Edited by
H. S. Muralidhara

Solid/Liquid Separation:

Waste Management and Productivity Enhancement

1989 International Symposium

BATTELLE PRESS
Columbus • Richland

Library of Congress Cataloging-in-Publication Data:

Solid/liquid separation: waste management and productivity enhancement / edited by H. S. Muralidhara
ISBN 0-935470-54-9 : $87.50
1. Factory and trade waste. 2. Refuse and refuse disposal.
3. Separation (technology). I. Muralidhara, H. S.
TD897.5.S625 1990
628.4--dc20
Library of Congress Number: 89-17828
CIP

Additional copies may be ordered through Battelle Press, 505 King Avenue, Columbus, Ohio 43201, U.S.A., (614) 424-6393, or 1-800-451-3543.

Dedicated to my wife, Ponnamma,
and our children, Shilesh and Shubha.

PREFACE

Solid/liquid separation, a vital process in treating municipal and industrial waste, is undergoing increasing use in the chemical, coal, and mineral industries. Separation involves removing either the liquid or the solid from solid-liquid systems. As Professor Frank Tiller aptly pointed out in *The Crisis in Solid-Liquid Separation Technology*, most environmental problems can ultimately be solved by the clean separation of particulates or immiscible liquids from water.

Because wastes generated by industry must be rendered harmless before they are discharged into the environment, there is a growing interest in developing new processes that can *economically* ensure a cleaner environment. Coupled with a climate of increasing restrictions and fuel costs, new solid/liquid separation processes are being developed, modified, and refined to meet this demand.

Traditionally, the solid/liquid separation process has employed a single property, using such forces as gravity, vacuum, or pressure. Ultrasonic, acoustic, electric, and magnetic forces are also utilized, but commercial applications of these are somewhat limited. In an effort to develop improved processes, the scope of solid/liquid separation technology has broadened to include combined field separation technologies, which combine two or more processes or properties in a single operation. The interest in this concept is reflected in the volume of current technical literature and the increasing development of commercial processes.

The first conference, Recent Advances in Solid/Liquid Separation held in November 1986, focused on recent developmental processes--flocculation, coagulation, and cake deliquoring--and detailed various field forces interacting with the solid, liquid, or interface.

This second conference, in December 1989, highlights the interdisciplinary nature of solid/liquid separation processes: geochemistry, colloidal science, flow-through process media, and petroleum, mechanical, and chemical engineering. Also key to the focus of this conference are environmental issues, specifically the Superfund sites, the Garbage Barge, and the increasing denial of land disposal requests...disposal sources are being reduced. Because land disposal must be utilized most efficiently, the waste relegated to land disposal must be compacted by composting or some other means. Waste incineration is another promising management technology that not only compacts waste, but delivers energy as well.

Planning for this conference began in 1988, when Richard Denning of Battelle's Corporate Technical Development Department granted initial funding to identify needs of the conference and to refine its focus. These ideas were supported by Ben Maiden and Bill Huffman, both of Battelle, and Professors Tiller and Aarne Vesilind, who verified the importance of a second conference.

I then met with Joe Sheldrick of Battelle Press to plan the corresponding publication, and by 1989, Don McConnell, Don Frink, and Bob Giammar, all of Battelle, were lending their support. Our intent was to detail specifically the growth of R&D activity, especially in terms of hazardous waste management and treatment. In addition, to foster better communication and to realize greater productivity, the book features a crosstalk approach to represent the exchange of ideas, approaches, various needs, and promising strategies between researchers from the various disciplines involved and members of industry.

The conference was an international effort, established to disseminate and share ideas and, we hope, to hasten viable results. Because of the diversity of the topics, the organization of the book was challenging; however, the order of papers presented within mirrors the order in which they were presented at the conference.

I wish to acknowledge my gratitude to a number of people whose help was invaluable to the conference and to these *Proceedings*. Dr. Robert Olfenbuttel of Battelle, who assisted in initial planning of the conference and identified key speakers in hazardous waste management areas; Cherlyn Paul, also of Battelle, who edited and managed the production of the *Proceedings*; all authors who worked very hard to submit their papers on time; and my wife, Ponnamma, and children, Shilesh and Shubha, without whose cooperation this effort would not have been possible.

CONTENTS

*Asterisks indicate corresponding authors.

AUTHOR INDEX

Solid/Liquid
Separation:
Waste
Management
and
Productivity
Enhancement

PROCESS FUNDAMENTALS OF SLUDGE DEWATERING

Solid/liquid separation processes play an increasingly important role in waste treatment and allied industries. The solid particle to be treated is very fine and the nature of suspension it forms is generally non-Newtonian; the nature of the suspension is at once inorganic, organic, and microbiological. Hence, an understanding of the principles of colloid science is becoming exceedingly important. Because the subject of solid/liquid separation is interdisciplinary, a multidisciplinary approach is warranted to solve some complex environmental problems. For example, subject of flow-through porous media is studied by geochemists, petroleum geologists, and hydrologists, among others, yet the fundamentals of the process are not understood.

Bearing this in mind, the *Proceedings* will attempt to provide a scientific basis for some of the process mechanisms. Professor Vesilind, Duke University, has performed significant work in the area of sludge dewatering. His overview identifies the knowledge gaps and discusses channelling in batch thickening operations. Professor Chase of the University of Akron provides a good model to explain the transport characteristics during cake filtration, and Dr. Jennings of Sir J. J. Thompson Labs from the United Kingdom expounds a new phenomenon called electrodynamic banding concentration in colloidal suspension. Rheological aspects to determine the state of sludges and theory of thin cake filtration is discussed by Mr. Crawford and Dr. Cheng, respectively.

CHANNELLING AND THE DOUBLY CONCAVE FLUX CURVE IN BATCH THICKENING

P. Aarne Vesilind
Professor and Chairman
Department of Civil and
Environmental Engineering
Duke University
Durham, North Carolina 27706
(919) 684-2434

and

Gregary N. Jones
Project Engineer
CH2M Hill
Rocky Point Centre, Suite 350
Tampa, Florida 33607

ABSTRACT

Channelling, which occurs during batch thickening of concentrated slurries, has often been implicated as the cause of the doubly concave flux curve. Channelling supposedly allows water to escape the settling slurry at a high rate, thus increasing the interface settling velocity and the solids flux. However, research shows that the upper layers of the settling column are progressively diluted and that the channel zone occurs at the interface between the diluted slurry and the slurry in compression. As the solids compress and expel water, this water escapes by means of the channels. Thus channelling is not directly responsible for the doubly concave flux curve, and the curve in fact results from the incorrect interpretation of batch thickening data.

INTRODUCTION

Gravity thickening is a process of separating excess water from sludge solids for the purpose of reducing the volume of sludge, thus facilitating its subsequent processing or disposal. Separation is achieved by the downward movement of the solids relative to the liquid. That relative motion is caused by the greater gravitational force per unit volume of solids, which is in turn proportional to the solids density. This separation allows a more concentrated sludge to form on the bottom of the thickener, while a relatively particle-free supernatant collects at the top.

Gravity thickening is a cost-effective process for solid-liquid separation in water and wastewater treatment as well as in various mining industries. It is used in water and wastewater treatment to thicken sludges with original solids concentrations between 0.5 and 4 percent solids to concentrations from 3 to 15 percent solids by weight. Gravity thickening has a beneficial impact on the sludge treatment costs as well as the overall operating costs of a wastewater treatment plant. For example, primary sludge thickening aids anaerobic digestion by reducing the volume of influent sludge, thus increasing detention time and making the digester temperature easier to maintain. The higher solids concentration improves the pH stability of the sludge and, because of increased viscosity, reduces the tendency for stratification[10].

CHANNELLING DURING GRAVITY THICKENING

During ideal gravity thickening, liquid flows upward through aggregates and around individual particles and flocs in uniform seepage across the horizontal cross section of the thickening device. However, under certain conditions some sludges will release liquid in streams, which are large and few in number when compared with the flow paths developed during the former type of flow. In transparent settling columns, these streams can be seen as miniature volcanoes at the solid-liquid interface and as small streams along the outside of the thickening sludge. This behavior, commonly known as channelling, has been defined as the creation of flow paths in a thickening suspension on a scale much larger than the size of the solid particles themselves[5].

Fitch has described channelling as essentially the short-circuiting of liquid upward through a concentrated suspension. He went on to note that, if the solids are idealized as a porous medium, channelling lowers resistance to movement of the liquid[6]. One way to understand the reason for this is by using the Hagen-Poiseulle equation to approximate flow through a porous matrix as is done in some derivations of Darcy's law[1]. If a unit cross-sectional area of the porous matrix contains N capillaries of diameter δ, then the total flowrate, Q, through the cross-section is N times the flowrate for an individual capillary, Q_i, as calculated by the Hagen-Poiseulle equation for laminar flow through a capillary,

$$Q = N\,Q_i = N \frac{\pi \delta^4 \rho g}{128 \mu} \frac{dP}{dx} \qquad (1)$$

where dP/dx is the pressure gradient. If it is assumed that the flow area is constant for a given concentration of sludge ($N^2 \pi \delta^2 = A_f =$ constant), then the diameter in equation (1) can be replaced by a function of N and A_f to yield

$$Q = \frac{1}{N} \frac{A_f^2 \rho g}{8 \pi \mu} \frac{dP}{dx} \qquad (2)$$

Since the total flowrate is inversely proportional to the number of capillaries, a smaller number of large channels will result in greater flow for the same pressure drop as a large number of small channels. Of course, this simple analysis does not take into account how the tortuosity of the flow paths decreases the flowrate. But it can safely be assumed that tortuosity decreases with the number of flow channels, thus making the increase in flowrate with decreasing N even more pronounced.

Flux in thickening is defined as the product of a mass concentration and the settling velocity characteristic of that particular concentration, hence yielding a mass flowrate of solids per unit area per unit time. A flux curve is a plot of flux over a range of concentrations. Many equations attempt to define the relationship between interfacial settling velocity and concentration[3,8,13,19]. None are derived from purely theoretical considerations, and some are completely empirical. Nonetheless, for the most part they yield similar results and preference is generally dependent on the particular type of sludge under study. All these analyses of the variation of settling rate as a function of concentration yield essentially exponential decreases in velocity with increasing concentration, which translates into a singly concave flux curve. Such a flux curve would be of the type following the dashed portion of the curve in Figure 1. Considerable empirical evidence, however, suggests that the solids flux curve has a second local maximum, as shown by the solid line in Figure 1. This doubly concave flux curve defies the logic of exponentially decreasing velocity.

Discussions of channelling in past research have noted two different, though perhaps related, phenomena: (1) the actual appearance of channels (whether vertical channels, diagonal channels, or volcanoes at the solid-liquid interface) and (2) abnormally high interfacial settling velocities that result in the doubly concave flux curve. Because these two phenomena have frequently coincided during thickening research, a causative relationship between the two has generally been presumed.

The occurrence of channels and/or unexpectedly high interfacial settling rates has been observed as long as thickening research has been conducted. In a study of metallurgical pulps in 1916, Coe and Clevenger described "trunk channels" that release liquid at the top of the compression zone[2]. Additionally, their paper presented several flux curves that showed increases in flux occurring at the higher solids concentrations.

Research using phosphate mining slimes[15,16] has shown that the addition of heavy impurities (such as mine tailings) or the suction of air bubbles out of sludge can increase the interfacial settling rate. In this and other work with phosphate mining slimes[7,14], there is evidence of an induction period during which flocculation takes place to cause a sudden increase in settling velocity. Assuming the flocculation leads to a stable velocity that is truly characteristic of the particular concentration, this behavior should cause no undue concern. But if the user is inconsistent about the choice of the representative

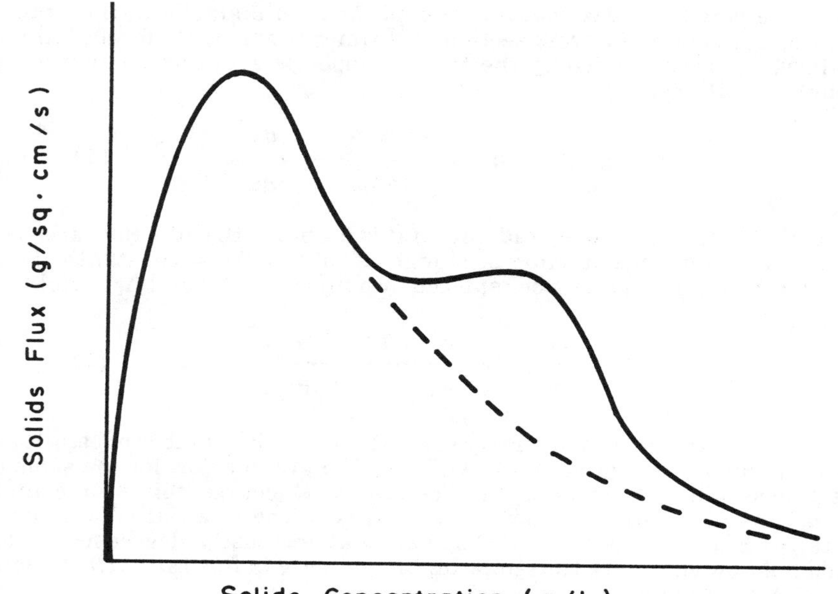

Figure 1. Doubly Concave Flux Curve

velocity, the false perception of a high settling rate being characteristic of a particular concentration may arise(18).

The appearance of channels and interface volcanoes in calcium carbonate sludge was recorded by Tory and Shannon(17), and the channels were described as "instabilities." Scott's work with calcium carbonate shows an abnormal plateau in the flux curve, which he attributes to channelling(11). Another doubly concave reduced flux curve was reported by Scott for silicon dioxide sludge(12) and Vesilind(19) reported a doubly concave curve for a chromium hydroxide suspension.

Two studies have found channelling in kaolin slurries. Cole noted interface volcanoes starting at an intermediate concentration, but quickly followed by the propagation of extremely dense sludge layers to the interface(3). Cole also found increasing settling rates late in batch tests, coincident with decreasing concentrations in the upper levels of the batch cylinders prior to passage of the interface. Dell and Kaynar(4) studied channelling in flocculated kaolin suspensions and calculated lower specific resistance to filtration for sludge in the channeled zone.

Some of the more dramatic examples of an abnormally fast interfacial settling rate for biological sludges are to be found in the works of Cole(3) and Vesilind(18). Cole found that a coarse floc structure developed during some experimental runs, which could cause differences of 500 percent or more in the observed settling velocity. The presence of the coarse floc structure increased with increasing cylinder diameter and decreasing sludge concentration. Vesilind also found this trend, called agglomeration, during his work, as well, and noted that the agglomeration did not necessarily occur throughout the entire height of the settling column, but would occur in a region and slowly move toward the top of the column until it reached the interface, causing an increase in interfacial settling velocity. Past examples of increased settling velocity directly imply increased flux rates.

If channelling occurs in continuous thickeners as well as in batch thickening columns, the increased flux rates due to channelling would make it possible to process more sludge, process the same amount of sludge to a higher concentration, or simply build smaller thickeners.

On the other hand, if channelling does not occur in continuous thickening, and yet the increased interfacial settling rate at a particular concentration in a batch test is factored into the design, then extrapolating the batch test results to the design of continuous thickeners will lead to undersizing. Though it can be said that the safety factors applied to such designs would cover the error introduced by channelling, the elimination of some fraction of the safety factor is not a desirable circumstance.

The objective of this study was to try to understand why channels occur in batch thickening tests, and how their onset can effect determinations of sludge flux rates.

EXPERIMENTAL EQUIPMENT AND DESIGN

Based on previous work in which sludges exhibited channelling, two different calcium carbonate sludges were chosen for this study. Due to the variabilities over time and difficulties in handling, waste-activated sludges were not considered, and it was believed that the application of the preliminary results with simpler sludges to the thickening of biological sludges would be a logical future step. Problems with variability and the complex nature of waste-activated sludge would not be resolved if the activated sludge was dosed with formaldehyde[9], and activated sludge is susceptible to the sudden formation of large floc agglomerates and the concomitant increased settling velocity noted by Cole[3] and Vesilind[18]. Such agglomeration would have overshadowed any effects caused by channelling.

The thickening experiments were performed in large transparent cylinders as shown in Figure 2. The sludge was pumped into the cylinders from the bottom and allowed to thicken without stirring. The interface drop and channels were observed through the cylinder walls.

OBSERVATIONS AND RESULTS

Several interesting observations were made on the channels that formed during the batch thickening runs. All of the channels were small (4 cm or less in length and about 0.1 cm wide), and they moved slowly upward as a fairly even layer over the course of the experiment. Figure 3 depicts the channel zone just as it meets the solid-liquid interface. As shown, two sizes of channel formed. The more obvious channels were typically from 1.5 and 2.0 cm long. Parallel to these "large" channels were smaller channels that were only 0.5 cm in length, originated about 1.0 to 1.5 cm above the beginning of the large channels, and terminated at the same height. Both types appeared randomly scattered around the cylinder wall, except that more small channels formed where greater distance separated the larger ones. Small particles could be seen being transported upward through the large channels. As the initial concentration of the thickening suspension was increased, there was a tendency for the channel zone to appear suddenly at higher levels in the thickening column, as opposed to being first noted very near the base of the column.

Since the solid-liquid interface fell and the channel zone rose, the two would eventually meet and volcano-shaped cones would form at the interface at the mouth of each channel. When the supernatant was clear enough, these volcanoes could be seen across the entire interface, not just at the walls. Also, when all thickening had ceased and the supernatant was very clear, dead volcanoes could be seen to be randomly located on the interface. No pattern could be discerned, nor did the volcanoes appear to have any preference or dislike for wall locations. The random location of volcanoes across the solid-liquid

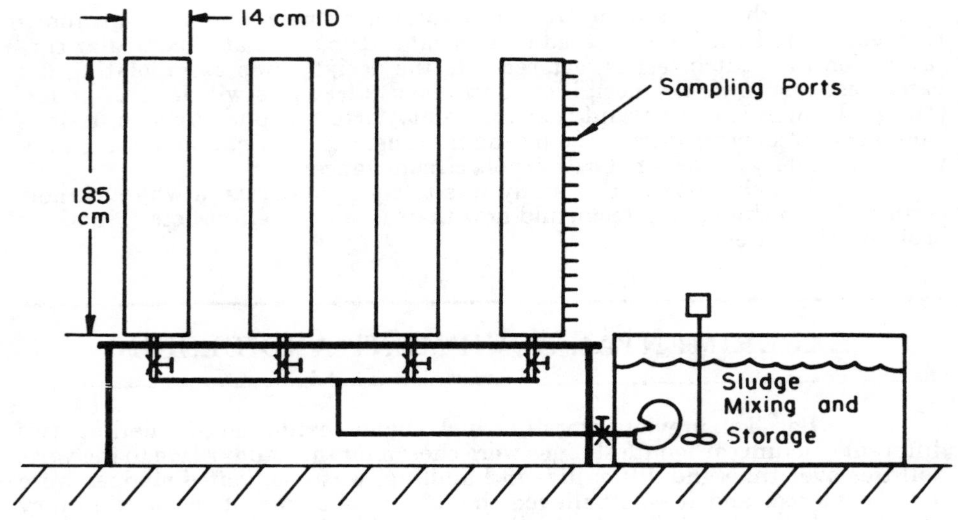

Figure 2. Experiment Set-Up

interface indicated that the channels were located all across the horizontal cross-section of the sludge layer, rather than only at the walls.

Differences were also noted among the relative sizes of the volcanoes formed as the channel zone reached the solid-liquid interface. Some of the higher initial concentration runs (approximately 500 g/L) built large volcanoes of approximately 6.5 cm diameter and 1.5 cm height. These can be compared with a typical cone of 0.3 to 1.0 cm diameter and 0.1 to 0.3 cm height during

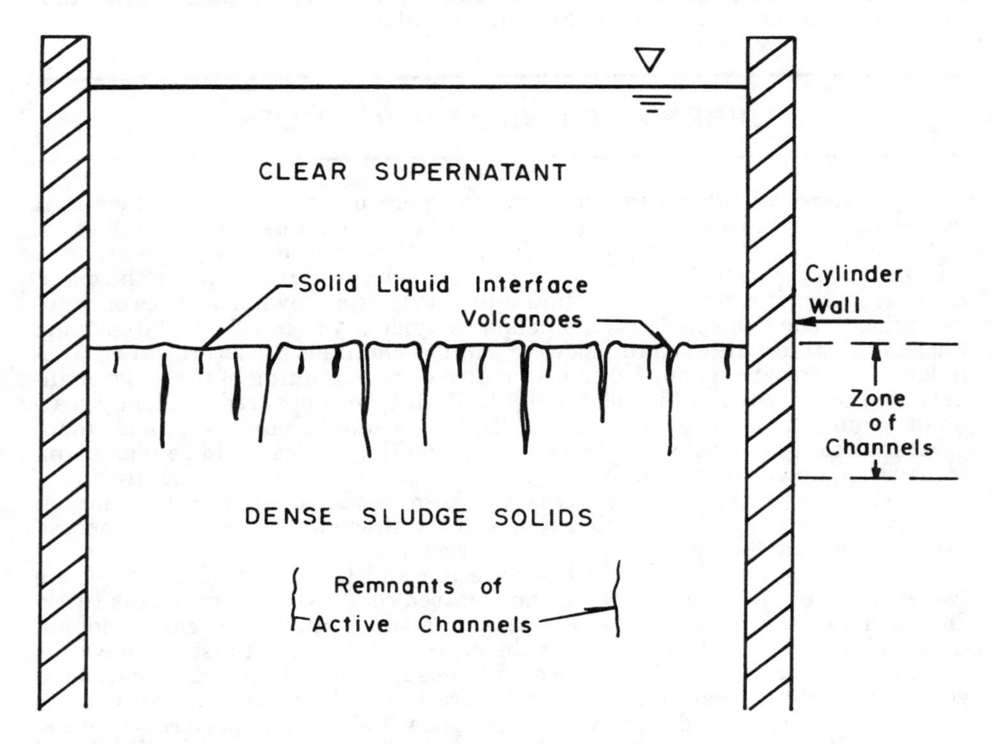

Figure 3. General Appearance of the Channel Zone

runs with lower initial concentrations. The formation of larger volcanoes in higher initial concentration runs indicated that greater amounts of solids were being carried out of the sludge beds. This would be caused by a higher fluid velocity in the channel, which would in turn be caused by greater compression taking place in the sludge bed.

It is also important to note that these channels were not large channels that reached from the compression region to the solid-liquid interface shortly after beginning the test. Instead, they occurred in a zone of very short length and therefore could not be directly responsible for the transferal of fluid from the compression region to the interface until very late in the experiments.

In every experiment, the meeting of the interface and the channel zone was quickly followed by a decrease in interfacial settling velocity. This sharp change in settling velocity indicated that the channel zone marked the transition from a free zone settling regime to much slower compression settling. Spot checks of the solids concentration around the channelled zone from 7 different experiments (with initial concentration ranging from 300 to 650 g/L) shown in Figure 4, reveal that a large gradient in concentration occurs in the vicinity of the zone and verifies that a change in the type of settling must take place as the zone reaches the interface.

Figure 4 also shows that the solids concentration at the top of the channel zone appears to be about 320 g/L regardless of initial solids concentration, and increases through the channel zone.

Six batch thickening tests were run with very low amounts of solids in the column to examine the effect of solids on the final average settled sludge concentration. The results indicate that the solids tend to settle normally to a

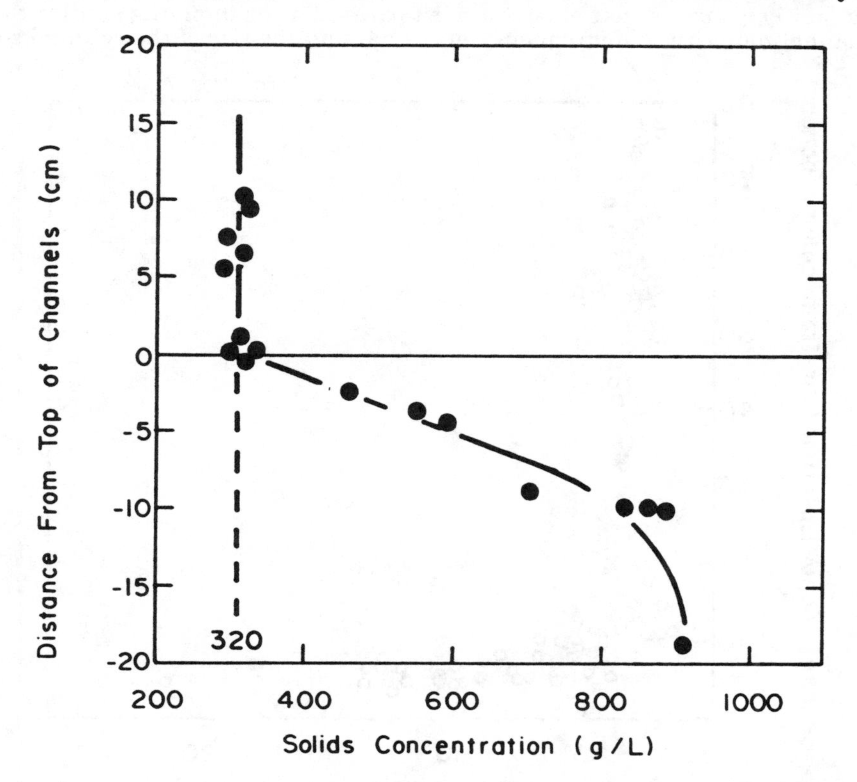

Figure 4. **Concentration Measurements at Various Vertical Distances from the Top of the Channel Zone. The Data are for 7 Tests with Initial Concentrations Ranging from 300 g/L to 660 g/L**

packing density of 572 g/L, and that higher concentrations are the result of compressive stresses due to higher solids beds. So the concentration range that Figure 4 defines for the channel zone could not be considered compression, but rather must be a very thick, hydraulically supported region. The vertical channels would then result from buoyant instability caused by dilution of the region just above the compression zone.

The data in Figure 4 also reveal that the concentration in the zone settling region, which is normally expected to be constant over the course of a run, is in fact decreasing. Concentrations near the channel zone were as much as 330 g/L below initial suspension concentration. Two possible explanations for this decrease in concentration were (1) flocculation resulting in differential settling and (2) dilution of the upper levels by fluid released from sludge compressing below the channelled zone. The formation of channels across a specific concentration gradient and the fact that that gradient was smaller than the initial sludge concentration in many of the experiments also explains why channels were initially noted later in a run and higher in the column in certain runs. If the sludge had to dilute before channels could form, the necessary dilution would occur well into the experiment.

Figure 5 is a composite of data from 15 batch thickening experiments with initial concentrations varying from 90 to 530 g/L. The channel location for these runs is plotted against the logarithm of time, thus eliminating minor data fluctuations. If channels appeared below a height of about 20 cm above the floor of the thickening column, this figure demonstrates that the channel zone is consistently at approximately the same location at the same time in runs with very different initial concentrations. Beyond that 20-cm level are a family of curves that appear almost parallel, thus implying similar rates of channel zone rise. The frequency of the data on the rise of the channel zone

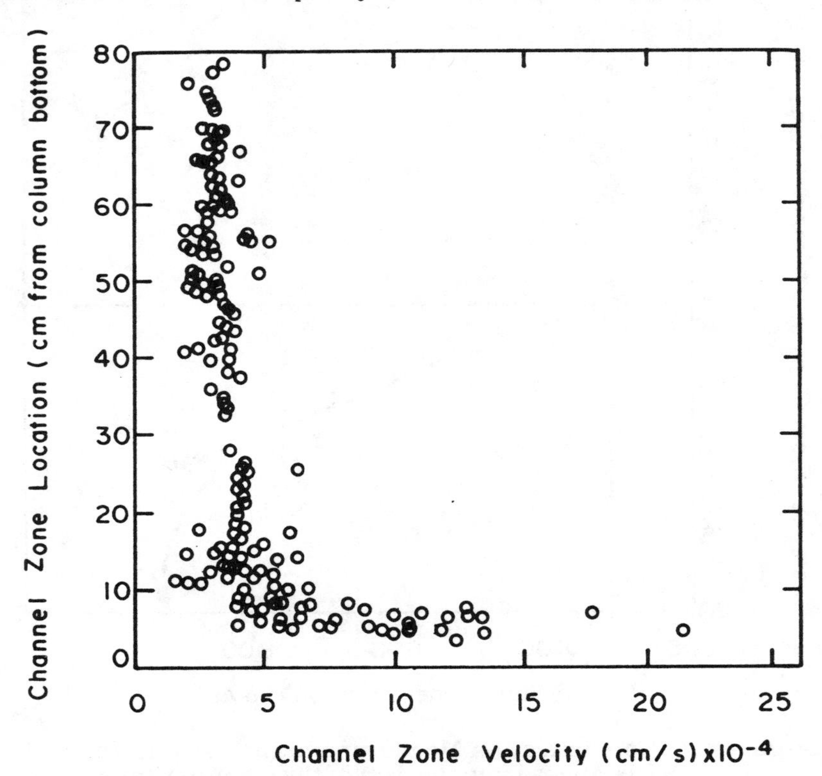

Figure 5. Channel Zone Location. Initial Concentrations from 92 g/L to 531 g/L

allows computation of the velocity of the zone between different heights. These channel rise velocities are plotted as a function of channel location in Figure 6. That figure shows that the channel zone rises at an almost constant rate, regardless of initial concentration, once it has reached a certain height in the sludge bed, indicating that an equilibrium is reached when the channels are formed at higher levels in the column. Normally in a batch thickening test, an equilibrium value for the rise rate of the compression zone (since the rise rate of the channel zone is equivalent) would be controlled by the settling velocity of the zone settling region. The fact that these experiments show a constant rise rate for a number of different initial concentrations, along with the previously noted dilution down to a specific zone concentration, implies that for this sludge the compression region strongly influences the zone settling region and, through that, the interfacial settling velocity.

It is interesting that the channels have a decreasing, rather than constant, rise velocity even in experiments with initial concentrations from 300 to 350 g/L. This is logical if channels merely serve to mark the transition from zone to compression settling. The compression zone should rise much faster at lower levels where lesser amounts of fluid are being expelled, but at higher depths, larger amounts of fluid are expelled in any time increment, making the upward progress of the zone much slower. The constant rise velocity would imply that some concentration in the upper portion of the compression zone limits the expulsion of the fluid, and this zone is not pushed to its full hydraulic capacity at the lower depths.

The testing program also examined interfacial settling velocities across a wide range of initial concentrations of homogeneous calcium carbonate

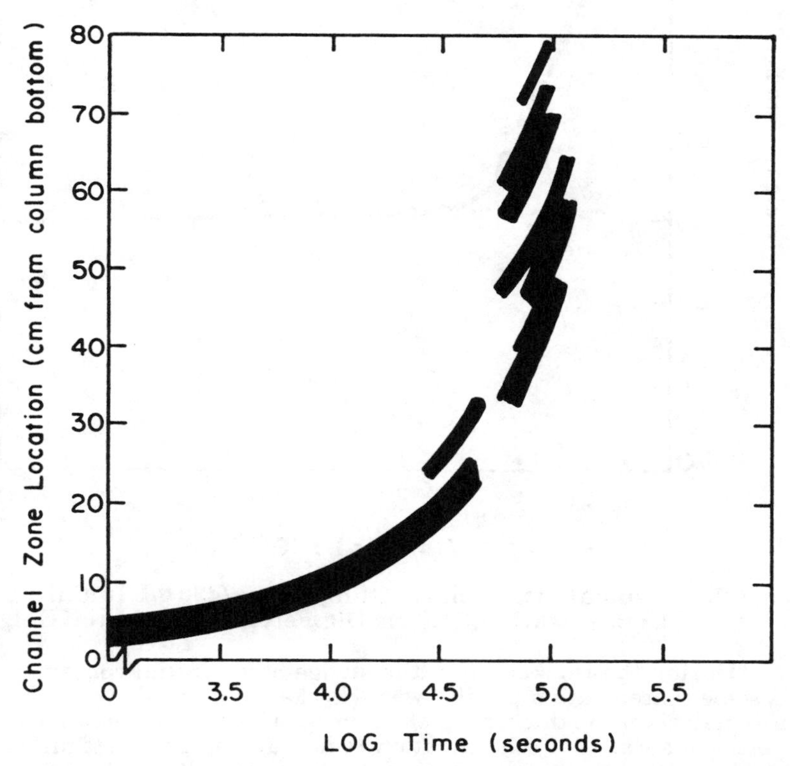

Figure 6. Relationship Between Channel Zone Upward Velocity and Its Location in the Cylinder. The Graph Summarizes Data from 15 Experiments, Initial Solids Concentration Ranging from 92 g/L to 531 g/L

sludges. The intent of this testing was to compare abnormally high velocities to the occurrence of channelling.

Interfacial settling velocities are normally determined by graphically fitting a straight line to the constant rate portion of the interfacial settling curve and using the slope of that line as the characteristic settling velocity of the initial concentration in the test. In this study, however, the velocity was calculated incrementally over the entire course of the run by simply dividing the drop in interface height by the elapsed time over which that drop occurred. Each value of incremental interfacial settling velocity was calculated across a number of points to smooth out fluctuations in velocity that would occur if two consecutive points were used (such fluctuations being caused by reading error when interface height measurements were made). Figure 7 shows a typical interface settlement curve, which certainly appears linear for the first 10^5 seconds. But Figure 8, a plot of the incremental velocity over time, shows the wide variation in velocity that actually occurs over the course of the run. Thus, what appears to be a straight line (and will have deceptively high R^2 if a least squares analysis is mistakenly applied to the data) is found to be almost continuously changing.

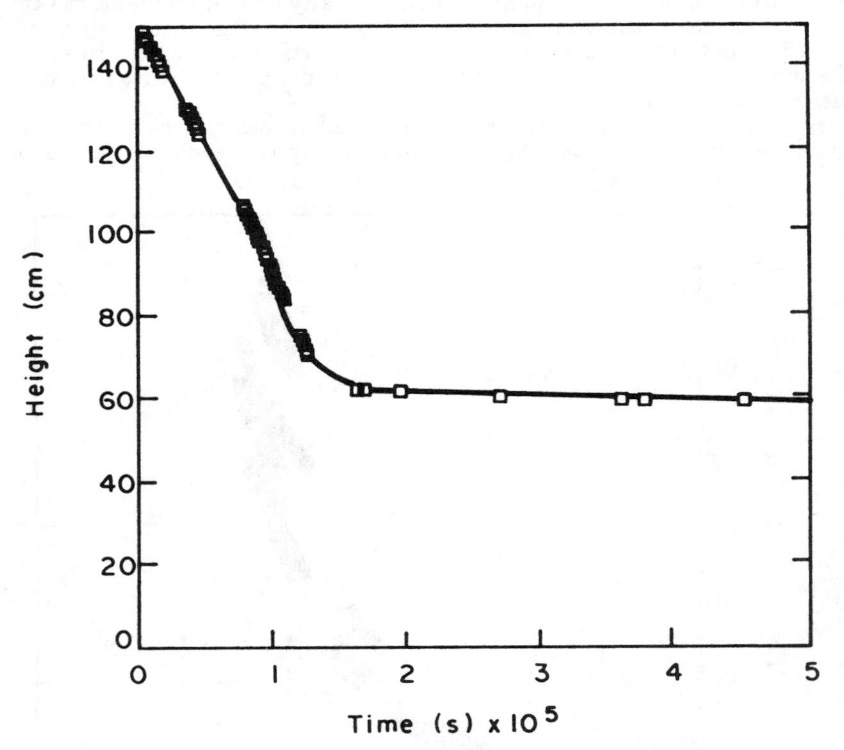

Figure 7. Typical Interfacial Settling Curve (430 g/L Initial Concentration, 14.0 cm Diameter, 150 cm Initial Height)

During the thickening of the sludge at an initial concentration of 520 g/L, some concentration profiles were checked and are shown in Figure 9. Standard analysis of the thickening of a homogeneous slurry requires that the slurry settle in a constant concentration region during zone settling or in an increasing concentration region during compression. It can be seen in the figure that the concentration profile for this sludge decreases with time. Such behavior might be thought to mean that dense impurities in the sludge were plunging through the calcium carbonate, but the sludge did not contain a large enough fraction of impurities to account for these changes in concentration.

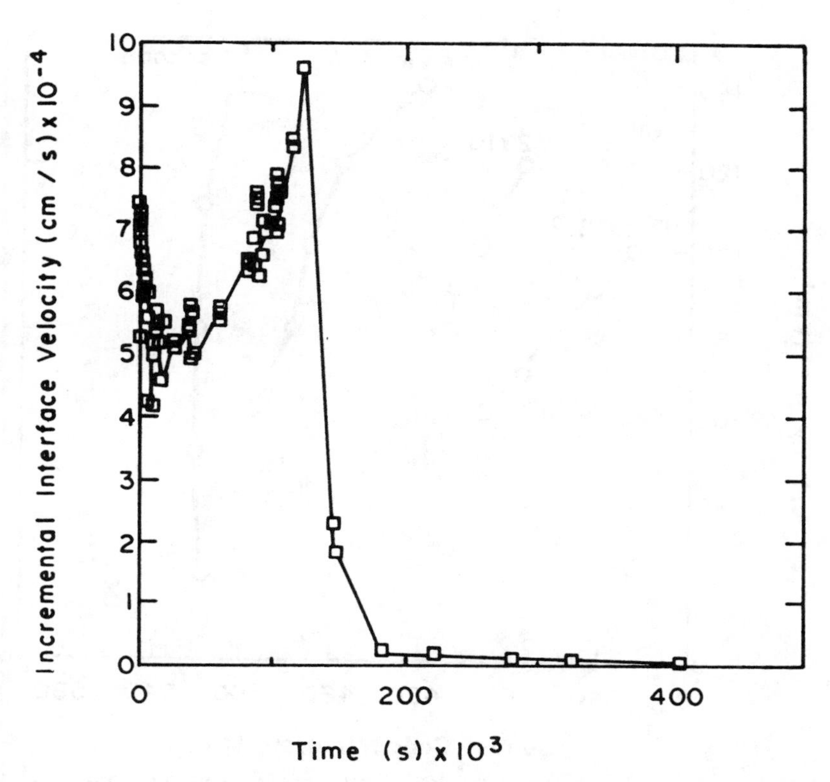

Figure 8. Example of Variation of Incremental Interfacial Velocity (430 g/L Initial Concentration, 14.0 cm Diameter, 150 cm Initial Height)

This particular experiment also exhibited a slight increase in settling velocity in the later part of the experiment. Even if the concentration had been lowered by differential settling of impurities, the remaining solids in the zone settling region should have been slowed because of loss of those impurities.

It is important to note that the initial drops in concentration did not have channel zones. Channels were only found around a concentration of about 320 g/L. The dilution occurred prior to the formation of any channels.

Since the channels exist across a consistent concentration zone, channelling prevents any zone, with concentrations higher than that at which the channelled zone terminates, from ascending at a higher rate than the upper surface of the compression region.

REEXAMINATION OF THE DOUBLY CONCAVE FLUX CURVE

The phenomenon that is commonly thought to be symptomatic of channelling is a doubly concave flux curve. The graphically determined interfacial settling velocities measured in this study using batch tests were multiplied by the initial concentration to yield the solids flux graph (Figure 10). From these data, it appears that the second concavity in the flux curve exists between 500 and 800 g/L, with a peak at about 650 g/L. To contrast with this normal method of measuring interface velocity, the initial velocity was determined, and the resulting flux is also plotted in Figure 10. Since this latter

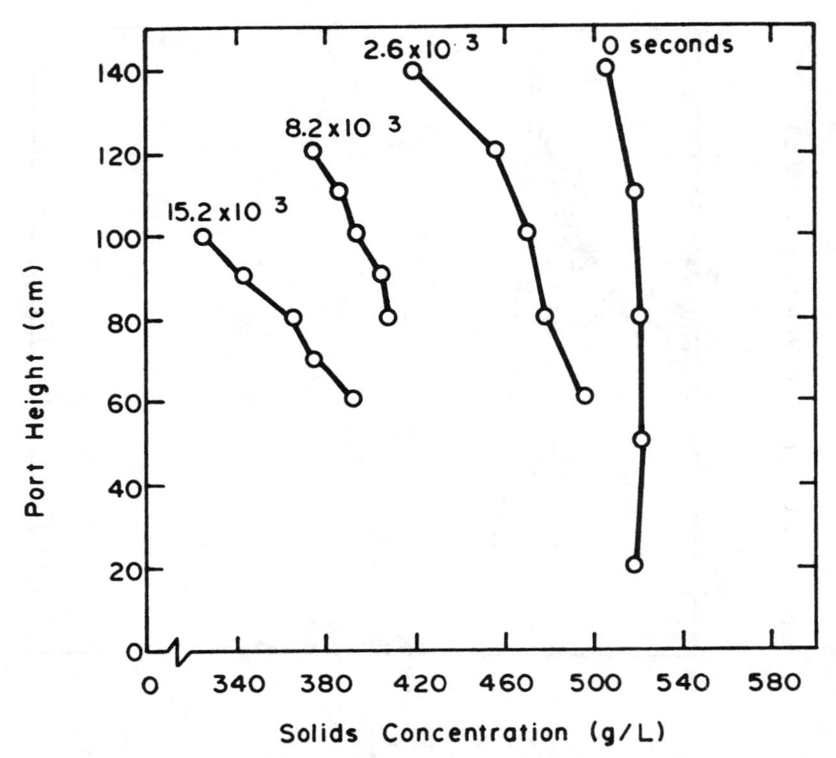

Figure 9. **Concentration During a Thickening Test (520 g/L Initial Concentration, 14.0 cm Diameter, 150 cm Initial Height)**

interface velocity is from early in each test, it could be argued that the flux thus calculated is more representative of a given concentration than are later interface velocity measurements that might include differential settlement of the solids and the progressive dilution of the upper sludge column. Since the flux data calculated by this second method do not demonstrate a second concavity, that concavity is not necessarily a true indication of abnormally high settling velocities across a range of initial concentrations. The second concavity is therefore an artifact of the particular method of determining settling velocity and cannot necessarily be considered representative of actual sludge thickening behavior. Further, the doubly concave flux curve does not result from channel formation. Rather, the progressive dilution of the upper sludge layers results in increased settling velocity and in the formation of channels. The channel region represents the transition between thickening and compression and is caused by the expulsion of water caused by the compression of the deeper sludge.

CONCLUSIONS

1. Channelling is a real phenomenon in batch thickening tests, and it can be induced and manipulated for a calcium carbonate slurry.

2. Channelling is probably not the cause of the double concavity in the flux curve.

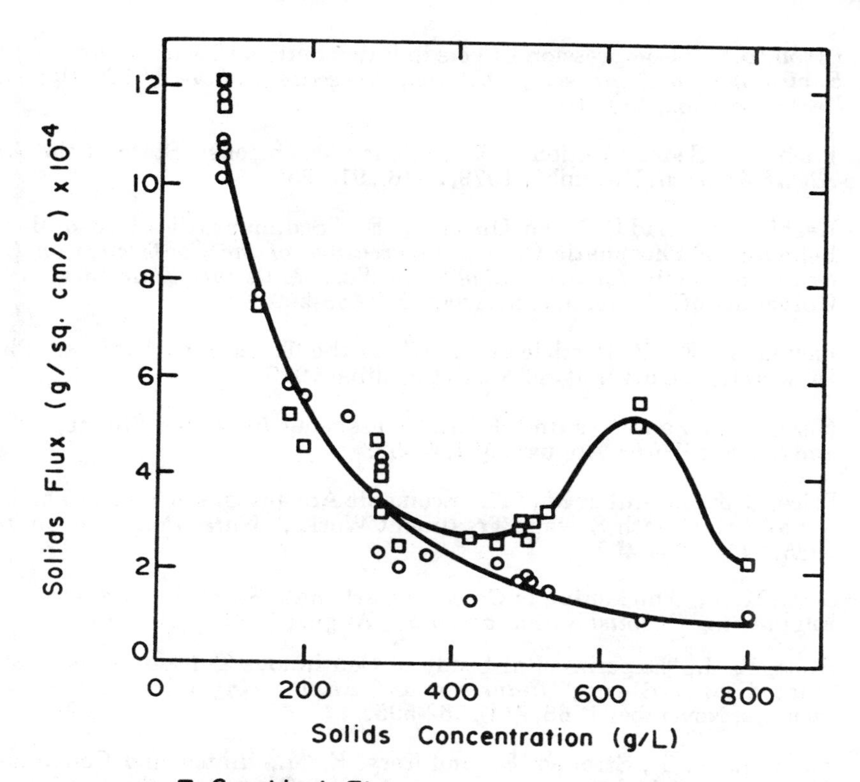

Figure 10. Comparison of Solids Flux Calculated from Graphical Velocity Using the Entire Interface Settling Curve to that Calculated from Initial Velocity (14.0 cm Diameter, 150 cm Initial Height)

3. The upper reaches of a thickening sludge in the batch mode are being progressively diluted by the expulsion of water from the lower sludge blanket.

4. Because of this dilution, the batch settling curve probably does not have a straight line "zone settling" portion, thus bringing into question the validity of designing continuous thickeners from batch tests.

REFERENCES

1. Bear, J., *Dynamics of Fluids in Porous Media*, New York, American Elsevier Publishing Company, 1972.

2. Coe, H. S. and Clevenger, G. H., "Methods for Determining the Capacities of Slime-Settling Tanks," *Transactions of the AIME*, 1916, *55*, 356-384.

3. Cole, R. F., "Experimental Evaluation of the Kynch Theory," Ph.D. Dissertation, University of North Carolina, 1968.

4. Dell, C. C. and Kaynar, M. B., "Channelling in Flocculated Suspensions," *Filtration and Separation*, July/August 1968, 323-327.

5. Dixon, D. C., "Compression Effects in Batch Settling Tests," *Journal of the Environmental Engineering Division, Proceedings of the ASCE*, 108 EE6, December 1982, 499-507.

6. Fitch, B., "Sedimentation of Flocculent Suspensions: State of the Art," *AIChE Journal*, November 1979, *25* (6), 913-930.

7. Keshian, B., Ladd, C., ad Olson, R. E., "Sedimentation-Consolidation Behavior of Phosphatic Clays," *Proceedings of the Conference on Geotechnical Practice for Disposal of Solid Waste Materials*, June 13-15, 1977, University of Michigan, Ann Arbor, MI, 188-209.

8. Lawler, D. F., "A Particle Approach to the Thickening Process," Ph.D. Dissertation, University of North Carolina, 1979.

9. Nissen, J. A. and Vesilind, P. A., "Preserving Activated Sludge," *Water and Sewage Works*, August 1974, 48-52.

10. Price, G. J. and Adler, M., "The Economic Advantages of Sludge Thickening at Avonmouth Sewage Treatment Works," *Water Pollution Control*, 1985, *84* (3), 394-402.

11. Scott, K. J., "Thickening of Calcium Carbonate Slurries," *Industrial and Engineering Chemistry Fundamentals*, August 1968, *7* (3), 484-490.

12. Scott, K. J., "Experimental Study of Continuous Thickening of a Flocculated Silica Slurry," *Industrial and Engineering Chemistry Fundamentals*, November 1968, *7* (4), 582-595.

13. Shannon, P. T., Stroupe, E., and Tory, E. M., "Batch and Continuous Thickening," *Industrial and Engineering Chemistry Fundamentals*, August 1963, *2* (3), 203-211.

14. Smellie, R. H. and LaMer, V. K., "Flocculation, Subsidence, and Filtration of Phosphate Slimes," *Journal of Colloid Science*, 1956, *11*, 720-731.

15. Somasundaran, P., "Thickening or Dewatering of Slow-Settling Mineral Suspensions," *Mineral Processing* (International Mineral Processing Congress, 13th, Warsaw), J. Laskowski, ed., Elsevier, Amsterdam, 1981, Volume 2, Part A, 233-261.

16. Somasundaran, P., Smith, Jr., E. L., and Harris, C. C., "Dewatering of Phosphate Slimes Using Coarse Additives," *Proceedings of the XIth International Mineral Processing Congress*, Instituto de Arte Minerali e Preparazione dei Minerali, Universito de Cagliari, 1975, 1301.

17. Tory, E. M. and Shannon, P. T., "Reappraisal of the Concept of Settling in Compression," *Industrial and Engineering Chemistry Fundamentals*, May 1965, *4* (2), 194-204.

18. Vesilind, P. A., "The Influence of Stirring in the Thickening of Biological Sludges," Ph.D. Dissertation, University of North Carolina, 1968.

19. Vesilind, P. A., *Treatment and Disposal of Wastewater Sludges*, Revised edition, Ann Arbor, MI, Ann Arbor Science Publishers, 1979.

SLUDGE AND WASTEWATER FILTRATION: A CONTINUUM ANALYSIS

George G. Chase, Assistant Professor, and
Max S. Willis and Ismail Tosun
Department of Chemical Engineering
The University of Akron
Akron, Ohio 44325-3906
(216) 375-7943

ABSTRACT

Stringent measures on environmental protection have increased the importance of sludge handling and disposal in which solid/liquid separations play a central role. The use of pressure filters in sludge dewatering has substantially increased over the last decade in the United States.

Filtration is generally considered an art rather than a science. Since the existing design methodologies are mainly empirical, design of filters is based either on previous experience or on a trial-and-error procedure. Current filter evaluation is based on *ad hoc* modeling of the processes.

This paper presents a continuum approach as an alternative to *ad hoc* modeling. This continuum approach is applicable to both cake filters and deep bed clarifiers. As an example, the continuum approach is applied to cake filtration, and the resulting working equation is compared with the two-resistance model currently used to evaluate filter cakes. The comparison indicates that the continuum approach provides novel methods for operating filters and increases attention to the cake/medium interaction.

INTRODUCTION

Growing demands for clean water and the increased reuse of wastewater has led to more stringent effluent standards for wastewater treatment plants[10]. Filtration is widely used in the treatment of municipal and industrial wastewater and sludges to clarify, dewater, and recover solids. Often filtration is the critical step in wastewater processing.

Generally solids are recovered via cake filtration, and clarification is usually achieved by filtration through granular (deep bed) filters. In cake filtration the solid particles are stopped by a primary medium, such as a cloth, and form a cake on the surface of the medium. In deep bed filtration the particles are retained on the surfaces of the granules within the depth of the bed.

The purpose here is to present a continuum approach as an alternative to the *ad hoc* modeling approach currently used in the analysis of filtration of sludge and wastewater. The *ad hoc* approach to filter design normally uses the specific resistance model for cake filtration and dewatering[4,6,16,18]. In the analysis of deep bed filters the *ad hoc* approach normally uses morphological and stochastic models[7,8,9,15]. In principle the continuum theory approach can be applied to both cake and deep bed filters.

The conceptual model or theory to be used in the analysis of filtration or water clarification must be selected prudently because these concepts not only govern the practical operation of process units, but also guide the research activities.

For example, the *ad hoc* approach to cake filtration identifies the specific cake resistance as the primary material parameter that controls the rate of filtration, and it ignores the resistance of the filter media as a significant operating factor. Consequently, considerable effort is directed toward modifying prefilt properties in an effort to obtain favorable values for the specific filtration resistance.

The continuum theory, on the other hand, directs attention to the selection of the filter medium and its interaction with the particles as the key factor affecting the filtration rate. This is more in harmony with actual practice because the lucrative market for manufacturers of filter media is evident.

MULTIREGION CONTINUUM APPROACH

From the point of view of the multiregion continuum approach to problems associated with processing multiphase systems, granular bed water clarification and filtration have more similarities than differences. In a deep bed filter, two multiphase regions interact at a visible discontinuity between the slurry region and the granular bed. In cake filtration, three multiphase regions interact at two distinguishable discontinuities: one is between the filter medium and the filter cake, and the other is between the filter cake and the slurry. This is schematically represented in Figure 1.

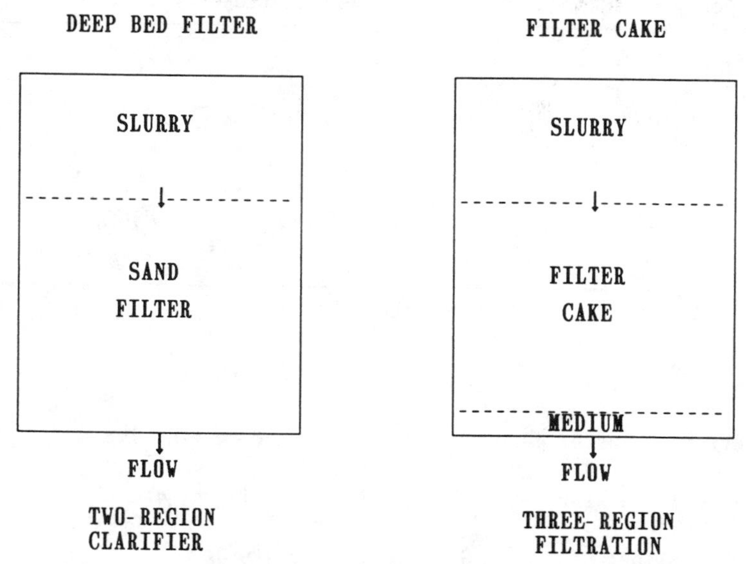

Figure 1. Schematic of the Two-Region Deep Bed Clarifier and the Three-Region Cake Filtration

The sand bed in the water clarification process and the filter medium in the cake filtration in Figure 1 are both multiphase regions governed by the same multiphase equations. In the clarifier, the medium region interacts with the prefilt; whereas in the filtration, the medium interacts with the filter cake.

In either case, the media capture the particles that migrate across the singularity that separates the media from the prefilt or the filter cake. As a result of this flux of particles into the media, the porosity diminishes, the surface area increases, and at constant fluid throughput, the local velocity, local drag, and pressure gradient all increase. This mechanism is one-sided, that is, the prefilt or the filter cake affects the media, but the media do not alter the prefilt or the filter cake. Since the clogging of the media does not alter either the prefilt or the filter cake, it is evident that the time-dependent condition of the media controls the overall throughput in the process.

Granted, the depth of penetration of the particles into the medium of the clarifier is much larger than the depth of penetration into the medium in the cake filtration, but nonetheless, it is evident that the mechanisms for both processes are exactly the same and that it does not make any difference whether the media clog as a result of an interaction with a slurry or with a filter cake. The throughput is diminished by the clogging of the media.

The design of these processes from a continuum point of view requires a formal procedure for developing continuum equations that both apply to multiphase regions and take into account the interaction between different multiphase regions. In addition, the continuum equations must account for any number of phases--interphase effects such as surface tension and heterogeneous reactions--and the presence of different chemical species in each phase. This formal procedure has been developed elsewhere[19], and the general equations, along with jump balances for interfaces and region singularities, are summarized in Table 1.

Table 1. General Property Balance Equations at the Homogeneous Scale, the Heterogeneous Scale, and the Region Scale in a Dispersed Multiphase Material

<table>
<tr><th colspan="2">Homogeneous Scale</th></tr>
<tr><td>Phase Property Balance</td><td>Interface Property Balance</td></tr>
<tr><td>$\frac{\partial \varphi_\alpha}{\partial t} + \nabla\cdot(\varphi_\alpha \mathbf{v}_\alpha) + \nabla\cdot\mathbf{i}_\alpha - (f_\alpha + g_\alpha) = 0$ (A)</td><td>$[\varphi_\alpha(\mathbf{v}_\alpha - \mathbf{v}_i) - \varphi_\beta(\mathbf{v}_\beta - \mathbf{v}_i)]\cdot\mathbf{n}_{\alpha\beta} + (\mathbf{i}_\alpha - \mathbf{i}_\beta)\cdot\mathbf{n}_{\alpha\beta} + \mathbf{i}_{\alpha\beta}\cdot\mathbf{n}_\sigma^\gamma + g_c^\delta = 0$ (B)

$g_c^\delta = \varphi_i(\mathbf{v}_i^\alpha - \mathbf{v}_i) + \varphi_i(\mathbf{v}_i^\beta - \mathbf{v}_i)$ (C)
$\mathbf{i}_{\alpha\beta} \equiv [\varphi_\alpha(\mathbf{v}_\alpha - \mathbf{v}_i) + \varphi_\beta(\mathbf{v}_\beta - \mathbf{v}_i) + (\mathbf{i}_\alpha - \mathbf{i}_i) + (\mathbf{i}_\beta - \mathbf{i}_i)]G_i$ (D)</td></tr>
<tr><th colspan="2">Heterogeneous Scale</th></tr>
<tr><td colspan="2">$\frac{\partial}{\partial t}\langle\varphi_\alpha\rangle + \nabla\cdot\langle(\varphi_\alpha \mathbf{v}_\alpha)\rangle + \nabla\cdot\langle\mathbf{i}_\alpha\rangle - (\langle f_\alpha\rangle + \langle g_\alpha\rangle) + (E_i^\alpha + I_i^\alpha + S_i^\alpha + G_i^\alpha) = 0$ (E)

Convective Transfer: $E_i^\alpha = \frac{1}{V}\sum_{\beta\neq\alpha}\int_{A_{\alpha\beta}} \varphi_\alpha(\mathbf{v}_\alpha - \mathbf{v}_i)\cdot\mathbf{n}_{\alpha\beta}\,dA$ (F)
Mechanical Transfer: $I_i^\alpha = \frac{1}{V}\sum_{\beta\neq\alpha}\int_{A_{\alpha\beta}} (\mathbf{i}_\alpha - \mathbf{i}_i)\cdot\mathbf{n}_{\alpha\beta}\,dA$ (G)
Property Slips: $S_i^\alpha = \frac{1}{V}\sum_{\beta\neq\alpha}\int_{A_{\alpha\beta}} \mathbf{i}_{\alpha\beta}^\alpha\cdot\mathbf{n}_\sigma^\gamma\,dA$ (H)
Heterogeneous Reaction: $G_i^\alpha = \frac{1}{V}\sum_{\beta\neq\alpha}\int_{A_{\alpha\beta}} g_c^\alpha\,dA$ (I)</td></tr>
<tr><th colspan="2">Region Scale (Labeled Regions 1 and 2)</th></tr>
<tr><td>Region Property Balance</td><td>Region Discontinuity</td></tr>
<tr><td>$\frac{\partial}{\partial t}\langle\varphi_o\rangle + \nabla\cdot\langle(\varphi_o \mathbf{v}_o)\rangle + \nabla\cdot\langle\mathbf{i}_o\rangle - (\langle f_o\rangle + \langle g_o\rangle) = 0$ (J)
$\sum_\alpha(E_i^\alpha + I_i^\alpha + S_i^\alpha + G_i^\alpha) = 0$ (K)</td><td>$[\langle\varphi_o\rangle_1(\langle\mathbf{v}_o\rangle - \mathbf{v}_s)_1 - \langle\varphi_o\rangle_2(\langle\mathbf{v}_o\rangle - \mathbf{v}_s)_2]\cdot\mathbf{n}_{12} + (\langle\mathbf{i}_o\rangle_1 - \langle\mathbf{i}_o\rangle_2)\cdot\mathbf{n}_{12} = 0$ (L)</td></tr>
</table>

In Table 1, equation (A) is a general property balance and equation (B) is an interface balance at the (single-phase) homogeneous scale for an α-phase in a multiphase material. The interface balance contains postulates that describe an arbitrary interface between any two phases. These postulates characterize the interface as a region that has one dimension at the molecular scale and two dimensions at the continuum scale. The boundaries of the molecular scale dimension are not parallel. This interface model appears as property slips (equation (D) of Table 1) at the continuum scale of the phases that adjoin the interface.

At the scale of local measurements in a multiphase domain, the general property balance (equation (E) of Table 1) accounts for the effects of the phases and interfaces. The excess terms--E_i^α, I_i^α, S_i^α, and G_i^α--in the property balance account for the convective, mechanical, geometric, and reactive effects of the interface on the α-phase properties, respectively.

When the heterogeneous scale balances are summed over all of the phases, the region property balance (equation (J) of Table 1) is obtained. The region balance reverts to the same form as that for a single phase because this general balance must be independent of scale. The sum of the excess terms (equation (K) of Table 1) properly zero as expected.

At the region scale, two distinct multiphase regions are separated by a singular surface. The region discontinuity balance (equation (L) of Table 1) accounts for property transfers across this discontinuity between regions, but does not contain any inter-region effects since none occur at region singularities.

The details of the interface model and theoretical development of the equations in Table 1 are provided by Willis *et al.*[19]. Specific property balances can be written for the different scale equations by substituting appropriate quantities from Table 2 for the equations in Table 1.

Table 2. Identification of the Property Functions that will Convert the General Property Balance to a Specific Property Balance for Mass, Momentum, Energy, and Entropy

General	Mass	Momentum	Energy	Entropy
φ_a	ρ_a	$\rho_a \mathbf{v}_a$	$\rho_a(U_a+\frac{1}{2}\mathbf{v}_a\cdot\mathbf{v}_a)$	$\rho_a S_a$
φ_i	ρ_i	$\rho_i \mathbf{v}_i$	$\rho_i(U_i+\frac{1}{2}\mathbf{v}_i\cdot\mathbf{v}_i)$	$\rho_i S_i$
$\mathbf{i}_a$	0	$-\underline{\mathbf{t}}_a$	$-\underline{\mathbf{t}}_a\cdot\mathbf{v}_a+\mathbf{q}_a$	$\mathbf{s}_a$
$\mathbf{i}_i$	0	$-\underline{\mathbf{t}}_i$	$-\underline{\mathbf{t}}_i\cdot\mathbf{v}_i+\mathbf{q}_i$	$\mathbf{s}_i$
f_a	0	$\mathbf{b}_a$	$\mathbf{b}_a\cdot\mathbf{v}_a+h_a$	$\rho_a m_a$
f_i	0	$\mathbf{b}_i$	$\mathbf{b}_i\cdot\mathbf{v}_i+h_i$	$\rho_i m_i$
g_a	0	0	0	$\rho_a \phi_a$
g_i	0	0	0	0
$\mathbf{i}^a_{a\beta}$	$\mathbf{j}^a_{a\beta}$	$\underline{\mathbf{t}}^a_{a\beta}$	$\mathbf{e}^a_{a\beta}$	$\mathbf{p}^a_{a\beta}$

APPLICATION TO CAKE FILTRATION

Figure 2 is a schematic of one-dimensional isothermal cake filtration. The filter cake is identified as Region 1, and the slurry is identified as Region 2. The cake/slurry boundary with its discontinuity in the local porosity is the singular surface between these two regions. The analysis presented here is an example of the application of the continuum approach to a multiphase process.

EQUATION OF CONTINUITY

Table 3 lists the assumptions and their impact on the equation of continuity. These assumptions not only simplify the equations but also impose restrictions on the solution.

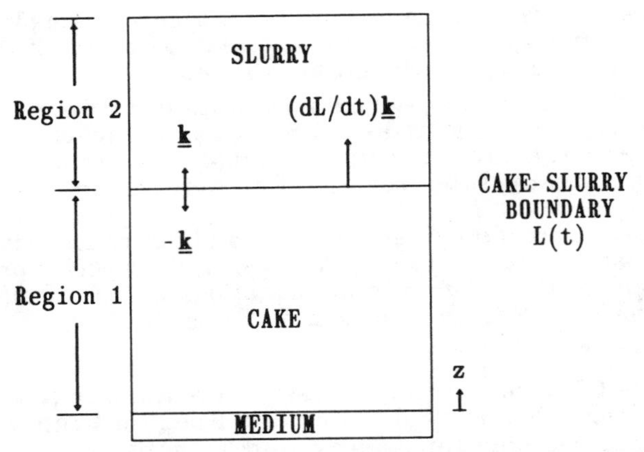

Figure 2. Schematic of a One-Dimensional Filtration with the Slurry Entering at the Top and Filtrate Leaving at the Bottom

Table 3. Assumptions and Their Impact on the Continuity Balances

ASSUMPTION	IMPLICATION ($j=\alpha$ or β)
• Incompressible α- and β-phases.	$\rho_j = \langle\rho_j\rangle^j_c = \langle\rho_j\rangle^j_{sl}$
• No radial variation in velocity.	$\langle \mathbf{v}_j\rangle^j = \langle v_j\rangle^j_z(-\mathbf{k})$
• No chemical reactions.	$g^\delta_c = G^j_i = 0$
• No mass transfer between phases.	$\rho_j(\mathbf{v}_j - \mathbf{v}_i) = 0$ on $A_{\alpha\beta}$
• No interfacial tension.	$G_i = 0$
• Slurry porosity is constant.	$\partial\epsilon_{jsl}/\partial t = \nabla\epsilon_{jsl} = 0$
• Interstitial velocities in the slurry are equal to each other.	$\langle \mathbf{v}_\alpha\rangle^\alpha_{sl} = \langle \mathbf{v}_\beta\rangle^\beta_{sl}$

CAKE CONTINUITY BALANCES

α-phase

$$\frac{\partial\epsilon_{\alpha c}}{\partial t} = \frac{\partial\langle v_\alpha\rangle_{zc}}{\partial z} \tag{A}$$

•β-phase

$$\frac{\partial\epsilon_{\beta c}}{\partial t} = \frac{\partial\langle v_\beta\rangle_{zc}}{\partial z} \tag{B}$$

•Region

$$\frac{\partial}{\partial z}\left[\langle v_\alpha\rangle_{zc} + \langle v_\beta\rangle_{zc}\right] = 0 \tag{C}$$

•Region Discontinuity [at z = L(t)]

$$\rho_\alpha\left[\langle v_\alpha\rangle_{zc} - \langle v_\alpha\rangle_{zsl}\right] + \rho_\beta\left[\langle v_\beta\rangle_{zc} - \langle v_\beta\rangle_{zsl}\right] = (\rho_\alpha - \rho_\beta)(\epsilon_{\alpha sl} - \epsilon_{\alpha c})\frac{dL}{dt} \tag{D}$$

The region equation (equation (C) in Table 3) is obtained by summing the continuity equations for both phases; it reveals that the sum of the phase average velocities depends only on time. When this result is combined with the boundary condition

$$\langle v_\alpha \rangle_{zc} = -\frac{1}{A}\frac{dV_F}{dt} \quad \text{and} \quad \langle v_\beta \rangle_{zc} = 0 \qquad \text{at} \quad z = 0 \tag{1}$$

and the fact that the filter assembly has a fixed volume, then we obtain

$$\langle v_\alpha \rangle_{zc} + \langle v_\beta \rangle_{zc} = -\frac{1}{A}\frac{dV_F}{dt} \tag{2}$$

which shows that at any position z in the filter cake, the total flow rate of both phases is equal to the volumetric flow rate of filtrate leaving the filter. The minus signs in equations (1) and (2) imply that the direction of flow is opposite to the positive direction of the z-coordinate.

The solid phase particles are actually stopped somewhere within the filter medium. However, this penetration depth is less than the accuracy of local measurements, and as a result, it is acceptable to use the boundary condition that the solid phase velocity is zero at $z = 0$.

The region discontinuity balance (equation (D) of Table 3) can be reduced to the form[3,19]

$$A\frac{dL}{dt} = G\frac{dV_F}{dt} \tag{3}$$

where

$$G \equiv \frac{1 - \epsilon_{\alpha s}}{\epsilon_{\alpha s l} - \epsilon^*_{\alpha c}} \tag{4}$$

accounts for the effect of the slurry porosity on the filter cake.

With the exception of the start of the filtration when clear fluid is displaced in the filter assembly with slurry, the term G is constant and equation (3) becomes

$$A\,L = G\,V_F \tag{5}$$

EQUATION OF MOTION

Combining the assumptions in Tables 3 and 4 with the general relation between stress and pressure, $\underline{t}_\alpha = P_\alpha \underline{I} + \underline{\tilde{t}}_\alpha$, and with equation (E) of Table 1 results in equations (A) and (B) of Table 4 for the α- and β-phase momentum balances. In these equations the drag force (F_{dz}) is the force between the phases due to momentum transfer, and $\langle \mathbb{P}_\alpha \rangle^\alpha$ is the piezometric pressure defined by

$$\langle \mathbb{P}_\alpha \rangle^\alpha \equiv \langle P_\alpha \rangle^\alpha + \langle \rho_\alpha \rangle g z \tag{6}$$

Table 4. Assumptions and Their Impact on the Momentum Balances

ASSUMPTION	IMPLICATION
• The only body force is the gravitational force.	$b_j = \rho_\alpha g$
• Inertial effects are insignificant	$\frac{\partial}{\partial t} <\rho_j v_j> + \nabla . <\rho_j v_j v_j> \simeq 0$
• No interfacial tension over $A_{\alpha\beta}$.	$P_\alpha \simeq P_\beta$ on $A_{\alpha\beta}$
• Pressure is continuous at the cake-slurry interface.	$<P_\alpha>^\alpha_c = <P_\alpha>^\alpha_{sl}$ at $z=L(t)$
• Viscous effects in the fluid phase are negligible.	$\underline{\tilde{t}}^\alpha_c \simeq 0$
• Dilute slurry, i.e., particles are not in contact with each other.	$\underline{\tilde{t}}^\beta_{sl} \simeq 0$

CAKE MOMENTUM BALANCE

• α-phase

$$\epsilon_{\alpha c} \frac{\partial <\mathbb{P}_\alpha>^\alpha_c}{\partial z} + F_{dz} = 0 \tag{A}$$

• β-phase

$$\epsilon_{\beta c} \frac{\partial <\mathbb{P}_\alpha>^\alpha_c}{\partial z} + \frac{\partial \tilde{t}^\beta_{zzc}}{\partial z} + \epsilon_{\beta c}(\rho_\beta - \rho_\alpha) g - F_{dz} = 0 \tag{B}$$

• Region

$$\frac{\partial <\mathbb{P}_\alpha>^\alpha_c}{\partial z} + \frac{\partial \tilde{t}^\beta_{zzc}}{\partial z} + \epsilon_{\beta c}(\rho_\beta - \rho_\alpha) g = 0 \tag{C}$$

Region Discontinuity [at $z = L(t)$]

$$\left[\tilde{t}^\beta_{zzc}\right]_{z=L} = 0 \tag{D}$$

CONSTITUTIVE RELATIONS

Examination of the equations in Table 3 and Table 4 reveals that the number of unknown functions exceeds the number of independent equations by two. Therefore, constitutive equations are required for the fluid phase drag (F_{dz}) and the solid phase stress ($\tilde{t}^\beta_{zzc}$).

Using the eight axioms (i.e., causality, determinism, equipresence, objectivity, invariance, neighborhood, memory, and admissibility) that must be satisfied by the constitutive relations, Willis *et al.*[19] show that

$$F_{dz} = \lambda \left[<v_\alpha>^\alpha_{zc} - <v_\beta>^\beta_{zc}\right] \tag{7}$$

$$\tilde{t}^\beta_{zzc} = \sum_{m=1}^{n} C_m(\epsilon^e_{\alpha c} - \epsilon_{\alpha c})^m \tag{8}$$

are admissible as constitutive relations. Here, λ is the resistance function that relates the drag force to the velocity difference, and $\varepsilon_{\alpha c}{}^e$ is the initial unstressed volume fraction of the α-phase in the cake. In postulating equation (8) the difference in volume fraction is considered to be the measure of deformation of the β-phase. It is important to note that $\tilde{t}^{\beta}_{zzc}$ is not a property of an individual solid particle, but rather, it is a property of the porous medium.

CORRELATION OF PROCESS PARAMETERS

The α- and β-phase continuity equations, equations (A) and (B) in Table 3, are explicitly dependent on time and contain only the volume fraction and velocity of the two phases. This means that the α- and β-phase velocity distributions are completely determined by the volume fraction distribution.

Complete specification of the velocity profiles by the continuity conditions leaves the α-phase equation of motion (equation (A) in Table 4) as a first-order differential equation. Therefore, the correlation of process parameters that specifies the filtrate rate for cake filtration does not require the simultaneous solution of continuity and motion equations. It can be obtained by evaluating the α-phase motion equation at the cake/septum interface, i.e., z = 0, where the region of highest drag is most likely to occur.

Substitution of equation (7) into equation (A) (Table 4) and the evaluation of the resulting equation at the exit of the filter cake where solid velocity is zero gives

$$\frac{dt}{dV_F} = \frac{(\lambda/\epsilon^2_{\alpha c})_{z=0}}{\Delta\mathbb{P}_c J_o A} L \tag{9}$$

The term J_o in equation (9) is the dimensionless pressure gradient at the cake/septum interface defined by

$$J_o = \left[\frac{\partial P^*}{\partial \xi}\right]_{z=0} \tag{10}$$

in which the dimensionless variables P^* and ξ are given as

$$P^* = \frac{\langle\mathbb{P}_\alpha\rangle^\alpha_c - \langle\mathbb{P}_\alpha\rangle^\alpha_c|_{z=0}}{\Delta\mathbb{P}_c} \tag{11}$$

$$\xi = \frac{z}{L} \tag{12}$$

Elimination of cake length (L) between equations (5) and (9) gives the correlation of process parameters as

$$\frac{dt}{dV_F} = \left[\frac{G}{\Delta\mathbb{P}_c A^2} \frac{(\lambda/\epsilon^2_{\alpha c})_{z=0}}{J_o}\right] V_F \tag{13}$$

When dt/dV_F is plotted versus V for a constant cake pressure drop filtration, equation (13) indicates that the variations in J_o and $(\lambda/\varepsilon_{\alpha c}^2)_{z=0}$ are responsible for the deviations from linearity. The interpretation of rate data by using equation (13) is given in detail by Willis *et al.*[19].

COMPARISON OF THE CONTINUUM AND *AD HOC* APPROACHES

A comparison of the working equations for the continuum and *ad hoc* approaches to cake filtration is shown in Table 5. The working equation listed for the continuum approach is a rearrangement of equation (13). This equation

Table 5. Comparison of Approaches to Cake Filtration

AD HOC APPROACH	CONTINUUM APPROACH
Working Equation	Working Equation
$\Delta P = (R_c + R_m) \frac{\mu}{A} \frac{dV_F}{dt}$	$\Delta \mathbb{P}_c = \left[\frac{G}{A^2} \frac{(\lambda/\epsilon_{ac}^2)_{z=0}}{J_o}\right] V_F \frac{dV_F}{dt}$
$R_c = \alpha^* \, cV_F$	
Design Parameters: A	Design Parameters: A
Operating Parameters: ΔP, V_F, t, c, R_m	Operating Parameters: $\Delta \mathbb{P}_c$, V_F, t, G, J_o, ϵ_a
Material Parameters: α^*, μ	Material Parameters: λ, C_m

relates the cake pressure drop ($\Delta \mathbb{P}_c$) to the filtrate rate and filtrate volume. The slurry porosity is accounted for by G, which is defined by equation (4). The dimensionless pressure gradient (J_o) defined by equation (10) accounts for the effect of the pressure gradient, or drag, on the deformation of the particulate phase.

The working equation for the *ad hoc* approach is the two-resistance model and is based on the heuristic analogy with Ohm's law[1,14] for the cake and medium resistances, R_c and R_m, in series. This equation relates the total pressure drop across the cake and medium (ΔP) to filtrate rate and volume. The cake resistance is assumed to be the product of the average specific resistance (α^*), the mass concentration of dry cake per volume of filtrate (c), and filtrate volume. This definition of α^* is analogous to the definition of electrical resistivity. Interpretation of filtrate rate data using this model depends on the empirical relation used for the specific cake resistance function (α^*)[3,12,13].

Traditionally the medium resistance in the two-resistance model is neglected, which implies that the total pressure drop is approximately equal to the cake pressure drop, $\Delta P \simeq \Delta \mathbb{P}_c$. However, recent literature[2,11] shows that the medium resistance increases during the filtration and may be a significant part of the total pressure drop.

As shown in Table 5, each approach has the same design parameter, the filtration area (A). The operating and material property parameters, however, differ. The material parameters identify the material that is being filtered. The resistance function (λ) for the continuum approach depends on the viscosity of the fluid phase and the surface area of the particulate phase. Also, the deformation of the solid phase, as described by the constitutive relation (equation (8)) and the material parameters (C_m) influence the dimensionless pressure gradient (J_o).

As stated in the introduction, the conceptual model or theory used in the analysis of filtration not only governs the practical operation of process units, but also guides the research activities. The *ad hoc* working equation provides three modes of operation[17]:

1. Constant pressure filtration. The pressure drop across the filter assembly is maintained constant by applying a constant pressure to the slurry and maintaining the filtrate outlet at a constant datum pressure.
2. Constant rate filtration. Constant rate is obtained by a positive displacement pump. The filtrate outlet is maintained at a constant datum pressure, and the inlet pressure varies according to the pressure drop across the filter assembly.
3. Variable pressure, variable rate filtration. This mode of operation is obtained by the use of a centrifugal pump. The filtrate

outlet is maintained at a constant datum pressure, and the inlet pressure varies with the varying flow rate and varying total pressure drop across the filter assembly.

The working equation for the continuum approach is based on the pressure drop across the filter cake and not the whole filter assembly. As such, this provides different modes of operation from that of the *ad hoc* approach. These modes of operation are[2,5]:

1. Constant cake pressure drop. Constant cake pressure drop is obtained by measuring the pressure drop across the filter cake and adjusting the applied pressure such that the cake pressure drop remains constant, but the applied pressure increases due to the increased pressure drop across the clogging filter medium.
2. Constant applied pressure, constant rate. In this mode of operation the applied pressure to the slurry is constant at the same time that a control valve in the outlet maintains a constant flow rate. This mode of operation recognizes the significance of the clogging of the filter medium and permits the pressure drop across the cake and filter medium to increase. This mode of operation is a direct result of the continuum analysis and the application of the solid phase stress constitutive relation, equation (8).

The *ad hoc* approach neglects the medium resistance and focuses on the specific filtration resistance (α^*) of the filter cake. Process research based on this model is directed only toward finding a prefilt preparation that will result in favorable values of the specific filtration resistance[6]. Only two controlled modes of operation are available, constant pressure and constant rate.

On the other hand, the continuum approach focuses on the clogging of the filter medium by the cake particles. This approach suggests that altering the aggregate structure of the solid particles by prefilt operations is not independent of the selection of the filter medium. Both must be taken into consideration to obtain favorable filtration rates. In addition, novel modes of operation that compensate for the clogging of the filter medium become apparent from the continuum approach.

CONCLUSIONS

It is evident from this review that multiphase continuum analysis can be used as an alternative to the *ad hoc* approach for the design of filters for wastewater and sludge treatment. It is also evident that the systematic approach normally reserved for single-phase systems can now be extended to dispersed multiphase material systems.

The continuum approach provides alternate modes of operation to those used with the *ad hoc* approach. The continuum approach also directs research into the interactions between the cake and medium, as well as alteration of the morphology of the particles by prefilt treatment.

The continuum approach is applied here to cake filtration. This same approach can be applied to the deep bed clarifier. The differences between the two applications are in the boundary conditions between the multiphase regions.

NOMENCLATURE

A	filter cross-sectional area
$A_{\alpha\beta}$	interfacial surface between the α- and β-phases
b_α, b_i	α-phase and interfacial region external supply of momentum
C_m	solid phase deformation coefficient, equation (8)

c mass concentration of dry cake per filtrate volume
E_i^α convective transfer of ψ_α across interface region (Table 1, equation (E))
$e_{\alpha\beta}^\alpha$ interfacial energy slip flux with the α-phase (Table 2)
F_{dz} z-component of the interfacial drag force
f_α external supply of property ψ_α
G ratio of cake to filtrate volume, equation (4)
G_i^α generation by chemical reaction in the interface (Table 1, equation (I))
G_i geometric factor associated with property slip in the interface
g gravity acceleration constant
g_α internal generation of property ψ_α
g_c^δ rate of property production per area $A_{\alpha\beta}$ (Table 1, equation (C))
h_α, h_i α-phase and interface region external supply of energy
$\underline{I}$ unit tensor
$\underline{I}_i^\alpha$ mechanical interaction between the α-phase and the interface
i_α, i_i surface flux vector of general properties ψ_α and ψ_i
$i_{\alpha\beta}$ general property slip in the interface region (Table 1, equation (D))
J_o dimensionless pressure gradient, equation (10)
$j_{\alpha\beta}^\alpha$ interfacial mass slip flux vector (Table 2)
k z-direction unit vector
L cake height
m_α, m_i α-phase and interface external supply of entropy
$n_{\alpha\beta}$ area unit normal vector of surface $A_{\alpha\beta}$
n_{12} area unit normal vector of the boundary between regions 1 and 2
n_σ^γ unit tangential vector of the interface directed into the γ-phase
P_α α-phase pressure
ΔP pressure drop across the filter assembly
$\Delta \mathbb{P}_c$ pressure drop across the filter cake
$<P_\alpha>_c^\alpha$ α-phase intrinsic average pressure in the cake
$<\mathbb{P}_\alpha>_c^\alpha$ piezometric pressure, equation (6)
$P_{\alpha\beta}^\alpha$ interfacial entropy slip flux
q_α, q_i heat flux vectors for the α-phase and interface
R_c, R_m cake and medium resistances in the two-resistance model (Table 5)
S_α, S_i α-phase and interface average specific entropy
S_i^α general property slip in the interface (Table 1, equation (H))
s_α, s_i α-phase and interface entropy flux
t time
$\underline{t}_\alpha, \underline{t}_i$ α-phase and interface stress tensors
$\underline{\bar{t}}_\alpha$ α-phase shear stress tensor
$\underline{\bar{t}}^\alpha$ α-phase averaged shear stress tensor
$\underline{t}_{\alpha\beta}^\alpha$ interface momentum slip with the α-phase
U_α, U_i α-phase and interface specific internal energies
V volume
V_F filtrate volume
v_α, v_i α-phase and interface velocities
v_i^α velocity of α-phase species in the interface region
v_s region boundary velocity
$<v_\alpha>_{zc}$ z-component of the α-phase averaged velocity in the cake
$<v_\alpha>_z^\alpha$ z-component of the α-phase intrinsic averaged velocity

Greek:
α^* average specific cake resistance in the two-resistance model
ε_α α-phase porosity (volume fraction)
ε_α^* cake averaged porosity
ε_α^e α-phase unstressed cake porosity
λ resistance function material parameter, equation (7)
μ fluid phase viscosity
ρ_α, ρ_i α-phase and interface region densities

$\langle \rho_\alpha \rangle^\alpha$	α-phase intrinsic phase averaged density
ψ_α	general property function

Superscripts:

α,β	α- and β-phase quantities

Subscripts:

α,β	α- and β-phase quantities
i	interface quantity
o	multiphase region, averaged quantity
$1,2$	multiphase regions 1 and 2
z	component in the z-direction
c,sl	quantity evaluated in the cake and slurry, respectively

Operations:

$\langle\ \rangle$	volume averaged quantity
$\langle\ \rangle^\alpha$	α-phase intrinsic volume-averaged quantity

REFERENCES

1. Almy, C. and Lewis, W. K., *J. Ind. Eng. Chem.*, 1912, *4*, 528-532.

2. Chase, G. G., Ph.D. Dissertation, University of Akron, January, 1989.

3. Cheremisinoff, N. P. and Azbel, D. S., *Liquid Filtration*, Ann Arbor Science, Massachusetts, 1983.

4. Christensen, G. L. and Dick, R. I., *J. Environ. Eng.*, 1985, *111*, 243-257.

5. Collins, R. M., Ph.D. Dissertation, University of Akron, January, 1988.

6. Dick, R. I. and Ball, R. O., *CRC Crit. Rev. Environ. Control*, 1980, *10*, 269-337.

7. Fan, L. T., Hwang, S. H., Chou, S. T., and Nassar, R., *Chem. Eng. Commun.*, 1985, *35*, 101-121.

8. Fan, L. T., Nassar, R., Hwang, S. H., and Chou, S. T., *AIChE J.*, 1985, *31*, 1781-1790.

9. Jackson, G. E., *CRC Crit. Rev. Environ. Control,* 1980, *10*, 339-373.

10. Kohn, P. M., *Chem. Eng. (N.Y.)*, 1977, *84*, 95-97.

11. Leu, W. and Tiller, F. M., *Sep. Sci. Technol.*, 1983, *18*, 1351-1369.

12. McCabe, W. L., Smith, J. C., and Harriot, P., *Unit Operations of Chemical Engineering*, 4th Edition, McGraw-Hill, New York, 1985.

13. Peters, M. S. and Timmerhaus, K. D., *Plant Design and Economics for Chemical Engineers*, 3rd Edition, McGraw-Hill, New York, 1980.

14. Ruth, B. F., *J. Ind. Eng. Chem.*, 1935, *6*, 708-723.

15. Tien, C., in *4th World Filtration Congress: Proceedings Part II*, R. Banbrabant, J. Hermia, and R. A.Weiler, eds., Royal Flemish Society of Engineers, Belgium, 1986.

16. Tiller, F. M. and Anantharamakrishnan, O. V., *J. Chem. Eng. Jpn.*, 1980, *13*, 380-385.

17. Tiller, F. M. and Crump, J. R., *Chem. Eng. Prog.*, 1973, *10*, 65-75.

18. Wilhelm, J. H., *J.-Water Pollut. Control Fed.*, 1978, *50*, 471-483.

19. Willis, M. S., Tosun, I., Choo, W., Chase, G. G., and Desai, F., in *Advances in Porous Media*, M. Y. Corapcioglu, ed., Elsevier, Amsterdam, in press.

OPTIMIZING POLYMER CONSUMPTION IN SLUDGE DEWATERING APPLICATIONS

P. M. Crawford
SCC Product Manager
Zenon Water Systems, Inc.
845 Harrington Court
Burlington, Ontario L7N 3P3
Canada
(416) 639-6320

ABSTRACT

As we move into the 1990s, upgrading wastewater treatment plants is becoming a very important consideration. Although upgrading can take many forms, one of the most economical is to optimize the performance of the existing equipment and structures. In the realm of sludge dewatering, also an increasingly important topic, an area that has received little attention in the past is the control of the sludge conditioning process prior to dewatering. In conjunction with the Wastewater Technology Centre in Burlington, Canada, Zenon Water Systems has developed the Sludge Conditioning Controller (SCC) to fulfill this need in the wastewater marketplace. Descriptions of both the hardware and software aspects of the SCC are presented, and typical operating performance of the microprocessor-based system is shown. Experience with full-scale systems has revealed that the benefits associated with the SCC far exceed the original objective of saving polymer. The others include automation of the dewatering device operation, increased capacity, and more uniform performance of the dewatering machine.

INTRODUCTION

Organic polymers have become the primary choice of conditioning agent for sludge dewatering operations. Until now, however, there has been no simple technique for automatically controlling the addition of polymer to the sludge to optimize performance of the dewatering machine. Several devices have appeared in the marketplace, but none to date have satisfactorily solved the problem. Some rely on an indirect measurement of sludge conditioning by analyzing filtrate quality, but these are subject to severe solids fouling. Another is capable of measuring the state of sludge conditioning on a labor-intensive manual basis, but this device does not provide closed-loop feedback control of the polymer feed pump. Yet another is applicable only to belt presses, and even then, only to certain designs. Zenon has overcome all of these deficiencies with the new Sludge Conditioning Controller (SCC).

The SCC is a unique, on-line, real-time sludge monitor and polymer flow rate controller developed in conjunction with Environment Canada's Wastewater Technology Centre (WTC). The relationship between the rheological characteristics of conditioned sludge and optimal polymer dosage was discovered by the WTC in the early 1970s[1,2]. At that time, however, process control hardware technology was not sufficiently advanced to allow production of a cost-effective controller. The proliferation of inexpensive microcomputers in the early 1980s, however, allowed development of the concept to the feasibility stage[3].

In 1984, Zenon began a federally funded project to develop the SCC from the prototype stage to a commercial device. The initial part of that work was a market study. Because that study identified a strong North American municipal market for polymer control technology, Zenon proceeded to the next phase of the project: development of rugged and inexpensive system for controlling polymer[5]. After this work was completed in late 1986, the first field trial installation of the SCC was made at Guelph, Ontario, on a 2.0-m belt press. During this initial operation, the performance of the SCC was compared with historical manual operating data. Based on about 4 months of operating data, the reduction in polymer consumption was about 24 percent[4].

During this first field experience, several significant improvements in ruggedness, reliability, and accuracy were made. All of these changes were incorporated into the production version of the SCC, which was launched into the marketplace in early 1987. Zenon is now actively marketing the SCC on a worldwide basis to both municipal and industrial plants.

THEORY

The Zenon Sludge Conditioning Controller uses proprietary process technology in combination with Zenon's own advanced microprocessor control technology. The SCC determines the state of sludge conditioning directly by measuring the intrinsic rheological characteristics of both conditioned and unconditioned sludge.

Rheology is the study of shear stress and shear strain behaviour of fluids. Most homogeneous materials develop shear stress in proportion to the flow rate or shear strain imposed on them. This kind of behaviour is called Newtonian.

When solid particles are added to the fluid, as in wastewater sludges, the behaviour is different. Fluid flow is inhibited by the particles until some minimum shear stress is applied. This kind of behaviour is called non-Newtonian. In addition, the more particles there are (i.e., the higher the solids concentration is), the higher the shear stress required to produce flow at the same rate. This relationship will be demonstrated more fully later.

When conditioning polymers are added to the sludge, the behaviour is different again. Fluid flow is again inhibited by the polymer chains until some minimum, but much higher, shear stress is applied. In addition, the shear stress declines with time to a lower value if the fluid flow is maintained constant. This kind of behaviour is called thixotropic.

Figure 1 shows schematically what the information that the SCC gathers looks like. On the graph, the torque or shear stress response of the sludge is plotted versus time. The "peak" is defined as the highest value of torque found for a given sample; the "base" is the average of the value of the data near the end of data collection. As shown, the peaks get higher in magnitude as polymer dosage increases. The lowest curve is typical of unconditioned sludges as it has no distinct peak in the data.

The SCC uses these kinds of information, gathered from sludge samples taken in real time, to determine the state of sludge conditioning. It then changes the polymer flow signal to adjust the amount of polymer added to the sludge.

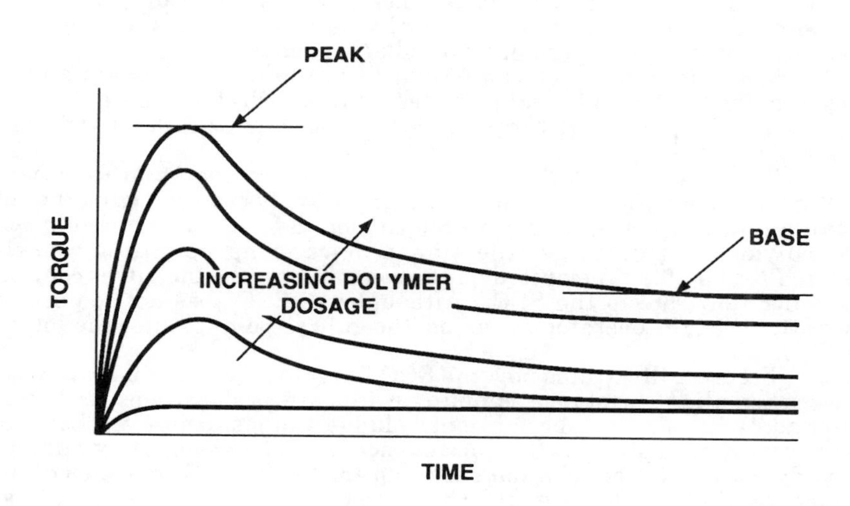

Figure 1. Shear Response of Conditioned Sludge

HARDWARE

The hardware for an SCC consists of four main components:

- Central control panel
- Local control station
- Sample vessel and sensor head
- Printer.

The central control panel (CCP) houses the single-board computer (SBC), sample vessel interfaces, control signal generators, and communications link. The SBC is the heart of the system. It controls all of the automatic sludge sampling and data collection, as well as making the control decisions on the polymer flow signal output. The sample vessel interfaces provide the necessary optical isolation between the SBC and the sampling system. The control signal generators also provide isolation between the CCP and the polymer pump controller. The communications link allows sending of the data and alarms from the SCC to a printer or computer. A single CCP can control the polymer flow rate for up to four dewatering devices both independently and simultaneously.

An operator interface is also provided on the front face of the CCP. It consists of a two-line LCD screen and a keypad to control the operation of the SCC.

The local control station contains the solenoid valves necessary to control both the air-operated pinch valves that control sludge sampling and the rinse spray nozzles that clean the sensor head rotor and level probes. In addition, it also has a keypad for the manual control of all sampling system functions.

The sample vessel is a PVC or stainless steel container with a capacity of about 10 L. It is piped into both the unconditioned and conditioned sludge lines and a drain via pinch valves that control the flow of sludge samples. All sludge flow in and out of the vessel is by gravity; no additional sludge pumping is required. The sensor head mounts on the sample vessel lid, which also serves as mounting for the rinse spray nozzles. The level probes, which sense when the vessel is full of sludge, and the rotor, which turns in the sludge sample, both extend through the lid into the vessel. Inside the sensor head is the electric motor, which turns the rotor and the optical sensor, which measures the torque or shear stress exerted on the rotor by the sludge sample.

The printer serves as an output device to record the data and any alarms from the SCC for archival purposes. If it is desirable to save this information in electronic form, it can be saved to a computer disk as well as, or instead of, being printed.

The SCC hardware is entirely automatic in operation from batch-sampling the sludge to self-cleaning after sample analysis to adjusting the polymer pump flow rate. The only requirement for the operator is to adjust the initial polymer dosage to get the sludge dewatering operation working properly. Pressing the tune button on the central control panel gives control of the polymer flow rate to the SCC. Although the SCC does not require any further attention, the operator can adjust the auto setpoint to alter the setpoint of the SCC.

Figure 2 illustrates how the SCC fits into a typical sludge dewatering flow schematic. The sludge sampling points are as shown upstream of the polymer addition point for unconditioned sludge and just before the belt filter press for conditioned sludge. Samples of each are taken separately and independent data are collected. The data are transmitted to the CCP, which decides to change the polymer flow signal. The system therefore operates as a classic closed-loop feedback control system with an additional feedforward input based on the unconditioned sludge sample.

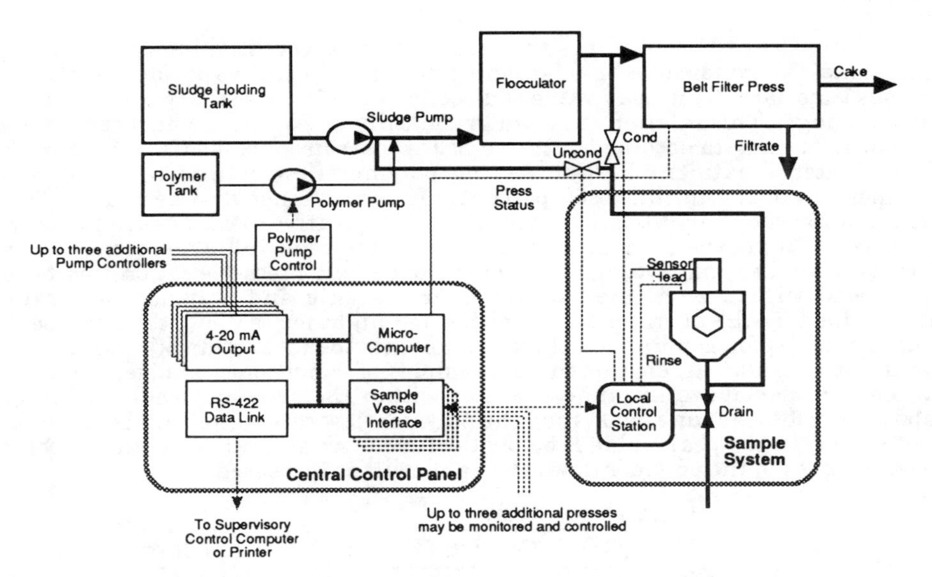

Figure 2. SCC Block Diagram

SOFTWARE

The most vital part of any computer-controlled system is the software. For the SCC, the software controls all aspects of the operation including flushing sample lines, filling the sample vessel, and changing operational parameters, as well as data collection, analysis, and output, and the control algorithm.

All parameters that control how the system will behave at a specific site can be changed using the built-in keypad operator interface. The values of these parameters are presented on the two-line LCD screen and can be viewed only, but not changed, unless the appropriate password is entered. Some of the parameters must be established only during the commissioning of the system and not adjusted again. Examples of these are the flushing and filling time parameters for the unconditioned and conditioned sludge samples, as well as the drain and rinsing times. Operational parameters such as controller gain, solids sensitivity, setpoint deadband, maximum polymer change, minimum and maximum polymer flow, and sample cycles per tune can be changed by the operator to customize the response of the system to the specific needs of the plant.

The data output consists of a summary for each sample cycle of the information collected during the calibration check in air, analysis of both the unconditioned and conditioned sludge samples, the present tune and auto setpoint values, and the most recent polymer flow rate signal. Each set of data is preceded by the date and time for reference purposes. The data are transmitted in ASCII format and serial form on an RS422, RS232C-compatible data link.

The main objective of the control algorithm is to maintain constant sludge conditioning with varying solids and dewaterability. To achieve this objective the SCC compares the present state of conditioning with the reference point of known satisfactory conditioning obtained during the tuning cycle. If the present state of conditioning is either satisfactory or better than satisfactory, then the algorithm directs the polymer flow signal to be reduced. If, on the other hand, the present state of conditioning is not satisfactory, then the algorithm directs the polymer flow signal to be increased.

The SCC determines the present state of conditioning by analyzing the sample of conditioned sludge and finding the peak value of the torque versus time data. This peak value is directly related to the state of conditioning of the sludge. This relation is shown graphically in Figure 3, which represents demonstration data obtained with the SCC sensor head at the City of Detroit's Wastewater Treatment Plant. As the polymer flow rate is reduced from 100 percent of the initial to 60 percent, the peak values also decrease. This information shows that the SCC sensor head is sensitive to changes in the polymer flow. From experience, we also know that the peak value represents a particular state of conditioning. Therefore, once belt press operation has been optimized with a particular sludge characteristic and polymer flow rate, determined by the operator to be neither too high nor too low, then the peak value obtained on conditioned sludge represents the ideal state of conditioning or the tune peak. If, on subsequent samples of conditioned sludge, the peak value is either above the tune peak, or within the setpoint deadband range above or below the tune peak, then the polymer flow rate signal will be reduced. Conversely, if the peak value is below the tune peak and outside of the setpoint deadband, then the polymer flow rate signal will be increased.

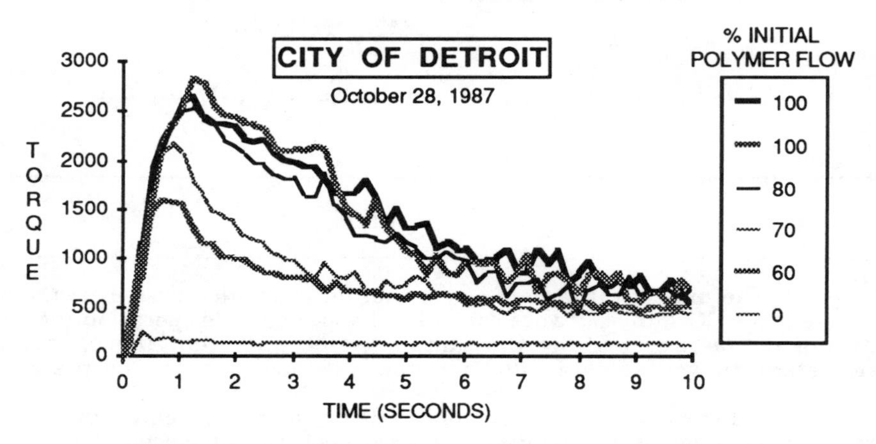

Figure 3. Typical SCC Demonstration Results

The preceding description assumes that only the polymer demand of the sludge changes, not in the solids concentration of the sludge. If the solids concentration changes, we again know from experience that the peak value of the conditioned sludge will also change. They are related such that a decrease in solids concentration will cause a decrease in peak value, even for the same state of conditioning. But using the previous approach, a decreasing or lower peak value would result in an increase in polymer flow rate. If the solids are decreasing, however, the polymer flow rate should be decreased, not increased. Therefore, additional information is required to determine whether the decrease in peak value is due to increased polymer demand or decreased solids concentration.

This information is obtained by analyzing an unconditioned sludge sample. The nature of the response of the SCC sensor head to unconditioned sludge is very different to that of conditioned sludge. Referring again to Figure 3, the curve for the 0 percent initial polymer flow rate is the typical response with no high peak value, simply a constant low-level torque value. The average value of the last two seconds of data is defined as the base value. By testing, we have found a relationship between the base value of the unconditioned sludge and the solids concentration. They are related such that decreases in solids concentration will cause decreases in the unconditioned base value. This is shown graphically in Figure 4. Therefore, the tune peak value can be adjusted in relation to the unconditioned base value to ensure that the

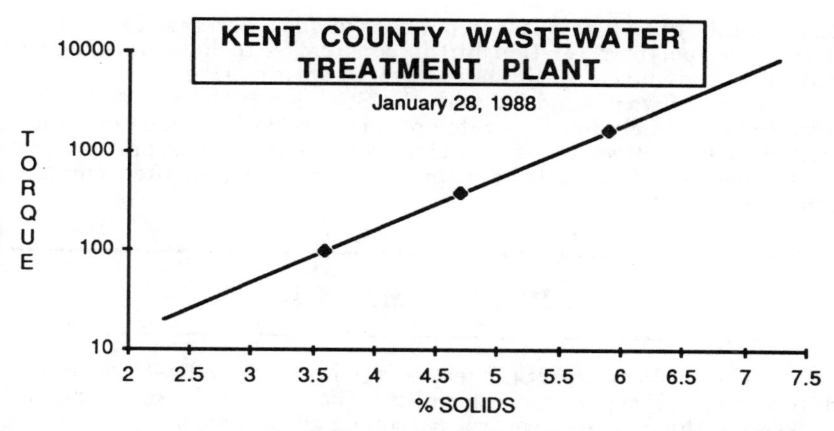

Figure 4. Typical Torque Versus Solids Concentration Results

polymer flow signal is always adjusted in the correct direction regardless of what happens to the solids concentration of the sludge.

For most sludges, up to a certain polymer dosage, the relationship between the conditioned sludge peak value and polymer flow rate is direct, that is, increasing the polymer flow rate with all other conditions held constant results in an increase in conditioned sludge peak value. Usually at some relatively high dosage in the overdose range, however, the relationship changes so that further increases in polymer flow rate cause the conditioned sludge peak value to decrease. The general nature of this more complete relationship is depicted in Figure 5. It is apparent that, using the control logic outlined above, once the polymer flow rate exceeded the maximum value on the Figure 5 curve, the system would be out of control. First the peak drops, and the algorithm responds by increasing the polymer flow rate. Then the peak drops again and polymer flow rate increases again. This process will continue until the maximum available polymer flow rate is reached. To prevent this unstable situation from occurring, the algorithm is continually testing for the relationship noted above in which increases in polymer flow rate result in peak values that are further away from the adjusted tune peak. When this condition is recognized, the algorithm continues to control, but now in the opposite sense; that is, if the peak is too low, the polymer flow rate is decreased, a change which should result in an increase in the peak for the next sample. This direction of control

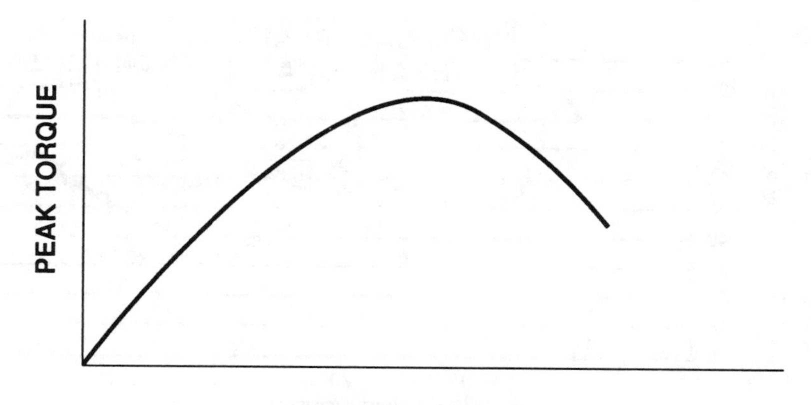

Figure 5. Typical Peak Torque Versus Polymer Dosage Relationship

action continues until a decrease in polymer yields a decrease in peak. At this point the peak/polymer relationship is considered to have returned to the normal state, and control continues in the original direction.

The software supplied with the SCC is complete, and the SCC will perform its task of polymer flow rate optimization without the addition of any other hardware or software. New versions of software with improved capability and ease of use are continually being prepared and released after complete field performance testing.

PERFORMANCE

The application of SCC technology is very broad with respect to both the dewatering device and the material to be dewatered. In addition to all belt press designs, the SCC can control polymer addition on belt thickeners, dissolved air flotation units, centrifuges, filter presses, and vacuum filters. Although developed initially for municipal wastewater sludges, the measurement technique is equally applicable to coal refuse slurries, chemical plant organic sludges, foundry inorganic sludges, and pulp and paper sludges, to name a few. In short, any dewatering or thickening operation that uses a flocculating agent for conditioning prior to treatment can benefit from the use of the SCC.

In municipal plant operation, the range of solids concentrations on which the SCC has worked successfully is from about 0.3 percent for waste-activated sludges up to about 7 percent for anaerobically digested sludges. In industrial applications, the lower limit is comparable, and although the sensor head carries no theoretical limitation, the practical limitation of being able to get the sludge samples to flow into the vessel governs the upper limit. Fortunately, this limitation is usually shared with the dewatering device, so that if the sludge will flow to the machine, it will flow to the SCC. As an example, in coal refuse applications, solids concentrations up to about 40 to 45 percent have been routinely sampled.

Over a dozen SCC systems are presently in the field in eight countries around the world. Typical operating data showing the changes in polymer flow rate signal with time are presented in Figures 6, 7, and 8. The savings indicated are specific to the time interval of operation, but are representative of normal operation. The average polymer savings range from about 15 to 40 percent, depending to some degree on the amount of overdosing experienced during manual operation.

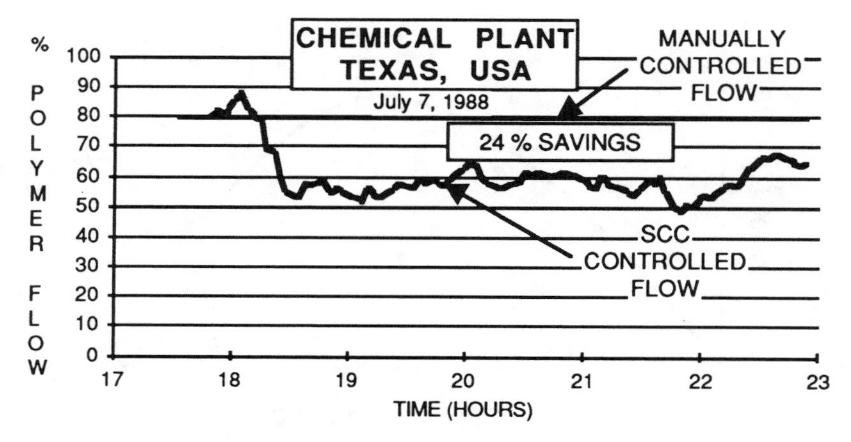

Figure 6. Typical Operating Results on Chemical Biowaste Sludge

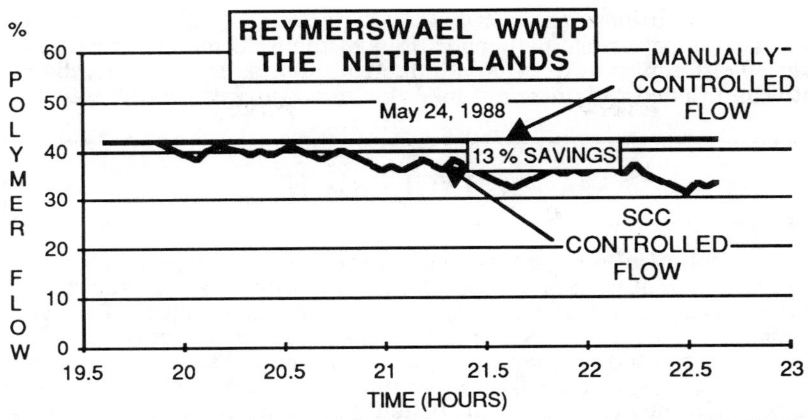

Figure 7. Typical Operating Results on Municipal Digested Sludge

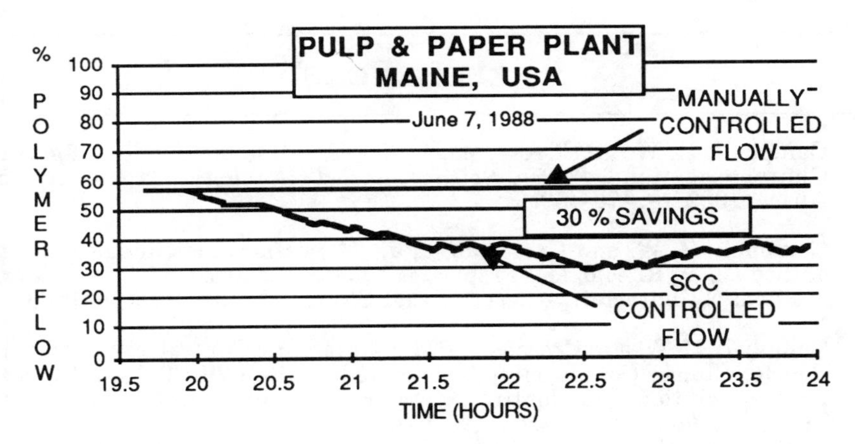

Figure 8. Typical Operating Results on Pulp and Paper Sludge

Based solely on polymer savings, the payback period for the SCC ranges from about 4 months to 4 years, with an average of about 1.5 years, depending mainly on the pre-SCC polymer consumption.

Additional benefits accrue with the installation of an SCC. The operator will gain the freedom to concentrate on maximizing the solids throughput on the dewatering device. Typically, these machines are mass flux limited: this means that if the solids concentration of the sludge decreases, the sludge flow rate can be increased to maintain the same solids loading. Normally, this is very difficult because the operator would have two variables to adjust, sludge flow rate and polymer flow rate. With the SCC controlling the polymer flow rate automatically, the operator can adjust the sludge flow to whatever is desirable. A future development of the SCC will be to use the solids concentration information obtained from the unconditioned sludge samples to control the sludge flow rate. By optimizing this aspect of the dewatering operation, an increase in total capacity can be obtained without resorting to the major capital investment of purchasing additional dewatering machinery.

Another benefit of using the SCC is the more precise control of cake solids that it provides because of the continuously optimized sludge conditioning. Since all forms of ultimate sludge cake disposal are cost sensitive to the cake solids concentration, this factor is very important with respect to savings and payback period.

In addition to the full-scale installations, Zenon and its overseas distributors have performed demonstrations of the SCC measurement technique at over 50 different municipal and industrial plants. In all cases, the response to both conditioned and unconditioned sludge samples was as expected.

CONCLUSION

Optimization is a vital part of the advanced management approach to wastewater treatment plant operations. Shrinking capital budgets force the need to upgrade wastewater treatment plants to operate more effectively without major investment. The SCC meets these requirements with an inexpensive microcomputer-based approach designed to implement sludge conditioning process control strategies reliably and economically on a wide variety of sludges and dewatering devices. The many benefits include automating the dewatering device operation, increasing capacity and more uniform performance of the dewatering machine, and savings in polymer costs.

REFERENCES

1. Campbell, H. W. and Crescuolo, P. J., "The Use of Rheology for Sludge Characterization," *Water, Science, and Technology*, Capetown, South Africa, 1982, *14*, 475-489.

2. Campbell, H. W. and Crescuolo, P. J., "Assessment of Sludge Conditionability Using Rheological Properties," presented at an EEC Workshop on Methods of Characterization of Sewage Sludge, Dublin, Ireland, 1983.

3. Campbell, H. W. and Crescuolo, P. J., "Automatic Control of Polymer Addition for Sludge Conditioning," presented at the IAWPRC Conference on Instrumentation and Control of Water and Wastewater Treatment and Transport Systems, Houston, TX, 1985.

4. Campbell, H. W. and Crescuolo, P. J., "Control of Polymer Addition for Sludge Conditioning: A Demonstration Study," *Water, Science, and Technology*, presented at the ICA session of the 14th IAWPRC Conference, Brighton, England, in press, 1988.

5. Campbell, H. W., Siverns, S. L., Crescuolo, P. J., and Harrison, E. J., "An Instrument for Automated Control of Sludge Conditioning," presented at the 9th AQTE Symposium on Wastewater Treatment, Montreal, Quebec, 1986.

THIN CAKE FILTRATION: THEORY AND PRACTICE

K.-S. Cheng, Ph.D.
Research Process Engineer
Millipore Corporation
80 Ashby Road
Bedford, Massachusetts 01730
(617) 275-9200

ABSTRACT

Thin cake filtration (TCF) is a useful technique for improving the filtration rate for a certain class of slurries. Contrary to conventional filtration in which cake deposits accumulate continuously during the filtration process, TCF controls cake growth by maintaining mechanical agitation close to the filter medium surfaces. The shear-compacted thin cake, only 2 mm thick, serves as a permanent precoat, delivering high rate and clear filtrates. Mechanical forces break down flocculated particulate structures, producing filter cakes with high solids content.

Although the effectiveness of TCF is well published and documented, the basic mechanisms are poorly understood. Current theories based on hydrodynamic models provide no more information than empirical models for curve fitting of experimental data. More importantly, they fail to explain why the filtration rate decreases irreversibly with increasing solids concentration. Our extensive experiences with TCF lead us to conclude that shear consolidation and particle blinding of the thin cakes provide a much better explanation of the mechanisms of TCF.

Practical applications of TCF for washing and filtering are discussed, and guidelines for applying TCF are described in detail.

INTRODUCTION

Thin cake filtration (TCF) is a relatively new concept for improving filtration rate. In conventional filtration, slurry flows perpendicular to the filter medium. The filter cake, composed of particulate deposits, becomes the resistance to flow. Increasing pressure is effective only for relatively incompressible slurries. But it also compresses the filter bed, reducing the size of the flow channels, thus offsetting some of the increased driving force. In extreme cases for very compressible materials, pressure has no effect on filtration rate.

Thin cake filtration, also called delayed cake filtration, cross-flow filtration, or dynamic filtration, tackles the root of the filtration resistance problem by limiting cake growth. Many variations of this concept have been implemented. In ultrafiltration, hydraulic forces are employed to pump slurries at high velocities to minimize cake buildup. Devices using vibration of the filter medium to shake off the filter cake have been built. Totally submerged rotary drums[1] with continuous cake removal have been tried. In the following, we will discuss a particular class of thin cake equipment that employs a rotating turbine blade to maintain thin filter cakes.

The advantage of thin cake filters are many: high rate/filter area, small physical size, clear filtrate, high cake solids, pumpable filter cake, and automatic and continuous operation. Its disadvantages are high cost/filter area and the potential danger of blocking, which will be explained below.

EQUIPMENT DESCRIPTION

Figure 1 shows a thin cake staged filter (artisan filter), consisting of alternating, rotating, and stationary filter elements. Slurry enters one end of the filter, and thick extrudate (cake) emerges from the opposite end. Since the highly concentrated cake is non-Newtonian, rheological properties of the filter cake are very important in determining the operating characteristics of the filter. The fixed plates are made in concentric circular patterns with a shaft passing through the inner radius. The turbines are composed of a solid disk and

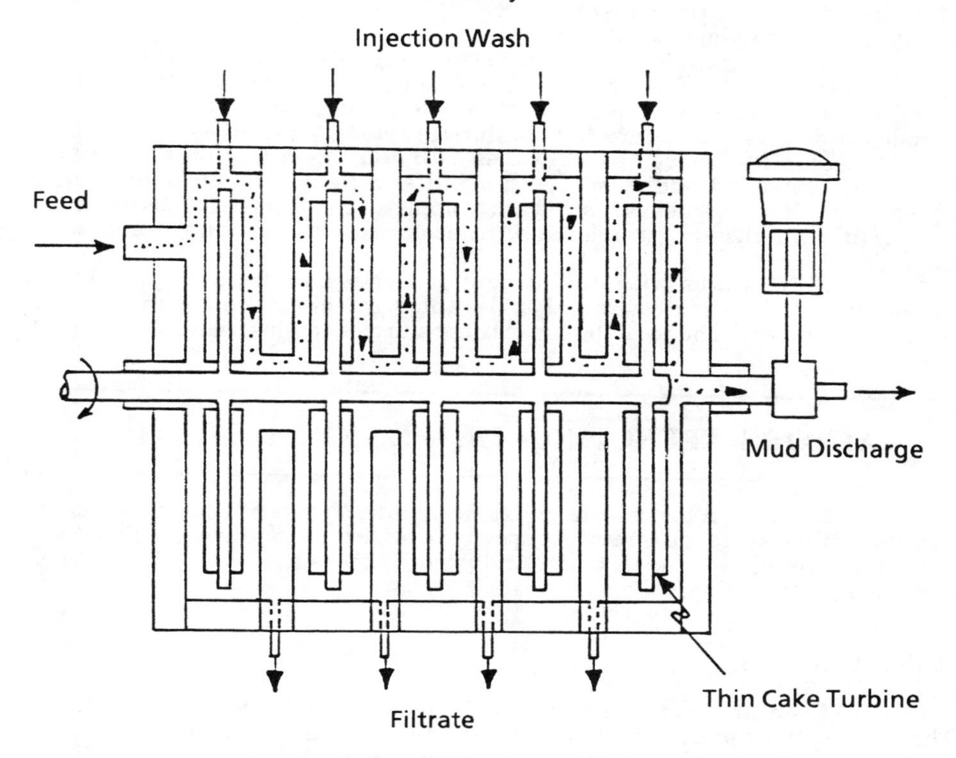

Figure 1. Thin Cake Filter

blades, rotating at a clearance of 2 mm from the filter medium. Rotating speed is kept low intentionally, at 175 rpm for a 2-ft-diameter plate, to avoid excessive abrasion of the blades. The concentrated paste is discharged through a modulated rubber pinch valve. Since viscosity is directly related to the power consumption of the drive motor, the solids content of the filter cake is kept relatively constant by monitoring and maintaining a constant level of torque.

There are many variations of this concept[1]. Earlier models of the thin cake filter (dyno filter) have a rotating filter plate instead of a turbine. Although providing more filter surfaces, these devices did not generate enough shear forces to keep the thickened cake moving, resulting in costly shutdown known as blocking. Filter blocking is a unique phenomenon in cross-flow filtration; in this phenomenon, the viscosity of the thickened slurry becomes so high that it loses the ability to flow. When that happens, the filter has to be shut down, completely disassembled, and cleaned, a time-consuming and labor-intensive task. Another variation of this concept uses a grooved rotating disk, which generates less shear force and is less effective than the turbine blades.

APPLICATION

Since the filter cake, a highly concentrated suspension, must be discharged as an extrudate, the rheology of the slurry is the most important consideration for TCF applications. Not all filter cakes can be extruded. Slurries containing substantial amounts of particles larger than 30 μ will not produce pasty filter cake for extrusion. Filter cakes containing large particles tend to plug the cake valve, consequently blocking the filter. Fine sand and grainy catalysts are examples of this category.

At the other end of the spectrum, slurries containing a large fraction of submicron particles deliver such a low rate (below 1 gal/hr/ft^2) that in most cases, TCF is not feasible economically. Blinding slurries such as oil waste and completely dispersed titanium dioxide slurries belong to this class.

Slurries producing filter cakes with dilating properties also will not filter well. When left undisturbed, dilated filter cakes have a moist and liquid appearance. But when shear is suddenly applied, the filter cake becomes dry and hard. Quicksand, starch, and ground calcium carbonate are some examples.

Filter cakes with thixotropic properties such as clay, calcium carbonate, metal hydroxides, metal oxides and pigment generally run very well in thin cake filters. The particle size of these slurries ranges from 0.8 to 10 µ.

FACTORS AFFECTING THIN CAKE FILTRATION

In addition to pressure and temperature, TCF is affected by a change in solids concentration, the rotating speed of the turbine, and the filter media. I will give a brief description of my industrial experience with each of these factors in the following.

RATE DECLINE

Even at steady state, thin cake filtration rate declines with time. The rate declines most quickly during the first few hours; thereafter, it continues to drop for the next few days, although at a much lower rate. The filtration rate normally will stabilize in about a week. A typical run is shown in Figure 2.

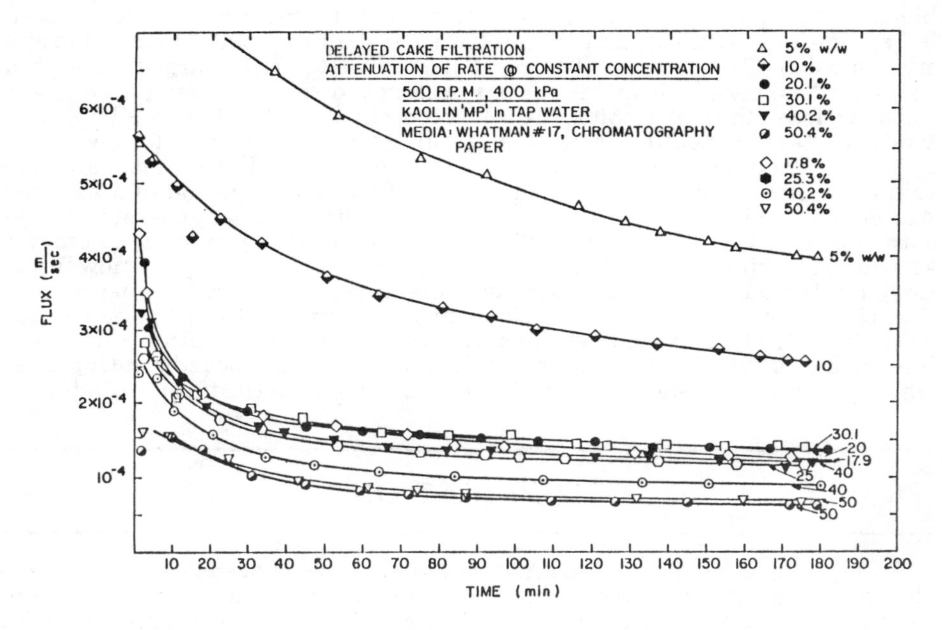

Figure 2. Filtrate Rate Versus Time

The decline is probably caused by continuous compaction of the thin cake by the turbine, particle blinding, and filter medium compaction. Backflush of the thin cake to regenerate the rate is only marginally effective if the medium is blinded.

Slurries with very narrow particle distribution exhibit the worst decline. Clay, calcium carbonate, and other materials with wider distribution fare better. Some aluminum hydroxide slurries stand out as exceptions--these slurries maintain a rate that never declines. The highly flocculated structure and complete absence of loose fine particles are probably responsible for this behavior.

SOLIDS CONCENTRATION

Filtration rate invariably decreases with solids concentration (Figure 3). Unlike other factors, its effect on rate is irreversible--i.e., once the thin cake is exposed to high solids concentration, the filtration rate remains depressed even if the solids concentration is reduced. In industry, we called it the "cake poisoning effect."

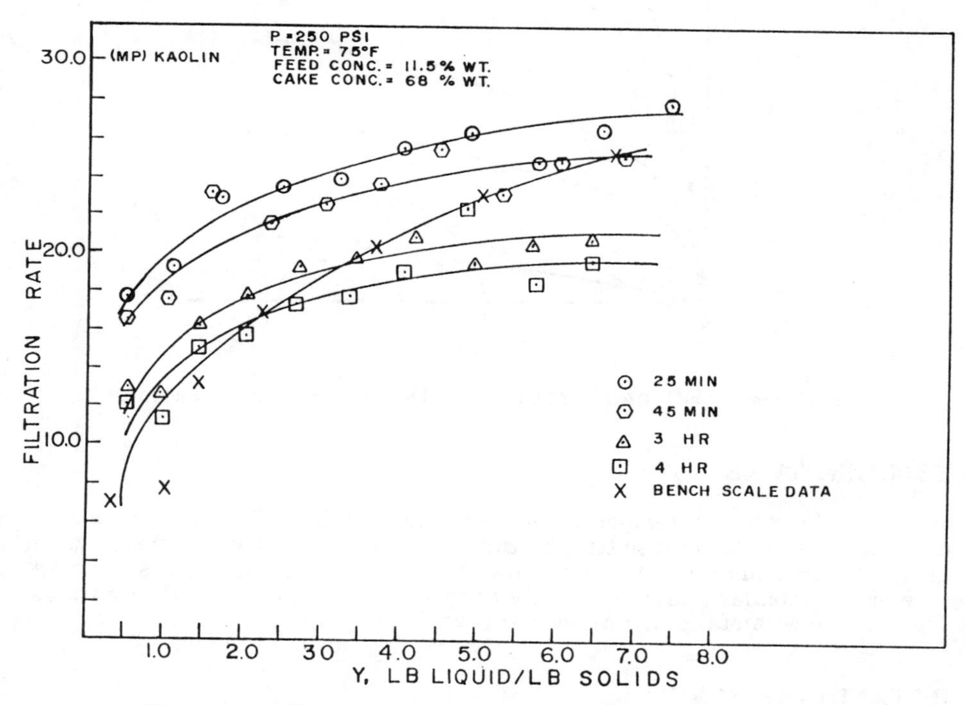

Figure 3. Filtrate Rate Versus Solids Concentration

PRESSURE

Filtration rate generally increases with applied pressure (Figure 4). However because of compression of the thin cake, the flow rate does not increase proportionally with pressure. For some materials, pressure produces a faster rate decline. In general, pressure produces mostly a transient effect that will dissipate in a couple of days. For example, increasing the pressure from 60 psig to 300 psig for kaolin slurries may bring 30 percent long-term improvement in rate.

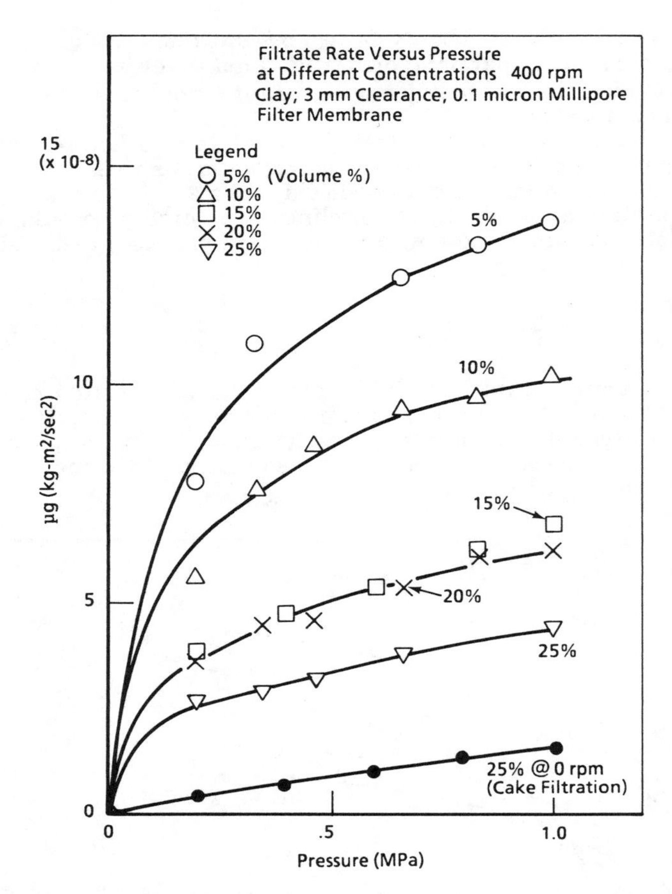

Figure 4. Effect of Pressure on Delayed Cake Filtration

TEMPERATURE

Increase in temperature generally increases filtration rate. Its effect is predictable because temperature lowers the viscosity of filtrate, which brings a corresponding increase in flow rate. In practice, the process temperature of a particular slurry is usually fixed because of process considerations. Therefore, few can take advantage of this variable.

ROTATIONAL SPEED

For most slurries, the filtration rate can be increased by raising the rotational speed of the turbines (Figure 5; the magnitude of improvement is, however, unpredictable. In general, less compressible slurries such as iron oxide, kaolin, and calcium carbonate respond positively to increasing speeds, whereas very compressible materials do not. In industrial environments, rotational speed is kept low to prevent excessive abrasion of the turbines. Furthermore, high rotational speeds cause heat buildup in filter cakes because of viscous dissipation.

It is not clear why rotational speed affects filtration rate. Very high speed can produce thinner cakes(1), but filtration rate improves even at moderate speed increases, which have no significant impact on cake thickness. Furthermore, unlike the effect of solids concentration, the effect of speed is

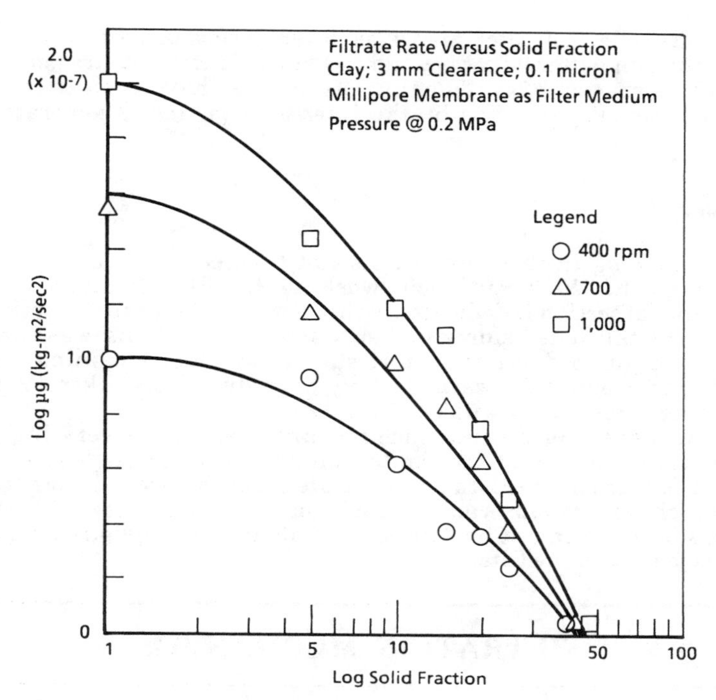

Figure 5. Filtrate Rate Versus Solids Fraction at 3-mm Clearance and 0.2 MPa

somewhat reversible, i.e., the rate will go up and down with rotational speed for many cycles.

A possible explanation for the effectiveness of speed is that the clearance between the blade and the filter cake increases at higher speeds, thus reducing the shear forces on the surface of the filter cake. Furthermore, the additional turbulence at higher speeds produces thinner cakes and decreases the probability that fine particles will penetrate the thin cakes.

CAKE THICKNESS

The thickness of the thin cake is determined by the speed of the turbines and by the clearance between the blade and the filter medium. For low rotational speeds, we found that a 2-mm clearance is the optimum. Reducing this clearance further will cause seizure of the turbines by the filter cake, and increasing this clearance will cause excessive cake buildup.

At higher rotational speeds, the clearance requirement can be relaxed. However, the filter will suffer increased abrasion and potential heat buildup in the filter cakes.

FILTER MEDIA

Thin cake filters have very peculiar requirements of the filter media. Most filter cloth or paper will not work because they cannot maintain rigidity and flatness over a 2-ft diameter. Filter felts usually are preferred because of their rigidity. Lately, certain newly developed reinforced filter cloths can be used also. For this reason, thin cake filters usually cannot take advantage of the latest developments in filter media.

The effect of filter media on filter performance is unpredictable. Some slurries run better with very tight medium (1 cfm); others perform much better with more open ones (30 cfm). Membranes have been tried, but they invariably wear off with time by the intense shear forces generated by the turbines.

WASHING

Washing in thin cake filters can be accomplished either continuously or by batch. With continuous washing, the filter is divided into three parts: the initial section is dedicated to thickening the feed to the optimum concentration, the thickened slurry is then washed by injecting wash water at a predetermined number of consecutive stages, and the remaining stages are dedicated to thickening the washed slurry to produce filter cakes. Washing in thin cake filters has been discussed elsewhere(2).

However, continuous washing cannot achieve a very high ratio of soluble removal because of the limited number of filter stages available for washing. For these slurries, the concentrated feed is recycled back to the feed tank, in which wash water is added continuously until the desirable amount of soluble has been taken out by filtration. This method obviously disables the continuous feature of the filter.

FILTRATION MECHANISM

Zhevnovatyi(7) probably was the first to investigate thin cake filtration. He investigated cake formation under shear forces produced by hydraulic means. He attempted to explain the data by applying boundary layer theory with suction to a monolayer of particles. The derived result, however, amounts to no more than a dimensional analysis of the problem.

Malinovskaya and Schevchenko(4) studied thin cake formation on the rotating filter plate. The filtration rate was found to reach a maximum with increasing rotational speed. They reasoned that the centrifugal forces generated by the rotational motion probably counter balanced the applied pressure at higher speeds. They proposed that specific filtration resistance of the filter cake increases at higher speed.

Other researchers, such as Porter(5) and Cheng(1), extended the concentration polarization model of the ultrafiltration model to explain thin cake filtration. The polarization model suggests that a layer of concentrate builds up at the filter medium interface, creating a concentration gradient across the boundary layer. As filtration brings a flux of particles to the surface, the thickness of the concentration layer builds up, slowing down the filtration rate until the rate is balanced by the rate of diffusion of particles back into the bulk.

This model, sometimes called particle polarization, works quite well with colloidal structures or molecules. With particles larger than 0.2 μ, this theory breaks down because the slow diffusion rate of larger particles cannot account for the much higher observed filtration rate. Porter(5) suggested that diffusion is augmented by other mechanisms such as radial migration. But radial migration is such a weak force, observable only in laminar flow, that this mechanism is a hypothesis at best.

Current research focuses on studying the flow patterns inside the filter(3).

The weakness of all of these existing theories is that they are inconsistent with one very important experimental observation: after the thin cake is formed, the filtration rate stays depressed even if the thickened slurry is purged and substituted with a dilute slurry. It is common knowledge among thin cake filter users that, unless the thin cake is removed physically, the filtration rate will remain low, even after a flush-out.

Therefore, any theory relating filtration rate with concentration difference must take into account the irreversible nature of the filtration rate. The boundary layer and particle polarization theories cannot stand up to this test: the premise that high concentration depresses the filtration rate implies that lowering concentration will increase the filtration rate.

A CONJECTURE

Thin cake filtration is a complicated phenomenon. Thin cake filters actually produce a new class of very high density solids suspension that cannot be produced by simple mixing. The shear forces in a thin cake filter are so strong that they break down floc structures to allow tighter particle packing. Until the basic factors can be better understood, attempts to describe this process mathematically will be futile. After years of experience and observation of thin cake filtration, I offer the following conjecture.

The irreversible nature of the filtration rate shows us that filtration resistance resides only in the thin cake and the filter medium. At low rotation speeds (below 200 rpm for 2-ft diameter under typical industrial operating conditions), the variations of cake thicknesses from stage to stage are minimal. Since the filtration rate decreases with increasing solids concentration and cake thickness remains relatively constant, the specific filtration resistance of the filter cake must increase with solids concentration.

To account for the increasing specific filtration resistance with concentration and its irreversible nature, I propose the concept called shear compression. The conjecture is that the thin cakes experience increasing shear forces at higher solids concentration because of higher slurry viscosity. The hydrodynamics theory of lubrication shows that very high normal forces can be generated by the flow of viscous fluid through small channels[6].

The intense shear forces and the enormous pressure forces generated by viscous flow through small crevices break down the particulate structure of the thin cake, creating increased resistance to flow. Since solids compression is irreversible, this explains why the filtration rate would not increase even after the solids concentration is reduced. Visual observation of the thin cake supports this theory as thin cakes appear to be highly compacted and brittle to the touch.

Tiller shows that cake consolidation is a gradual phenomenon, which probably explains the initial decline of the filtration rate within hours after start-up. The long-term rate decline is probably caused by continuous fine particle impingement, reducing the pore size for flow. The irreversible part of the rate decline is probably caused by severe compaction of the filter medium from the turbines and fine particle penetration into the medium matrix.

CONCLUSION

Thin cake filters represent the latest advance in solid/liquid separation technology. For a certain class of slurries, thin cake filters easily outperform conventional filtration equipment in terms of filtrate rate per square foot of filtration surfaces. But because thin cake filters are complex and expensive to build, each application must be examined in its own merits. Until thin cake filters can be built more inexpensively, their use will be limited to special applications.

The basic mechanisms of thin cake filtration are exceedingly complex. Efforts to understand the structure of the thin cake have been hampered by the thinness of the filter cake itself. At a thickness of only 2 mm, it would be extremely difficult to build a pressure probe to measure the hydraulic pressure drop inside the thin cake. Without this information about the local conditions, the thin cake phenomenon will remain as a black box. We can try to propose different hypothesis to explain it, but we will never be able to prove it.

REFERENCES

1. Cheng, K.-S., Doctoral Dissertation, University of Houston, 1979.
2. Cheng, K.-S., Tiller, F. M., and Bagdasarian, A., *Filtration and Separation*, 1982, March/April.
3. Lu, W. M. and Chuang, C. J., *J.Ch.E. Japan*, 1988, *21* (4), 368.
4. Malinovskaya, T. A. and Schevchenko, V. F., *Theoretichsekoi Tekhnologii*, 1973, *1*, 592.
5. Porter, M. C., *IEC Prod. Res. Dev.*, 1972, *1*, 2.
6. Schlichting, H., *Boundary Layer Theory*, McGraw Hill Book Co., 2nd Edition, 1960, 279.
7. Zhevnovatyi, A. I., Trans. from *Zhurnal PriKladnoi Khimmi*, Plonum Pub. Corp., NY, 1973, *48*, 334.

ELECTRO-OPTIC STUDIES OF CONCENTRATION BANDING IN PARTICULATE MEDIA

B. R. Jennings, R. Isherwood, and M. Stankiewicz
Optics Group
J. J. Thomson Physical Laboratory
University of Reading
Whiteknights
P.O. Box 220
Reading, Berkshire RG6 2AF, UK

ABSTRACT

A phenomenon has been observed in which particles in suspension associate into highly concentrated "bands" that arrange themselves perpendicular to the field lines of uniform electric fields. These bands concentrate with time and field strength and are optimized for given field frequencies. They only form in alternating electric fields and are in opposition to the well-known pearl-chain tendency for particles to associate along field lines in dc electric fields. Electric birefringence and electrophoretic light scattering measurements have been made to demonstrate the tendency of pearl chains to be associated with the electrical polarizability, whereas banding is predominantly associated with electrophoretic mobility of the particles. The generality of the effect is shown by its existence in various emulsion systems and rod and disk-like particles.

INTRODUCTION

It is well known that particles or droplets associate or coalesce in the presence of electric fields. This has been commercially exploited both in the design of electrostatic precipitators(4) for aerosol and airborne particles and in the use of electrostatic separators(5) that isolate immiscible fluids such as oil-water emulsions. In each case the phenomena must relate to the electrical properties of the particles or droplets in their suspending environment. Little is known of the details of the mechanism of association.

We have recently reported(9) some observations on a novel phenomenon in which kaolinite discotic clay particles in suspension can be swept into regions of high density in the presence of alternating electric fields. Unlike the established pearl-chain phenomenon(14,16), in which particles associate along the field lines in dc electric fields, the bands accompany alternating fields and only appear after a short period of time (typically of the order of minutes). The object of this paper is to describe these bands and the conditions under which they appear. A theoretical description of their origins indicates that chains and bands are associated with different electrical properties of the particles within the medium. Auxiliary electro-optical experiments using electric birefringence(7) and electrophoretic light scattering(2) that support the theoretical description are described. Results on a wider range of materials indicate the universality of the phenomenon.

EXPERIMENTAL

Direct observation of the banding phenomenon was made in a thin glass cell that held a pair of electrodes so as to give a horizontal electric field across the interspace. Strips of 0.1-mm stainless steel provided electrodes, which were isolated from the glass with rubber spacers. The field was applied across the 4-mm gap within the 23 × 0.9 mm cell. The cell was in the form of a pair of microscope-type slides, which were then mounted directly in a Karl Zeiss Jena "Ergaval" microscope. This was coupled directly to a CCTV camera whose output was displayed on a visual display unit. Video recording of the dynamic processes involved was thereby facilitated.

Using a range of electric generators available to this research team, both dc and ac fields of continuous or pulsed forms could be applied for up to 30 minutes' duration with fields of amplitude up to 400 V/cm and for various frequencies below 20 kHz. Observations were made on a variety of materials, falling broadly into two classes. The first was for aqueous suspensions of mineral particulates of various geometry. The second was for emulsions in which either water or oil were the predominant phase. In addition, the cover glass material was changed to include polished polymethyl methacrylate sheets or glass that had been coated with trimethylchlorosilane, which is a strongly hydrophobic material. The independence of the effects seen on the substrate material indicated that the phenomena to be described were not associated with the substrate surface.

ELECTRIC BIREFRINGENCE OBSERVATIONS

Under the influence of a pulsed electric field, anisotropic particles rotate into orientational order as their permanent or induced dipole moments couple with the applied field. Electric birefringence measurements are undertaken by the placing of the sample in a glass cell across which a pair of electrodes generate a transverse potential gradient. A light beam propagates between the electrodes after having been polarized at 45° to the field direction. The cell is followed by a quarter-wave plate whose fast axis is set parallel to the polarizing azimuth of the incident beam. Light then falls on a polarizing analyzer, which is set in quadrature azimuth to that of the polarizer. Light only penetrates this optical arrangement if the sample is isotropic and non-birefringent. Upon the induction of orientational order, any optical anisotropy associated with the particle gives rise to a collective birefringence (Δn), which can be related to the intensity of the light signal that penetrates the analyzer and falls upon a photodetector. Details of the optical array, its adjustment and manipulation can be found elsewhere(15).

Induced birefringence is wavelength dependent and is often described in terms of a Kerr constant(1):

$$K = \operatorname*{Lt}_{E \to 0} \left(\Delta n / \lambda E^2\right) = \frac{2\pi c}{15 n \lambda} \Delta g \left\{ \frac{\mu^2}{k^2 T^2} + \frac{\Delta \alpha}{kT} \right\} \quad (1)$$

Here E is the applied field strength, c is the concentration of the suspension of refractive index n, and Δg is the anisotropy in the optical polarizabilities associated with the major and minor axes of the particles. The parameters k and T are the Boltzmann factor and the absolute temperature, respectively. Equation (1) is written for dc fields. For alternating fields, the polarizability is insensitive to frequency until very high frequencies ($>10^6$ Hz), whereas the permanent moment (μ) is frequency dependent and exhibits a frequency dispersion phenomenon. For particles of droplets of the order of 1 μ in size, an ac field of 1 to 10 kHz is sufficient to eliminate the μ contribution. Hence $\Delta\alpha$. the difference in the polarizabilities (α_i), with $i = 1$ or 2 can be obtained. Here α is the electrical polarizability associated with the major (subscript 1) or minor (subscript 2) geometric axes of a particle.

Following the termination of the pulsed field, the birefringence decays in a multi-exponential manner for which the effective time constant defines a rotary relaxation time, τ_r. Particle size can be evaluated from τ if the shape of the particle is assumed. In general(17),

$$\tau_r = \frac{\pi \eta l^3}{18 kT \psi} \quad (2)$$

where η is the solvent discosity, l the major particle dimension, and ψ varies with the shape. Typically it has the value of 1/3 for spheres, is approximately 4 for rods, and varies with the axial ratio for spheroids.

A typical example of a transient response is given in Figure 1. In this case, the amplitude of the effect can be used to evaluate the droplet polarizability, and the rate of the decay is a measure of the droplet size via the parameter, τ_r.

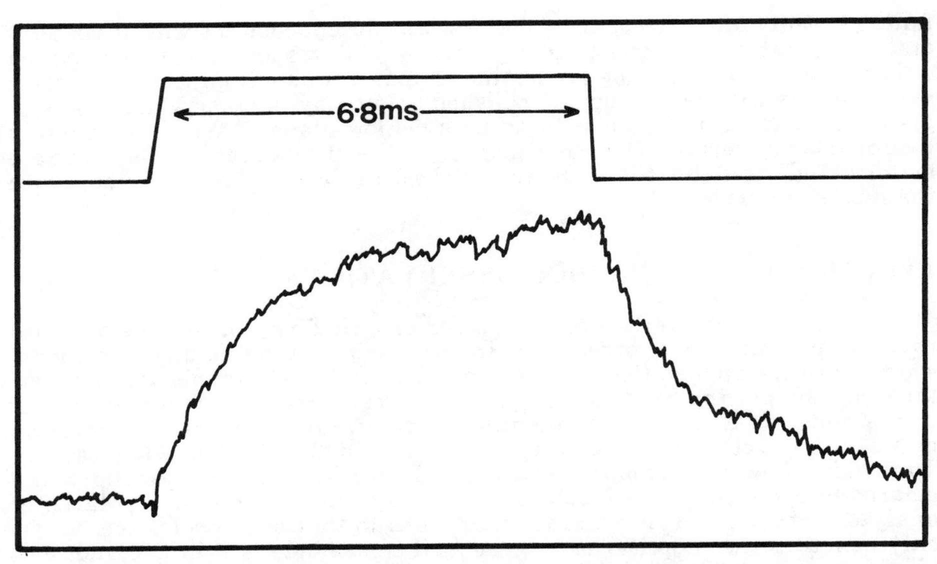

Figure 1. Transient Electric Birefringence of Water Droplets in Kerosene. The applied dc field pulse of 9,000 V/cm was applied for 6.8 μs duration; time runs from left to right. The amplitude of the induced optical birefringence was $\Delta n = 1.3 \times 10^{-9}$.

ELECTROPHORETIC LIGHT SCATTERING

The light scattered by a dilute suspension over a long time period, while assuming a constant average intensity, does in fact fluctuate in intensity as particles undergo translatory motion between chaotic random collisions. Any temporal structure in the flicker pattern can be analyzed to measure this translatory diffusion. The procedure is to count optically the scattered photons over short equi-interval time periods and to autocorrelate these signals together. The period of delay (τ) in the autocorrelation is accompanied by a diminution of the correlation function in a Lorentzian manner. The time characteristics of this decay follow the form(2)

$$C(\tau) = \underset{\substack{\theta \to 0 \\ \tau \to 0}}{Lt} \quad A \exp\left(- 2D_t K^2 \tau\right) + 1 \tag{3}$$

where

$$K = \frac{4\pi}{\lambda} \cdot \sin \theta/2 \tag{4}$$

A is an optical constant, and D_t is the translatory diffusion coefficient of the particles. It can also be related to the major dimension of the particles using equations of the form(18)

$$D_t = \frac{kT}{3\pi\eta} \left\{\frac{\psi'}{l}\right\} \tag{5}$$

where ψ' is also a shape factor. In this case, it is unity for a sphere.

Should the particles be forced to undergo translatory diffusion due to electrophoretic mobility, then the correlation function becomes regenerative and oscillates. To achieve this experimentally, the optical scattering signal must be mixed in a heterodyne procedure with a sample of unscattered light. In that case the correlation function becomes[2]

$$C(\tau) = A' \exp\left(- D_t K^2 \tau\right) \cos \left\{\left(\bar{K} \cdot \bar{v}\right) \tau\right\} \tag{6}$$

where A' is a modified optical constant and v is the particle translational velocity.

Using the fact that the electrophoretic mobility,

$$u = \frac{v}{E} \tag{7}$$

then the period of oscillation of the correlation function is given by[19]

$$T = \lambda \left\{uE \sin \theta\right\}^{-1} \tag{8}$$

where θ is the scattering angle to the forward direction of the incident light beam.

The apparatus for measuring this effect consists of a helium-cadmium or argon laser that projects a blue light beam of wavelength λ through a collimating optical system into the center of a glass cell that holds a few milliliters of the sample. The light scattered at an angle θ is detected by an optical limb, which can be adjusted to variable θ. This limb consists of a lens and a well-defined pinhole to define the optical beam reaching the photocathode of a high sensitivity photon-counting photomultiplier. The signal from this is fed directly to an autocorrelator whose output is available for data analysis or visual display. Full details of the method of measurement have been given elsewhere[8]. A typical example of the response for the samples studied here can be seen in Figure 4 of reference 8 where the field-induced oscillations in the correlation function can be easily seen. For the various samples studied in this work, Figure 2 presents the results of the analysis of such oscillating correlellograms, and displays the data according to equation (8) so that values of the electrophoretic mobility (u) can be determined.

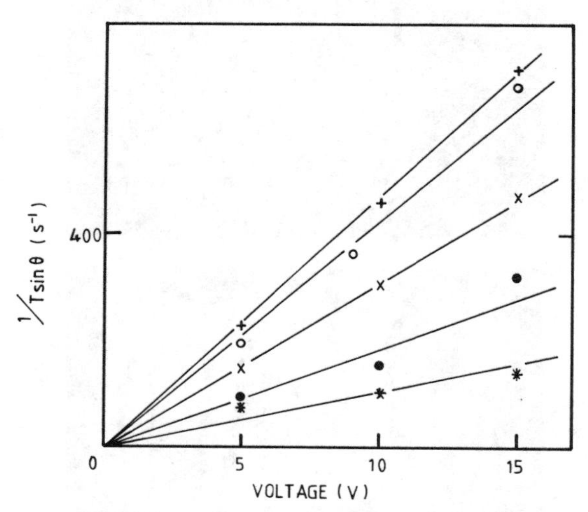

Figure 2. Field Strength Dependence of the Oscillation Period of the Correlation Function. Data for + kerosene; x kaolinite; x crocidolite asbestos suspension; • amosite asbestos and * polytetrafluoroethylene spherical latex. All samples were in aqueous dispersion. The electric voltage was applied across a gap of 1.6 mm. See equation (8) for the significance of the plot.

BANDING OBSERVATIONS

The majority of microscopic observations were made on a suspension of kaolinite clay particles whose dimensions were approximately discotic with 0.5 μm diameter and 0.1 μm thickness. The sample was obtained from Messrs. English China Clays International of Cornwall, UK. The data were obtained on media at neutral pH and 2.5×10^{-3} g/mL concentration. Fields were applied up to 250 V/cm. In all the figures shown here, the electric field vector was from left to right at constant uniform magnitude.

Figure 3 shows the result of an applied electric field of 100 Hz frequency and 125 V/cm continuously applied to the sample. Here, the initial uniform distribution of particles throughout the medium slowly gives way to local clustering of the particles, which concentrate into regions that become essen-

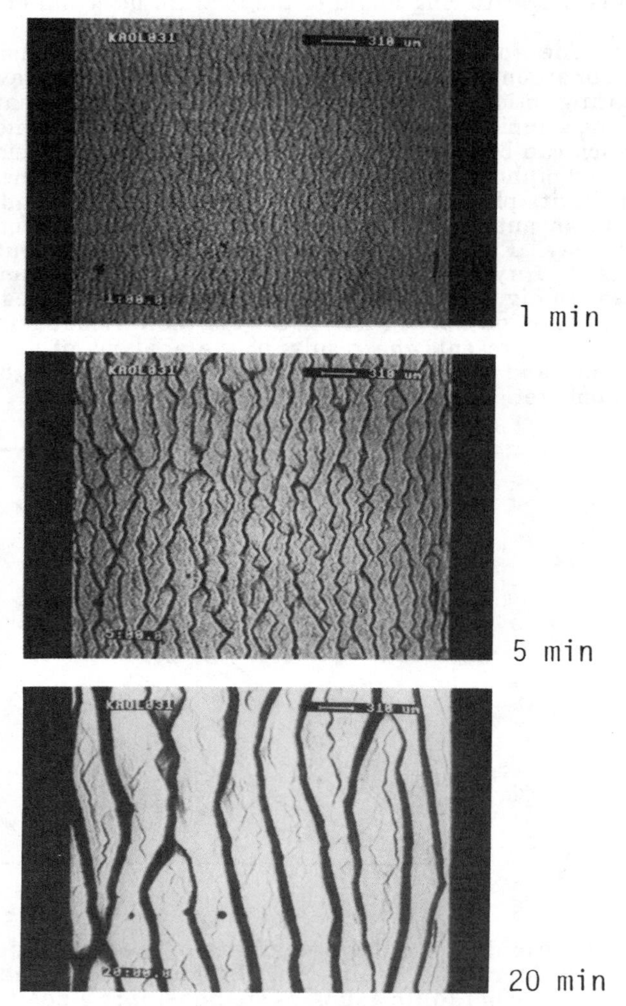

Figure 3. Electrodynamic Concentration in a 0.25 Percent Kaolinite Suspension in Water. The applied field was left to right. Data for E = 200 V/cm and 500 Hz frequency after time intervals from the left of 1, 5, and 20 minutes, respectively.

tially linear and perpendicular to the field direction. These regions are characterized by the following features:

1. They are essentially perpendicular to the field vector.
2. They exhibit kinks of approximately 50° for low fields, straightening toward a 70 to 80° inclination at high fields.
3. The particles move within these kinked regions in a manner that is essentially clockwise in regions kinked to the left and anticlockwise in those inclined to the right.
4. The effect is frequency dependent and optimizes for frequencies of the order of 50 to 100 Hz.
5. The effect is strongly field dependent and is increasingly enhanced with field strength until chaotic turbulent motion eliminates the order.
6. Banding is most predominant at the bottom of the cell, which is presumably the region where the strongest particle concentration is encountered.
7. The number of bands decreases as the effect proceeds, as the concentration in these bands increases.

It is apparent that these bands are regions of extremely high local particle concentration. The process represents a very efficient, electrically driven, dynamic separation procedure when encountered with noncoalescing particles. In such cases, the particles flow within local regions. It is evident that these regions ought to represent a focus for serious study because of their scientific interest and their commercial potential for achieving efficient, electro-dynamic coalescence and precipitation. Finally, attention is drawn to the fact that the bands associate in a direction essentially perpendicular to that associated with the conventional static field pearl-chain formation and should not be confused with that mechanism, which has been extensively studied in the past.

THEORETICAL INTERPRETATION

We have recently considered the origins of the concept and described the relevant theoretical equations fully elsewhere[11]. A summary follows.

PEARL-CHAIN FORMATION

As early as 1936, Krasny-Ergen[12] considered the energy required to bring a sphere of radius r and electrical polarizability α from infinity to a distance R from an identical sphere placed in an electric field E. The direction of approach along R is assumed to be at variable angle ϕ to the field direction. Within the field, both particles polarize and give rise to mutual attraction between extremities of the particles along the field direction. Such mutual attractions eventually lead to linear associations of the particles. These result in the "pearl-chain" or "rouloux" formations often described in the electrostatic literature[13,14,16] for particles that are separated such that R is $\gg$ r. The potential between the pair of particles becomes

$$U = -\varepsilon_r \left(\frac{r}{R}\right)^3 \alpha \, (3\cos^2\phi - 1) E_0^2 \qquad (9)$$

Here ε_r is the relative permittivity of the particles. The equation shows that the tendency to associate depends upon R, r, ϕ, and E. Furthermore, the following factors are known. Firstly, optimal interparticle attraction occurs for $\phi = 0$, which is *along* the field vector lines. Secondly, the particles positively repel each other in the direction of $\phi = 90°$ and exhibit neutrality when $3\cos^2\phi = 1$ and ϕ is approximately 55°. Hence, for such a mechanism, particles tend to associate along the field lines and to strongly oppose association perpendicular to these lines. The mechanism is induced by the particle polarizability α, and is the key to pearl-chain formation, whose frequency dependence is reflected in the frequency dependence of the polarizability as described by a Debye-type[6] dispersion where

$$\alpha = \alpha_0 \left\{ 1 + (\omega\tau_r)^2 \right\}^{-1} \quad (10)$$

Here α_0 is the dc field polarizability and τ_r the rotary relaxation time of the particles. This relates to the value in equation (2). If required, equation (9) can be further simplified by expressing the polarizability of a sphere as

$$\alpha_0 = 4\pi\varepsilon_0 r^3 \quad (11)$$

where ε_0 is the permittivity of free space. One can see that with ac fields the effect is strongly diminished due to a frequency-dependent term in equation (10). Pearl chains are thus predominantly encountered under dc electric fields.

DYNAMIC BANDING

In incompressible fluids, moving particles experience the Bernoulli effect. This is the tendency for local pressure reductions to occur as particles move within a fluid medium. Particles tend to attract laterally under such motion. In principle, therefore, particles that are being displaced electrophoretically in a uniform parallel field will, under favorable conditions, experience lateral attraction. This is enhanced by electro-osmosis in which the support fluid moves to fill the space vacated by translating particles. Electro-osmosis is thus the inverse effect of electrophoretic mobility. The interaction of two such particles moving under the influence of an *alternating* electric field has been considered, in principle, as early as 1933 by Bjerknes *et al.*[3]. Reconsidering their equations leads to the following expressions for spheres which are widely separated such that $R \gg r$, under an alternating electric field of amplitude E_0 and frequency ω. We have shown elsewhere[11] that, for submicron particles in water at frequencies less than 1 MHz, the potential between two such particles separated by distance R under angular azimuth ϕ to the field vector direction, is given by

$$U = \frac{\pi\rho_0}{2} \left(\frac{r}{R}\right)^3 r^3 \frac{u^2 E_0^{\,2}}{(1 + \omega^2 \tau_h^{\,2})} (3 \cos^2\phi - 1) \quad (12)$$

Here ρ_0 is the density of the medium and τ_h relates to the translatory diffusion coefficient expressed in equation (5).

From equation (12) we note the following. First, there is a tendency for particles to attract under the conditions of $\phi = 90°$, to repel strongly for $\phi = 0°$, and to be neutrally interactive for $\phi = 55°$. This is exactly in opposition to the conditions for pearl-chain formation. Second, the effect is strongly dependent upon the particle surface charge exhibited not through polarizability but through electrophoretic mobility, which is a surface charge-related parameter[10]. Again, the effect increases with field strength but is frequency dependent and limited. Equation (12) is thus the key to understanding band formation, which depends upon the particles having surface charge and being mobile within the medium under alternating electric field excitation.

ADDITIONAL COMMENTS

Many experiments have been reported on the observation of pearl chains in dc electric fields. Due to the ability of such chains to form conducting pathways and arcs between electrodes along the field lines, it appears that the majority of experimenters did not investigate the influence of alternating fields, nor did they in general leave their fields on for sufficient time to observe the electrostatic banding phenomenon. The phenomenon occurs essentially in regions of high concentration, such as toward the bottom of the cell with particulate suspensions and toward the top of the cell with separating dispersions. Whereas in many colloidal media the electrophoretic mobility and the electrical polarizability may be strongly influenced by the electrical double layer surrounding charged particles, nonetheless, it appears that banding is strongly associated with net particle surface charge, which would influence the electrophoretic mobility alone. The tendency of the two effects to compete in their angular behavior explains why the two phenomena are rarely seen together.

Two further experimental observations are worthy of report. First, to find banding and pearl chains together one must look for materials that are characterized by both significant α and u. Observations must be made under alternating fields of relatively low frequency. Pearl chains favor dc fields, whereas bands are predominantly a dynamic behavior. In Figure 4 a high magnification is shown of band formation in an emulsion of 50 drops of decane in 1 L of water. Under the conditions shown, one can see the tendency of chains to be swept into bands. There is no droplet motion in this case because the

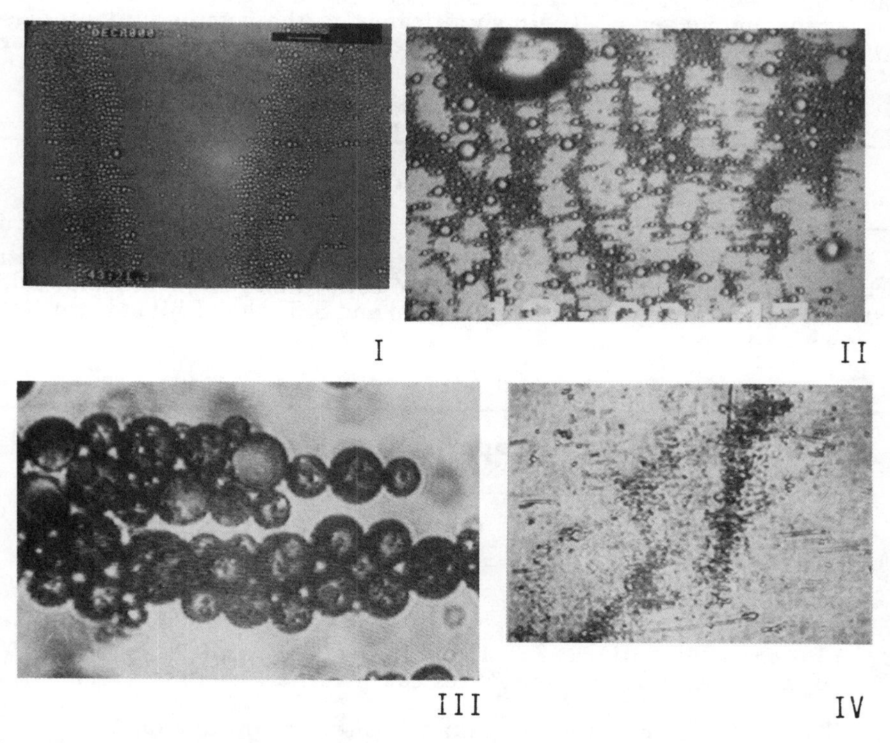

Figure 4. Band Formation in a Variety of Media

I. High magnification of a decane in water medium. Note the presence of the pearl chains within the bands; E = 150 V/cm and 20 Hz.

II. A kerosene in water emulsion; E = 300 V/cm and 50 Hz.

III. A crude oil in water emulsion; E = 300 V/cm and 10,000 Hz.

IV. A dispersion of asbestos fibers in water; E = 600 V/cm and 500 Hz frequency.

droplets have a tendency to associate and coalesce and are therefore relatively "sticky." They are able to be "held" within the 54° inclination without continuous motion. The figure does exhibit band formation and pearl-chain formation in quadrature directions.

Second, this figure also illustrates the same tendency of bands and chains to form together in a range of materials. The first example is a kerosene in water emulsion; the second is a crude oil in water emulsion. In both cases the same two physical phenomena are manifest. Finally, to illustrate the universality of the effect, the final frame of the figure indicates an early stage of band formation in a very dilute suspension of asbestos particles in water. These are fibrous particles, as compared with the discotic particles of the kaolinite. Bands thus form independently of particle morphology.

A third and final factor is simply to conclude that the tendency to form electro-dynamic bands follows exactly the sequence of electrophoretic mobility values indicated in Figure 2. It was impossible to see banding in PTFE suspensions, whereas kaolinite was the most ready of the materials studied to exhibit bands. In addition, it was also confirmed that pearl chains form most readily in those media with the highest electrical polarizability.

It is the authors' belief that these regions of high, local electrically induced concentration ought to be the object of far greater study for an understanding of the basic physical parameters responsible for electrically induced interaction phenomena. Their significance to the commercial processes involved in emulsion separations and in electrostatic precipitators and separators is self-evident.

ACKNOWLEDGMENTS

This work was supported by extramural awards to BRJ by Messrs. British Petroleum Ltd. Under this scheme both RI and MS held Fellowships. Extensive discussions with both D. Graham and S. Taylor of BP Ltd. are gratefully acknowledged.

REFERENCES

1. Benoit, H., *Ann. Phys. (Paris)*, 1951, *6*, 561.

2. Berne, B. J. and Pecora, R., "Dynamic Light Scattering," Wiley, New York, 1976.

3. Bjerknes, V., Bjerknes, J., Solberg, H., and Bergerson, T., "Hydrodynamique Physique," Presses Universitaires, 1934, Volume I.

4. Böhm, J., "Electrostatic Precipitators," Elsevier, Amsterdam, 1982.

5. Clayton, W., "Theory of Emulsions and Their Technical Treatment," C. G. Sumner revision, J. &A. Churchill, London, England, 1954.

6. Debye, P., "Polar Molecules," Dover Publications, New York, 1929.

7. Fredericq, E. and Houssier, C., "Electric Dichroism and Electric Birefringence," Clarendon Press, Oxford, England, 1973.

8. Isherwood, R. and Jennings, B. R., *SPIE*, 1984, *492*, 116.

9. Isherwood, R., Jennings, B. R., and Stankiewicz, M., *Chem. Eng. Sci.*, 1987, *42*, 913.

10. James, A. M., *Surface and Colloid Science*, R. J. Good and R. S. Stromberg, eds., Plenum, New York, 1979, Volume II, 121.

11. Jennings, B. R. and Stankiewicz, M., *Proc. Royal Soc. A.*, in press.

12. Krasny-Ergen, W., *Ann. Phys.*, 1936, *27*, 459.

13. Kruyt, R. and Vogel, G., *Kolloid Zeit*, 1941, *95*, 2.

14. Muth, E., *Kolloid Zeit*, 1927, *41*, 97.

15. Parslow, K., Jennings, B. R., and Trimm, H., *Europ. J. Phys.*, 1984, *5*, 88.

16. Pearce, C.A.R., *Brit. J. Appl. Phys.*, 1954, *5*, 136.

17. Perrin, F., *J. Phys. Radium*, 1934, *5*, 497.

18. Perrin, F., *J. Phys. Radium*, 1936, *7*, 1.

19. Ware, B. and Flygare, W., *Chem. Phys. Lett.*, 1971, *12*, 81.

SLUDGE AND NOVEL DEWATERING PROCESSES

It is most desirable to develop new and improved dewatering methods for reducing the moisture content in sludges since landfilling and incineration are becoming increasingly expensive. For example, urban areas such as Philadelphia, New York, and Los Angeles have virtually no landfill space left, and transportation distances and costs to acceptable landfills are increasing. The incentive to come up with newer dewatering methods to alleviate some of these problems is enormous.

Dr. Meijer, Technical Director for the Netherlands Wastewater Authority, gives a very interesting paper on how a small country in Europe is trying to cope with an insurmountable sludge problem. Papers by Mr. MacConnell of Montgomery Consulting and Mr. Gustafson of Hach Co. address some of the practical problems and solutions encountered in a wastewater treatment plant. Dr. Martel of the U.S. Army delineates a freeze/thaw process, Dr. Mayer of Du Pont discusses new developments in plate and frame filters, and Mr. Smith of United Coal describes a high-gravity centrifugal dryer.

STATE OF WASTEWATER TREATMENT INDUSTRY IN THE NETHERLANDS

H. A. Meijer
Technical Director
Hollandse Eilanden en Waarden Water Authority
P.O. Box 469
3300 Al Dordrecht
The Netherlands
31 78 141288

ABSTRACT

The Netherlands is a small, densily populated country in the delta of the international rivers Rhine, Meuse, and Scheldt with special water quality problems. The Pollution of Surface Waters Act, which became effective in 1970, provided authorities with the legal power and financial means to abate water pollution. In the last 20 years an ambitious sewage treatment plant investment programme was executed.

The load of surface waters with oxygen-consuming substances is reduced from 40 million to 5 million population equivalents. Different wastewater treatment systems used in the Netherlands are mentioned. A peculiar underground two-stage sewage treatment plant (Rotterdam-Dokhaven) is described in greater detail.

Sludge disposal is a matter of concern, since the central government has announced that within a few years conditions for agricultural application of sewage sludge and other forms of reuse will be made very stringent. Incineration of sludge seems to be the only alternative method of disposal.

From 1995 on phosphate removal of effluents will be implemented at sewage treatment plants according to the Rhine Action Programme. Consequences of nutrient removal for treatment costs are described.

Attention is given to the Applied Research Programme, Future Generation Waste Water Treatment Works.

INTRODUCTION

The Netherlands is a wet country, short of water.

Looking at the yearly rainfall of 750 mm and the situation in the delta of the rivers Rhine, Meuse, and Scheldt, this phrase seems to be incorrect. The phrase becomes more credible when the word pure is added. Pure surface water is indeed very exceptional in my country.

As obvious from a map of western Europe (Figure 1), the rivers in the Netherlands are part of international riverbasins, where more than 100 million people live, work, and pollute the streams. Many chemical and mining industries are present in the catchment areas of these riverbasins. In the Netherlands alone, where another 14.8 million people live in an area of only 37,000 km^2, the attack on water quality is heavy.

Geography is a critical factor, as well. Figure 2 shows that approximately 25 percent of the land lies below sea level. Without the protection by dikes and the continuous pumping of water, the lower parts of the country would have disappeared into the sea long ago.

"God made the world, the Dutch made Holland." The Dutch have practised land reclamation for a thousand years. The names "Amsterdam" and "Rotterdam" refer to the period when we built dams and sluices. After the invention of the windmill and the steam engine, bigger and deeper lakes were reclaimed. After the 1953 disaster when part of the country was flooded and 1,800 people were drowned, high-tide protection became first priority. The Delta plan was developed, and after ratification by the Houses of Parliament, the Delta project was executed (Figure 3).

The Delta works were accomplished over 30 years. Dams or high-tide barriers enclose the estuaries in southwest Holland. The Delta works protect the low polders in the western part of the Netherlands against flooding. The Delta works are very expensive: for instance, the storm surge barrier dam in the Easter Scheldt, operational since October 1986, costs Dfl 8,000 million, $3,500 million (U.S.).

A negative side effect of the closing of estuaries by dams and barriers is the creation of lakes with stagnant water. These stagnant waters are

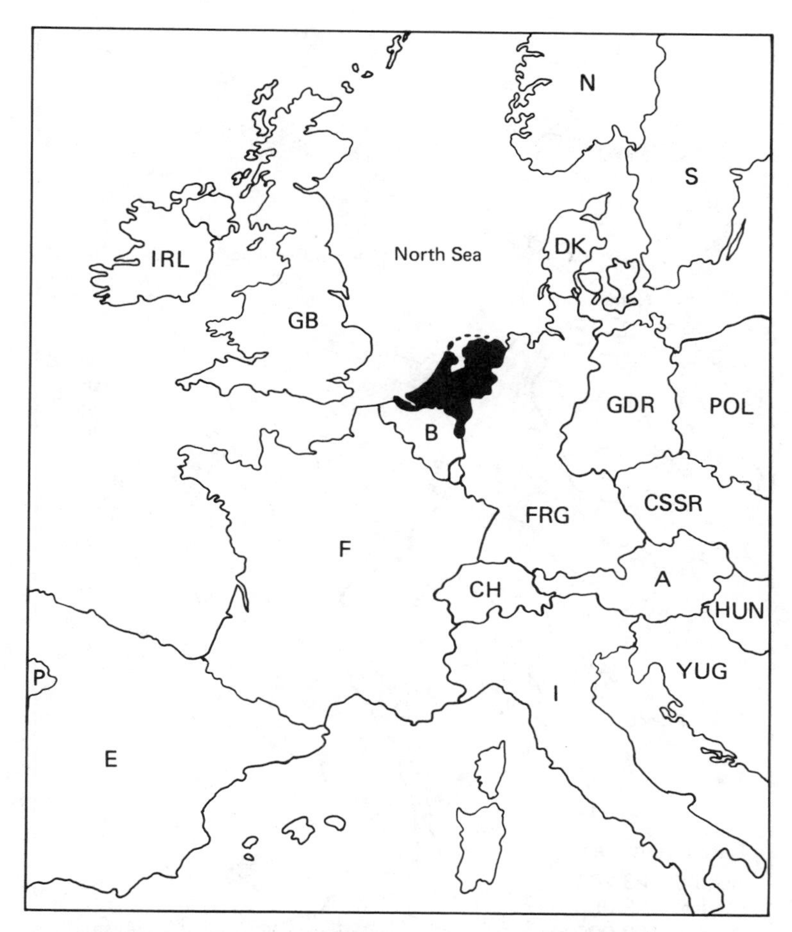

Figure 1. Indication of the Location of the Netherlands and the Catchment Areas of the Rivers Rhine, Meuse, and Scheldt[1]

sensitive for eutrophication caused by high nutrient concentration of the incoming water. The river silt is contaminated with phosphates, heavy metals compounds, and nondegradable organic substances (PCBs, polycyclic aromates, and pesticides). Thick layers of polluted sediments damage fish and fish consumers. Restoration of these lakes is currently being studied. A big problem is how to dispose of the contaminated sediments. Locations for land disposal are scarce: city councils have not been very cooperative.

The rising of the sea level by the greenhouse effect will make future adjustments of the heights of the dikes inevitable. So the conservation of the Netherlands is a never-ending job.

WASTEWATER MANAGEMENT IN THE NETHERLANDS

After World War II, the Dutch government turned its attention to reconstruction. Until the end of the 1960s, water pollution was not seen as a problem, but as the price of growing prosperity, and the treatment of urban and

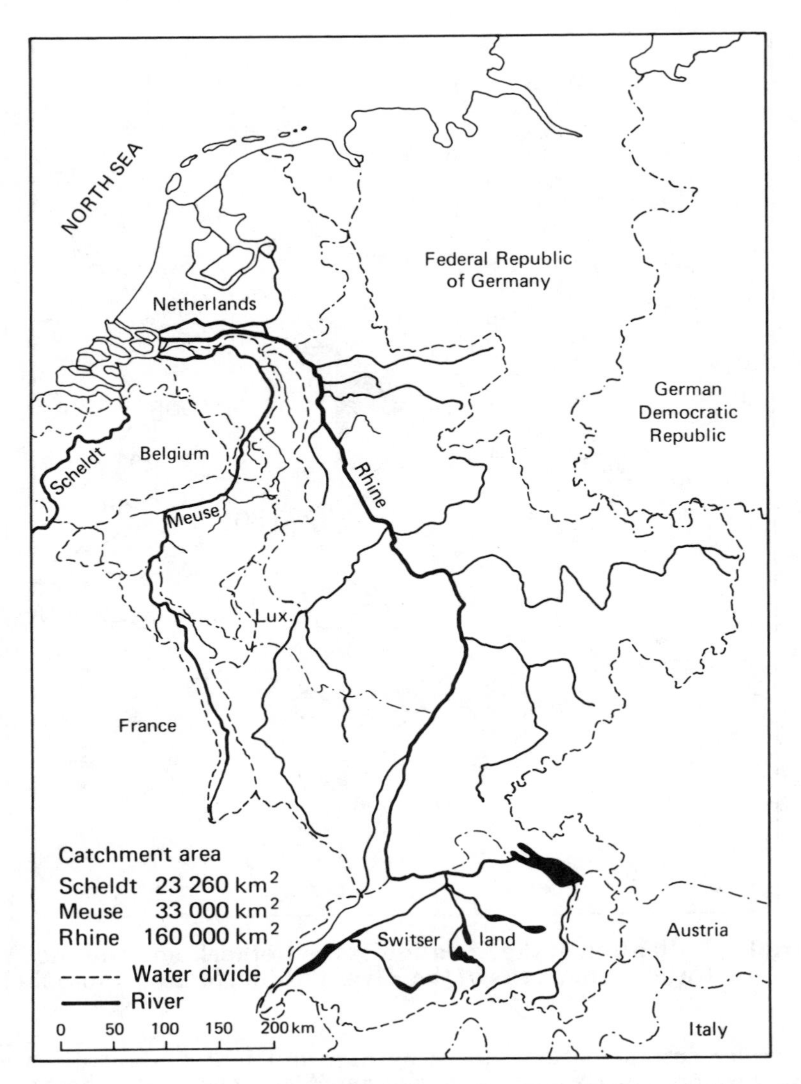

Figure 1. Indication of the Location of the Netherlands and the Catchment Areas of the Rivers Rhine, Meuse, and Scheldt(1) (Continued)

industrial effluent was incidental in the Netherlands. Rachel Carson's book Silent Spring made people all over the world conscious of the dangers of polluting the environment. It is not surprising that in a country with so much water--very dirty water in many places--abatement of water pollution was first tackled. Table 1 illustrates the water pollution situation in 1969.

Although several towns had sewage treatment plants, a structural approach of the water quality was missing. It sometimes happened that sewage was treated downstream, but not upstream, so the efforts of the town downstream were completely ruined by its neighbors.

Conditions for an efficient approach to water pollution abatement are a strong legal base. The Pollution of Surface Waters Act, which came into operation in December 1970, provided authorities with legal power and financial means to fulfill their tasks. The main points of the Pollution of Surface Waters Act are:

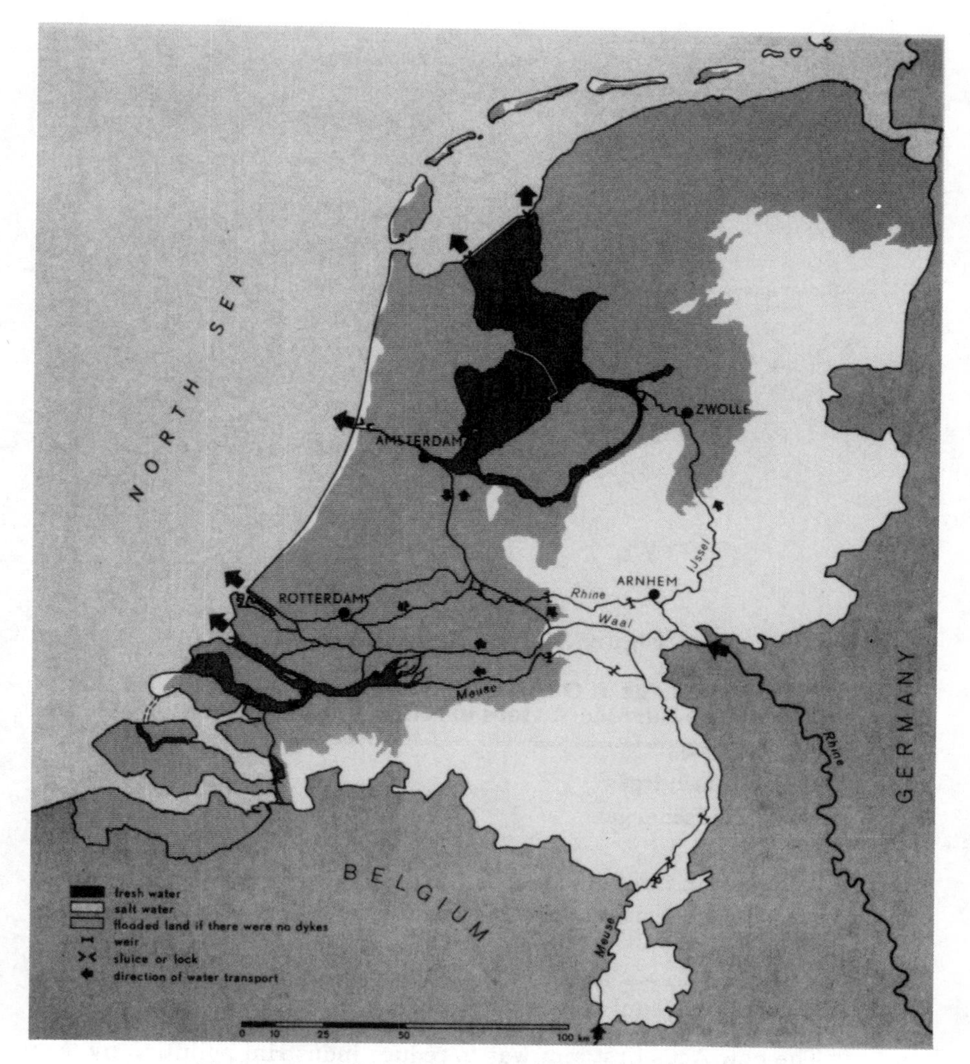

Figure 2. Water Systems in the Netherlands

1. Every discharge of wastewater into surface water is subject to discharge consent from the water management agency. The agency has power to stipulate the quality and quantity of the effluent to be discharged.
2. Each discharger of wastewater into surface water or a sewer has to pay a pollution levy according to the "polluter pays" principle.

Meijer(2) contains more detailed information about the Pollution of Surface Waters Act.

From 1970 on, water pollution control in the Netherlands was regulated by agencies (Figure 4), in most cases existing or newly formed water authorities. Water authorities are public bodies, operating on the same level as municipalities. The same agencies regulate flood control and drainage of polders. Characteristic for the water authorities is their own tax system. Some water authorities in the western part of the country have a history of more than 700 years.

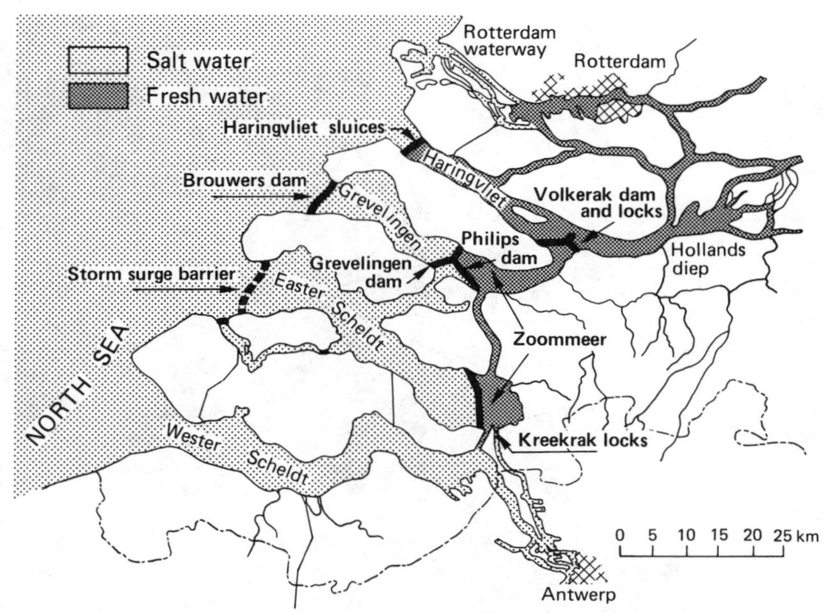

Figure 3. The Delta Plan(1)

Table 1. The Discharge of Oxygen-Consuming Substances and the Load into Surface Waters in Million Population Equivalents*

Domestic discharges	12.5
Industrial discharges	33.0
Total	45.5
Eliminated in treatment plants	5.5
Load of oxygen-consuming substances into surface waters	40.0

* 1 population equivalent = 54 g BOD per day.

The new Act's first aim was to reduce industrial pollution by legal measures; its second was to reduce the load of communal wastewater by building and operating sewage treatment plants. Top priority was given to reducing the load of oxygen-consuming substances from the effluents to restore the oxygen balance in the surface waters. Removal of phosphates was foreseen in a later stage, in connection with the reduction of phosphates from detergents and agriculture. Very important for a country situated at the delta of international rivers is acting in concert with Switzerland, France, West Germany, and Belgium.

To demonstrate the relative contribution of discharges from Dutch sources to water pollution, Table 2 shows both the load into surface waters and the influx via the frontier-crossing rivers Rhine and Meuse for a number of substances. The figures in this table underline the major pollution contribution to Dutch surface waters from the Rhine and Meuse. Dutch efforts against river pollution can succeed only if the other countries along the Rhine and Meuse show an equal diligence in pursuing their pollution abatement programmes.

The pollution load of the river Rhine with respect to oxygen-consuming matter, as well as cadmium, mercury, and other nondegradable substances, has decreased (Figure 5a). The phosphate concentration has diminished in recent years, but concentrations as high as 0.6 mg P/L are still quite common.

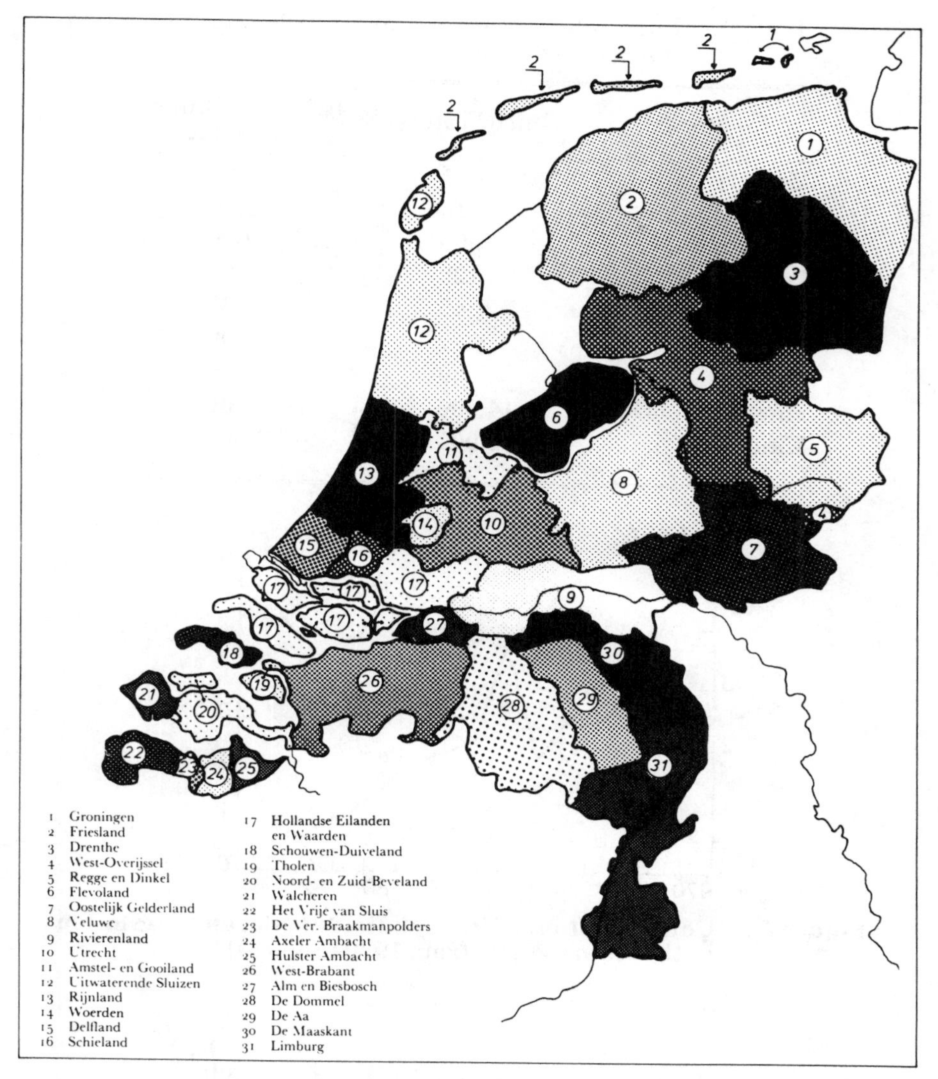

Figure 4. Regional Water Quality Authorities in the Netherlands as of January 1, 1981

Nitrogen compounds concentrations are still increasing, and the sodium chloride situation is still without any prospect of success. The French alkali mines have been reluctant to solve their waste problem. Figure 5b shows the increase of chloride in Rhine water.

In 1987 the Rhine states undersigned the Rhine Action Programme to reduce the load of phosphates from different sources by 50 percent by 1995. Requirements for phosphates will be sharpened in the near future, and nitrogen compounds will be limited. In Table 3 phosphate limits for new and existing sewage treatment plants are given. These values were recently published by the Dutch Environmental Protection Ministry. Deviations from these limit values are allowed only when the overall P-removal efficiency of all treatment plants under a given water quality agency is at least 75 percent.

Since 1975 within the European Community (EC) several regulations on water pollution control have sought to harmonize the water quality policy in the different member countries. The EC has prepared guidelines on

Table 2. Load into Surface Waters in 1978 and 1980 (tonnes per annum)

	Dutch Waters	Influx via Rhine and Meuse
Mercury	1	20
Cadmium	15	200
Zinc	900	10,000
Copper	120	1,300
Nickel	45	1,000
Chromium	110	2,600
Lead	200	1,800
Phosphate (as P)	14,000	57,000
Mineral oil	8,000	23,000

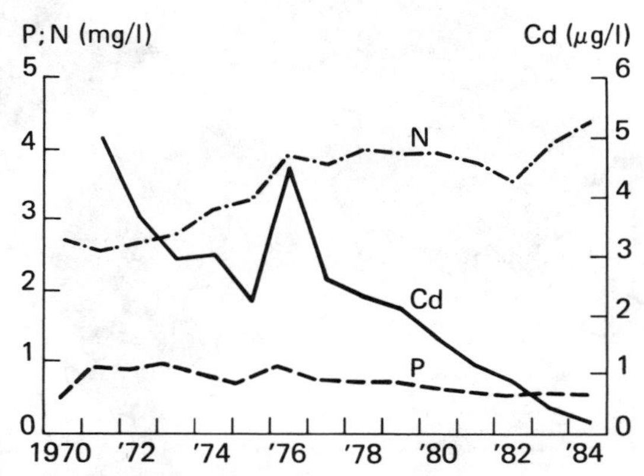

Figure 5a. Concentration of Phosphate, Nitrate, and Cadmium in the Rhine Water from 1970 to 1984(1)

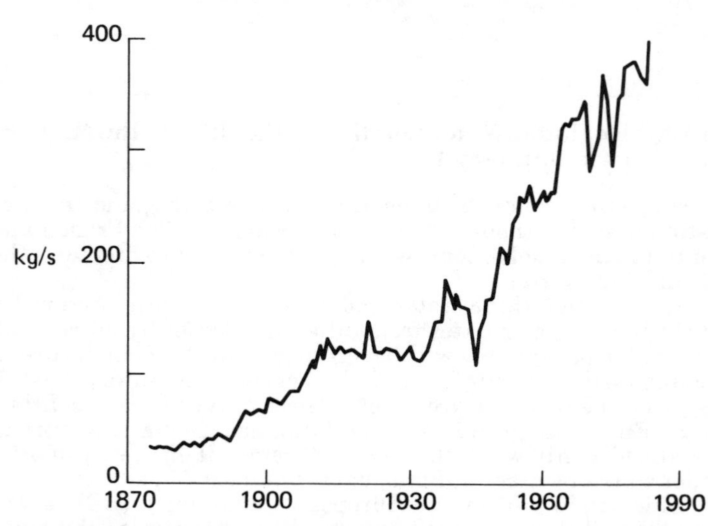

Figure 5b. Chloride Load of the River Rhine Since 1880(1)

the quality of surface water for drinking water production, on swimming water quality, on fishing water quality, on emission reduction of so-called "Black List" substances, and on agricultural use of sewage sludge. The member countries have the duty to insert these guidelines in their environmental legislation. In 1983 Dutch legislation was adapted to EC guidelines.

Table 3. Phosphate Limit Values for Communal Treatment Plants

Sewage treatment plants cap. >100,000 p.e.	max. 1 mg P/L
Sewage treatment plants cap. > 20,000 p.e.	max. 2 mg P/L
Sewage treatment plants cap. < 20,000 p.e.	max. 2 mg P/L

EXECUTION OF THE WASTEWATER TREATMENT PROGRAMMES

The Dutch government set 1985 as the completion date for the building of all treatment plants for the reduction of oxygen-consuming substances. Figure 6 shows the progress that has been made with the building of wastewater treatment plants. The strong rise of the capacity of secondary treatment plants after 1970 is obvious. Table 4 compares the water pollution situation in 1988 with the situation of 1969 (Table 1).

The investment programme for sewage treatment plants was not completely executed in the planned period. The economic recession from 1982 to 1986 caused an inevitable delay.

The number of sewage treatment plants in 1969 was 420, with a total capacity of 7.6 million population equivalents (p.e.) (primary treatment plants only); in 1988 the number of plants increased to 485. Many new biological plants were built, and 130 primary treatment plants were closed. Sewage treatment plants usually serve more than one municipality to control the number of effluent discharging points. Sewage of a town or village without a suitable receiving stream is pumped to a central sewage treatment works situated near a better effluent recipient.

WASTEWATER TREATMENT SYSTEMS

Table 5 portrays the wastewater treatment situation in the Netherlands in 1988.

For the higher capacities (above 100,000 p.e.) one-stage activated sludge plants are most common. Sludge is anaerobically digested, and methane gas is used for electrical energy production.

Oxidation ditches are very popular in the Netherlands; since the invention by the Dutch scientist Dr. Pasveer in 1953, more than 200 oxidation ditches have been built. The first ones were used in small towns, but the introduction of Carrousel type ditches with vertical surface aerators increased the system's capacity to 200,000 p.e. and, for industrial wastewater, to more than a million p.e. Sludge of oxidation ditches is aerobically stabilized during the aeration process. Energy consumption is higher than conventional activated sludge systems, effluent quality is excellent, and operation is rather simple.

In the last decade, several two-stage biological treatment plants have been built in the Netherlands. A prominent example is the sewage treatment plant built in Rotterdam in a disused harbour called Dokhaven. The Dokhaven underground sewage treatment works came on line in 1987. The

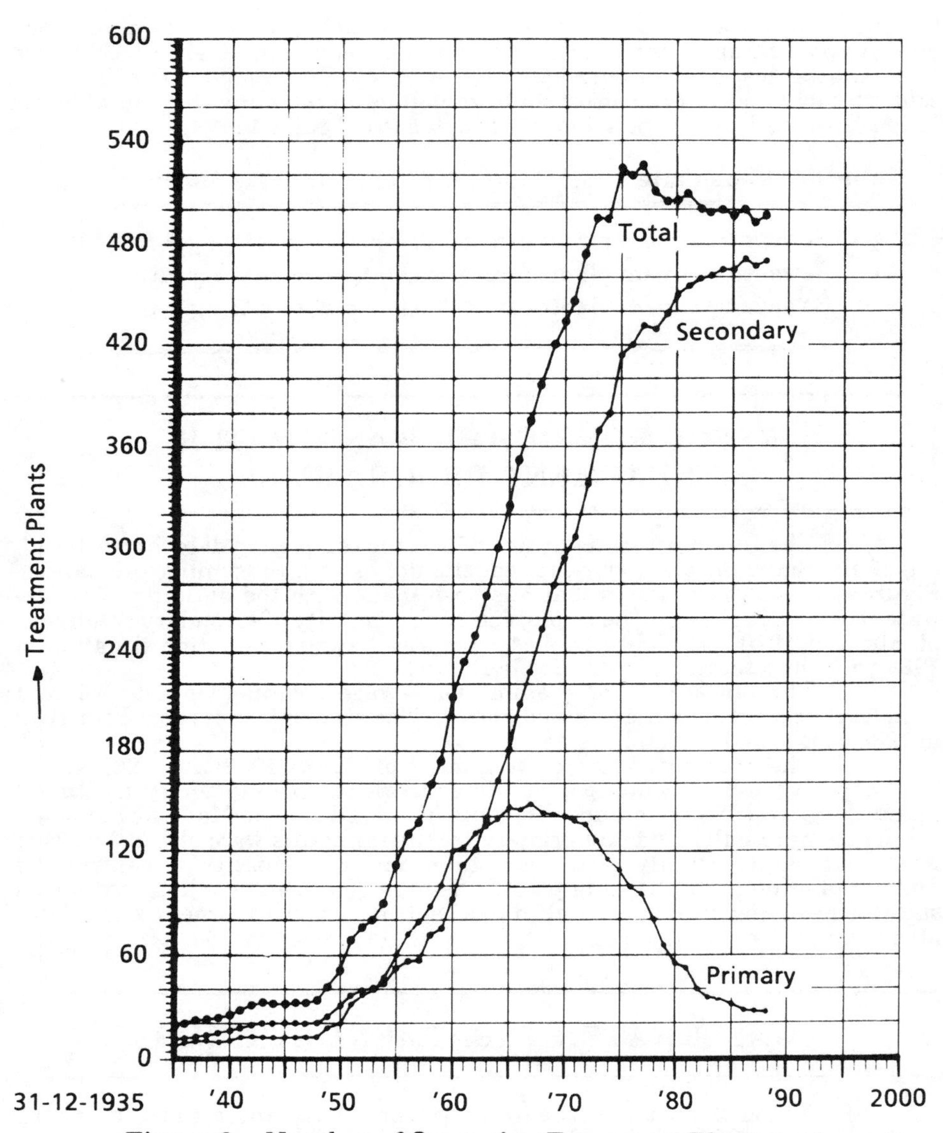

Figure 6. Number of Operating Treatment Plants and Their Design Capacities from 1935 to 1989

two-stage activated sludge plant serves a population of 470,000 and is completely covered. The first-stage aeration tanks have air diffusers, and the second-stage use surface aerators. A three-step wet chemical scrubber purifies the foul air at each stage. Since the plant is in the middle of a housing estate, it was not possible to discharge ventilation air on the spot. Of the purified air (mixed with clean air), 240,000 m^3/hr is released through a 50-m-high chimney stack at the sludge-handling site, which is 600 m from the Dokhaven plant. The hydraulic capacity is 19,000 m^3/hr, and 470,000 p.e. equates with 25,000 kg BOD/day.

The layout of the sewage treatment plant is shown in Figure 7; note how the rectangular tanks minimize the plant's volume.

The choice of the treatment system, i.e., the A-B process of Prof. Dr. Ing. B. Böhnke of the Technical University of Aachen (FRG), was made after

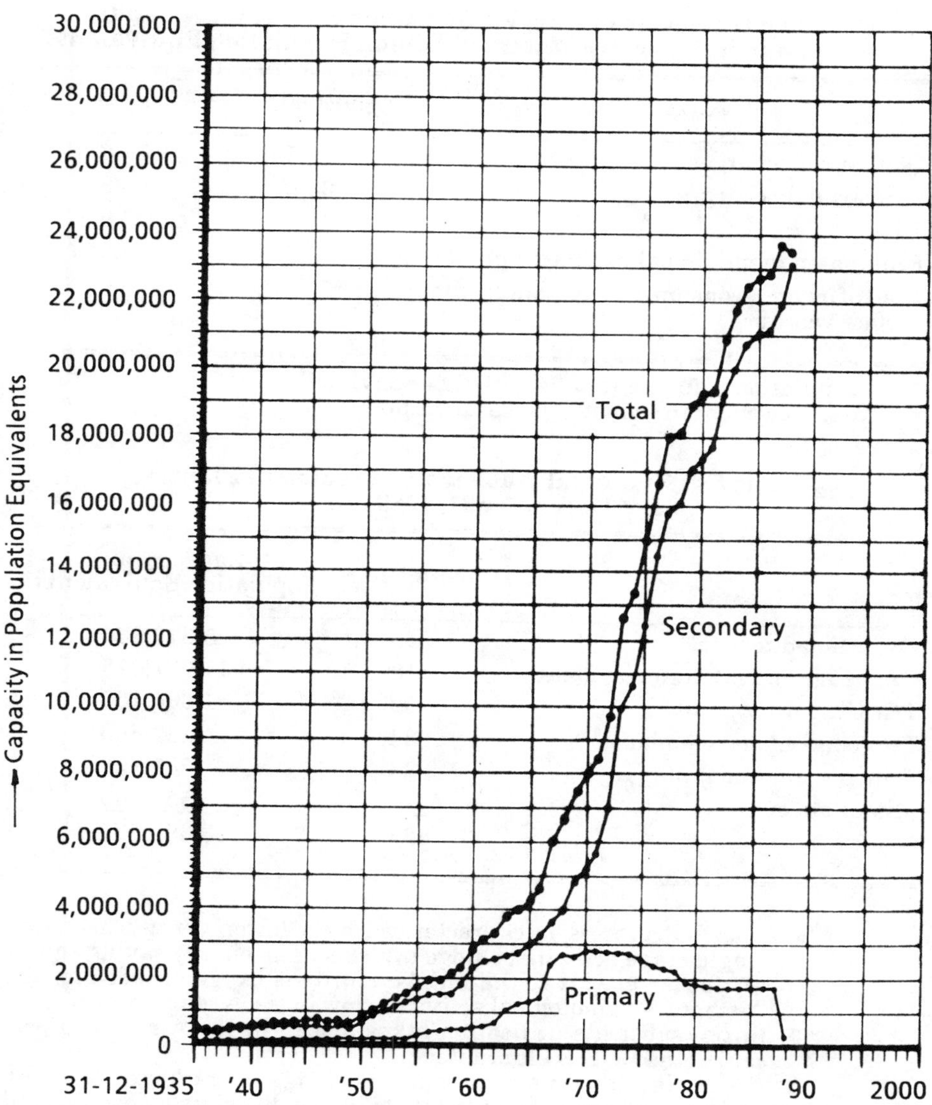

Figure 6. Number of Operating Treatment Plants and Their Design Capacities from 1935 to 1989 (Continued)

comparing several treatment systems and pilot-plant experiments. The most important reasons for this decision were:

- The high sludge loading in the A-stage of the activated sludge process (2.7 kg BOD/kg MLSS, day) and the allowable high surface loading of the sedimentation tank of 3.0 m/hr due to the sludge volume index of 40 to 60 mL/g, meet very well the limited space available for the construction of the treatment plant in the former harbour.
- The absence of primary sedimentation not only saves room, but also provides the opportunity to aerate the raw sewage almost directly after its arrival at the plant. Bad odour compounds formed by putrescent processes during transportation are oxidized or driven out before the sewage enters the first settling tank.

Table 4. The Discharge of Oxygen-Consuming Substances and the Load into Surface Waters in Million Population Equivalents*

	1969	1988
Domestic discharges	12.5	14.8
Industrial discharges	33.0	8.8**
Total	45.5	23.6
Eliminated in municipal treatment plants	5.5	18.6
Load of oxygen-consuming substances into surface waters	40.0	5.0

* 1 population equivalent = 54 g BOD per day.
** After treatment in industrial treatment plants.

Table 5. Municipal Wastewater Treatment Plants as of December 31, 1988

	Number	Capacity in Population Equivalents
Oxidation ditches	245	5,676,330
One-stage activated sludge plant	118	11,081,025
Trickling filter plants	52	1,604,100
Two-stage biological plants	34	4,132,600
Primary treatment plants	25	341,820
Miscellaneous	11	556,500
	485	23,392,375

- The A-B process is characterized by a lower energy use than single-stage activated sludge systems. The energy saving can be attributed mainly to the small quantity of oxygen needed for the A-stage. The biological activity is low in the A-stage, and only an initial substrate utilisation takes place. In the A-stage 60 percent of the BOD is eliminated.

At the sludge treatment site surplus sludge is thickened to a dry solids content of 4 percent and digested at 32 C in two digesters by compressed gas mixing. The digested sludge, with a dry solids content of 3 percent, is dewatered with Pennwalt/Sharpless centrifuges to a dry solids content of 20 to 23 percent. The dewatered sludge is open-air dried from 4 to 8 weeks to a shear strength of 10 to 20 kN/m^2 before being dumped in a landfill. Digestion gas is used for power generation, providing 35 percent of the energy consumption of the sewage treatment works (Table 6).

The sewage treatment works are controlled by a central computer system. Automation made it possible to operate the works with a staff of 25, covering daytime hours only for 38 hours per week. In the nights and weekends two operators are on call.

The Rotterdam-Dokhaven sewage treatment plant investment costs were Dfl 300 million.

With the execution of the building programme of the treatment plants, nearly all wastewater in the Netherlands is biologically treated. These first-generation treatment plants only begin to address the problem: the quantity of sewage sludge is growing, and its disposal will be the next hurdle.

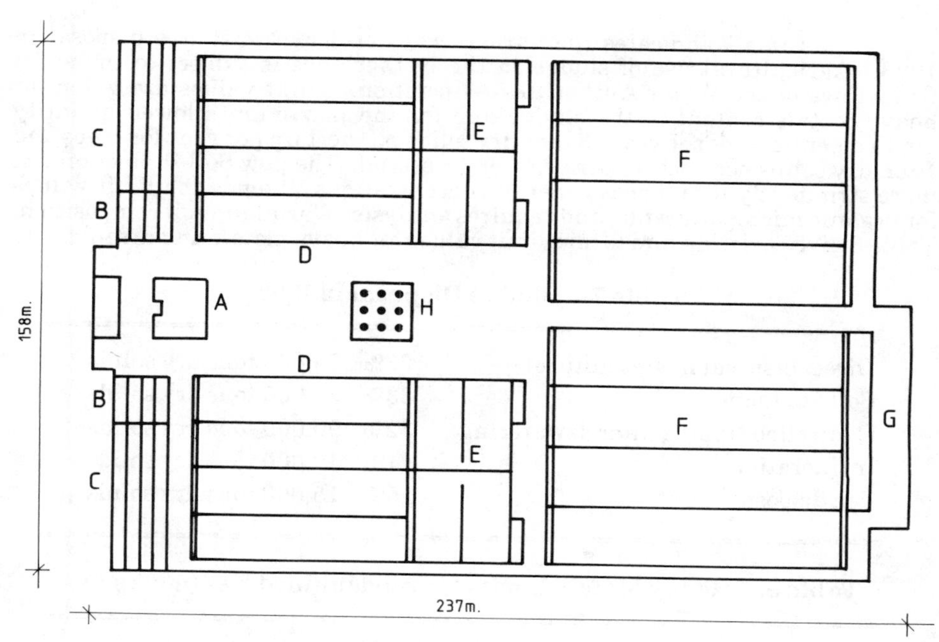

Figure 7. Rotterdam-Dokhaven Sewage Treatment Plant(3)

Table 6. Treatment Results at the Dokhaven Plant, Mean Values over 1 Year (1988/1989)

	COD, mg/L	BOD, mg/L	N-kj, mg/L	P_{tot}, mg/L
Influent	435	167	46	8.2
After A-Stage	170	57	35	5.4
Efficiency %	61	66	25	35
After B-Stage	54	9	19	3.5
Efficiency % after A+B stages	90	97	65	59

SLUDGE TREATMENT AND DISPOSAL

Domestic sludge production in the Netherlands was 280,000 tons dry solids in 1988. Phosphate removal from effluents could increase the yearly amount of domestic sludge by 100,000 tons; 400,000 tons is expected in 2000.

Table 7 indicates the various ways of domestic sludge disposal in 1988. Agricultural use of sludge in the Netherlands is subject to the strict Guidelines of the Water Authorities Association. Limit values are given for heavy metals content of the sludge and for the maximum allowed quantity applied per hectare per year: 2 tons dry solids per hectare per year for crops and 1 ton dry solids per hectare per year for grassland. The new Soil Protection Act more stringently limits heavy metals concentrations, extends the limit values for organic micropollutants, and requires analysis of farmland soil composition. Table 8 gives existing and future limit values for heavy metals and arsenic.

Table 7. Sludge Disposal in 1988

Direct disposal in agriculture	26%	75,000 tons dry solids
Soil production	33%	90,000 tons dry solids
Controlled tipping after dewatering	32%	90,000 tons dry solids
Incineration	4%	10,000 tons dry solids
Sea disposal	5%	15,000 tons dry solids

Table 8. Heavy Metals Limits for Agricultural Use (mg/kg ds)

	Limit Values Until January 1, 1991	Limit Values from January 1, 1991 Until January 1, 1995	Limit Values Indicative from 1995 on
Zinc	2,000	1,400	300
Copper	900	425	75
Lead	500	300	225
Chromium	500	350	75
Nickel	100	70	38
Cadmium	5	3.5	1.25
Mercury	5	3.5	0.75
Arsenic	25	25	

Sharpening the regulations for the agricultural use of sewage sludge will result in a significant reduction in the quantity of reused sludge. Other ways of disposal have to be found. The same situation will occur with the reuse of sludge in compost and soil production. New legislation will also end this disposal method within a few years. After 1990 disposal of sewage sludge into the North Sea, currently 180,000 tons per year, will be forbidden.

For the short term buying dewatering equipment and dumping seems to be the only solution. Over the long term, incineration to reduce the quantity of the remaining solids must be considered. In both cases dumping locations are inevitable. Environmental impact procedures make the creation of new controlled tipping sites a time-consuming activity.

Better technical solutions other than incineration have to be developed, but so far no replacement for incineration has been found. Studies carried out in the Netherlands on the Vertech system and the Carver-Greenfield process have not proven that these systems prevail over incineration. Better dewatering devices are also needed, since most of the commercially available machines cannot meet the standards for autothermic incineration or the minimum dry solids content required for monodeposition.

Battelle's electroacoustic dewatering investigations on Dutch sewage sludge were promising: without filling agents, dry solids contents of between 38 and 45 percent were obtained. With belt presses or centrifuges, dry solids contents of not more than approximately 20 percent are manageable. The new Dutch regulations on sludge disposal have consequences for wastewater treatment costs. Agricultural use of sludge costs circa Dfl 250 per ton ds. Sludge incineration costs at least Dfl 750 per ton ds, an increase of Dfl 500. In 1988, 75,000 tons of sludge were used for agricultural purposes, for a total of Dfl 37.5 million (75,000 × 500).

CONSEQUENCES OF P AND N REMOVAL FOR WASTEWATER TREATMENT COSTS

New regulations on phosphorus (P) removal and nitrogen (N) reduction will also lead to higher wastewater treatment costs.

Until the agreement between the Rhine and North Sea countries on phosphate and other pollutants limitation was reached, residential and industrial sewage was generally treated at low-rate biological treatment plants. Standard requirements are <20 mg/L for BOD and N-Kj, and <30 mg/L for suspended solids. Figure 8 shows the layout of a conventional activated sludge plant with a capacity of 90,000 p.e. and a hydraulic capacity of 3,500 m^3/hr.

Legend.

1. Screens
2. Primary settling tank
3. Aeration tank
4. Final settling tank
5. Sludge digestion tank
6. Sludge thickener
7. Gasholder
8. Sludge dewatering

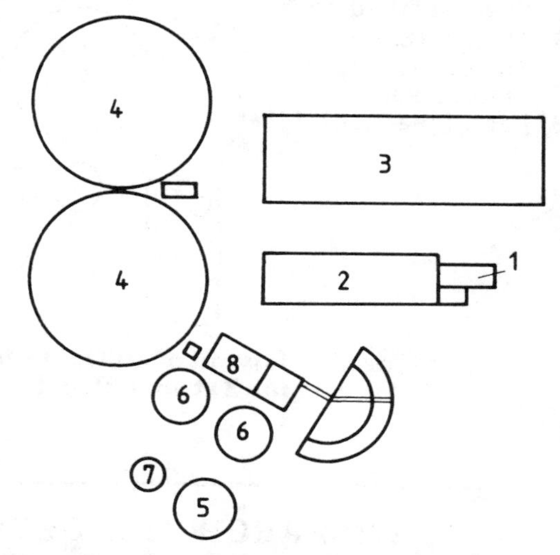

Figure 8. Layout of a Conventional Activated Sludge Plant

In the first stage of the Rhine Action Programme the P standard will be 2 mg/L, and the nitrogen standard, expressed in N-total, will be 20 mg/L. With simultaneous P removal and a doubling of aeration tank volume (BOD-load halved from 0.15 to 0.074), these standards can be reached. It will be far more difficult to reach the standards of 1 mg/L for P and 10 mg/L for N-total currently discussed. Another doubling of aeration tank volume will be necessary for meeting the N-total limit and for P removal, as will simultaneous precipitation via post treatment with filters.

A recent (not yet published) study in the Netherlands has shown that these stringent effluent standards can best be reached by designing an oxidation type of treatment plant using post-precipitation of phosphate. When incineration of the sludge is foreseen, primary sedimentation and digestion of sludge have no advantages.

Figure 9 shows the layout of an oxidation ditch type of plant for P and N removal to 1 mg/L and 10 mg/L, respectively.

The before-mentioned study demonstrated that the extra costs for extreme P and N removal are high: investment costs increase 60 percent and yearly costs, 50 percent.

In the Netherlands two new techniques for post P removal are developed. DHV Consultants, Amersfoort, has developed a pellet reactor system for P removal based on a physicochemical process in which phosphates react with lime and crystallize on grains of sand to form pellets in a fluidised bed reactor. A demonstration plant is in operation at the sewage treatment plant at Westerbork.

Smit Nijmegen precipitates phosphate with lime. By mixing the Ca-P precipitate with magnetic ferric oxide, P removal in a magnetic field is made possible. A demonstration plant for this high-gradient magnetic separation system is under construction.

It is the intention of both systems to reuse the phosphate residue as fertilizer or as raw material for P production.

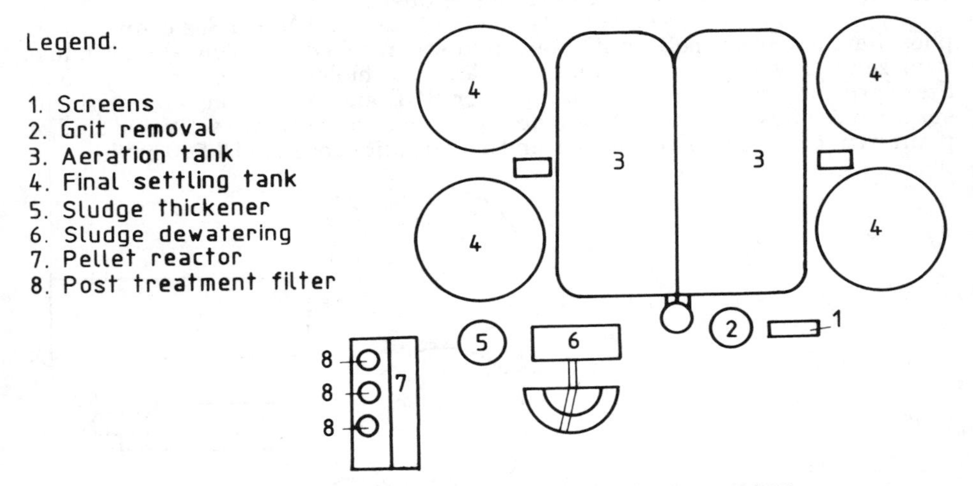

Figure 9. Layout of an Oxidation Ditch Type of Plant for Extreme P and N Removal

RESEARCH AND DEVELOPMENT ON WASTEWATER TREATMENT SYSTEMS

Applied research on wastewater treatment in the Netherlands is coordinated by the STORA Foundation. STORA was founded by the state, provinces, and water authorities in 1972 to stimulate and coordinate applied research in the field of wastewater and sludge treatment. Water quality control was later added to its charge.

In 1988 STORA and the Ministry of Transport and Public Works agreed on an investigation programme called "Future Generation Waste Water Treatment Works" (RWZI 2000). The RWZI 2000 Programme will provide the wastewater industry with the new technology needed between 2000 and 2020, when the present generation of treatment plants is due to be renovated.

For investigations from 1988 to 1992, Dfl 10 million was spent. STORA has no research facilities. Studies and investigations are executed by consulting engineers, universities, institutes, and private companies.

Examples are the above-mentioned studies by DHV and Smit Nijmegen on phosphate removal and by Battelle on EAD sludge dewatering.

The main points of the RWZI 2000 Programme are:

- Wastewater treatment
  - Anaerobic pretreatment of domestic sewage
  - Evaluation of A-B two-stage system
  - Deep shaft systems
  - Sludge on carrier-activated sludge systems
  - Phosphate removal, various systems; pellet reactor, magnetic separation, biological P removal
- Sludge production and sludge treatment
  - High-performance bioreactor
  - Thermophilic digestion
  - Vertech wet air oxidation
  - Multiple evaporation Carver-Greenfield process
  - Electroacoustic dewatering
- Fundamentals of sludge/water separation.

With the exception of the study on the fundamentals of sludge/water separation, most studies are critical evaluations of known systems or principles.

CONCLUSIONS

The enforcement of the Pollution of Surface Waters Act in 1970 was the start for a strong attack on water pollution in the Netherlands.

The choice for a decentralised approach by water authorities with their own financial resources and legislative support has been successful. Water pollution by industries and municipalities has been greatly reduced. Towns and villages, even in rural areas, are almost completely serviced by sewer systems connected to sewage treatment plants.

The quality of surface water in the Netherlands is nevertheless inadequate in many places, especially in the low parts of the country. The main reason is the high nutrient concentration, causing eutrophication problems in stagnant waters. Nonpoint sources of nutrients include runoff from agricultural land, release from river sediments and acid rain, and very important for the Netherlands, the boundary crossing rivers. All complicate any abatement strategy.

Only by international treaties can the nutrient load of the rivers decrease: the trend in this direction has been set by the Rhine Action Programme of 1987. The Dutch water authorities, provinces, and central government have signed a document stating that from 1995 on, phosphates from effluents will be reduced by 75 percent.

In the next 10 years the cost of sewage treatment in the Netherlands is expected to rise by 50 to 100 percent. Sewage treatment plants will have to include facilities for nutrient removal. More expensive sludge treatment systems are necessary, and the restoration of lakes and water courses contaminated by sediments must be financed.

REFERENCES

1. Colenbrander, H. J., c.s., CHO.TNO., *Water in the Netherlands 1986*.

2. Meijer, H. A., "Industrial Effluent Control in The Netherlands and Legislation in this Field in Some Other European Countries," *W.P.C.*, 1983, *2*, 213-226.

3. Meijer, H. A., "Rotterdam-Dokhaven Sewage Treatment Plant: A Large Sewage Treatment Plant in the Midst of a Developing Residential Quarter," *Wat. Sci. Tech.*, 1988, *20* (4/5), 267-274.

CENTRIFUGE THICKENING: FULL-SCALE TESTING AND DESIGN CONSIDERATIONS

Gary S. MacConnell and David S. Harrison
Senior Engineer and Principal Engineer
James M. Montgomery Consulting Engineers, Inc.
250 North Madison Avenue
Pasadena, California 91109
(818) 796-9141, x6607

and

Kenneth W. Kirby and Hansong Lee
Process/Startup Engineer
Bureau of Sanitation
Hyperion Division
and
Farhad Mousavipour
Project Engineer
Bureau of Engineering
Hyperion Division
City of Los Angeles
Playa Del Rey, California 90291

ABSTRACT

The City of Los Angeles installed four high-capacity thickening centrifuges at the City's Hyperion Treatment Plant in 1988. The purpose of this project was to provide interim waste-activated-sludge thickening capacity and to provide a full-scale test for the comparison of two manufacturers' high-capacity centrifuges. An existing building was retrofitted to house the interim waste-activated-sludge thickening facility. The facility has an effective capacity of 1,800 gallons per minute (gpm) with three of the four units in service.

The process and equipment designs of the two types of centrifuges are compared. The systems that comprise the interim facility are presented: centrifuge feed, centrifuge, polymer, thickened sludge transfer, and centrate disposal. Centrifuge facility design considerations are discussed.

The results of a long-term performance test are presented and discussed. The effects of the operational parameters of the two centrifuges are evaluated. These operational parameters include sludge feed rate, differential speed, polymer addition, and sludge discharge area for one of the machines. The performance of the machines is based on solids capture efficiency, centrate quality, and thickened sludge concentrate. Power consumption, machine reliability, and wear and corrosion are evaluated for each centrifuge.

INTRODUCTION

The City of Los Angeles' Hyperion Treatment Plant currently processes approximately 400 million gallons per day (mgd) of wastewater; 180 mgd receives secondary treatment by the conventional activated sludge process. The City was under court order to provide full secondary treatment by 1998. The high-purity oxygen-activated sludge process (HPOAS) was chosen to provide full secondary treatment when the plant was upgraded and expanded to a liquid treatment capacity of 450 mgd in 2010.

This increase in secondary treatment will significantly increase the volume of activated sludge that must be wasted to 11.5 mgd in 2010. This waste-activated sludge, with a solids concentration of approximately 0.5 to 0.7 percent, requires thickening to approximately 5.0 percent solids prior to anaerobic digestion.

In 1988, the City installed four high-capacity thickening centrifuges with a design capacity of 600 gpm each to provide interim waste-activated-sludge thickening capacity and to provide a full-scale test for comparing two manufacturers' centrifuges. An existing building was retrofitted to house the interim waste-activated-sludge thickening facility. The facility was comprised of the four centrifuges and necessary appurtenances such as pumps, piping, and controls. The interim facility was designed to have three operational units for a combined capacity of 1,800 gpm. Design factors and facility modifications necessary during start-up are presented.

The results of this full-scale test are discussed in detail, with emphasis on the performance of each manufacturer's centrifuges under different operating conditions. In addition to the performance of each unit, relative electrical, operational, maintenance, and chemical requirements are compared. The two types of centrifuges are also compared with respect to design, materials of construction, and operation. A complete analysis of the two manufacturers' high-capacity thickening centrifuges is presented based on the full-scale test.

All types of centrifuges, regardless of their design and configuration, use the basic principles of sedimentation for liquid/solids separation[2]. The theory of sedimentation was best described by Stokes' law, which describes the velocity of a spherical particle settling in a liquid under a gravitational field as presented below:

$$V_g = [(\rho_s - \rho_l)\, d^2\, g]/18\,\mu \qquad (1)$$

where:
- V_g = settling velocity of the particle due to gravity
- ρ_s = density of the solid
- ρ_l = density of the liquid
- d = diameter of the particle
- μ = viscosity
- g = gravity constant (9.81 m/sec^2, 32 ft/sec^2).

Centrifugation increased the settling rate by increasing the gravitational force. The multiple increase in the gravitational field produced by the centrifuge was directly related to the speed of rotation as:

$$G = (w^2 r)/g \qquad (2)$$

where:
- G = multiple of gravitational field
- w = speed of rotation
- r = radius of rotation.

Stokes' law may thus be modified to apply to centrifuges as follows:

$$V_c = \{[(\rho_s - \rho_l)\, d_p^2]\,(w^2 r)\}/18\,\mu \text{ or} \qquad (3)$$

$$V_c = \{[(\rho_s - \rho_l)\, d_p^2]\, G\}/18\,\mu$$

where:
- V_c = settling velocity of the particle due to equivalent gravitational force from centrifuge.

Two centrifuge manufacturers, using the above principles in the design of their machines, each provided two centrifuges to the City of Los Angeles. Both manufacturers' machines were designed to thicken waste-activated sludge (WAS) produced at the Hyperion Treatment Plant.

WAS is excess biomass that is produced in the biological treatment process for municipal sewage. It is composed primarily of water and organic biomass with relatively little inorganic material. The specific gravity of WAS is generally about 1.005, and the specific gravity of the sludge solids averaged about 1.25. Due to the relatively low specific gravities of WAS and WAS solids, this material did not separate as readily as other sludges that contain more dense inorganic material.

The two manufacturers are Humboldt-Wedag and Sharples. Each has a horizontal solid-bowl, scroll-type centrifuge that was designed to achieve similar process performance with continuous feed. However, there are distinct differences in the design, the approach to thickening, and the operation of the two units. The design criteria for the Humboldt-Wedag and Sharples units are presented in Tables 1 and 2, respectively. The similarities and differences between these two machines are discussed below.

THE HUMBOLDT-WEDAG CENTRIFUGE

The design of this machine was cocurrent. The centrifuge received feed sludge and thickened it along the same direction as the flow through the bowl (Figure 1). Centrate traveled in the opposite direction to the thickened sludge and to the overflow weirs at the head (or feed end) of the machine. A horizontal conical conveyor, installed concentrically within the bowl, rotated at a slightly higher speed than the bowl and conveyed thickened sludge to the tail or discharge end of the bowl, where it exited through small nozzles and ports. The differential speed between the conveyor and bowl was controlled by driving the conveyor with a Viscotherm hydraulic backdrive unit. The system was operated so that a preset differential speed was maintained. The hydraulic con-

Table 1. Humboldt-Wedag Centrifuge Design Criteria

Model	BC5-2
Type	Co-Current
Design Capacity	600 gpm
Bowl Speed	1385 - 1600 rpm
Main Drive	400 Horsepower with hydraulic turbo-coupling
Bowl Diameter	44 inches
Bowl Length	165 inches
Pond Depth Setting	8 inches
Bowl Material	Carbon steel with coal tar epoxy coating
Conveyor Material	Carbon steel with coal tar epoxy coating
Bowl Construction	Fabricated from steel components
Abrasion Protection:	
Feed Components	Adiprene
Conveyor Blades	Ceramic
Sludge Discharge	Eutalloy
Lubrication:	
Type	Circulating oil
Reservoir	20 gallons
Motor	1/2 horsepower
Conveyor:	
Drive	Rotodiff 112D
Control Unit	Viscotherm B30-9624
Motor	40 horsepower

veyor drive rotodiff allowed for the stepless variation of the differential speed between the bowl and the conveyor.

The pond depth was determined by adjustable weir plates on the centrifuge discharge end of the machine. The bowl and conveyor of the Humboldt-Wedag machine were fabricated of carbon steel, and nonwear surfaces were coated with coal-tar epoxy.

The feed components were protected from abrasion and erosion with adiprene; the conveyor blades were coated with eutalloy; and sludge discharge components, including the face of the conveyor blades and bowl beach, were protected with ceramic tiles.

The Humboldt-Wedag centrifuge was considered a low-speed machine since its bowl speed of 1,385 to 1,600 rpm was lower than other machines (including the Sharples machine, which may reach a maximum speed of 2,000 rpm). However, the bowl diameter of 44 in. and the bowl length of 165 in. were greater than the Sharples machine. This related to a greater effective volume than the Sharples machine, given the same pond depth, and a greater retention time within the machine at the same flow rates. The theory behind the low-speed machine was that the greater bowl volume and longer retention time can compensate for the lower G forces generated by the machine. Referring to the modified Stokes' equation for centrifuges, the lower settling velocity (V_c) of a particle in the low-speed machine, in theory, can be equalized with a longer retention time to provide the same performance as a high-speed machine.

Table 2. Sharples Centrifuge Design Criteria

Model	PM 95, 000 AD
Type	Counter current
Design Capacity	600 gpm
Bowl Speed	Maximum 2000 rpm
Main Drive	300 horsepower with variable frequency drive
Bowl Diameter	40 inches
Bowl Length	140 inches
Pond Depth Setting	11.73 inches
Bowl Material	Stainless steel
Conveyor Material	Stainless steel
Bowl Construction	Centrifugally cast and machined
Abrasion Protection:	
Feed Components	Replaceable tungsten carbide tile
Conveyor Blades	Replaceable tungsten carbide tile
Sludge Discharge	Replaceable tungsten carbide tile
Lubrication:	
Type	Circulating oil
Reservoir	30 gallons
Motor	1/2 horsepower
Conveyor:	
Drive	DC backdrive system
Control Unit	SCR controller
Motor	50 horsepower

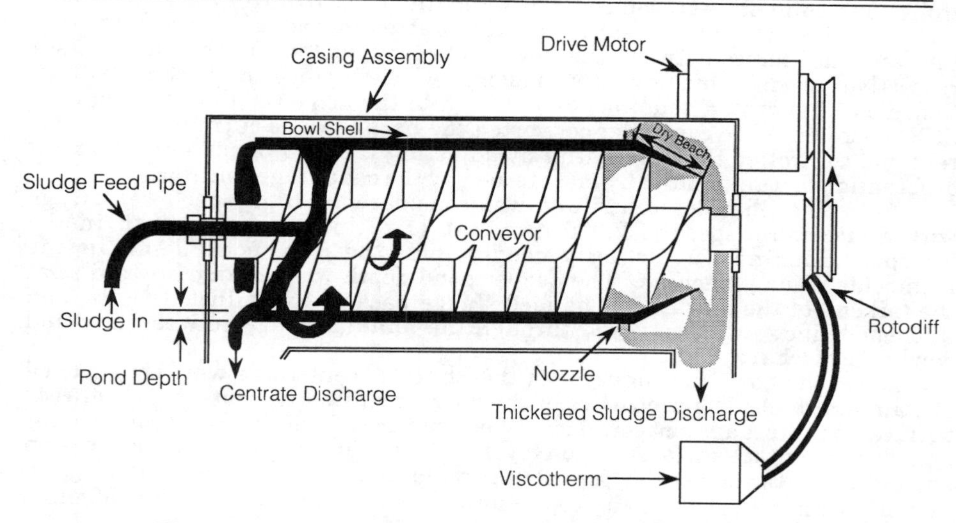

Figure 1. Humboldt-Wedag Centrifuge Schematic

THE SHARPLES CENTRIFUGE

This machine was a countercurrent design. The feed sludge entered and the thickened sludge existed at the same end of the unit (Figure 2). Liquid sludge entered and thickened sludge exited at the tail end of the machine where the main drive and motor were located. Centrate flowed along the axis of the machine and exited at the centrate discharge or head end of the machine. The pond depth was controlled by a replaceable disc-shaped weir plate at the head end of the machine.

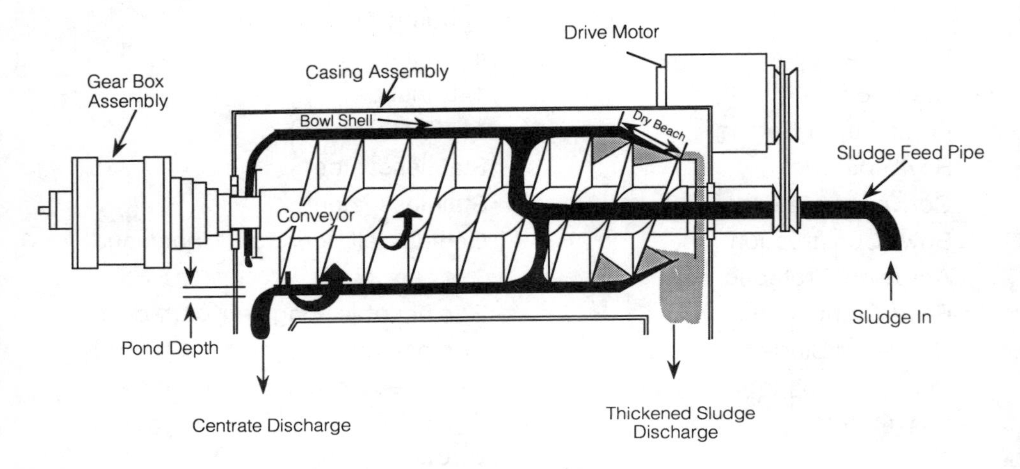

Figure 2. Sharples Centrifuge Schematic

The Sharples conveyor was similar to the Humboldt-Wedag conveyor. However, the Sharples conveyor has parallel vanes along the axis of the conveyor, interior to the conveyor blades at the centrate end of the machine. These vanes assist in clarifying the centrate. The conveyor in the Sharples centrifuge rotates at a slightly lower speed than does the bowl. The differential speed in the Sharples machine was achieved by a dc motor backdrive and controlled by a solid-state silicon-controlled rectifier (SCR) controller.

The backdrive system essentially acted to brake the conveyor since the fractional forces between the bowl and conveyor tended to equalize their respective speeds. This system was energy efficient since the backdrive functioned as a generator, sending current back to the ac power lines through the SCR controller. The controller converted the backdrive dc current into ac current and controlled the differential by adjusting the flow of current. A preset differential was automatically maintained by the centrifuge controls.

The Sharples centrifuge was considered a high-speed centrifuge with a maximum speed of 2,000 rpm. At 11.73 in., the pond depth in the Sharples machine was actually deeper than the 8-in. pond depth in the Humboldt-Wedag machine. The deeper pond depth was obtained by a disc at the tail end of the machine. The feed sludge backed up against the disc, and thickened sludge was conveyed through the annular space between disc and bowl to be discharged.

The bowl and conveyor of the Sharples centrifuge were constructed of stainless steel. The properties of the metal required no coating on nonwear surfaces to protect against corrosion. Wear surfaces, including feed components and sludge contact surfaces of conveyor blades, were protected from abrasion and erosion with replaceable tungsten carbide tile. The bowl was centrifugally cast of stainless steel and balanced to form a monolithic structure free of joints to prevent areas of high stress.

WAS CENTRIFUGE THICKENING FACILITY DESIGN

The interim WAS centrifuge thickening facility was composed of several systems as schematically presented in Figure 3. These subsystems included centrifuge feed, centrifuge, polymer, thickened sludge transfer, and centrate disposal. Each system is described below.

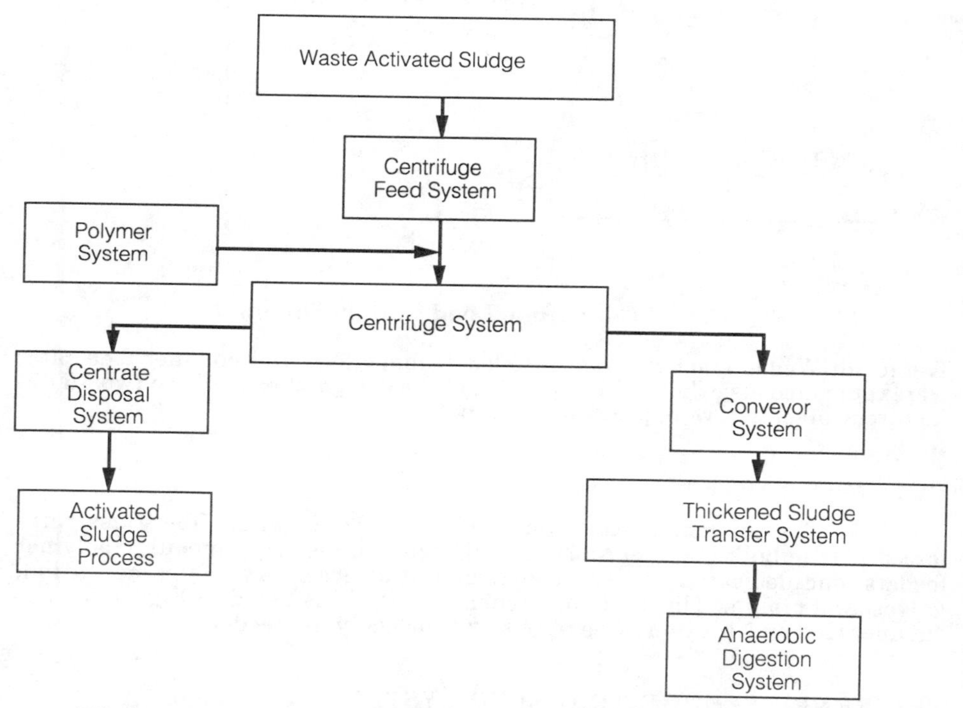

Figure 3. Centrifuge Thickening Facility Process Schematic

CENTRIFUGE FEED SYSTEM

The centrifuge feed system was divided into four subsystems, with one subsystem per centrifuge, as schematically presented in Figure 4. The subsystem for each centrifuge consisted of one 30-hp recessed impeller centrifugal pump with a rated capacity of 500 to 1,000 gpm; variable frequency drive; magnetic flow meter; and accessory piping, valves, instrumentation, and controls. This system and the flow to each centrifuge was controlled from the system control panel.

CENTRIFUGE SYSTEM

The centrifuge system consisted of four centrifuges, including drives, backdrives, oil-to-water cooling units, local control panels, and motor starters. The desired operating conditions of each centrifuge were controlled and monitored from its local control panel.

The adjustable operating parameters included bowl speed, conveyor speed, and differential speed. In addition, a discharge nozzle (port) area for the

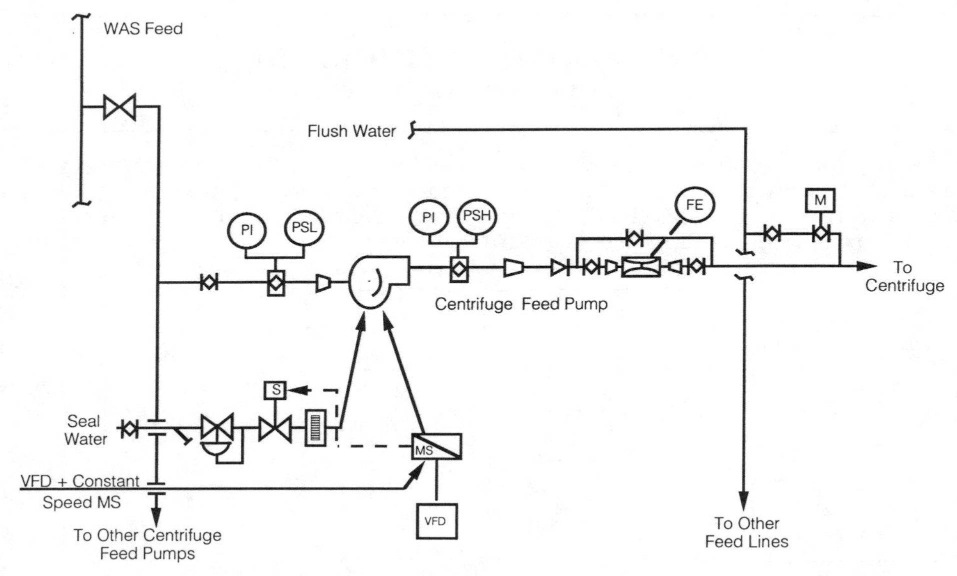

Figure 4. Centrifuge Feed System Schematic

Humboldt-Wedag machine was included. Sludge feed and polymer feed rates were controlled elsewhere. The specifics of the design of each of the two manufacturers' machines were previously discussed.

POLYMER SYSTEM

The polymer system consisted of a 10,000-gallon fiberglass reinforced plastic bulk polymer storage tank; four high-energy modular polymer feeders, one dedicated to each centrifuge; and accessory instrumentation, valves, and piping. This system is schematically illustrated in Figure 5. The polymer feed and dilution rates were set at each polymer feeder.

THICKENED SLUDGE TRANSFER SYSTEM

The thickened sludge transfer system is schematically illustrated in Figure 6. Two sets of two-screw conveyors transferred thickened WAS (TWAS) from each of the two manufacturers' centrifuges to two separate 4,000-gallon bolted steel tanks. TWAS was withdrawn from the two storage tanks through a common header by one of two 20-hp progressive cavity pumps with a rated capacity of 50 to 300 gpm. The TWAS was then pumped through a common line to the plant's anaerobic digesters. Each pump was equipped with a hydrostatic variable speed drive.

Each pump's discharge rate may be controlled in the manual mode or automatically adjusted to maintain a stable tank level. In addition to the equipment mentioned above, accessory controls, magnetic flow meter, valves, and piping made up the TWAS transfer system.

THE CENTRATE DISPOSAL SYSTEM

The centrate disposal system consisted of four interconnected 12-in. gravity lines, complete with valves, as illustrated in Figure 7. Due to the retrofit of the existing facilities, a portion of the gravity lines would have full pipe flow or would be surcharged with an inverted siphon condition. The centrate drained to the plant's activated sludge aeration inlet channels.

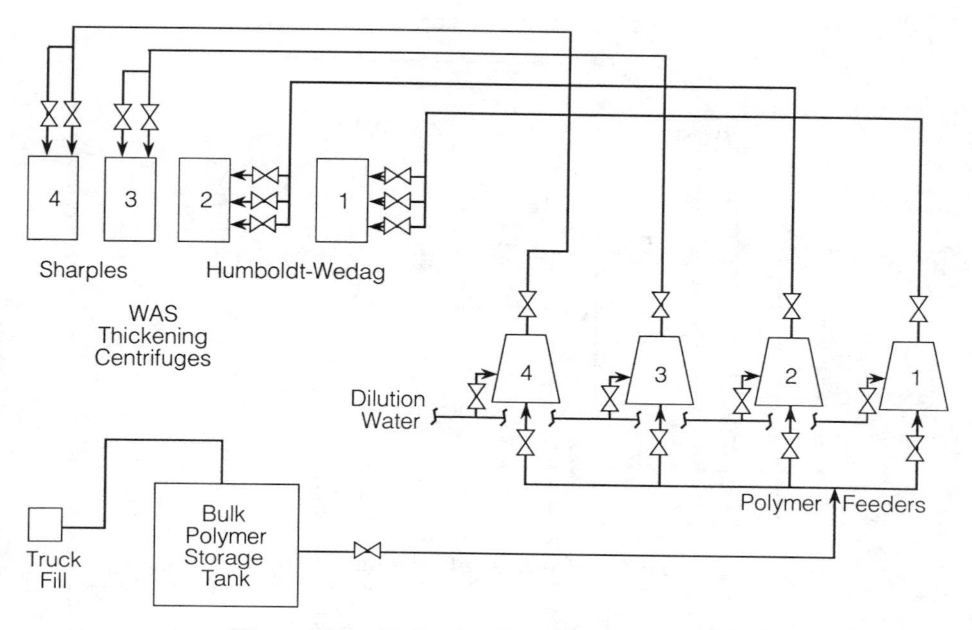

Figure 5. Polymer System Schematic

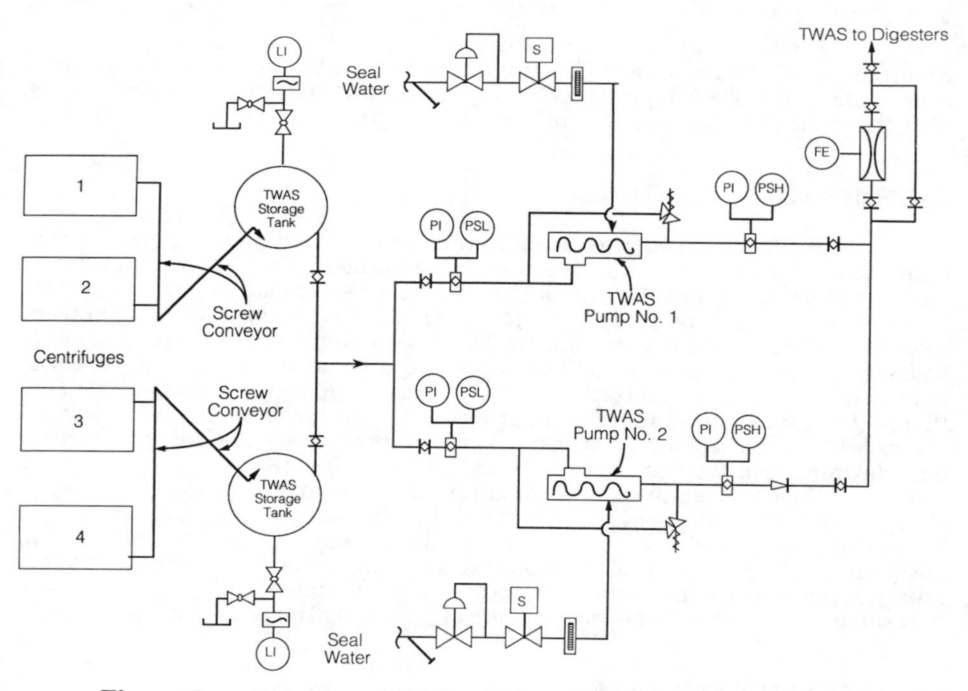

Figure 6. Thickened Sludge Transfer System Schematic

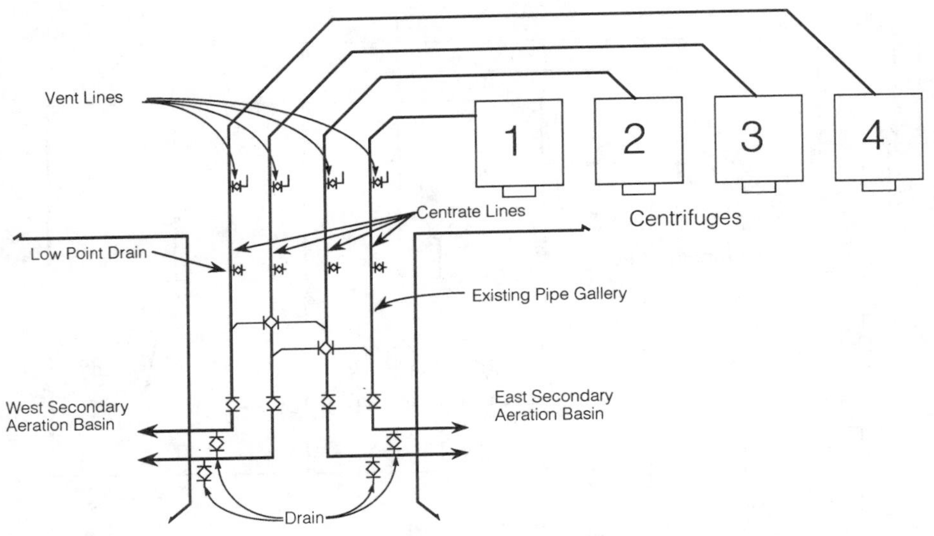

Figure 7. Centrate Disposal System

SYSTEMS MODIFICATION

During start-up of the facility, two problems were encountered that required modifications to the facilities. These two problems included high head conditions at the TWAS pumps and restricted flow through the centrate lines. Mitigative measures were taken to correct these problems as discussed below.

HIGH HEAD CONDITIONS

Discharge pressures encountered at the TWAS pumps were approximately two times greater than the calculated values. The design values were based on standard hydraulic calculations for water and modified for sludge with curves from Metcalf and Eddy[3], which were based on sludge type and percent solids. Research[1,4] revealed that the use of these and similar curves to equate fractional losses of water to various types and concentrations of sludges could result in gross and significant errors. It is recommended that engineers thoroughly investigate the physical properties of a plant's thickened sludge or slurry prior to the design of a transfer system since these materials can exhibit non-Newtonian properties.

Wastewater sludges above a certain concentration, specific for each sludge, exhibit the rheological properties of a thixotropic material, which can be approximated as a Bingham plastic. Where site-specific analysis of sludge is not permissible prior to construction, a conservative design is recommended. This problem was mitigated by tying into a larger diameter abandoned pipeline that ran parallel to the same location and abandoning the original line.

RESTRICTED CENTRATE LINES

During the start-up of the centrifuges, flow was restricted in the centrate lines and would back up into the centrifuges. This would occur in some instances with flow rates as low as 250 gpm, even though the centrate lines were designed for an excess of 1,000 gpm. Test risers that were comprised of short lengths of pipe and a ball valve were installed in one of the centrate lines, and it was discovered that a combination of foam and air was restricting flow through the pipe by reducing the effective diameter of the pipe. The foam problem was greatest when excessive polymer was being added to the primary clarifier and the plant was experiencing *Nocardia sp*. problems, two well-documented sources of foam in treatment plants.

This problem was mitigated by venting the centrate lines in several locations to a level above the centrifuge for air release and adding a spray system to the centrate hopper, complete with chemical defoamer injection capabilities. These modifications were successful, and the spray system has worked without the need for defoamer. It was recommended that centrate systems be conservatively sized and well vented for air release. In addition, spray systems should be installed similar to that described above to mitigate the occurrence of foam.

PURPOSE OF FACILITY AND FULL-SCALE TEST

The interim WAS thickening facility had the dual purpose of providing interim WAS thickening capacity to the Hyperion Treatment Plant and providing for the full-scale testing of two manufacturers' centrifuges. Following the facility's start-up, each machine was required to meet the guaranteed performance standards as presented in Table 3. Following the completion of the performance testing of each machine, a 6-month full-scale test was begun in March 1989.

Table 3. Guaranteed Performance Standards

Sludge Feed	0.5 to 0.7% solids by weight
Feed Rate	600 gpm
Cake Solids	4.0% solids by weight
Solids Capture	85.0%
Polymer	None

The purpose of the long-term test was to collect data to be used in the evaluation of each manufacturer's machine. The data collected will be used to determine centrifuge performance under various operating parameters, as well as electrical, operational, maintenance, and chemical requirements, for facilities using each of the two manufacturers' equipment.

The various tests performed on each manufacturer's centrifuges and their results are presented. Humboldt-Wedag centrifuge number 1 and Sharples centrifuge number 4 were used to represent each manufacturer's centrifuges. Each of the tests, with the exception of the performance test with polymer addition, were performed side-by-side and at the same time so that both machines received identical feed sludge for each set of samples taken. Feed sludge concentrations ranged from 4,920 mg/L to 5,880 mg/L. However, for the majority of the tests, the feed sludge concentration was above 5,600 mg/L.

RESPONSE TIME TEST

This test was conducted to determine the time it took for each centrifuge to reach equilibrium following a deliberate change of an operational parameter. The goal of this test was to determine the required grace period between an operational change and the initialization of sample collection. This test was performed without polymer and at differential speeds of 7.5 rpm for the Humboldt-Wedag and 3.2 rpm for Sharples.

Feed sludge, TWAS, and centrate samples were collected and solids concentrations analyzed for both machines at 10-minute intervals following instantaneous sludge feed flow changes of 500 to 700 gpm and 700 to 500 gpm. The results of this test are graphically presented in Figure 8. For both machines, equilibrium was reached within approximately 15 minutes after increasing the sludge feed from 500 to 700 gpm. When the sludge feed rate was reduced from 700 to 500 gpm, both the TWAS concentration and capture rate equalized within approximately 15 minutes. However, the centrate took 30 to 40 minutes before it reached concentration equilibrium. Therefore, a grace period of 30 to 40 minutes is recommended following an operational change.

PERFORMANCE TEST WITHOUT POLYMER

This test was conducted to establish machine performance without polymer based on operational modifications and adjustments. The goal of this test was to determine performance trends as a result of operational changes and establish optimum operational parameters. Various differential speeds were evaluated for each machine at differing feed rates. During this test, the WAS had a sludge volume index (SVI) of 130 and an average solids concentration of 5,800 mg/L.

The Humboldt-Wedag machine was tested at differential speeds of 3, 6, 9, and 11, and at flow rates of 400, 550, 700, and 820 gpm. The bowl speed of 1,600 rpm, the pond depth of 8 in., and two 10-mm diameter nozzles with a total discharge area of 78.5 mm^2 were constant throughout the test. The same differential speeds and flow rates were tested for the Sharples machine. However, the bowl speed of 1,995 rpm and the pond depth of 11.73 in. were constant throughout the test. Nozzle size was not a variable with the Sharples machine.

The results for the Humboldt-Wedag machine are presented in Figure 9. This figure clearly shows that capture efficiency decreased as the flow rate increased. This corresponded to the degradation of centrate quality (higher solids concentration) as the flow rate increased. The optimum differential speed appeared to be between 6 and 9. However, due to the relatively flat curves, differential speed was not an operational control parameter, as stated by the manufacturer. Both capture efficiency and centrate quality degrade at both the low and high ends of the differential speed setting. At the lower differential speeds, the sludge was not conveyed to the discharge ports quickly enough, which caused the TWAS layer to increase, resulting in solids being discharged with the centrate.

At the higher differential speeds, centrate quality also degraded possibly as a result of excessive turbulence. With respect to TWAS solids, the results indicated that there was an optimum flow rate (at approximately 700 gpm) for a given set of conditions. At flow rates above and below this optimum flow rate, the TWAS solids concentration was lower.

This information clearly showed that the differential adjustment had no useful effect on improving centrifuge performance or controlling TWAS concentration or centrate quality. The manufacturer had stated that the primary control parameters were flow rate and sludge discharge nozzle area. The data obtained verified that the flow rate was indeed a major control factor for centrifuge performance. The sludge nozzles were not available to be changed during this testing. Further testing will be performed to document performance with various sludge discharge nozzle areas. It should be noted that the nozzle can be changed only when the centrifuge was at rest. The operation to change the nozzles was quite simple, requiring approximately 15 to 30 minutes after the centrifuge has been stopped and flushed.

The data collected on the Sharples machine, as presented in Figure 10, illustrates an interesting phenomenon. The capture efficiency performance curves for all four flows converged at a differential speed of 6.0 rpm. This indicated that, under the feed conditions present during the test, 87 per-

RESPONSE TIME DETERMINATION (Humboldt 1)
Flowrate changed from 500 to 700 gpm

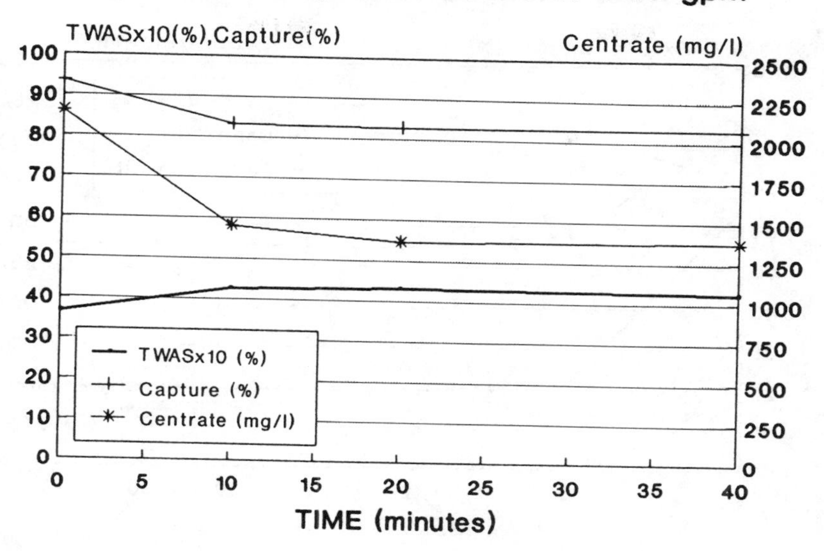

RESPONSE TIME DETERMINATION (Humboldt 1)
Flowrate changed from 700 to 500 gpm

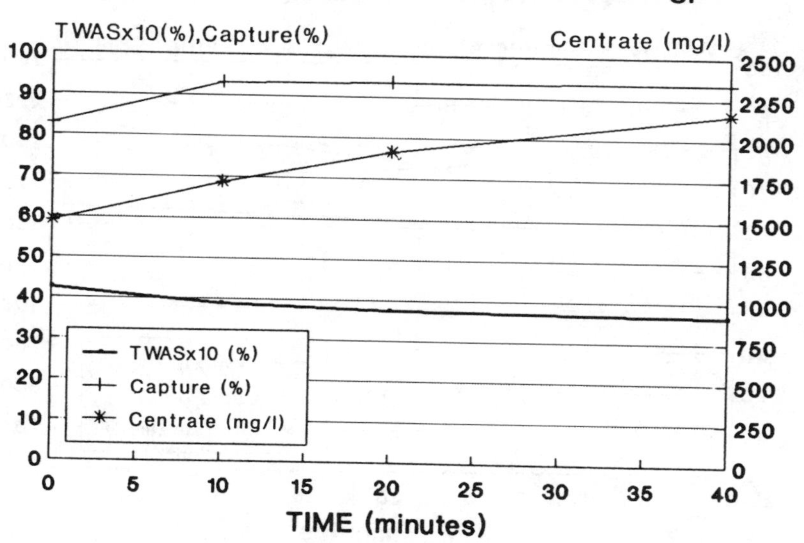

Figure 8. Centrifuge Response Time

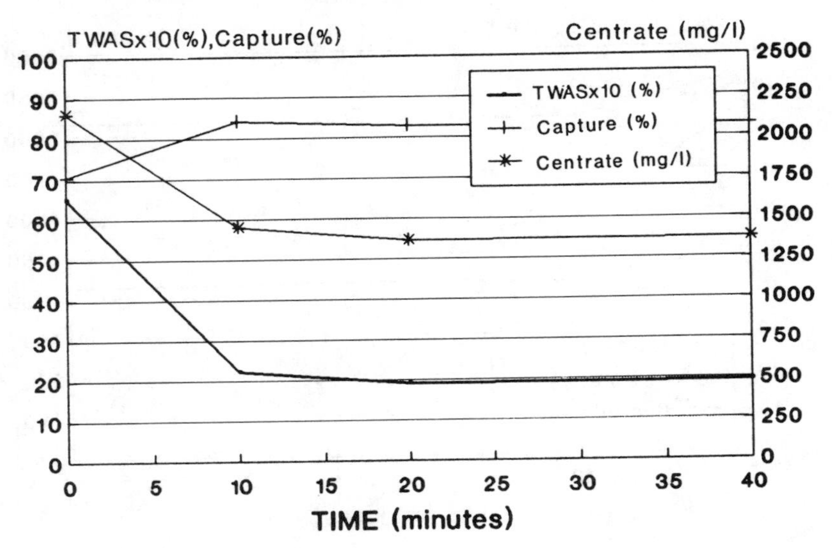

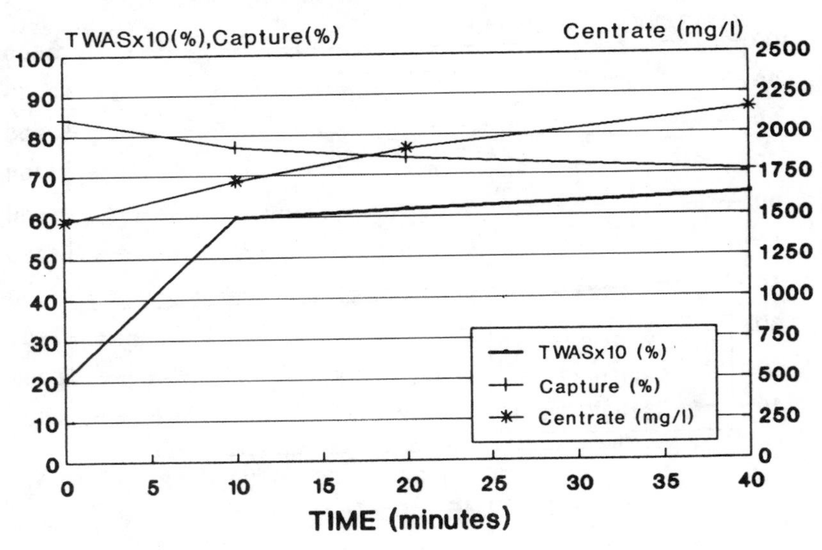

Figure 8. Centrifuge Response Time (Continued)

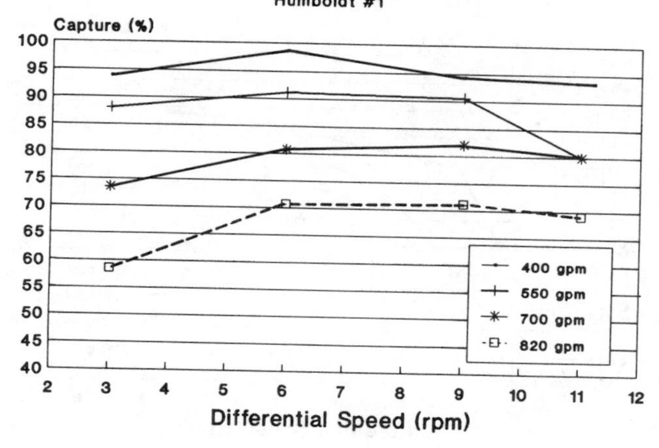

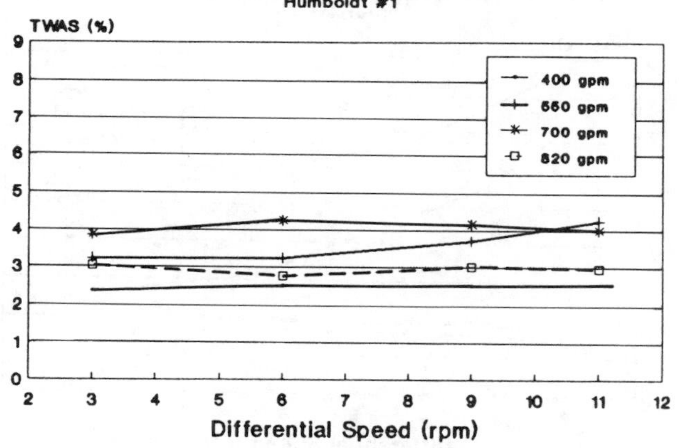

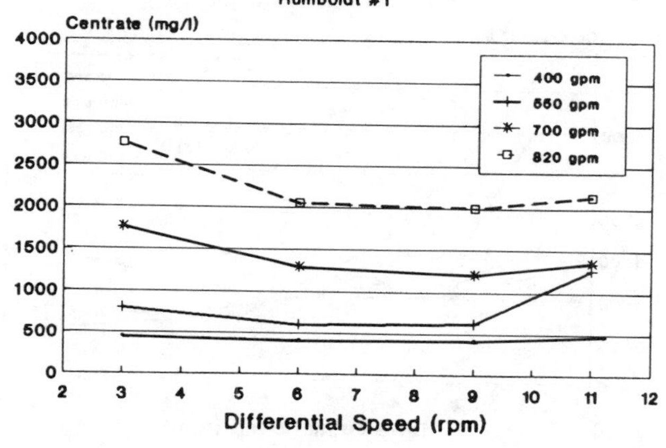

Figure 9. Humboldt-Wedag Performance

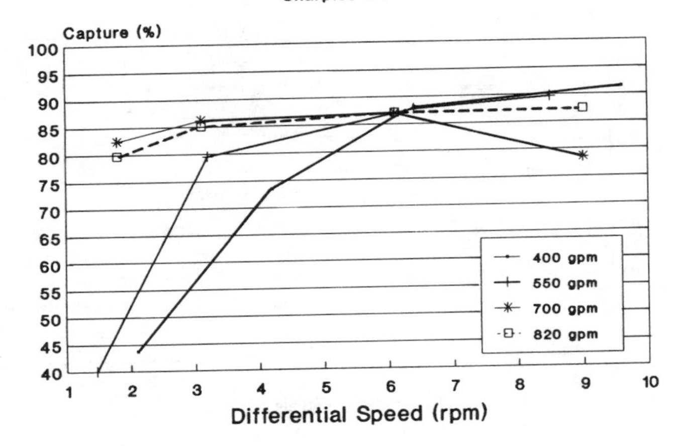

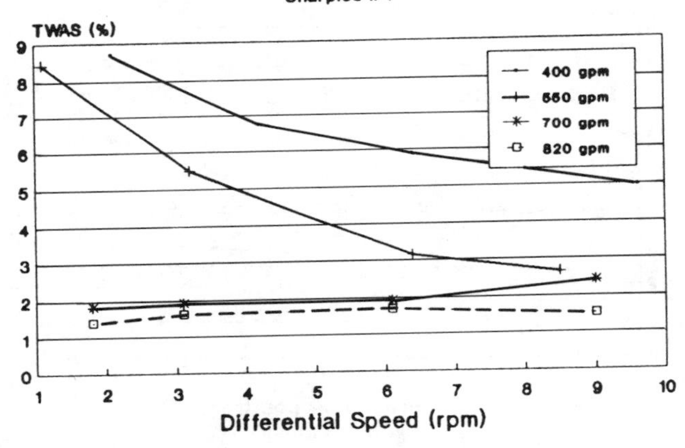

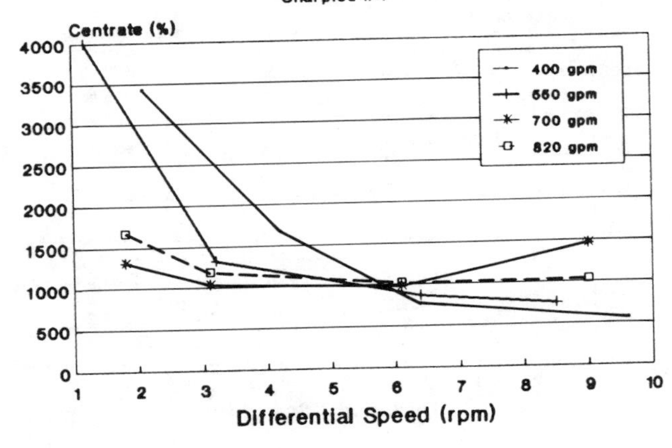

Figure 10. Sharples Performance

cent capture efficiency was obtainable at flows between 400 and 820 gpm. At flows of 550 gpm or lower, differential speed has the largest influence on performance.

At flows of 700 gpm or greater, the maximum TWAS concentration obtained was approximately 2 percent total solids, and the differential speed had very little effect on performance. This indicated that, somewhere between 550 and 700 gpm, the centrifuge ceased to perform effectively without polymer addition. However, at flows of 550 gpm or below, TWAS concentrations were produced from 3 to 8.5 percent total solids. This was possible by adjusting the differential speed electronically from the local control panel (LCP) while the centrifuge was in operation.

As expected, the TWAS concentration decreased and the centrate quality improved as the differential speed was increased. When the differential speed was decreased, the converse was true. Also, the TWAS concentration and centrate quality were closely related. As the TWAS concentration increased, the centrate quality degraded and vice versa.

PERFORMANCE TEST WITH POLYMER

This test was conducted to establish machine performance with polymer addition at various rates and other operational modifications and adjustments. The goal of this test was to determine performance trends as a result of operational changes and to establish optimum operational parameters with polymer additions. Diatec 4240A, a high molecular weight, cationic polymer at 3 percent active neat liquid, was used for all polymer tests.

The Humboldt-Wedag centrifuge was tested at feed rates of 600, 650, 700, and 800 gpm with polymer feed rates ranging from 0 to approximately 4.0 pounds of polymer per ton of sludge solids (lb/ton). The differential speed was set at 6.8 rpm, based on the previous test without polymer.

The results of this test on the Humboldt-Wedag machine are graphically presented in Figure 11. The curves indicated that the capture efficiency was relatively insensitive to polymer addition since only small improvements of approximately a 5 percent increase in capture efficiency were noted at flow rates of 700 gpm or less when polymer was added at approximately 2.0 lb/ton. Polymer dosages above 2.0 lb/ton yielded no significant additional capture. The capture efficiency increased noticeably with increased polymer dosage at sludge feed above 700 gpm up to 4.0 lb/ton. Testing has not been conducted above 4.0 lb/ton.

Polymer addition showed little effect on TWAS concentration at any of the flow rates tested. The TWAS concentration increased as the flow rate increased within the ranges tested. The most significant improvement occurred from 600 to 650 gpm.

Polymer reduced the solids concentration of the centrate by approximately one-half at a polymer dose of approximately 2.0 lb/ton for all sludge feed rates. Centrate quality appeared to level off at a concentration of approximately 250 mg/L; additional polymer did not improve the centrate quality. Additional nozzles were not available to incorporate this operational parameter into the test. However, they will be tested at a later date.

The results confirmed Humboldt-Wedag's claim that their machine was relatively insensitive to operational change; this was demonstrated by their consistent performance under varying operational conditions that included feed rate and polymer addition.

Polymer addition can marginally improve the quality of the centrate and capture efficiency but had little effect on TWAS concentration.

The Sharples machine was tested at feed rates of 600, 700, and 800 gpm and at polymer rates ranging from 0.0 to approximately 2.8 lb/ton of dry solids. Various differential speeds were tested ranging from above 0.0 to approximately 10.0. Unlike the Humboldt-Wedag machine, polymer addition drastically improved the performance of the Sharples machine. However, for each polymer dose at a set feed rate, there is an optimum differential speed at

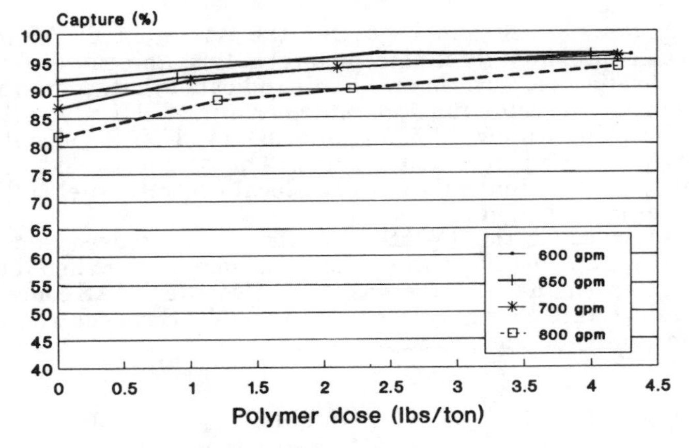

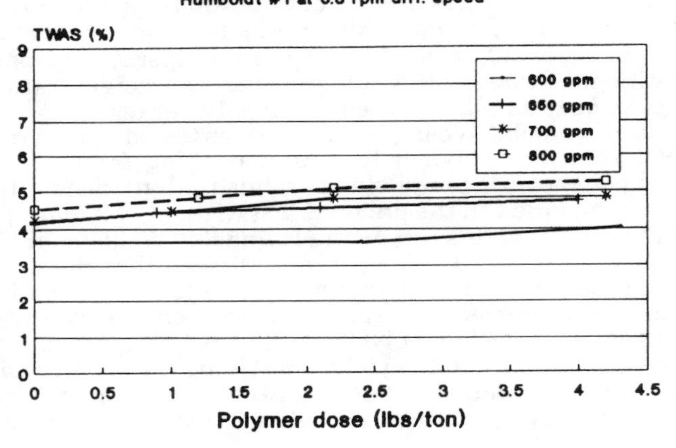

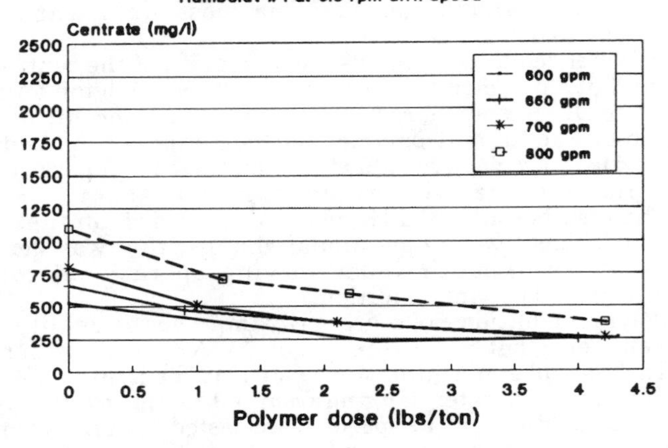

Figure 11. Humboldt-Wedag Performance with Polymer

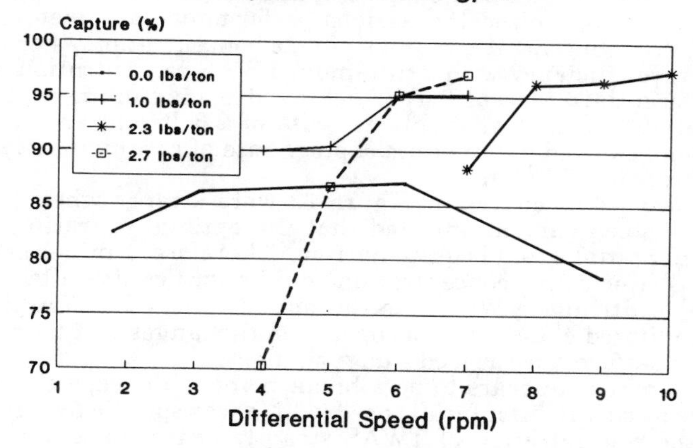

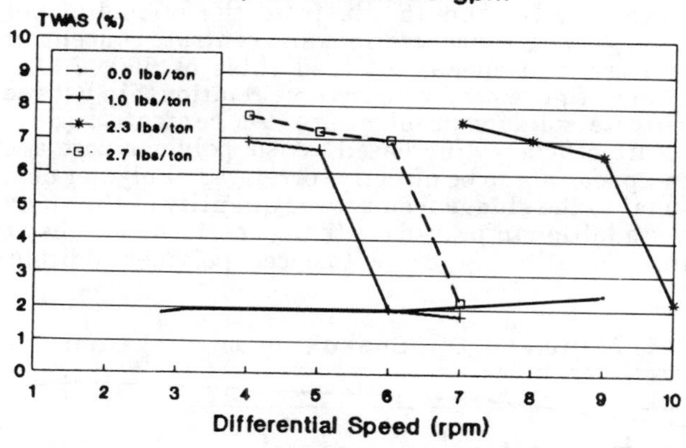

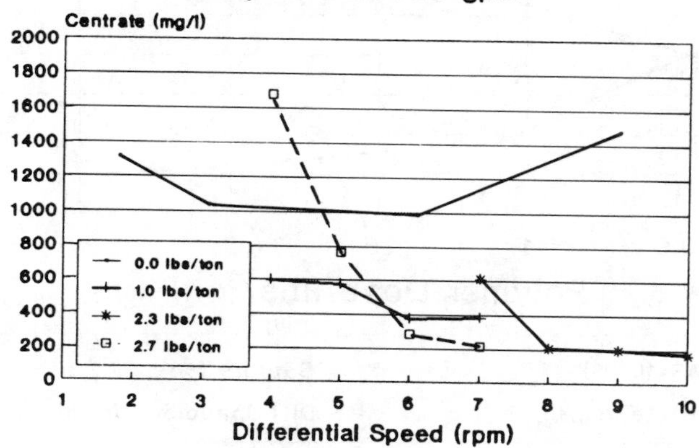

Figure 12. Sharples Performance with Polymer at 700 gpm

which the centrifuge performance is optimized. This concept is graphically presented in Figure 12. This figure clearly shows that, at a feed rate of 700 gpm and polymer addition rates of 0.0, 1.0, 2.3, and 2.7 lb/ton of dry solids, specific differential speeds optimized the various performance parameters of capture efficiency, TWAS concentration, and centrate concentration. For instance, to optimize capture efficiency with a minimum TWAS concentration of 7 percent solids, a differential speed of 8 rpm achieved a maximum capture rate of approximately 97 percent at a polymer rate of 2.3 lb/ton; and a differential speed of 6 rpm achieved a maximum capture rate of approximately 95 percent at a polymer rate of 2.7 lb/ton.

The various curves in Figure 12 were steeper when polymer was added. These steep curves indicated that the optimal operating range with respect to differential speed is quite narrow. Therefore, a minor change to the feed sludge SVI or solids concentration could significantly affect the performance of the centrifuge. With this concept in mind, the Sharples machine should be monitored closely so that operational changes can be made to optimize machine performance as feed sludge changes.

There also appears to be a break point with respect to differential speed. Above a critical differential speed (which was specific for each operating condition), the concentration of TWAS steadily degraded to where the centrifuges were of minimal value. This may result from a break in the sludge seal between the bowl and conveyor, or possibly from excessive turbulence within the centrifuge.

Figures 13, 14, and 15 illustrate the effects of polymer dose on capture efficiency, TWAS concentration, and centrate concentration at the best or optimum differential speeds, at feed rates of 600, 700, and 800 gpm, respectively. These figures show an inverse relationship between capture efficiency and centrate solids concentration. As centrate solids concentration decreased, capture efficiency increased. Also, polymer dose and optimum differential speed appear not to be directly correlated. Polymer dose requirements were proportional to the sludge feed rate or quantity of the sludge solids. Also, limited polymer addition improved centrifuge performance; beyond this threshold, performance actually degraded with excess polymer addition.

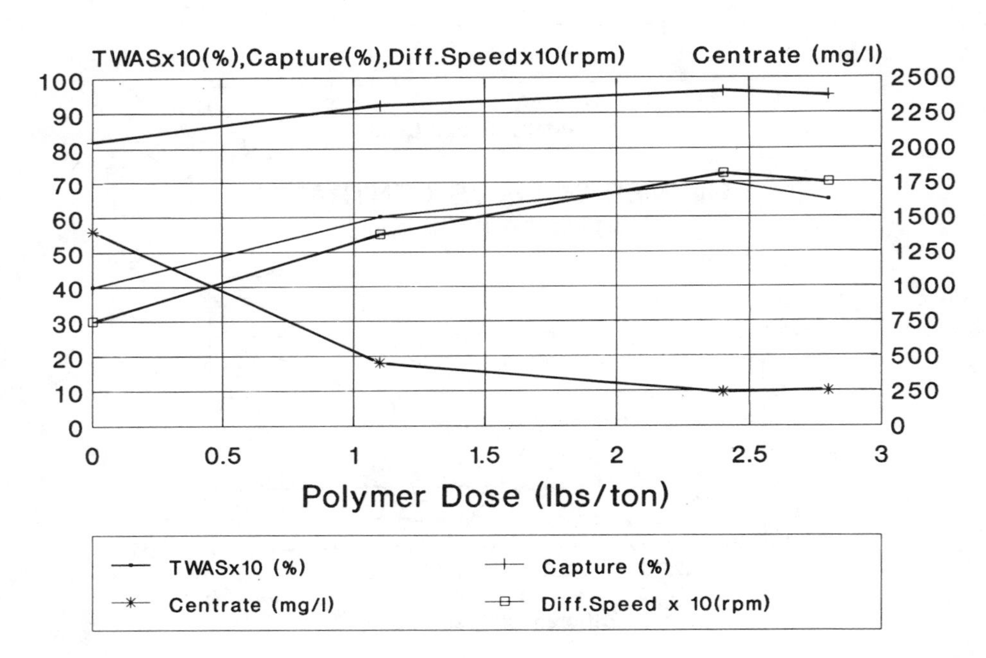

Figure 13. Optimum Sharples Performance with Polymer at 600 gpm

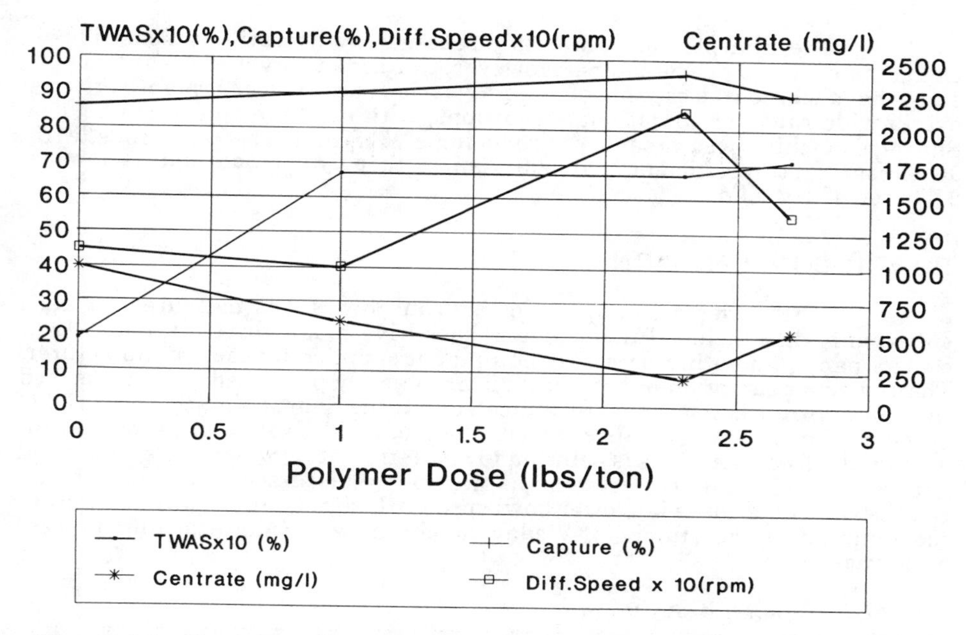

Figure 14. Optimum Sharples Performance with Polymer at 700 gpm

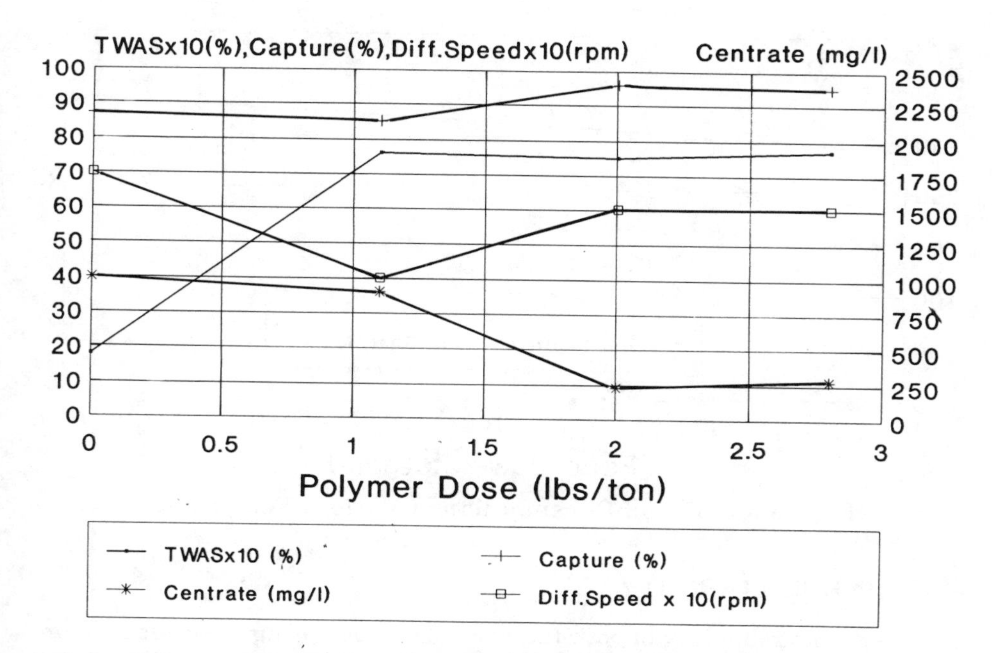

Figure 15. Optimum Sharples Performance with Polymer at 800 gpm

The Sharples machine, as discussed, was relatively sensitive to change in operating parameters. However, this sensitivity was advantageous, since the machine was capable of meeting various desired performance criteria over a wide range of operational conditions, with machine adjustments. The Sharples machine was capable of producing TWAS at 7.5 percent solids, with 95 percent capture efficiency, at 800 gpm, with polymer addition as low as 2.0 lb/ton of dry solids.

POWER CONSUMPTION

The power consumption of both machines was measured for each entire unit that included main motor, backdrive motor, lube oil motor, local control panel, and other accessory equipment supplied by each manufacturer. The power consumption of both units is graphically presented for various feed rates in Figure 16. These results indicated that the Sharples machine was 30 to 50 percent more energy efficient than the Humboldt-Wedag machine, dependent on the feed rate. This is believed to result from its smaller drive motor and energy efficient backdrive system, as previously discussed. Testing is being conducted to determine if a lower bowl speed will significantly lower the power consumption of the Humboldt-Wedag machine, while maintaining process performance.

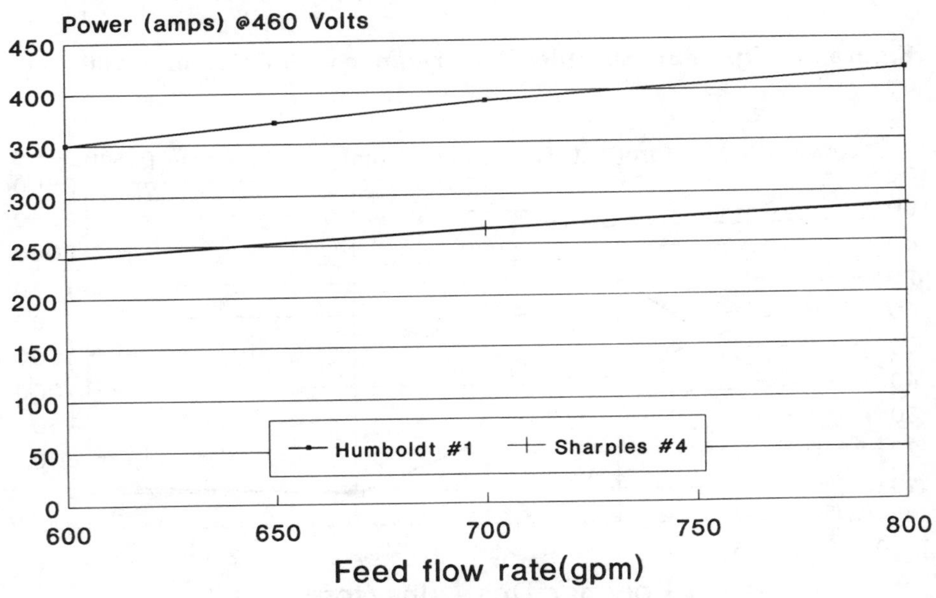

Figure 16. Power Consumption at Various Feed Rates

MACHINE RELIABILITY

The reliability of both the two Humboldt-Wedag centrifuges (numbers 1 and 2) and the two Sharples centrifuges (numbers 3 and 4) was measured in number of occurrences and times out of service (days) and is presented in Figure 17. The data presented were conservative, as machines were not penalized for time out of service when an accessory piece of equipment that was not supplied by either manufacturer was down. In addition, machines were not penalized for occurrences that could have been mitigated by operational staff. Downtime, therefore, includes only equipment malfunction or component failure. Causes of downtime are summarized by manufacturer in Table 4.

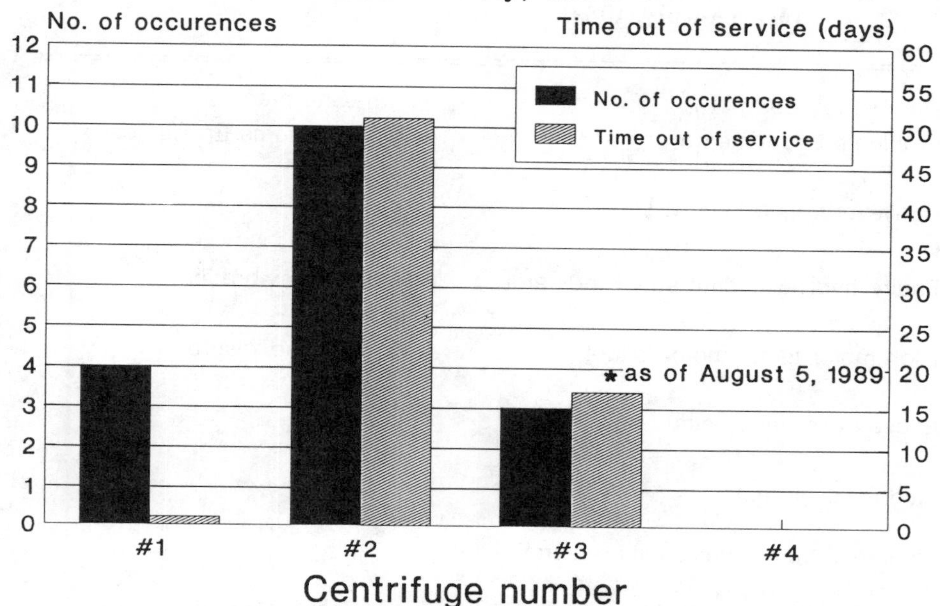

Figure 17. Unscheduled Centrifuge Downtime (March-July, 1989)

WEAR AND CORROSION

No extensive wear was noticed on any of the machines. Corrosion was noticed on both Humboldt-Wedag machines in the form of rust on steel components that include the bowl, conveyor, and casing. Although corrosion does not currently create a problem, its long-term implications are unknown. No noticeable rust was found on the Sharples machine.

SUMMARY AND CONCLUSIONS

The start-up and performance testing of two manufacturers' centrifuges at the City of Los Angeles' Hyperion Treatment Plant provided valuable information and data on the use of centrifuges for thickening municipal waste-activated sludge. The information and data collected identified key system design considerations, established process performance criteria for both centrifuges, and provided insight into the operation of both machines.

Two conditions that required mitigative measures during start-up included (a) excessive head loss in the thickened sludge transfer lines and (b) restricted flow through the centrate lines. Investigation into the cause of the excessive head loss in the thickened sludge line concluded that thickened sludge is a non-Newtonian, thixotropic material that may be modeled as a Bingham plastic. Since much erroneous information has been published with respect to sludge head loss, engineers are advised to use site-specific data on sludge head loss (when available) or thoroughly research sludge head loss and provide a conservative design when site-specific data are not available.

Detailed investigation revealed that the combination of air and foam can greatly reduce the effective capacity of the centrate piping. To mitigate this potential problem, centrate lines should be conservatively sized and well vented. A water spray system with defoamer capabilities is also recommended to break up foam in the centrate hopper of centrifuges.

Table 4. Unscheduled Downtime Caused by Centrifuge Malfunction (March-July, 1989)

Humboldt-Wedag Problems Encountered	Sharples Problems Encountered
Conveyor bound to bowl [1]	PLC problems[3]
Faulty hydraulic differential indicator	High vibration
High motor temp. motor failure	Choke failure[4]
Loose main drive belts	
Hydraulic oil leak[2]	
Hydraulic pump motor failure alarm	
High lube oil temp. alarm.	

1: The recurring problem required that the bowl and scroll assembly be removed from the carriage three times. The third time it was discovered that a scraper mechanism was not built within proper tolerances. The faulty assembly was modified, and the problem appears to be resolved.

2. Several other oil leaks have been experienced with machine number 1 and machine number 2, but have not been large enough to shut down the unit.

3. The PLC (Programmable Logic Controller) lost some values stored in memory registers.

4. The dc Choke has been sent for repairs. The total downtime is undetermined. as the unit has been out of service It was 14 days as of 8-5-89.

The performance test and evaluation included several aspects. It was concluded that a grace period of 30 to 40 minutes is required, after an operational change, to allow the centrifuge to reach equilibrium before samples are collected. The performance test of both the Humboldt-Wedag and Sharples centrifuges, with and without polymer, yielded considerably different results with respect to each centrifuge. The two main operational variables for the Humboldt-Wedag machine were sludge feed rate and nozzle size. An increase in feed rate resulted in a decrease in both the capture efficiency and centrate quality. With respect to TWAS concentration, there appears to be an optimum feed rate for a given set of conditions. A feed rate above or below this optimum setting results in less concentrated TWAS. Differential speed has little effect on machine performance, which confirmed the manufacturer's contention that it is not an operational control variable. Surprisingly, the addition of polymer was only marginally successful in improving the machine's performance with respect to capture efficiency and centrate quality. No significant improvement was noticed in TWAS concentration with polymer additions. Nozzle size could not be tested as an operational control variable since additional nozzles were not available during the test. This parameter will be tested at a later date.

The Humboldt-Wedag machine was operationally stable: its performance did not change significantly with operational modifications. In effect, operational adjustments were not required with minor changes to the feed sludge SVI and solids concentration. Performance adjustments are made by changing the nozzle size. This machine did not require a great deal of operator attention.

The testing of the Sharples machine showed that sludge feed, differential speed, and polymer addition were all effective operational parameters. An increase in the differential speed resulted in an increase in both the capture efficiency and centrate quality, with a decrease in TWAS concentration.

The addition of polymer drastically increased the effective capacity of the centrifuge and improved all measurable performance parameters. Optimum polymer dose was proportionate to the sludge feed rate or quantity of total solids. Polymer addition above the optimum dose resulted in a decrease in machine performance. For each operational condition with polymer, there was an optimum differential speed. A maximum or breakpoint differential speed exists for each operational condition. Above this speed, the performance of the machine steadily degrades.

The Sharples machine is relatively sensitive to operational conditions. A minor change in the feed sludge SVI or solids concentration can significantly affect the machine's performance. Therefore, the machine must be monitored and operational adjustments made to optimize the machine with changes in sludge feed conditions. The sensitivity of the machine is beneficial in that the machine may be adjusted to a desired level of performance. The centrifuge can be operated over a wide range of feed rates and at various performance criteria.

The Sharples machine was 30 to 50 percent more energy efficient than the Humboldt-Wedag machine, dependent on the sludge feed rate. The Sharples machines were more reliable than the Humboldt-Wedag machines when measured as a group. However, the most reliable Humboldt-Wedag machine was more reliable than the least reliable Sharples machine.

Corrosion was noticed on Humboldt-Wedag components. No corrosion was noticed on the Sharples machines.

DISCLAIMER

The authors of this paper and their respective organizations do not endorse any of the products mentioned herein. The data and results presented in this paper are based on site-specific conditions in a controlled setting. The authors firmly believe that all data and results are capable of duplication under identical conditions and use of the same equipment. However, although the performance of the equipment presented in this paper is expected to be representative of similar conditions and equipment, identical data and results are not anticipated. Readers considering using centrifuges for sludge thickening are encouraged to pilot test units using their own site-specific sludge, where possible, for collection of data more applicable to their own situation.

REFERENCES

1. Carthew, G. A., Goehring, C. A., and Van Teylingen, J. R., "Development of Dynamic Head Loss Criteria for Raw Sludge Pumping," *J. Water Pollut. Control Fed.*, 1985, *55*, 472-483.

2. Letki, A. G., "Dewatering Fine Particles Using Centrifuges," presented at the Consolidation and Dewatering of Fine Particles Conference, University of Alabama, August 10, 1982.

3. Metcalf and Eddy, Inc., *Wastewater Engineering: Treatment Disposal Reuse*, McGraw-Hill, 2nd Edition, New York, NY, 1979, 588-595.

4. Mulbarger, M. C., Copas, S. R., Kordic, J. R., and Cash, F. M., "Pipeline Friction Losses for Wastewater Sludge," *J. Water Pollut. Control Fed.*, 1981, *53*, 1303-1313.

TEN-MINUTE DIGESTION

David E. Gustafson
Hach Company
P.O. Box 389
Loveland, Colorado 80539-0389
(303) 669-3050

ABSTRACT

Findings of a recent study conducted to establish the accuracy and precision of the patented Hach Digesdahl™ system by comparative analysis of a reference standard material indicate the Hach system deserves consideration as the method of choice. Reference sewage sludge of domestic origin, supplied by the Commission of the European Communities, was analyzed in numerous European laboratories. Many analytical methods for final development were used. Analysts at Hach Company used the Digesdahl system for sample preparation and colorimetric methods for final analysis of the same sludge material. Digesdahl results, comparable to certified and indicative values obtained from the Community Bureau of Reference, show the Digesdahl system is a complete and practical system for sample preparation.

INTRODUCTION

A new and innovative digestion procedure developed at the Hach Company replaces the conventional nitric acid digestion. The conventional digestion method described in the U.S. Environmental Protection Agency's (EPA) March 1979 procedural manual, titled "Methods for Chemical Analysis of Water and Waste," involves a troublesome digestion requiring a lot of experience, time, and patience. After the conventional digestion has been completed, total Kjeldahl nitrogen (TKN) cannot be analyzed. Briefly, the conventional procedure is to transfer a sample into a Griffin beaker and add 3 mL of concentrated nitric acid. The next step--placing the beaker on a hot plate inside a fume hood and evaporating the sample to near dryness without boiling--requires full and careful attention. The beaker is cooled before addition of another 3 mL of concentrated nitric acid. After digestion is complete, a small amount of hydrochloric acid solution (1:1) is added to dissolve any precipitates.

A major disadvantage of this conventional method is the failure to analyze TKN--an extremely important determination for farmers who apply sludges to their fields. A second major disadvantage is the constant monitoring required to avoid sample boiling and total dryness.

Research in Hach laboratories has been concentrated on determining metals and minerals with the Digesdahl apparatus, rather than the conventional nitric acid digestion described above. Digesdahl apparatus results were comparable to the certified and indicative values obtained by the Community Bureau of Reference analysts.

PROCEDURE

INSTRUMENTATION

After choosing a number of methods, European analysts applied various digestion techniques and instruments in their studies. Methods included:

European Instrument Techniques

AAS	atomic absorption spectrometry
AFS	atomic fluorescence spectrometry
DDC	diethyl dithio carbamate
DPASV	differential pulse anodic stripping voltammetry

CVAAS	cold vapor atomic absorption spectrometry
ES	emission spectrometry
ET-AAS	electrothermal atomic absorption spectrometry
HMDE	hanging mercury drop electrode
ICP	inductively coupled plasma emission spectrometry
IDMS	isotope dilution mass spectrometry
INAA	instrumental neutron activation analysis
MIBK	methyl isobutyl ketone
MS	mass spectrometry
RNAA	neutron activation analysis with radiochemical separation
SSMS	spark source mass spectrometry
XRF	X-ray fluorescence spectrometry

European Digestion Techniques

Digestion with HNO_3/HF at 160 C for 30 minutes
Digestion with HNO_3 followed by $HCLO_4$ at 220 C for 10 hours
Digestion with HNO_3 at 150 C for 2.5 hours
Digestion with aqua regia at 150 C for 2 to 5 hours followed with boric acid
Diqestion with $HCLO_4/HF$ at 160 C till dry (3 times)
Pressure digestion with HNO_3 followed by HF treatment
Predigestion at room temperature for 14 hours with aqua regia;
boiling temperature for 2 hours; HF treatment
Digestion by evaporation of HNO_3/H_2SO_4 (twice); HF-treatment (3 times);
HNO_3 (3 times)

Hach scientists used a Hach DR3000 Spectrophotometer--microprocessor-operated with a single-beam spectrophotometer, a double-pass grating monochromator, and a 1-in. sample cell. Preprogrammed calibrations were used without manual best-fit calibration curves. The Hach Digesdahl apparatus, consisting of a heater assembly, heat shield, fractionating column and filter pump (aspirator), was used for sample digestion. The fractionating column fits into the mouth of the digestion flask; hydrogen peroxide addition is controlled with a pump and timer. A digest suitable for direct analysis is produced without distillation, solvent extraction, or the use of catalysts. Digestion time, usually less than 10 minutes, far surpasses the 2 to 6 hours required for the conventional nitric acid/hydrochloric acid digestion. Some digestions used by the European laboratories take longer than 10 hours.

REAGENTS

Digestion Reagents

1. Sulfuric Acid (concentrated analytical reagent grade, Hach Cat. No. 979-09)
2. Hydrogen Peroxide Solution, 50 percent (Hach Cat. No. 21196-49)

Procedural Reagents

Reagents used by the Hach scientists are listed below:

1. Aluminum Testing	(Hach Cat. No. 22420-00)
2. Copper Testing	(Hach Cat. No. 14188-66)
3. Iron Testing	(Hach Cat. No. 854-66)
4. Lead Testing	anion exchange method
5. Manganese Testing	(Hach Cat. No. 22433-00)
6. Nickel Testing	(Hach Cat. No. 22426-00)
7. Nitrogen Testing	(Hach Cat. No. 21194-49)
	(Hach Cat. No. 21994-49)

8. Phosphorus Testing (Hach Cat. No. 2125-68)
9. Zinc Testing (Hach Cat. No. 22448-00)

SAMPLE PREPARATION

The following sample preparation was completed by Community Bureau of Reference analysts. The reference material was sterilized at 130 C for 2 to 8 hours, followed by grinding in a hammer mill with 2-mm openings for quick through-put and short contact time with the mill. The sample was sieved through 90-µm screen, blended for 10 days in a special blending drum, and evaluated for batch homogeneity before bottling. To ensure homogeneity before sampling, the analyst added solid polytetrafluoroethane (PTFE) balls of 17 mm diameter to each bottle.

Glassware was cleaned according to Sampling and Storage section instructions in the DR3000 Spectrophotometer manual.

Digesdahl apparatus, instead of the conventional nitric acid digestion or other digestions familiar to the Europeans, was used by Hach scientists to digest 15 replicates of the Community Bureau of Reference (BCR No. 144) Sewage Sludge Domestic sample. Following the sequential steps listed below, analysts:

1. Transferred a 0.5000-g standard reference sludge sample into a 100-mL digestion flask.
2. Added 4.0 mL of sulfuric acid, concentrated, to the digestion flask.
3. Turned on the fume eradicator and ensured suction to the fractionating head was in effect. Turned the temperature dial to a heat setting of 440 C (825 F).
4. Placed the flask weight and then the fractionating head on the flask. Placed the flask on the heater and heated the sample for 6 minutes.
5. Pumped 10 mL of 50 percent hydrogen peroxide to the charred sample by setting the hydrogen peroxide pump and timer to 4 minutes for the actual peroxide pumping.
6. Boiled off excess hydrogen peroxide for two more minutes after addition of hydrogen peroxide was complete. This can be set automatically by presetting the hydrogen peroxide pump and timer to a post-heat setting of 2 minutes.
7. Took the flask and fractionating column off the heater to cool before the fractionating column was removed from the digestion flask. Placed the fractionating column into its holding receptacle.
8. Diluted the digest with approximately 70 mL of deionized water.
9. Placed the digestion flask into a boiling water bath for 10 minutes.
10. Cooled to room temperature and diluted to the mark with deionized water. Inverted several times to mix.
11. Adjusted the pH of each aliquot taken for analysis as follows:
    a. Pipetted the appropriate analysis volume into the selected volumetric flask as stated below in Table 1.
    b. Added one drop of 2,4-Dinitrophenol Indicator Solution (Hach Cat. No. 1348-37) to each aliquot. When analyzing for total Kjeldahl nitrogen, a TKN Indicator Solution (Hach Cat. No. 22519-26) was used in place of the 2,4-Dinitrophenol Indicator. Aliquots smaller than 1.0 mL did not require pH adjustments.
    c. Added 8 N Potassium Hydroxide (KOH) Standard Solution, swirling between each addition, until the first, persistent pale yellow color formed (pH 4). When using TKN Indicator, the color change is from clear to the first appar-

ent pale blue color (pH 4). Initially an 8 N KOH solution was used, followed by a 1 N KOH solution as the yellow end point was approached.

d. Diluted to the appropriate volume with deionized water. The sample was then ready for colorimetric determination by specific procedures, as can be seen in Table 1.

Table 1. Colorimetric Procedures

PARAMETER	ALIQUOT DIGESTION	ALIQUOT ANALYSIS	DR-3000 METHOD NO.	METHOD
Aluminum	0.5 g	0.3 mL	2	Aluminon
Copper	0.5 g	1.0 mL	17	Bicinchoninate
Iron	0.5 g	0.1 mL	26	1,10-Phenanthroline
Lead	0.5 g	1.0 mL	100	Porphyrin
Manganese	0.5 g	3.0 mL	30	PAN
Nickel	0.5 g	3.0 mL	33	PAN
Nitrogen	0.5 g	0.4 mL	42	Nesslers
Phosphorus	0.5 g	0.1 mL	51	Ascorbic Acid
Zinc	0.5 g	0.5 mL	61	Zincon

RESULTS

Results for aluminum, copper, manganese, nickel, nitrogen, phosphorus, silver, and zinc were compared with the certified and indicative values obtained by the European analysts. The comparison of the certified data versus the values determined by the Digesdahl system and the comparison of the indicative values versus the determined values by the Digesdahl system can be seen in Tables 2 and 3, respectively.

Table 2. Results of Certified Values

ELEMENT	CERTIFIED VALUES mg/ Kg	DIGESDAHL VALUES mg/Kg
Copper	713 ± 30	713 ± 26
Lead	495 ± 19	428 ± 34
Manganese	449 ± 13	437 ± 16
Nickel	942 ± 22	877 ± 12
Zinc	3143 ± 103	3331 ± 180

Table 3. Results of Indicative Values

ELEMENTS	INDICATIVE VALUES mg/Kg	DIGESDAHL VALUES mg/Kg
Aluminum	18500 ± 900	13738 ± 2036
Iron	45660 ± 750	45219 ± 672
Nitrogen	----------	21801 ± 348
Phosphorus	20040 ± 150	19738 ± 443

DISCUSSION

Results obtained in most of the domestic origin sludge reference material tests were quite comparable. Certified and indicative values of the unknown sludge have been established by comparative testing; therefore, an analyst cannot be 100 percent certain of the actual value. This type of comparison study is one of the tools an analyst can use to be more certain of results. However, to provide a high degree of confidence in accuracy of results, primary standards and standard additions should be used. Primary standards are not available for all types of sample environments, so the Hach Company will synthesize a primary standard sludge for analysis from an array of organic and inorganic materials with known amounts of nutrients and metals.

CONCLUSION

The accuracy and precision of the Hach Digesdahl system have proven. The system achieves results similar to findings reached in much longer and more complicated procedures when comparisons of certified and indicative values are made. Based on this study, the Hach Digesdahl system would be the method of choice because of analysis speed, reduced system costs, and option of analyzing for minerals and metals with the same sample digest.

SLUDGE DEWATERING BY NATURAL FREEZE-THAW

C. James Martel
Environmental Engineer
U.S. Army Cold Regions
Research and Engineering Laboratory
72 Lyme Road
Hanover, New Hampshire 03755-1290

ABSTRACT

It is a well known fact that sludges are easily separated into solid and liquid fractions by freezing and thawing. Water and wastewater treatment plants in cold climate areas can take advantage of this process by freezing and thawing sludge during the winter and summer seasons in a new unit operation called a sludge freezing bed. The purpose of this study was to measure the dewaterability of freeze-thaw conditioned sludges and measure how well they drain at various depths. Typical water treatment, anaerobically digested and aerobically digested, sludges were tested. The main conclusion of this study was that up to 2.0 m of these sludges could be applied to a freezing bed.

INTRODUCTION

The solid and liquid fractions in an aqueous sludge can be separated by freezing. Separation is achieved during the formation of ice crystals, which can only be grown with pure water molecules. All other molecules including sludge solids are rejected to crystal boundaries where they are compressed and/or dehydrated. When the sludge is thawed, the ice crystals melt and the water drains away leaving the solids in a highly dewatered state. This process seems to work on all types of aqueous sludges.

Although this process has been known for several decades, it has not been widely used to dewater sludges from water and wastewater treatment plants. Many have attempted to build a mechanical device to do the freezing and thawing, but none have been commercially successful. The main reasons are the large amounts of electrical energy needed to freeze and thaw the sludge and the structural failure of freezing equipment from the expansion of ice. However, this process can be used in cold regions where nature can do the freezing and thawing. A unit operation that maximizes the use of natural freeze-thaw has been proposed by the author in a previous report(4). This unit operation is called a sludge freezing bed.

The size of a freezing bed will depend on the volume of sludge generated by the plant and the depth of sludge that can be frozen and thawed. The sludge volume can be estimated based on wastewater characteristics and flow. The design depth can be calculated from models based on local climatic conditions(5). These models predict that the design depth will be greater than 1.0 m in most cases. For example, the predicted design depths for freezing beds in Hanover, New Hampshire, and Fairbanks, Alaska, are 1.2 and 1.5 m, respectively.

No information could be found in the literature on whether sludges frozen to these great depths in a freezing bed would drain and dry after freeze-thaw conditioning. To answer this question, a series of dewaterability tests were conducted at the U.S. Army Cold Regions Research and Engineering Laboratory in Hanover, New Hampshire. These tests included specific resistance and capillary suction time tests, drainage tests at depths up to 220 cm, and drying tests. Three common sludges were used in these tests: a water treatment sludge, an anaerobically digested wastewater sludge, and an aerobically digested wastewater sludge. Natural freezing was simulated by freezing in a coldroom at -10 C for 7 days. Thawing was conducted at room temperature. All analyses were performed in accordance with *Standard Methods*, 15th Edition (1980). The initial physical characteristics of each sludge are shown in Table 1.

Table 1. Initial Physical Characteristics of Sludges

Sludge Type	Total Solids, %	Volatile Solids, %	Specific Gravity
Water Treatment	0.9	52	1.005
Anaerobic	8.7	48	1.006
Aerobic	1.5	77	1.005

LABORATORY TESTS

Specific resistance tests were conducted using the procedure developed by Adrian *et al.*[2]. Whatman No. 5 paper was used to filter water treatment and aerobically digested sludge samples, and Whatman No. 1 paper was used for the anaerobic sludge. The solids content (after drainage) was determined by dividing the total mass of cake solids by volume of filtrate generated at the end of test, as suggested by Christensen and Dick[3].

The results of the specific resistance tests before and after freeze-thaw are shown in Table 2. The specific resistance of water treatment sludge decreased slightly after freeze-thaw, whereas it increased for both wastewater sludges. An increase in specific resistance indicates a degradation in dewaterability, which is contrary to observed results. This same anomaly was observed by Adrian and Nebiker[1], who concluded that specific resistance may not be a valid dewaterability parameter for freeze-thaw conditioned sludges because of the rapid settlement of solids onto the filter paper. Apparently, these solids can blind the filter paper when a vacuum is applied, causing abnormally high specific resistances.

Capillary suction time (CST) values for each sludge were measured using the Komline-Sanderson Capillary Suction Timer. Five tests were conducted on each sludge to determine an average value. Based on a statistical analysis of preliminary data, this value should be within ±10 percent of the true mean at the 95 percent confidence limit.

Unlike the specific resistance tests, the CST tests showed a definite improvement in sludge dewaterability after freeze-thaw. For example, the water treatment sludge had a CST of 32 seconds before freezing and only 6 seconds after freezing (see Table 2). Similar reductions were obtained with the anaerobically and aerobically digested sludges. Thus, the CST test appears to be better than the specific resistance test for evaluating the dewaterability of freeze-thaw conditioned sludges.

Table 2. Dewaterability of Sludge Before and After Freeze-Thaw

	Specific Resistance (s^2/g @ 38.1 cm Hg)		Capillary Suction Time (s)	
Sludge Type	Before	After	Before	After
Water Treatment	4.8×10^9	6.0×10^8	32	6
Anaerobic	1.5×10^9	1.7×10^9	212	57
Aerobic	2.1×10^9	1.3×10^{10}	37	11

To determine the effect of freezing on filtrate quality, the filtrate obtained during the specific resistance tests was analyzed for 5-day biochemical oxygen demand (BOD), chemical oxygen demand (COD), turbidity, and pH. Examination of these data (see Table 3) indicates that freeze-thaw degrades the quality of the filtrate. Turbidity, BOD, and COD concentrations were all higher in the filtrate from the freeze-thaw conditioned sludges. The reason for this degradation may be the release of interstitial waters, which contain higher concentrations of dissolved materials than the surrounding free water. Similar results were obtained by Rush and Stickney(7), who concluded that BOD, COD, total organic carbon (TOC), and total phosphorus concentrations in the filtrate were three to six times higher than those in raw sewage. However, because the volume of filtrate is relatively small, the impact of returning it to the head of the plant should be minimal.

Table 3. Filtrate Quality Before and After Freezing

	BOD (mg/L)		COD (mg/L)		Turbidity (NTU)		pH	
Sludge Type	Before	After	Before	After	Before	After	Before	After
Water Treatment	--	--	--	--	1.4	2.7	5.3	5.2
Anaerobic	117	459	500	1064	58	72	8.2	7.8
Aerobic	51	65	117	230	3.4	43	7.4	7.5

DRAINAGE TESTS

Drainage tests were conducted at four depths ranging from 30 to 220 cm. Clear cast acrylic columns, assembled as shown in Figure 1, were used in these tests. A pair of columns were used for each depth. The odd-numbered column of each pair received unfrozen sludge, and the even-numbered column received an equal amount of frozen sludge. All eight columns were mounted against a white backboard for easier observations of sludge depth.

Sludge was not frozen inside the columns because the expansion force of ice could cause them to fracture. To circumvent this problem, sludge was frozen in the form of individual cakes. These cakes were then stacked in the even-numbered columns to the desired depth. This freezing technique closely simulated the layer freezing method used in a sludge freezing bed.

A thin section of one of the sludge cakes is shown in Figure 2. This picture clearly shows the separation of this sludge cake into zones of clear ice and solids. The zones of clear ice along the sides and bottom are an indication that the freezing rate was slow enough for solids to be rejected ahead of the advancing ice front. The freezing rate was slower in these areas because they were insulated by the polystyrene container used to freeze the sludge cakes. Vertical clear zones can also be seen in the interior of the sludge cake. These zones appear to be individual ice crystals formed by a more rapidly advancing ice front. Rapid freezing occurred from the top downward because the top surface was directly exposed to ambient air. In this case, the solids were rejected to ice crystal boundaries. However, whether the solids were rejected along a front or between ice crystal boundaries, the effect was the same, i.e., a separation of solid and liquid fractions.

As the cakes in the columns thawed, large angular fragments were observed falling to the bottom, where they accumulated to form a loose aggregate. Overlying the solids aggregate was the relatively clear supernatant. When the drain valve was opened, the overlying supernatant rapidly drained through this aggregate even though some consolidation was observed. In contrast, the unfrozen sludge did not drain well. Solid particles in the unfrozen

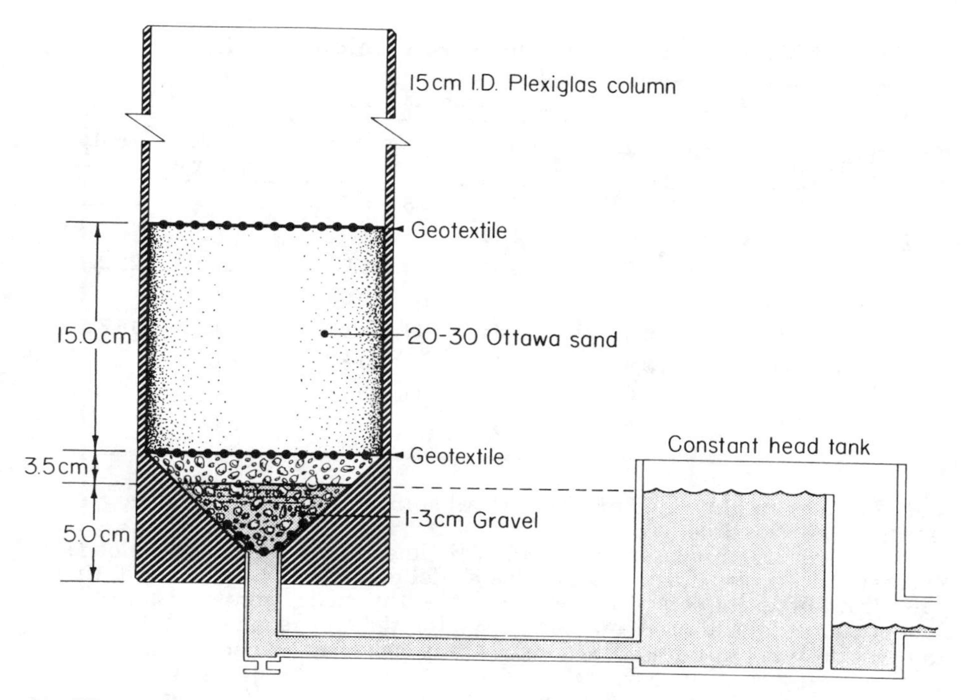

Figure 1. Acrylic Column Assembly Used During Drainage Tests

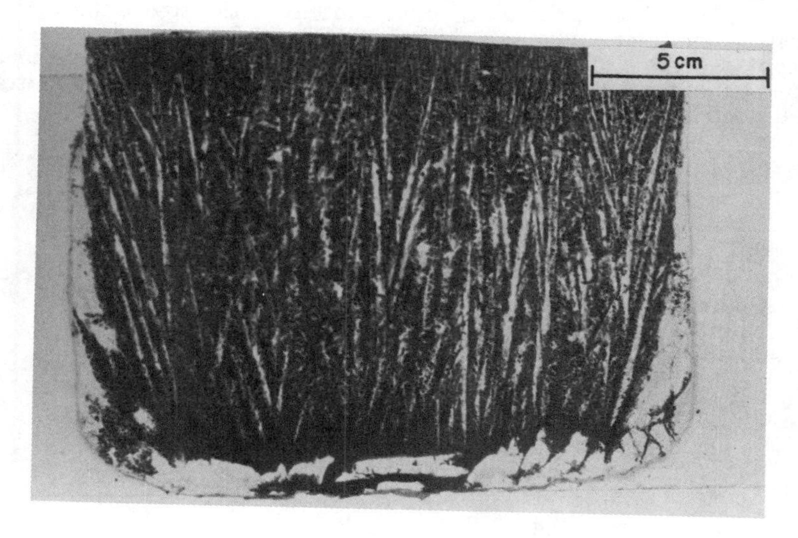

Figure 2. Section of Sludge Cake

sludge remained finely divided and formed an increasingly dense blanket that became nearly impermeable. These observations were common to all three sludges tested.

The results of the drainage tests clearly show the remarkable improvement in drainability after freeze-thaw. As shown in Table 4, all columns containing the freeze-thaw conditioned sludges drained in minutes, compared with days for the equivalent depth of unfrozen sludge. Thus, it is reasonable to expect that up to 2.0 m of these sludges could be applied to a freezing

Table 4. Drainage Times from Column Studies

Column Number	Water Treatment Sludge Depth, cm	Water Treatment Drainage Time, min	Anaerobic Sludge Depth, cm	Anaerobic Drainage Time, min	Aerobic Sludge Depth, cm	Aerobic Drainage Time, min
1	220	21,350.0+	207	28,935.0+	200	21,225.0+
2	220	6.5	200	111.0	200	18.5
3	120	21,350.0+	100	11,595.0	100	21,225.0+
4	120	5.0	100	9.0	100	5.0
5	60	5,780.0	54	11,595.0	54	16,875.0
6	60	3.8	54	3.8	54	2.2
7	30	4,290.0	31	11,595.0	31	7,275.0
8	30	1.8	31	3.0	31	1.8

bed. Considering the rapid drainage of all sludges at this depth, even greater depths may be possible.

An attempt to predict drainage times using a model developed by Nebiker *et al.*[6] was unsuccessful. The model predicted much longer drainage times than actually occurred. For example, the model predicted a drainage time of 5 days for 2.0 m of anaerobically digested sludge; the actual drainage time was only 111 minutes. Presumably, this lack of agreement resulted from a physical change in sludge particle size due to freeze-thaw. Instead of a suspension of finely divided particles, which the model assumes, the sludge was converted to a slurry of settleable particles in water.

After drainage, the total solids content of the freeze-thaw conditioned water treatment sludge ranged from 28.2 to 31.6 percent and averaged 30.3 percent (see Table 5). An increase in solids from 0.9 to 30.3 percent means that 97 percent of the water was removed from the original sludge. In contrast, the average solids content in the drained unfrozen sludge was only 9.8 percent.

Table 5. Total Solids Content After Drainage of Freeze-Thaw Conditioned Sludges

Column Number	Total Solids Content (%) Water Treatment	Anaerobic	Aerobic
1	--	17.2	5.9
2	30.1	37.0	16.6
3	--	21.2	6.8
4	28.2	36.3	17.5
5	10.4	23.0	6.0
6	31.6	33.0	16.7
7	9.3	19.1	7.3
8	31.2	34.2	15.6

The average solids content in the freeze-thaw conditioned anaerobic sludge after drainage was 35.1 percent. This increase represents a 75 percent removal of water. The unfrozen sludge contained an average of 20.1 percent solids.

From Table 5 the average solids content in the freeze-thaw conditioned aerobically digested sludge was 16.6 percent, which means that 91 percent of the water in the original sludge was removed. The average solids content in the odd-numbered columns containing the unfrozen sludge was only 6.5 percent.

Based on these results, only the freeze-thaw conditioned aerobically digested sludge would require further dewatering by drying. Both the water treatment and anaerobically digested sludges had solids contents greater than 30 percent, which is high enough for immediate removal with mechanical equipment(9). The reason for the lower solids content in the aerobically digested sludge is not clear. Perhaps the reason is related to the greater volatile solids content of aerobically digested sludges (see Table 1). Rush and Stickney(7) noted a similar result in experiments conducted with both digested and undigested sludges. The digested sludges, which contained less volatile solids, dewatered more efficiently than the undigested sludges.

DRYING TESTS

Drying tests were conducted on the aerobically digested sludge to see how long it would take to dry to a 20 percent solids content. These tests were conducted in the same columns used during the previous drainage study. Periodically, core samples were taken with a device made from a 2.54-cm-diameter brass pipe. The pipe was slowly pushed into the sludge layer until it contacted the underlying geotextile sheet. This technique provided a good vertical composite sample of the sludge.

The solids contents in each column during the drying tests are shown in Table 6. Generally, the solids dried more quickly in the columns containing less sludge. For example, after drying for 405 hours (almost 17 days), the solids content in Column 8 was 45.9 percent. During this same period, the solids content in Column 2 was 20.7 percent. Originally, the depth of sludge in Columns 2 and 8 were 30 cm and 200 cm, respectively. The drying time needed to achieve 20 percent solids in Columns 8, 6, 4, and 2 were approximately 66, 165, 335, and 405 hours, respectively. From this data it seems reasonable to expect that anaerobically digested sludge will dry to a manageable solids content within 2 to 3 weeks.

Table 6. Drying Rate Data for Freeze-Thaw Conditioned Aerobically Digested Sludge

Drying Time (h)	Total Solids Content (%)			
	Column 2	Column 4	Column 6	Column 8
0	16.6	17.5	16.7	15.6
66	17.0	18.0	18.4	19.3
165	23.6	20.1	23.6	23.4
236	19.3	18.5	26.6	21.6
335	19.7	21.4	34.6	38.1
405	20.7	21.6	36.2	45.9
550	16.9	21.1	--	--
715	19.2	24.2	--	--

CONCLUSIONS

The main conclusions of this study are:

1. Capillary suction time is a better indicator than specific resistance for evaluating the dewaterability of freeze-thaw conditioned sludges.

2. The depth of sludge that can be applied to a freezing bed is not limited by drainability--at least to a depth of 2.0 m. Since all sludges drained easily at 2.0 m, even greater depths may be possible.

3. The solids content in freeze-thaw conditioned water treatment and anaerobically digested sludges is high enough to allow mechanical removal immediately after drainage is complete. Further drying is not needed. However, aerobically digested sludge may require a 2- to 3-week drying period before it can be removed.

REFERENCES

1. Adrian, D. D. and Nebiker, J. H., Report No. EVE 13-69-1, 1969, Civil Engineering Department, University of Massachusetts, Amherst, MA.

2. Adrian, D. D., Lutin, P. A., and Nebiker, J. H., Report No. EVE 7-68-1, 1968, Civil Engineering Department, University of Massachusetts, Amherst, MA.

3. Christensen, G. L. and Dick, R. I., *ASCE Journal of Environmental Engineering*, 1985, *111* (3), 258-271.

4. Martel, C. J., Ph.D. Thesis, Colorado State University, October 1987.

5. Martel, C. J., *ASCE Journal of Environmental Engineering*, 1989, *115* (4), 799-808.

6. Nebiker, J. H., Sanders, T. G., and Adrian, D. D., In Proceedings of the 23rd Annual Purdue Industrial Waste Conference, Purdue University, Lafayette, IN, 1968.

7. Rush, R. J. and Stickney, A. R., Report No. EPS 4-WP-79-1, Wastewater Technology Centre, Environment Canada, Burlington, Ontario, 1979.

8. *Standard Methods for the Examination of Water and Wastewater*, 15th Edition, American Public Health Association, Washington, DC, 1980.

9. Water Pollution Control Federation, *Sludge Dewatering*, Manual of Practice 20, Washington, DC, 1983.

TYVEK® MICROFILTRATION OF HAZARDOUS WASTEWATERS

Ernest Mayer
Engineering Department Louviers 1359
E. I. du Pont de Nemours, Inc.
P.O. Box 6090
Newark, Delaware 19714

ABSTRACT

Du Pont has developed and recently commercialized a new filter media based on Tyvek® flash spinning technology. This new medium has an asymmetric pore structure, a greater number of submicron pores, and a smaller average pore size(2-4). As a consequence, it has superior filtration properties, longer life, and in many instances can compete with microporous membranes, PTFE laminates, and various melt-blown media. When coupled with an automatic pressure filter (APF), it provides an automatic wastewater filtration process that is a dry-cake alternative to conventional crossflow microfilters and ultrafilters. This Tyvek®/APF process has proved extremely useful in filtering heavy metal and other hazardous wastewaters to meet strict EPA NPDES discharge limits. Specific examples and actual case histories will be highlighted to illustrate its benefits. In addition, this Tyvek®/APF technology has recently been selected for EPA's SITE-3 program, which is also discussed.

TYVEK® T-980 MEDIA

Du Pont has recently commercialized a new filter medium based on Tyvek® flash spinning technology. It has an asymmetric pore structure, a greater number of submicron pores, and a smaller average pore size(2-4). As a consequence, it has superior filtration properties and longer life, and in many instances it can compete with microporous membranes, PTFE laminates, and various melt-blown media(4). Its key property is its tight pore structure at a very low cost compared to competitive products. Table 1 outlines the medium's cost per gallon of waste filtered and shows that Tyvek® T-980 grade, which is the lowest basis weight manufactured (0.9 oz/yd^2) and the only grade evaluated here, is very cost effective. In most applications, the T-980 grade was sufficient so the added cost for a higher basis weight grade was not warranted. For example, Tyvek® T-980 produced slightly poorer effluent quality than the "standard" 0.45 µ microporous membrane at a fraction of the cost; equivalent effluent quality to the PTFE laminate at a fraction of the cost; and much better effluent quality than typical 1- and 5-µ melt-blowns at equal or significantly lower cost (depending on application). Tests with actual wastes showed the

Table 1. Media Cost per Gallon Waste Filtered

			Actual Costs/Gallon Filtrate (cents/gal)*			
Media	Nom. Rating (µm)	Overall Filtrate Quality	60 ppm ACFTD**	600 ppm ACFTD	Low-Level Radioactive Waste	Lead-Bearing Waste
Microporous Membrane	0.45	Excellent	6.7	29	69	14
Tyvek® T-980	1	V. Good	0.5	1.3	1.7	0.3
PTFE Laminate	0.8	V. Good	8.2	41	63	12
Melt-blown PP	1	Good	0.7	2.4	6.5	1.3
Melt-blown PP	~5	Poor	0.3	1.5	4.0	0.8

*Based on actual media costs, flux rates, and measured life cycles.

**AC Fine Test Dust (ACFTD) challenge tests.

greatest cost benefit (Table 1). A similar cost benefit was obtained in operation with a low-level radioactive plating waste at the Savannah River Plant[5,6]. This installation realized almost a $200 M annual savings when Tyvek® T-980 medium was used with a more efficient filter aid. An added benefit of the Tyvek® is its superior strength compared with microporous membranes and the PTFE laminate medium. This strength permits its use in robust automatic pressure filters[8]. The high strength[11] coupled with the tight, ~1-μ nominal pore structure[4] led to Tyvek®'s use in the EPA Superfund SITE program[7].

OBERLIN AUTOMATIC PRESSURE FILTER (APF)

The Tyvek® T-980 medium requires a suitable filter housing. The Oberlin Filter Company's automatic pressure filter was chosen for its simplicity and fully automatic operation[9]. The Oberlin APF has other advantages, namely:

- Completely automatic, unattended operation save for Tyvek® roll replacement and chemical treatment makeup
- Enclosed operation for safety and handling of hazardous wastes
- Fairly high operating pressure (up to 60 psig)
- Completely in-line treatment chemical addition (i.e., filter aids and polymer flocculants)
- Automated shutdown flushing capability
- Cake washing capability to remove hazardous filtrate, if required
- Automatic, positive dry cake discharge
- Direct submicron filtration without the need for further downstream processing
- Reliable, low-maintenance performance
- Completely automatic safety interlocks and enclosures and/or purging, if required
- Explosion-proof design, if required
- Completely integrated pumping system(s), if required
- Dirty medium take-up and doctoring, brushing, or washing, if required (automatically accomplished during take-up)
- Standard PLC control.

TYVEK®/OBERLIN APF COMBINATION

Thus, the Tyvek®/Oberlin APF combination has some unique advantages, especially its completely automatic, submicron, low-cost filtration, and its dry cake discharge[7]. This dry cake discharge feature is precisely why the resultant cakes pass the modified EPA "Paint Filter Test" for land disposal[10] and in some instances pass the new EPA TCLP (Toxic Characteristic Leaching Procedure) test for hazardous components[1]. Dry cake/submicron filtration in one operation is why the Tyvek®/Oberlin APF combination was selected at Savannah River over conventional crossflow microfilters and ultrafilters. In plating waste treatment the simple, one-step Tyvek®/Oberlin APF combination replaced the conventional three-step clarifier/overflow sand filter/underflow recessed filter press process. However, the purpose of this paper is to highlight case histories where the Tyvek®/Oberlin APF filtration process has been successfully applied. The technology is most suitable for hazardous wastewater where the solids loading is not too high. Examples include contaminated groundwater, plating wastewaters, low-level radioactive wastes, plant equipment/floor washings, cyanidic wastes, plant wastewaters that contain heavy metals, and metal grinding wastes[11].

SAVANNAH RIVER SIMULATED WASTE

Table 2 summarizes a series of Oberlin bench-scale tests with simulated plating wastewater (to duplicate existing plating line effluents without uranium). The Tyvek® T-980 medium matches all competitive media in rate and filtrate turbidity, but produces a drier cake because of better air flow due to less media blinding. Tyvek® provides excellent cake release and the lowest (15 mg) "Δ weight gain" after the cake is removed. Furthermore, the Tyvek® T-980 is much stronger than the competitive media so it could be rerolled in the Oberlin APF without a carrying belt. All the others required careful handling. This feature simplified the operation and reduced maintenance.

Table 2. Savannah River Simulated Waste* Testing in Oberlin Pressure Filter

Media**						
Type	Nom. Rating (μ)	Δ Wt. Gain (Mg.)	Cake Release	Cake % Solids	Filtrate Turbidity (NTU)	Avg. Rate (gfd)
Microporous Membrane	0.45	50	Good	30	0.46	1050
Tyvek® T-980	1	15	Excellent	34	0.29	1070
PTFE Laminate	0.8	60	Fair	28	0.51	1120
Melt-blown PP	1	560	Fair	32	0.49	1040
Melt-blown PP	~5	445	Good	29	5.0	1150

* Simulated waste with 600 ppm metal hydroxide solids (except uranium) plus 3:1 ratio of Celite 577 filter aid.

** Δ Wt. Gain is media weight gain after cake removed; and cake release judged visually after bending media to simulate discharge around roll.

ACTUAL SAVANNAH RIVER OPERATION

Table 3 details actual Savannah River Plant wastewater treatment specifically aimed at aluminum and uranium removal[5,6]. The data were obtained from two Tyvek® T-980/Oberlin APF units that had been operating for about 3 years. Aluminum forming and metal finishing operations generate a high content of solids, aluminum, and turbidity. These solids can be reduced below the National Pollution Discharge Elimination System (NPDES) limits with the T-980/Oberlin combination at very high (1,200 gal/ft^2/day) rates. This rate is about seven times higher than the ultrafilter (UF)/reverse osmosis (RO) system (200 gal/ft^2/day) originally considered. Pilot testing also showed that the UF/RO system repeatedly fouled with this waste, requiring aggressive cleaning agents, which significantly added to the waste volume.

ELECTRONICS MANUFACTURING PLANT WASTEWATER

This plant's effluent exceeded the local sewer authority's lead discharge limit. As a consequence, the plant was mandated to cease discharge and dispose of their wastewater in an off-site hazardous waste landfill at a $0.55/

Table 3. **Actual Savannah River Plant Operation with Tyvek® T-980/Oberlin Pressure Filter**

Property	Raw Waste	T-980/ Oberlin APF	NPDES Limits *
Turbidity (NTU)	110	0.32	--
TSS (ppm)	687	1.4	31
Aluminum (ppm)	127	0.95	3.2
Lead (ppm)	1.6	0.2	0.43
Zinc (ppm)	0.5	< 0.1	0.32
Copper (ppm)	2.0	< 0.1	0.21
Uranium (ppm)	2.3	0.01	0.5
Capacity (gpd)	35,000	60,000	--
Flux (gfd)	--	1,200	--

*Actual State discharge permit values.

gallon cost. The Tyvek® T-980/Oberlin combination was installed instead of a crossflow microfilter because its dry cake feature significantly reduces waste volume. The T-980/Oberlin units repaid their cost in 3 months of operation based on disposal cost savings alone.

Table 4 details tests similar to those with the Savannah River plating waste. Tyvek® T-980 again compares favorably with the competitive media in rate and filtrate turbidity (except for the 0.45-µ microporous membrane). However, Tyvek® T-980 produces the driest cake (75 percent solids), gives excellent cake release, and has the lowest media "Δ weight gain" (0 mg). Again, the Tyvek® T-980 medium was the only one that could withstand the rigors of Oberlin APF rerolling without a support belt.

Table 5 details actual plant operation and compares the Tyvek® T-980 effluent with the plant's discharge limits. As shown, the T-980/Oberlin units reduced the effluent total suspended solids (TSS) and lead levels to well below the plant's required discharge limits at very high (1,500 gal/ft^2/day) flux ratios. In addition, the Oberlin APF routinely achieved dry cakes of >60 percent solids that could be disposed of in a RCRA-approved landfill (at significant cost savings compared with the concentrate from the crossflow microfilter).

ELECTRONICS PLANT CARTRIDGE REPLACEMENT

This plant's effluent had to be polished by absolute 0.45-µ cartridge filters to meet heavy-metal discharge limits. Cartridge costs exceeded $1,200

Table 4. Electronics Wastewater Testing* in Oberlin Pressure Filter

Type	Media** Nom. Rating (μ)	Δ Wt. Gain (Mg.)	Cake Release	Cake % Solids	Filtrate Turbidity (NTU)	Avg. Rate (gfd)
Microporous Membrane	0.45	20	V. Good	63	0.47	1730
Tyvek® T-980	1	0	Excellent	75	2.0	2160
PTFE Laminate	0.8	20	Excellent	72	1.4	2050
Melt-blown PP	1	50	Fair	64	2.6	1900
Melt-blown PP	~5	135	Good	63	2.6	1900

*Actual electronics manufacturing wastewater that contains about 6500 ppm heavy metals (Pb, Ba, etc) plus 1950 ppm Superaid filter aid (0.3 ratio) and 50 ppm high-charge cationic polymer flocculant (Praestol K-122L).

**Δ Wt. Gain is media weight gain after cake removed; and cake release judged visually after bending media to simulate discharge around roll.

Table 5. Actual Electronics Manufacturing Plant Wastewater* with Tyvek® T-980/Oberlin Pressure Filter

Property	Raw Waste	T-980/ Oberlin APF	Discharge Limits**
Turbidity (NTU)	>1000	3.4	N.R.
TSS (ppm)	6500	6.0	26
Lead (ppm)	500	<0.25	0.7
Barium (ppm)	~ 100	<1	N.R.
Capacity (gpd)	2500	7500	---
Flux (gfd)	---	1500	---
Cake % Solids	---	>60	---

* Includes floor washings also

** N.R. = not regulated.

daily plus significant labor charges. A Tyvek® T-980/Oberlin APF unit was installed at significant cost savings compared with these cartridges or with the crossflow microfilter that was also considered. Payback was on the order of 3 months; the Oberlin APF produced dry cakes suitable for landfilling off-site, and significant labor savings resulted.

Table 6 demonstrates that the Tyvek®/Oberlin system easily met the TSS and lead discharge limits at very high (1,500 gal/ft^2/day) flux. Cakes were also quite dry at ~50 percent solids.

Table 6. Electronics Plant Cartridge Filter Replacement* with Tyvek® T-980/Oberlin Pressure Filter

Property	Raw Waste	T-980/ Oberlin APF	Discharge Limits**
Turbidity (NTU)	2400	0.43	N.R.
TSS (ppm)	5500	1.0	20
Lead (ppm)	2300	0.18	0.7
Capacity (gpd)	2000	11,000	---
Flux (gfd)	---	1500	---
Cake % Solids	---	50	---

* Includes floor washings also ** N.R. = not regulated.

MUNITIONS PLANT WASTEWATER

This plant was faced with severe RCRA restrictions on heavy metals discharge. They hired an environmental engineering firm to solve their waste problem, namely to remove all metals (primarily lead, antimony, and barium) to well below discharge limits and to produce dry cakes that could be hauled off-site to an RCRA-approved landfill. This firm opted for the conventional cross-flow microfilter/press approach, which was expensive, required manual labor, and was prone to fouling. I proposed the less costly, single-step, Tyvek® T-980/Oberlin approach that was installed about a year ago. Table 7 shows that effluent TSS, lead, antimony, and barium levels are well below the discharge limits. In addition, the system demonstrated very high capacity (15,000 gal/day), very high flux (3,700 gal/ft^2/day), and fairly dry (40 percent solids) cakes. The unit operates at 200 F, which would have been troublesome for the cross-flow filter and explains the somewhat higher effluent lead level.

CLARIFIER UNDERFLOW HEAVY METALS REMOVAL

This plant was faced with a land ban of their main clarifier underflow sludge (~2 percent solids) because it did not pass the EPA "Paint Filter Test"[10]. This clarifier treated the entire plant effluent and removed primarily lead, zinc, and copper. To satisfy the RCRA land ban restrictions, the plant hired an expensive ($400/day) mobile dewaterer who used a manual, recessed-chamber filter press that produced sloppy cakes. These cakes had to be shovelled into dumpsters for off-site disposal. The Tyvek®/Oberlin combination was installed about 6 months ago. Table 8 shows excellent effluent quality at very high (10,000 gal/day) capacity and 900 gal/ft^2/day flux. The cakes were sufficiently dry (50 percent solids versus 40 percent requirement) to pass the "Paint Filter Test," were acceptable to the hauler and landfill operator, resulted in a significant cost saving to the plant. The Tyvek®/Oberlin system operates automatically at significant labor savings compared with the manual press.

Table 7. Munitions Manufacturing Plant Wastewater* with Tyvek® T-980/Oberlin Pressure Filter

Property	Raw Waste	T-980/ Oberlin APF	Discharge Limits**
Turbidity (NTU)	>1000	0.8	---
TSS (ppm)	1100	0.5	30
Lead (ppm)	540	1.8	5.0
Antimony (ppm)	4.4	0.4	1.0
Barium (ppm)	50	0.6	50
Capacity (gpd)	3000	15,000	---
Flux (gfd)	---	3700	---
Cake % Solids	---	40	---

* Includes floor washings also

** N.R. = not regulated.

Table 8. Clarifier Underflow Heavy Metals Removal* with Tyvek® T-980/Oberlin Pressure Filter

Property	Raw Waste	T-980/ Oberlin APF	Discharge Limits**
Turbidity (NTU)	>1000	1.0	N.R.
TSS (ppm)	17,000	2.5	20
Lead (ppm)	40	<0.01	5.0
Zinc (ppm)	410	0.2	5.0
Copper (ppm)	1050	1.2	5.0
Capacity (gpd)	6800	10,000	---
Flux (gfd)	---	900	---
Cake % Solids	---	50	40

* Existing plant clarifier underflow sludge was previously hauled off-site to hazardous landfill.

** N.R. = not regulated.

GROUNDWATER BARIUM REMOVAL

Contaminated leachate from a landfill required silt and barium removal to prevent fouling in downstream organics removal equipment. A Tyvek®/Oberlin system was chosen for this application because of its automatic operation and dry cake feature. Table 9 shows that TSS and barium levels are well below required limits at an extraordinarily high (2,700 gal/ft²/day) flux. This was substantially higher than the flux obtained from the ultrafilter also considered for this application.

Table 9. Contaminated Groundwater Barium Removal* with Tyvek® T-980/Oberlin Pressure Filter

Property	Raw Waste	T-980/ Oberlin APF	Discharge Limits**
Turbidity (NTU)	>1000	9.0	N.R.
TSS (ppm)	1200	6.6	20
Barium (ppm)	500	6	50
Flux (gfd)	---	2700	---

* From hazardous landfill leachate.

** N.R. = not regulated.

BATTERY MANUFACTURING HEAVY METALS REMOVAL

This plant consistently exceeded its permitted discharge limits from its conventional clarifier/underflow press system and considered an overflow polishing sand filter. Simultaneously, they heard of our Tyvek® T-980/Oberlin APF technology and decided to evaluate it as a polishing filter. Testing showed that it could replace the entire clarifier/press installation. Table 10 shows excellent metals removal, TSS reduction, and excellent effluent turbidity (0.5 NTU). Flux is quite high at 2,400 gal/ft²/day, and capacity from the single Tyvek®/Oberlin system exceeds 60,000 gal/day. These benefits are in addition to dry cakes that could be easily disposed of in an RCRA-approved landfill.

SUMMARY

These cases and actual operating applications demonstrate the utility of the Tyvek® T-980/Oberlin APF technology to remove heavy metals at very high flux rates and to simultaneously produce dry cakes that pass the EPA "Paint Filter Test." This technology is quite competitive when compared with microfilter cartridges, crossflow microfilters, and ultrafilters. In some

Table 10. Direct Filtration for Heavy Metals Removal from a Battery Manufacturing Plant* with Tyvek® T-980/ Oberlin Pressure Filter

Property	Raw Waste	T-980/ Oberlin APF	Discharge Limits**
Turbidity (NTU)	175	0.5	N.R.
TSS (ppm)	1600	<1	20
Nickel (ppm)	30-50	<0.10	2.27
Cadmium (ppm)	15-40	<0.05	0.4
Zinc (ppm)	0.2-1.0	<0.10	1.68
Cobalt (ppm)	0.4-1.5	<0.05	0.22
Capacity (gpd)	40,000	60,000	---
Flux (gfd)	---	2400	---

* Replaced clarifier and underflow press.

** N.R. = not regulated.

instances, the Tyvek®/Oberlin system can replace the conventional three-stage metal treatment process of clarifier, underflow filter press, and overflow polishing sand filter. Thus, the waste engineer/consultant now has a simpler, one-step process for treating hazardous metal-bearing wastewaters.

REFERENCES

1. Federal Register, June 13, 1986, Volume 51, No. 114.

2. Lim, H. S. and Mayer, E., Paper presented at 1987 Membrane Technology Conference, Boston, MA, October 22, 1987.

3. Lim, H. S. and Mayer, E., Paper presented at the International Technical Conference on Filtration and Separation, Ocean City, MD, March 22, 1988.

4. Lim, H. S. and Mayer, E., *Fluid/Particle Separation Journal*, March 1989, *2* (1), 17-21.

5. Martin, H. L., Paper presented at 10th Annual AESF/EPA Conference on Environmental Control for the Metal Finishing Industry, Orlando, FL, January 23-25, 1989.

6. Martin, H. L., Gurney, P. K., and Fernandez, L. P., Paper presented at 8th Annual AESF/EPA Conference on Pollution Control for the Metal Finishing Industry, San Diego, CA, February 11, 1987.

7. Mayer, E., Request for Proposal, SITE-3 Solicitation, Demonstration of Alternative and/or Innovative Technologies, "Groundwater Remediation via Low-Cost Microfiltration for Removal of Heavy Metals and Suspended Solids," February, 1988.

8. Mayer, E., "New Trends in SLS Dewatering Equipment," *Filtration News*, 1988, May/June, 24-27.

9. "Oberlin Pressure Filter," Oberlin Filter Company's Bulletin, February, 1988.

10. Sell, N. J., *Pollution Eng.*, August, 1988, 44-49.

11. "Tyvek® Engineered Specifically for Filtration," Du Pont Tyvek® Bulletin E-24534, 1988.

DECANTING AND FILTRATION IN A HIGH-GRAVITY CENTRIFUGAL DRYER

Lloyd B. Smith
Vice President, Product Development
and
Dr. Richard A. Wolfe
President
Coal Technology Corporation
103 Thomas Road
Briston, Virginia 24201
(703)669-6515

ABSTRACT

To address the need for an energy efficient and environmentally acceptable clean coal fines dryer, a new dewatering method has been developed that employs decanting in combination with very high gravity centrifugal filtration in a batch process. The new process uses a novel system for the prevention of filter blinding. It employs the combination of a carbon fiber composite bowl and a highly efficient, variable frequency, alternating current drive to conserve the energy used in accelerating and decelerating the bowl. A full-scale prototype is under test. Commercialization is planned in the near future.

INTRODUCTION

A study was made in 1981 to determine the most satisfactory processes available for dewatering the froth product from coal processing plants. This very fine coal is being produced in increasing quantities because of the use of modern mining and coal processing methods, and at the same time, the coal consumers are becoming more sophisticated in their determinations of the efficiencies they may gain from the use of dryer coal. None of the existing dewatering methods was deemed to be completely satisfactory, so a program of research and development was initiated.

The criteria adopted for the program were:

1. The new system should be capable of drying the coal fines to the order of 10 percent surface moisture.
2. The subject coal should be 1.2 mm (28 mesh) × 0 size with at least 35 percent minus 44 microns and contain no more than 15 percent ash.
3. The new system should be capable of drying only the coal fines of 0.45 mm (100 mesh) × 0 if necessary without larger coal particles being needed in the process.
4. The system should be efficient in its use of energy.
5. There should be no detrimental environmental effect.
6. The system should present an attractive economic structure.

The energy efficiency criterion guided the work toward a mechanical system and, in turn, to centrifugation. However, conventional centrifugal dryers were found to be unable to meet all of the criteria, and so a study was begun into state-of-the-art centrifugal drying.

PROGRAM HISTORY

During some early bench-scale tests, it was found that higher levels of artificial gravity would be needed than could be safely achieved in commercially available full-sized solid bowl centrifuges. It was also found that available basket centrifuges could not achieve either the speed or the productivity necessary to meet the criteria of the program. It was then decided to attempt to develop an entirely new process and apparatus. The criteria for the new system were developed after extensive bench-scale testing:

1. A batch machine should be used.
2. The system should be capable of achieving 4,000 G in commercial size.
3. It should be able to handle high levels of vibration at full speed without problems.

4. It should use both decanting and filtration so that both final surface moisture and productivity would be adequate.
5. It should be commercially practical and durable.

In order to determine the feasibility of using centrifugation for drying coal fines to 10 percent surface moisture in a commercially practical apparatus, extensive bench-scale testing was conducted. The initial determinations were made using a 25.4-mm (1-in.) diameter centrifuge operated at 2,094 rad/s (20,000 rpm) and producing 5,683 G. With this small device, it was found that a surface moisture with the subject coal fines of under 15 percent could be achieved in about 60 seconds of centrifugation. Although this did not fully demonstrate the goal of 10 percent surface moisture, it did indicate that centrifugation was worth further investigation.

A study of the results obtained with the 25.4-mm (1-in.) centrifuge led to a projection that a 127-mm (5-in.) centrifuge operating at 1,242 rad/s (11,864 rpm) and inducing 10,000 G could fully meet the goal. Such a machine was built and tested. It did achieve 10 percent surface moisture with the subject coal. It also yielded much valuable information about the problems that would be encountered in a scale-up. The filter used was of broadcloth supported by several layers of 1.2-mm (16-mesh) screen. It was found necessary during these experiments to make the inner surface of the centrifuge conical to bias the filtrate to move axially toward the portion of the bowl in which the exit ports were located.

Experiments with the 127-mm (5-in.) centrifuge showed possible problems in obtaining adequate productivity if all of the water must move through the bed of coal by centrifugal filtering. With this in view, a device was built in which coal could be compressed to the pressures expected within a full-sized centrifugal dryer and then water could be forced through it with sufficient hydrostatic pressure to produce conditions similar to those expected in commercial high-gravity centrifuges. Experiments with this apparatus proved that the centrifugal dryers would not be feasible unless some of the water was removed by means other than filtration.

Because of the urgent need in the coal industry for a solution to the coal fines drying problem, it was decided at this point to scale up immediately to a nearly full-sized prototype. The resulting machine was designated Prototype 1. This machine was designed and built at a time when the process had not been fully developed. It was, therefore, deemed prudent to provide for speed and gravity levels greater than projections from bench-scale work had indicated would be necessary. With this in view, the original nearly full-scale centrifuge had a 0.66-m (26-in.) diameter stainless steel bowl capable of very high speed operation by virtue of a special stress-relieving construction. The centrifuge was also fitted with custom-designed hybrid hydrostatic/hydrodynamic bearings capable of operating at up to 377 rad/s (3,600 rpm), if necessary.

Robotics systems were provided to, in turn, insert and retrieve a filling tube and cutting device. The centrifuge incorporated a system for changing the natural frequency of the bowl suspension and bearing housing as a part of the operating cycle. Thus, it could support badly out of balance loads of coal at high speeds without damaging any of its components.

Initial testing confirmed that allowing all of the water to exit through the filter was impractical. Productivity in such a case would have been unacceptably low. It was decided that a combination of decanting during the early part of each cycle and then removing the remaining water by high-gravity centrifugal filtration must be used. This method was found to be successful. The decanting process occurred immediately after the bowl was filled. It was also found that a squeegee operation on the inner surface of the load using a plastic blade attached to the fill tube improved the process by removing clays that collected on the inner surface of the newly formed load. Using this machine, a data base involving approximately 1,800 tests was assembled. Studies of this data base concluded that a commercial size centrifuge employing the method demonstrated in Prototype 1 had an acceptable probability of success.

As plans were laid for commercialization, some new technologies were beginning to be sufficiently developed so that they could be incorporated into a pre-production prototype of the new dryer. They were (1) carbon fiber composite construction for strengthening and lightening the bowl and (2) highly efficient regenerating all-solid-state variable frequency alternating current drive for reducing energy consumption. Since the new drive technology involved no in-house development, the decision was made to accept it without any testing for inclusion in the commercial prototype. The lightweight carbon fiber composite bowl, on the other hand, had serious ramifications branching into the vital areas of vibration dynamics and the filtration process. A 0.723-m (28.5-in.) carbon fiber composite bowl was, therefore, constructed and installed in Prototype 1.

When the new bowl was installed, it brought to light two problems that had to be resolved before the system could be commercialized. One of these was a mechanical dynamics problem. The very light composite bowl had a dramatically reduced gyrodynamic effect that left the bearing reactions with unbalanced loads at an unacceptable level. This problem was solved by the development of a device designated as a neutralizer ring. This heavy metal ring around the mouth of the bowl (for which patents have been applied) has the effect, if properly designed, of preventing any appreciable radiation of vibration to the machine frame when unbalanced loads are being processed. The second problem was the exact detailed design of the nonblinding filter system that had been pioneered in the 0.66-m (26-in.) stainless steel bowl. The manner in which this problem was solved will be treated later in the Discussion section.

As the test work on Prototype 1 reached the point where the probability of commercial readiness was indicated, a license was obtained and plans were made to produce a pre-production prototype.

Computer studies indicated that the economics were sound and the dynamic considerations were met with a 0.914-m (36-in.) diameter fully automatic centrifuge operating with a 145-second cycle time and producing 7.26 metric (8 U.S.) tons/hour of dry coal from froth flotation product. Such a machine has been built. It was designated Prototype 2 and it is under test in a coal plant at this time. It is illustrated in the schematic diagram on Figure 1.

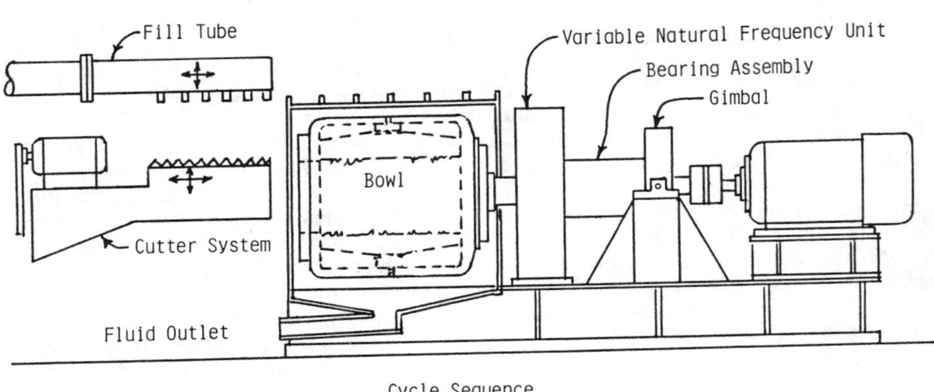

Cycle Sequence

Operation	Time Span	Ending Bowl Speed rad/s	Ending Bowl Speed RPM
Fill & Load Forming	60 sec	147	1400
Acceleration	15	294	2800
Dwell	15	294	2800
Deceleration	23	47	450
Cutting Out	22	21	200
Reset	10	147	1400
	145 sec		

Figure 1. Schematic of High-Gravity Centrifugal Dryer

The scale-up from 0.723 m (28.5 in.) to 0.914 m (36 in.) was not without its problems. Learning to build a load of 294 kg (650 lb) of coal fines in less than 20 seconds has required some development. The design of a quick change production filter has also required careful attention. The change from a single filter, as used in Prototype 1 and as illustrated in Figure 1, to a two-filter system, as is actually used in Prototype 2, introduced some complications relative to the orifice area required for successful filter operation. The problems were worked out, and Prototype 2, at this writing, is under development to bring it to production readiness. However, variables remaining in this unusual process will result in many years of development to optimize the system.

DISCUSSION

MECHANICAL SYSTEM

In order to achieve safely the required 4,000 G level, and after preliminary work with special stainless steel bowls, a carbon fiber composite bowl construction was adopted. Because the carbon fiber composite has an endurance limit more than three times that of stainless steel and because it weighs only 1/5 as much, a strength/weight ratio improvement of at least 15/1 over stainless steel was achieved. This accomplished two very important ends: the torque requirement and the resulting energy consumption were held to a minimum, and the useful life of the bowl was greatly extended.

As stated above, the beds of coal that form in a batch centrifuge are inherently unbalanced. To overcome this obstacle, a special suspension system was developed and patented. The new system changed the natural frequency and, thus, the stiffness during each cycle. For high-speed operation the gimbal mounted system, shown in schematic on Figure 1, is supported on air bags in the variable natural frequency unit to produce a soft ride with a low natural frequency. For cutting the dried coal out at low speed, the natural frequency is automatically and quickly raised so that the stiffness produced is appropriate for the cutting out function(1). This system, together with the neutralizer ring, which keeps the dynamic axis of the bowl coincident with that of the gimbal, enables the centrifuge to operate with out-of-balance loads without transmitting any excessive vibration to the structure on which the machine is mounted.

DESCRIPTION OF PROCESS

The schematic drawing of Prototype 2 in Figure 1 illustrates the full-sized pre-production pilot. This machine is equipped with a programmable controller that modulates the speed of the main drive and times the insertion and other functions of the fill tube and cutter system.

At the beginning of each 145-second cycle, the fill tube is inserted while the bowl is accelerated to 147 rad/s (1,400 rpm). At this time, a fine coal slurry, thickened to 50 percent coal solids, is introduced to the bowl, and the bed-building process begins. Figure 2 illustrates how the bed of solids is built rapidly and without blinding the filter. Figure 2 shows the feed tube in place and the slurry being introduced at 31.5 L/s (500 gpm). The rapid flow of slurry is used both to increase productivity and to cause turbulence on the inner portion of the slurry while the load is being solidified. This turbulence keeps the particles of clay and extremely fine coal in suspension and prevents them from moving toward the filter and blinding it. As the load is formed, a portion of the moisture is decanted and removed from the mouth of the bowl. This process also further benefits the coal by reducing the amount of clay in the load.

To render the nonblinding filter system effective, it is also necessary to restrict the outflow of water through the filter while the bed of coal is forming. Figure 2 shows that the filter medium consists of a 0.13-mm (120-mesh)

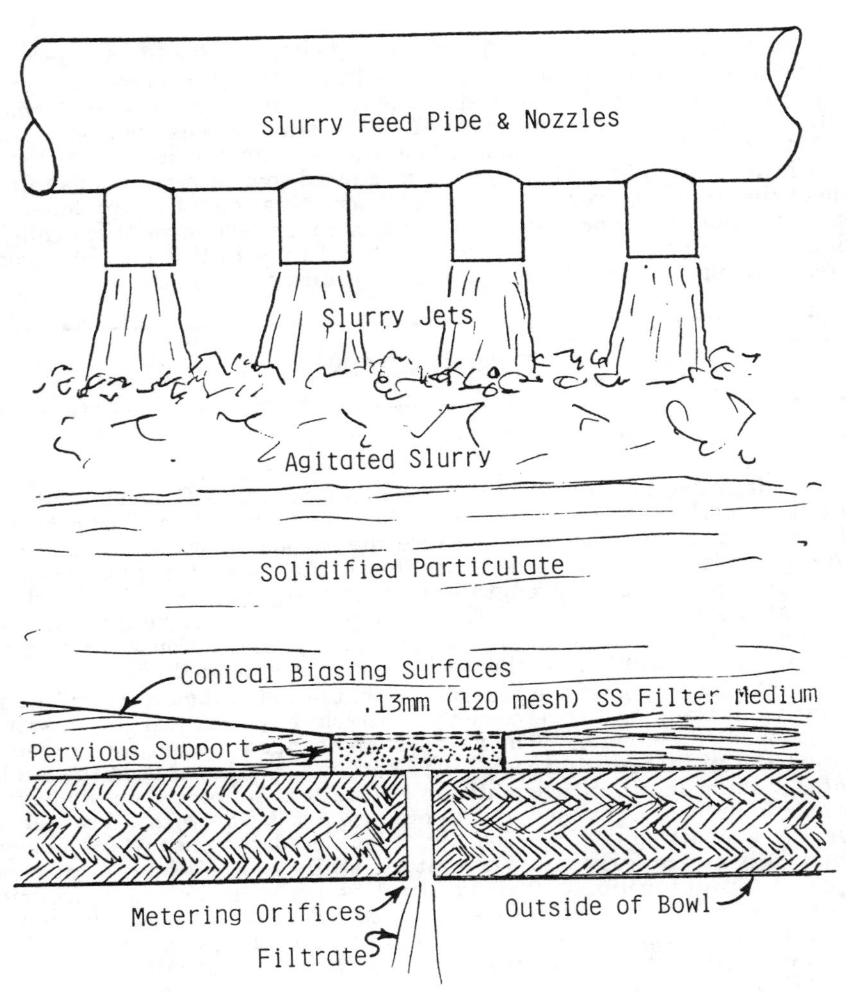

Figure 2. Schematic of Nonblinding Filter System

stainless steel screen filter supported by a pervious mass of coarse screen to support the medium against the 5.861×10^7 dyne/cm^2 (850 psi) pressure that is encountered from the coal under 4,000 G. The metering holes directly under the pervious support allow the filtrate to escape at a carefully controlled rate. This prevents the filtrate from moving toward the filter at a sufficiently rapid rate to take the clays and ultra-fine coal out of the turbulent suspension and move them toward the filter before the load is formed. The result is that most of the clay remains on the inner surface of the load of coal and the filter does not blind. A last step during the load-building process is the application of a squeegee-like plastic wiper blade attached to the fill tube (not illustrated). This wipes clean much of the clay, which collects on the inner surface of the load, from the inner surface of the load and disposes of it in the water that is decanted and removed from the mouth of the bowl.

It should be noted that conical biasing surfaces are provided within the bowl to induce the water to move toward the filters. It is also of interest that during the filling process, the coarse particles settle out very rapidly, leaving a higher percentage of fines near the inside diameter of the newly formed load. The 147-rad/s (1,400-rpm) filling speed produces 1,003 G. This is the order of the drying speed of most commercial centrifuges. However, this level of gravity is insufficient for achieving 10 percent surface moisture with 0.44-mm (100-mesh) × 0 coal fines. Therefore, after the load of coal is formed, the bowl is accelerated quickly to 293 rad/s (2,800 rpm) and allowed to dwell briefly at this 4,000 G level to remove more water by high-gravity filtration. The bowl is then decelerated to 47 rad/s (450 rpm) and the cutting out of the coal begins. Cutting is done using a special high-speed screw conveyor that removes 650 pounds of dry coal fines from the coal in less than 30 seconds. At this point the cutter is removed and the fill tube is reinserted to make ready for the next cycle.

IMPORTANT VARIABLES

At first, the conclusion might be drawn that the larger the bowl, the greater the productivity would be. It is true that larger bowls produce deeper beds and greater weights of coal per load, and that a given final moisture for coal fines can be achieved with lower gravity levels in a larger bowl. A summary of test findings concerning bowl diameter versus required gravity is given on Figure 3. However, even though a given final moisture level can be achieved with less gravity in a larger bowl, the greater bed depth produces more unit pressure on the outer layers of coal and, as a result, the coal becomes less pervious. The result is a requirement for greater residence times and, thus, longer cycle times and lower productivity. Figure 4 shows the percent of voids in dried clean coal fines under varying levels of unit pressure. The unit pressures developed in the three sizes of centrifuges shown in Figure 3, when each achieves 10 percent surface moisture with the subject coal, are given in Figure 4. The percent voids versus pressure curve in Figure 4, when viewed in light of the three sizes of centrifuge bowls shown in Figure 3, shows that the deeper the bed depth, the less pervious the coal will be and the greater will be the required residence time. This effect is obviously compounded by the greater physical distance that the water must transverse when exiting through the load in larger bowls.

A computer program was developed to determine the optimum bowl diameter for the full-sized production prototype now under test. This program takes into account, among other variables, that

1. The inertia of the bowl (assuming all dimensions are proportional) and hence the power required in the bowl drive, varies directly as the fifth power of the bowl diameter.
2. The weight of coal dried per cycle increases only as the third power of the bowl diameter.
3. The residence time increases logarithmically as the bowl diameter increases, but so does the hydrostatic head in the bed of coal when under enhanced gravity.

The appropriate bowl diameter for drying the subject coal fines was determined to be 0.914 m (36 in.) inside. This appears to have been an acceptable choice.

Among the more sensitive variables in the system is the orifice area in the filter. An extensive data base was developed in the 0.723-m (28.5-in.) centrifuge, designated Prototype 1, concerning the orifice area versus final surface moisture and filter blinding characteristics. As is shown on Figure 5, the ideal orifice area was found to be 64.52 cm^2 (10 $in.^2$). In smaller areas, the final surface moistures were elevated. In larger areas, the moistures were higher, and if the orifice area reached the order of 129 cm^2 (20 $in.^2$), danger of blinding became a factor.

The flow rate of the slurry during filling is also critical. If the rate of flow from the filling nozzles is allowed to drop from the ideal 8.2 L/s (130 gpm) in Prototype 1 to the order of 3.15 L/s (50 gpm), there is danger of blinding the filter. In this case the lack of turbulence resulting from the slower flow can no longer maintain the clays and extremely fine coal in suspension near the inner

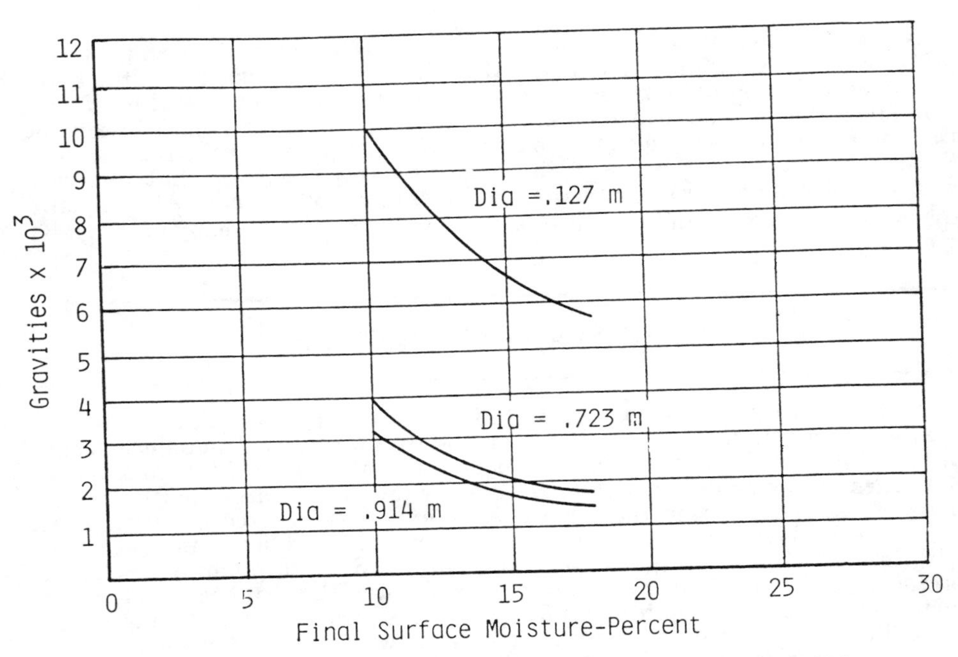

Felationship of Coal Bed Thickness to Centrifuge Dia.

Bowl Diameter		Coal Bed Thickness	
Meter	inch	mm	Inch
.127	5	25.4	1
.725	28.5	134.6	5.3
.914	36.0	190.5	7.5

Figure 3. Gravity Versus Moisture Versus Bowl Diameter Curves

surface of the coal bed as the load is formed. The result is that the clays and extremely fine coal are conducted through the bed by the filtrate, and they will collect on the filter medium. The result will be blinding.

When a load of dried coal is cut out of the centrifuge, it is cut in such a manner that a thin filter bed of coal is left in the bowl. For this reason, there is no wear on either the filter or the bowl. The cutter blade is set to cut within 1 in. of the filter and biasing cones. The dried coal, as a result of the high pressures to which it is subjected during the process, is slightly fused, and for this reason, it is somewhat less inclined to dust than thermally dried coal.

ECONOMIC ADVANTAGES

The selling price of the new centrifugal dryers has not been determined at this writing; however, there is evidence that the dryers can be installed at competitive first costs. Considering this, the operating cost structure could become the deciding factor in economic feasibility. Since the operating energy consumption is a primary cost factor, the use of the composite bowl combined with the variable frequency, alternating current drive has put the new dryers in a favorable economic position.

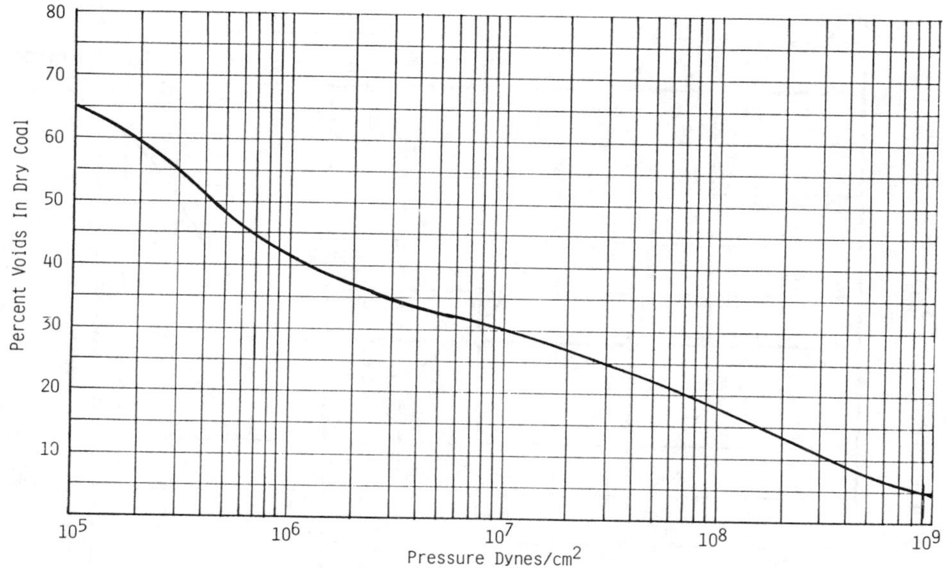

Figure 4. Pressure Versus Percent Voids in Dry Fine Coal

To operate with a 145-second drying cycle, which is necessary to achieve 7.26 metric (8.0 U.S.) tons/hour productivity, a maximum of 608.4 metric (600 U.S.) horsepower is employed during acceleration to and deceleration from the top drying speed. Figure 6 shows the energy use profile for an operating cycle of Prototype 2. The electronic inverter units, which convert alternating current to direct current, and then in turn convert the energy back to alternating current of a different frequency, were custom designed for this application. The drive can supply any level of torque at any appropriate frequency in response to the programmable controller that governs the sequences of the drying cycle. It also acts as a highly efficient generator, when required, to recover energy.

As shown in Figure 6, power is drawn from the power lines throughout much of the cycle. However, during deceleration, the drive is in a regenerating mode. This means that all but the heat losses in the inverter and the motor, a maximum of 15 percent of the power employed, is returned to the power lines for credit. The net result is that a motor, which employs up to 608.4 metric (600 U.S.) horsepower during portions of the cycle, only consumes a long-term average of less than 65 metric (64 U.S.) horsepower.

The very low energy consumption places the new centrifugal dryer system in a competitive position with other available methods. This energy consumption level is approximately the same as that for vacuum disk filters of the same capacity.

CONCLUSIONS

The high-gravity, batch type centrifuge is a viable system for the drying of coal fines produced by froth flotation in coal process plants. The new drying method may perform well in applications, other than coal fines drying, in which the particulate to be dewatered retains its integrity of shape under fairly high pressures and where an excessive amount of very fine leaf-like structures is not present in the product. Further, the new drying technology may be capable of development and application at much higher levels of gravity for products meeting the above criteria and with sufficient value to warrant the development.

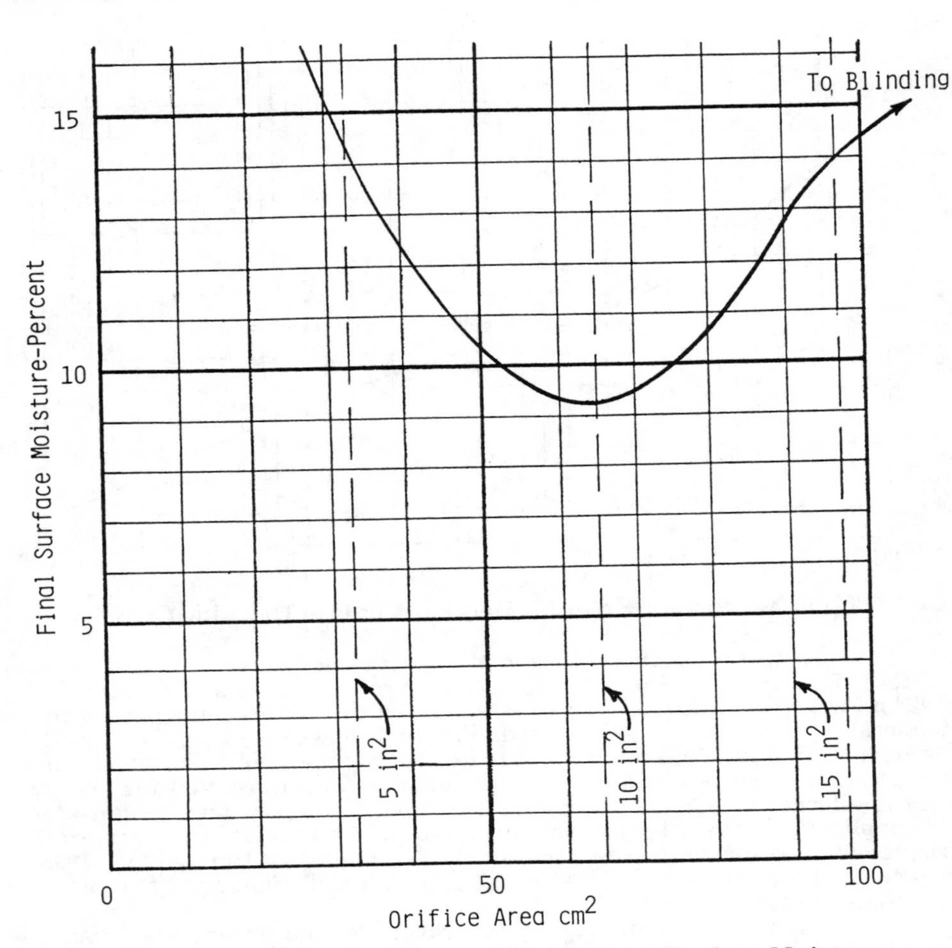

Figure 5. Filter Orifice Area Versus Final Surface Moisture

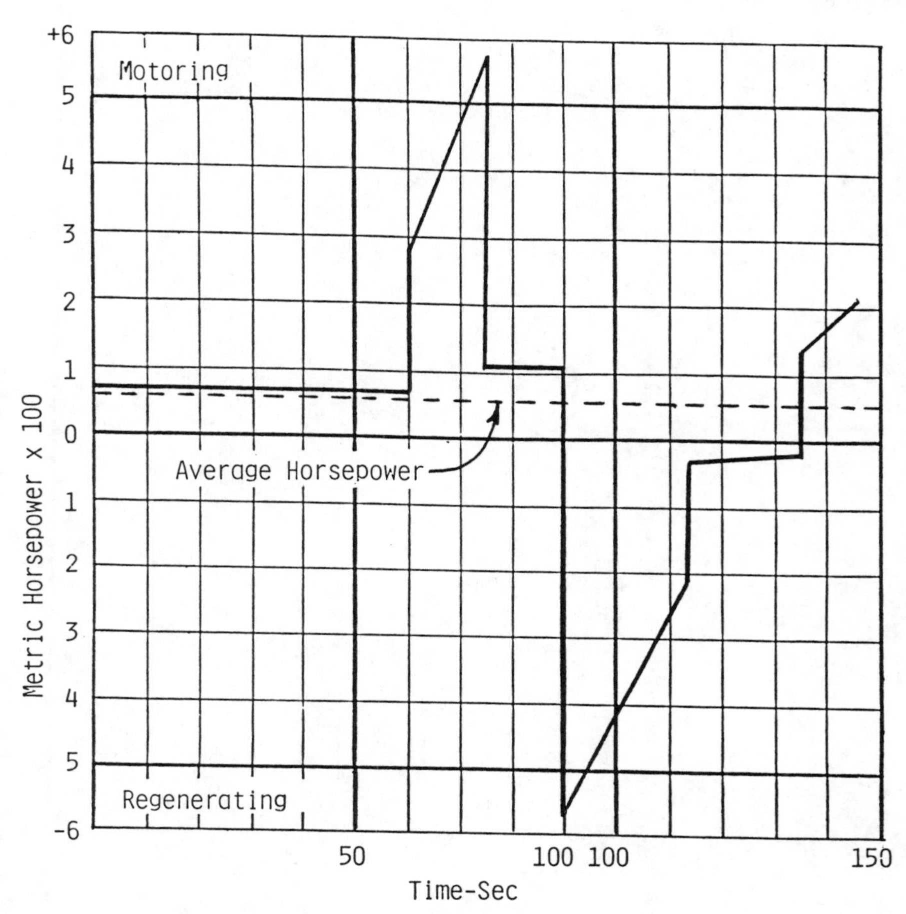

Figure 6. Power Versus Time Profile of One Cycle

REFERENCE

1. Smith, L. B., U.S. Patent No. 4639320, January 27, 1987, Method for Extracting Water from Solid Fines or the Like.

INDUSTRIAL COAL AND MINERAL WASTES

The United States is expected to spend about $23 billion by 1990 to attack some of the pervasive hazardous waste problems. Proportional spending might be expected from other industrialized countries in Europe and elsewhere in the world. Because hazardous waste problems place a severe burden on the environment, they should be addressed with the utmost seriousness by both corporate and federal agencies.

From a technological standpoint, many technologies and novel approaches have the potential to reduce waste or treat it in a satisfactory manner to achieve levels of compliance required. A number of options are available for waste disposal--for example, landfills and incineration--but all would be enhanced by a reduced volume of waste. Therefore, separation technology is vital to successful waste treatment.

Heavy metals in industrial wastewater come from a variety of industries such as steel, metal finishing, and electroplating. However, electroplating and metal finishing produce the largest amount of discharges. Dr. Peters from Argonne Lab provides an interesting paper on the performance of some conventional techniques in metal-laden wastewaters. Dr. Venkateswar of Ebasco, Professor Cheng of the University of Pittsburgh, Dr. Stana of IMC Fertilizers, and Mr. Hampel of Battelle Frankfurt discuss a number of very interesting case studies involving a variety of materials--waste acid, metals in mine drainage, fine coals, and treatment of dredged materials to produce ceramic pellets.

The keynote address by Dr. Daley of Chemical Waste Management throws new light on what solid/liquid separation can do to alleviate some treatment and disposal problems, and Dr. Harry Freeman, U.S. Environmental Protection Agency, addresses the EPA's role in helping the industry to solve some of these problems.

FORCES CONVERGE TO EXPLODE WASTE APPLICATIONS FOR SEPARATION TECHNOLOGY

Peter S. Daley, Ph.D., P.E.
Director, Research and Development
Chemical Waste Management, Inc.
Geneva Research Center
2000 South Batavia Avenue
Geneva, Illinois 60134

INTRODUCTION

It's an exciting time in the separation technology business. Four forces are creating enormous potential for separation methods' innovation in the fast-growing waste industry. The forces?

- Pressing needs to protect resources
- New, fast analytical chemistry methods
- New approaches and materials that will let us do new things
- Cheap, powerful computers are readily available.

Let's take these one at a time.

WASTE TREATMENT LEADS THE NEEDS

In the water treatment area alone, the number of facilities the nation will build or rebuild in the next decade is many tens of thousands. Pretreatment rules to protect publicly owned treatment works affect thousands of wastewater generators. New regulations governing lagoons used to treat, store, or dispose of wastewater are leading thousands of managers to buy new plants or, *even better*, recover water for reuse. We'll build thousands of treatment systems for sites contaminated by chemical wastes. And leachate from thousands of landfills will be treated to correct past shortcomings in siting, design, operation, and closure.

On the water supply side, growing concerns about drinking water quality are generating demand for better ways to remove trace contaminants at both treatment plants and the point of use.

A thousand here, a thousand there, "and pretty soon we're talking real money," drawing on the words of Tip O'Neil, former Speaker of the House. These "thousands" are not hyperbole; the treatment plants will be built; many will use advanced separation technologies.

RESOURCE RECOVERY NEEDS

Beyond water treatment, resource recovery as a waste minimization technique is perhaps the fastest growing segment of the waste business. Resource recovery is synonymous with separation technology. Here's a short list of some of the resource-recovery and waste-separation applications under development at Waste Management, Inc.

- Water recovery by evaporation with catalytic oxidation
- Metal recovery from electroplating and electric arc furnace wastes
- Hydrofluoric acid recovery from quartz crystal manufacturing
- Acid recovery from spent pickle liquor
- Plastic recovery from household wastes.

The driving force for these programs is not philanthropy; it's hard economics. Costs of disposal processes are rising. More treatment is required. Standards are higher. Land disposal space is less available. Recovered products' values, waste disposal cost avoidance, and risk reduction combine to make options attractive today that were rejected only a few years ago.

ANALYTICAL METHODS: A TWO-EDGED SWORD

Analytical chemistry is simultaneously making separations tougher and easier. On one end the ever-improving methods have found contaminants that raise concerns: things we didn't worry about in our ignorance, like trihalomethanes, selenium, dioxin, beryllium, boron, chromium(VI), radium, and most of the heavy metallic elements. We've concluded as a society that we should design plants better to control these contaminants, now that we know they're there. On the problem-solving end, analytical chemists can now provide real-time performance data for process design and control; data we've never been able to get before.

NEW METHODS AND MATERIALS

New separation methods are everywhere, and old ones are being revitalized through use of new materials. Membrane systems (e.g., reverse osmosis, electrodialysis, ultrafiltration) are finally finding broad application. Magnetic, microwave, and other electromagnetic systems are becoming practical as capital and operating costs fall. Enzyme methods are opening yet more treatment possibilities. Enzymes are being developed to produce valuable organic chemicals from sewage sludge. Advanced separation methods will be used both to produce the enzymes and purify the waste-derived chemicals.

Even lowly distillation is finding new applications in acid recovery, sludge dewatering, and water recovery. At the other thermal extreme, freezing is being more broadly applied in dewatering, concentration, and purification.

COMPUTERS OPEN NEW DOORS

The role of computers? For equipment design and process control, of course, but also in expert systems to speed process selection and design. And for scaleup so we can do process design lab work in hours, not days, using liters, not tons, of test materials. What about using efficient batch and plug flow methods instead of plain-vanilla continuous systems? Filter presses, a batch separation system, seem to be pushing continuous belt presses out in many applications because they can do a better job. All batch processes are more practical because computer control systems make them easy to run. And what about computer methods to mass produce the thousands of separation systems we're going to build?

AN EXAMPLE FROM CHEMICAL WASTE MANAGEMENT

What does this have to do with Waste Management, Inc.? Why is a garbage man here telling you this? Because we're at the focal point. We recover products, purify water, and clean up waste sites and sewer discharges; our business, perhaps more than any other, must react to the improvements in analytical chemistry.

I would like to discuss a whole list of first class separation technology development programs growing in response to these powerful driving forces, but clearly that cannot be. So I'll take just one and use it as an example to demonstrate waste industry needs and applications. The system I'd like to focus on is a Chem Waste Management, Inc., process called PO*WW*ER™.

PO*WW*ER™ is a very powerful wastewater purification process, hence its name. Figure 1 is a PO*WW*ER™ system flow chart. At its heart is a still; almost any type could be used, but we'd focused our work on forced-circulation evaporators because they can remove almost all the water from the brine residuals.

In the PO*WW*ER™ system the distillation step separates the nonvolatile contaminants from the wastewater. The steam from the still is super-

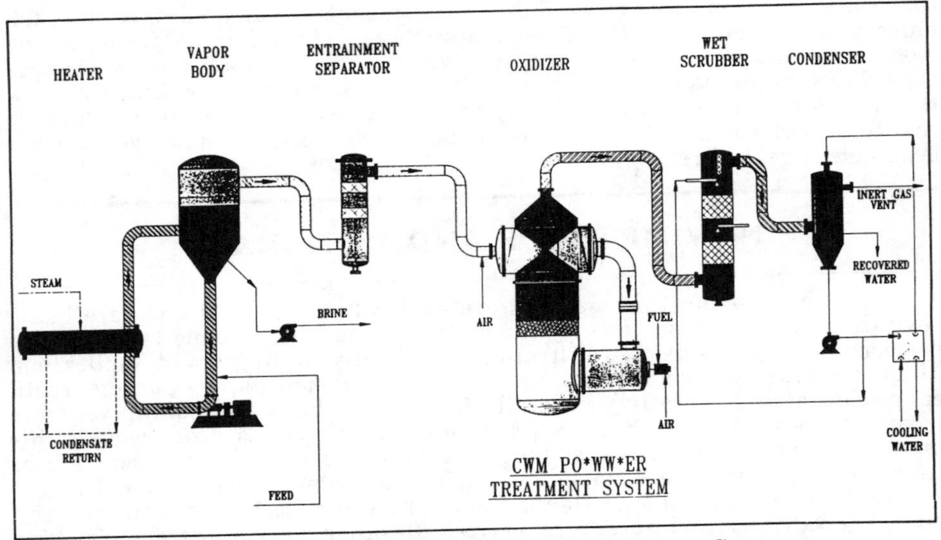

Figure 1. CWM PO*WW*ER™ Treatment System

heated to about 300 C and passed through a catalytic converter to destroy most volatile contaminants (see tabulation below). The catalytic oxidizer produces more steam, carbon dioxide, sulfured oxide, and mineral acid gases. The latter two are removed from the steam in a superheated-lime fluidized bed. The steam is condensed and the water further purified, if necessary, for reuse.

PO*WW*ER™ Separation/Destruction Steps

- Initial distillation separates
  - All inorganic salts and metals
  - Very low volatility organic materials (BP > ~500 C)
- Catalytic oxidation destroys
  - Volatile organic compounds
  - Ammonia and volatile amines
  - Hydrogen sulfide and volatile mercaptans
  - Other volatile organic materials
- Lime fluid-bed scrubbing separates and neutralizes
  - Mineral acids
  - Nitric and sulfuric acid
- Condensation separates
  - Carbon dioxide
  - Other noncondensable gases
- Final polishing (reverse osmosis, electrodialysis, double distillation, etc.) separates
  - Entrained carryover particles
  - Condensed gases.

The end result of all this is the production of very pure water from virtually any wastewater at a remarkably low cost, 3 to 10 cents per gallon. Clearly, the PO*WW*ER™ process has many variants that can make it more cost-effective for specific applications; what I've described is the brute-force, universal system.

The four driving forces I mentioned at the beginning (needs, new analytical methods, new materials, and cheap computers) have combined to make PO*WW*ER™ commercially viable today, whereas it wasn't 5 years ago. Let's look at these one at a time.

PO*WW*ER™ is a relatively high-cost system. The cost is justified by its versatility and its ability to produce reusable water at cost comparable to the sum of treatment costs by other methods and costs avoided in securing supplies of increasingly scarce water elsewhere. This recycling feature is a major factor in PO*WW*ER's™ commercial attractiveness. In another vein, one of the principal cost savings associated with PO*WW*ER™ is that it produces very little sludge requiring land disposal because no bulk precipitation agents like lime are added. In alternative conventional processes the rising cost of sludge disposal is making them increasingly unattractive.

New analytical methods have heavily influenced PO*WW*ER™. They have helped create the need by producing a demand for high-purity treatment processes, they have helped us speed development and demonstrate its effectiveness, and they have lowered the cost of commercial systems because they can be used to provide fast product quality measurements.

New materials are playing roles in several areas. Corrosion is a major problem in evaporators; we will be employing some very advanced alloys to keep it under control. More exciting is our current work, which we believe, will lower system costs by 20 to 30 percent through the use of advanced, high-temperature polymers in evaporator construction. In addition, we will employ new membranes in ultrafiltration, reverse osmosis, and electrodialysis polishing systems.

We've used computers extensively to speed system development, reduce development costs, and manage the process. In our commercial system, programmable logic control allows us to adapt the process quickly to changing feed character. This is essential if we are going to get the high utilization we need to make the system cost effective in large-volume markets.

THE FUTURE HOLDS MANY OPPORTUNITIES, BUT BE CAREFUL

It should be clear that the forces making PO*WW*ER™ a success will influence a wide range of separation and treatment technologies. In developing the PO*WW*ER™ system, we set our goals early, then looked at the new tools that were available. We specifically looked at traditional methods that might be viable if we could overcome their weaknesses. One result of this approach was our conclusion that membrane separation methods were being underutilized in our industry. This is still the case; there are more and more opportunities for clever separation scientists to apply their knowledge successfully to waste management problems.

I encourage you all to think hard about the four forces that have opened up unprecedented opportunities for separation technology applications in the waste industry--growing needs, better analytical methods, new materials, and cheap computers--and imaginatively apply your knowledge to our problems. No doubt, numerous successes are waiting that will make our world a better place.

What should you look for? Try these targets:

- Processes that can effect broad separations with few steps
- Processes that don't create new or bigger waste problems
- Processes that have inherently small air emission problems.

If you can bring these three attributes together with modest cost performance, you will have a winner. If you don't succeed on all three of these points, your chance of success in waste applications is small. Marginal separations and multiple steps force complexity and produce excessive labor and quality control costs. Public objections to the creation of yet more waste problems will leave you perpetually picking up loose ends, and public objections to air emissions are an almost certain kiss of death.

Good luck to each of you. I know there are the seeds of many valuable separation methods here today. I can safely speak for the entire waste services industry when I say we are greatly in need of better separation technologies and will listen to anyone with ideas.

NEW TECHNOLOGIES: THE POLLUTION PREVENTION PERSPECTIVE

I. J. Licis and H. M. Freeman
U.S. Environmental Protection Agency
Risk Reduction Engineering Laboratory
26 West Martin Luther King Drive
Cincinnati, Ohio 45268

BACKGROUND

Within the United States and other industrial nations throughout the world, there is a growing commitment to encouraging the development and adoption of clean technologies and clean products as a means of reducing environmental problems. In the United States, the term most often used to describe such approaches to environmental improvement is "pollution prevention," i.e., reduction and recycling to reduce the amount and toxicity of contaminants and thus the need for pollution controls.

Reflecting the importance placed on pollution prevention in the United States, on January 7, 1989, the U.S. Environmental Protection Agency (EPA) issued a policy statement committing itself to instituting pollution prevention throughout all of its media-specific programs (air, water, and land). It stated that:

> "EPA believes that developing and implementing a new multi-media prevention strategy focused primarily on source reduction and secondarily on environmentally-sound recycling, offers enormous promise for improvements in human health protection and environmental quality and significant economic benefits."

Pollution prevention approaches to environmental protection include

- Eliminating pollutants by substituting nonpolluting chemicals or products (e.g., material substitution, changes in product specifications), or altering product use
- Reducing the quantity and/or toxicity of pollutants generated by production processes through source reduction and process modifications
- Recycling of waste materials (e.g., reuse, reclamation).

Pollution prevention approaches do not include treatment and disposal methods.

A 1987 National Academy of Sciences report entitled "Multi-media Approach to Pollution Control" indicated that the transfer of pollutants from one medium to another is a major problem and that it is essential for the EPA to take a multimedia focus that takes into consideration a pollutant's impact on all media. The Pollution Prevention program being implemented by the EPA takes this approach.

A DISCUSSION OF TERMS

Prior to the coining and eventual adoption of the term "pollution prevention," there was a continuing discussion of what should be the proper descriptive term for those processes that resulted in less disposable waste. One term championed primarily by the Office of Technology Assessment was "waste reduction," defined as in-plant practices that reduce, avoid, or eliminate the generation of hazardous waste, which was defined as all nonproduct hazardous outputs from an industrial operation. The term used in the Hazardous and Solid Waste Amendments of 1984 and generally used by the EPA was "waste minimization." Waste minimization is defined as the reduction, to the extent feasible, of hazardous waste that is generated or subsequently treated, stored, or disposed of. It includes any source reduction or recycling activity undertaken by a generator that results in either the reduction of total volume or quantity of hazardous waste, or the reduction of toxicity of the hazardous waste, or both, so long as such reduction is consistent with the goal of minimizing present and future threats to human health and the environment. Earlier interpretation by the EPA of this definition included treatment options such as volume reduction. However, after much discussion at many conferences, the interpretation was modified to include only source reduction and recycling.

Neither of these terms really address the aspect that is becoming increasingly important: the characteristics of the actual product being produced. There are many examples of clean processes producing products that themselves may very well produce environmental problems when they are consumed or introduced into a waste stream somewhere. For instance, pesticides that may be produced by an efficient process with very few emissions or waste are eventually applied directly to the land and may show up in runoff from the treated field. A pollution prevention analysis of this process would incorporate the potential pollution from the product as well as from the process.

All of these terms are still in use. However, "waste minimization" has generally come to be recognized as a term dealing with industrial hazardous process waste reduction, and "waste reduction" has evolved into a term to represent the reduction of hazardous, as well as nonhazardous, waste. In this paper we will use "pollution prevention" as the more inclusive term.

WASTE MANAGEMENT HIERARCHY

Over the past 5 years, the EPA has attempted to redirect the nation's pollution control strategy toward pollution prevention by adopting a waste management hierarchy that placed priority on pollution prevention. The first two elements in the hierarchy, depicted in Table 1, focus on source reduction and recycling. This hierarchy has been widely cited as guidance for determining environmentally preferable waste management options, the idea being that options in the top sectors were preferable to those falling in the categories near the bottom of the list. This hierarchy is also useful in providing criteria for evaluating new products and processes from a pollution prevention perspective.

Table 1. EPA Waste Management Hierarchy

- **Source Reduction.** The reduction or elimination of waste at the source. Source reduction measures include product and process modifications, feedstock substitution, improvements in feedstock purity, housekeeping and management practice changes, increases in the efficiency of equipment, and recycling within a process.
- **Recycling.** The use or reuse of waste as an effective substitute for a commercial product or as an ingredient or feedstock in an industrial process. It includes the reclamation of useful constituent fractions within a waste material or the removal of contaminants from a waste to allow it to be reused.
- **Treatment.** Any method, technique, or process that changes the physical, chemical, or biological character of any waste so as to neutralize it; recover energy or material resources from it; or render it nonhazardous, less hazardous, safer to manage, amenable for recovery outside a process, amenable for storage, or reduced in volume.
- **Disposal.** The discharge, deposit, injection, dumping, spilling, leaking, or placing of waste into or on any land, water, or the air.

THE EPA'S POLLUTION PREVENTION RESEARCH PROGRAM

The Fiscal Year 1989 Environmental Protection Agency Appropriations Act required the Agency to submit a multiyear, multimedia pollution prevention research plan to Congress. This plan has been prepared and is presently undergoing review and clearance procedures. However, regardless of the final details of the report, we plan to structure our FY 1990-1992 pollution prevention research program according to the six fundamental goals discussed in the report:

1. Stimulate the development and use of products that result in reduced pollution. This goal focuses on the pollution prevention problems related to the use and disposal of specific products. Although products are often considered to include only manufactured items, as it is used here, the term also includes chemicals used in manufacturing processes and service industries; packaging for parts, commodities, and manufactured items; fluids and gases used as solvents, carriers, refrigerants, coatings, and lubricants; and additional items of commerce. These materials often are not viewed as wastes or industrial discharges, but do in fact impact the environment and pose a risk to health.
2. Stimulate the development and implementation of technologies and processes that result in reduced pollution. Numerous pollution prevention opportunities exist in manufacturing, mining, agricultural, and service processes. This goal addresses the need to focus research activities on these processes to enable broad-scale reduction in pollution generation.
3. Expand the reusability and recyclability of wastes and products, and the demand for recycled materials. Research is needed to improve the reusability and recyclability of wastes and products, and to increase the capacity of and demand for recycled materials in production processes. Such improvements will prolong the useful life of materials and reduce the environmental impacts of wastes and pollutants from all waste streams.
4. Identify and promote the implementation of effective nontechnological approaches to pollution prevention. This research area includes socioeconomic and institutional factors that motivate behavior and foster changes in behavior as they relate to incentives for adopting pollution prevention techniques. Research is needed to understand the roles of nontechnological factors in implementing pollution prevention approaches and their impact on the effectiveness of pollution prevention programs.
5. Establish a program of research that will anticipate and address future environmental problems and pollution prevention opportunities. Research is needed to assist the EPA in anticipating and responding to emerging environmental issues and to evaluate new technologies that may significantly alter the status of pollution prevention programs in the future. A flexible program is needed for conducting research that may affect long-term pollution prevention program directions and objectives. This research program will enable the EPA to anticipate and potentially prevent future environmental problems. In addition, this program provides the Agency with the ability to address emerging issues that will shorten the time between detection of a new environmental problem and the EPA's effective response.

6. Conduct a vigorous technology transfer and technical assistance program that facilitates pollution prevention strategies and technologies. It is imperative that the results of research investigations conducted under this program or by industry and academic research programs are communicated to appropriate audiences.

For the purpose of our discussion at this symposium on solid/liquid separation, I think it is useful to highlight certain parts of our pollution prevention research goals. To begin, the first goals address products rather than processes. This is a departure from past EPA research agendas that tended to address waste from production processes as the top priority. This shift in emphasis is a recognition that all products eventually become wastes and that it is important to consider ultimate disposal problems during the design steps for the products, especially for such waste streams as municipal solid waste.

Second, even though we are emphasizing cleaner products, we will continue to encourage clean production technologies by encouraging generators to develop and adopt the types of new processes being featured in this symposium. In addition, priority is given to technologies that represent improvements in the manufacturing processes as opposed to improvements in the end-of-the-pipe treatment processes. This is not to say that improvements in treatment are not important, but that priority consideration for research support will be given to those processes that come closer to producing no waste, rather than those more efficiently treating waste streams. Most of the processes being discussed in this symposium would be equally applicable to production processes and treatment processes.

Third, the goal that addresses increased recyclability of products and wastes is intended to influence product design considerations. Clearly, such considerations are becoming more important. The waste disposal capacity problem will not go away. Therefore we must, and in many instances do, take recycling into consideration as part of product design. This consideration has a direct effect on the field of technology of solid/liquid separation. Close coordination is needed between those designing the new separation processes and those producing products and the resulting waste streams.

FUTURE PLANS

We believe that the pollution prevention perspective will continue to be a strong force in future Agency action. The Agency's commitment to the concept was spelled out in recent technology by the Agency's Administrator, William K. Reilly, before a committee of the U.S. House of Representatives on May 25, 1989:

> "Existing environmental statutes, and our national pollution-control programs of the past twenty years, have emphasized containment and treatment of pollutants after they are produced. The widespread use of air pollution scrubbers, wastewater treatment plants and the recent introduction of land disposal restrictions are, in fact, significant accomplishments that show our nation's commitment to a clean environment. Nevertheless, our efforts have focused on end-of-the-pipe treatment processes. The limitations of environmental protection from that aspect only are becoming increasingly apparent. There is a growing recognition that traditional approaches--which stress treatment and disposal after pollution has been generated--have not adequately dealt with existing environmental problems. Nor will they provide an adequate basis for dealing with emerging problems such as global warming, acid

rain, and stratospheric ozone depletion. EPA believes that further improvements in the environmental quality will be best achieved by preventing the generation of pollutants that may be released to the air, land, and water by eliminating or reducing them at their source and encouraging environmentally-safe recycling of those which cannot be eliminated. We need to supplement our efforts with a new strategy--one that couples conventional controls and vigorous enforcement of our current laws, with pollution prevention. We must cut down on the amount of toxics and other pollutants being generated."

DISSOLVED AIR, INDUCED AIR, AND NOZZLE AIR FLOTATION SYSTEM PERFORMANCE COMPARISON FOR PRECIPITATION/FLOTATION OF METAL-LADEN WASTEWATERS

Robert W. Peters, Ph.D., P.E.
Waste and Safety Engineering Section
Energy Systems Division
Argonne National Laboratory
9700 South Cass Avenue
Argonne, Illinois 60439
(312) 972-7773

and

Gary F. Bennett, Ph.D., P. Eng.
Department of Chemical Engineering
The University of Toledo
2801 West Bancroft Street
Toledo, Ohio 43606

ABSTRACT

The performance of combined chemical precipitation/air flotation treatment has been addressed for two wastewaters employing three different air flotation systems: dissolved air flotation (DAF), induced air flotation (IAF), and nozzle air flotation (NAF). Two wastewaters, one synthetic and one industrial, were used in these performance assessments. The synthetic wastewater used Jayfloc-806 as the chemical collector, and the industrial wastewater used Jayfloc-824 as the collector. The synthetic wastewater was capable of removal efficiencies exceeding 96 percent. The DAF system provided the best metal removal, and the NAF system, the poorest metal removal. For the industrial wastewater, removal efficiencies were lower, on the order of 75 percent. For lower collector dosages, use of the DAF system resulted in the lowest residual concentrations, whereas at higher dosages (about 5 mg/L), use of the IAF system resulted in the lowest concentrations. Enhancement ratios were much greater for the industrial wastewater (compared with the synthetic wastewater).

BACKGROUND

Industrial sources of heavy metals in industrial wastewaters include the steel and metal-finishing industries, mining, electroplating, photography, tanning, and printing/dyeing; the manufacture of paint products, fertilizers, insecticides/pesticides, paper, fiber, and electronics products; cooling water and pipe corrosion; handling of metal-containing petroleum-based products; and battery manufacturing. Among these sources, the electroplating and metal-finishing industries produce more individual wastewater discharges than any other industrial category[4]. A similar list of major industrial sources of pollution was compiled by Dean *et al.*[7] and included processes associated with paper and pulp, organic and inorganic chemicals, fertilizers, petroleum, motor vehicles/aircraft plating and metal finishing, textile mill products, basic steel works foundries, basic nonferrous metal works foundries, flat glass, cement, leather tanning and finishing, asbestos products, and steam generation power plants. The number of industrial sources identified indicates that heavy metals are employed in a number of widely diversified fields.

Chemical and electrochemical methods are employed in the electroplating, metal-finishing, and allied industries for the purposes of protecting or decorating metal surfaces[11]. Most of these processes are followed by rinsing operations to remove the excess chemicals and other waste materials from the treated surfaces, giving rise to industrial waste effluents. In particular, pickling and electroplating operations give rise to high waste metal concentrations. For example, a typical wastewater from sulfuric acid pickling of copper and brass is reported to contain 250 to 300 mg/L of H_2SO_4, 60 to 90 mg/L of soluble copper (Cu), 20 to 30 mg/L of soluble zinc (Zn), and 10 to 15 mg/L of suspended solids[11].

The number of firms in the United States that are engaged wholly or partially in electroplating or other metal-finishing operations exceeds 13,000[17]. These firms discharge their spent process waters either to waterways or to publicly owned treatment works (POTWs)[17]. The pollutants contained in these process wastewaters are potentially toxic. The Clean Water Act of 1977 (Public Law 95-217) provides that the water discharges must be treated to oxidize cyanides, reduce hexavalent chromium (Cr^{+6}), remove heavy metals, control pH, and reduce oil concentrations, all to acceptable levels, even for wastewater discharges into sewers. Problems in achieving the required levels of heavy metals for wastewaters usually occur when: (1) the flow of wastewater exceeds that allowed in the guidelines, (2) mixtures of heavy metals are to be removed, or (3) the precipitated metal salt is difficult to separate from the wastewater by conventional gravity settling.

Currently, removals of oils and heavy metals from industrial wastewaters are generally accomplished separately using completely different treatment processes. The usual method of oil removal involves simple American Petroleum Institute (API) gravity separation, possibly followed by air flotation. Bennett[2] has provided a detailed discussion on the principles and applications of air flotation for oil removal. Heavy metals are generally removed by hydroxide precipitation accomplished through addition of an alkali such as lime or caustic soda to precipitate the metals as metal hydroxides, followed by flocculation, sedimentation, and/or filtration. Chemical precipitation is the most widely used industrial technique for removal of heavy metals from solution[7,13]; nearly 75 percent of the electroplating facilities employ a precipitation technique to treat their wastewaters[14]. The solubility of all metals is a strong function of the solution pH. To remove the heavy metals from wastewaters, the pH is adjusted to the point at which the metals exhibit their minimum solubilities. The pH for minimum solubility is different for each metal species.

The process of flotation involves four basic steps: (1) bubble generation in the wastewater being treated, (2) contact between the particles (or oil droplets) and the gas bubbles, (3) attachment of the particles to the bubbles, and (4) rising of the bubble-particle combination to the surface. Particles to be floated should be, to some extent, water-repellent or hydrophobic. This property can be enhanced through the addition of chemical reagents (flotation reagents or chemical collectors), which can be weak acids, bases, or their salts. These collectors are heteropolar with two functional ends; the ionic end adsorbs to the particle surfaces, and the organic end provides the hydrophobic surface to the particle.

Flotation equipment for wastewater treatment can be broadly classified into two major categories, depending on how the air is introduced into the system: dissolved air flotation (DAF) and induced air flotation (IAF). DAF is the conventional air flotation system used in industry. In this process, very fine gas bubbles are generated by reducing the pressure of a wastewater stream that has been exposed to air at elevated pressures. Typically, the applied pressure is in the range of 4 to 80 psig. The pressurized gas is released into the flotation cell, where small bubbles nucleate and form from the supersaturated solution. The bubbles attach to and become entrapped by suspended materials and rise to the surface, where they are removed.

IAF processes generate and discharge the air bubbles into the liquid using high-speed rotating impellers or diffusers, or by homogenization of a gas/liquid stream. In IAF processes, the gas is induced into and dispersed throughout the flotation cell at ambient pressure. Performances of both the DAF and IAF processes for oil removal have been reviewed by Bennett[2]. Churchill and Tacchi[6] reported that DAF and IAF processes resulted in effluent oil concentrations of about 25 and 10 mg/L, respectively. These values corresponded to removals of 89 and 93 percent for the DAF and IAF processes, respectively. One major difference in these two processes involves the size of the air bubbles produced. In DAF systems, bubble sizes are typically in the range of 10 to 100 μm; in IAF systems, the bubbles are generally about an order of magnitude larger[6].

Another recent air flotation technique involves nozzle air flotation (NAF), which uses an eductor or an exhauster as a gas aspiration nozzle to draw air into recycled treated wastewater, which in turn is discharged into the flotation vessel. Bennett[2] has also discussed the principles and application of NAF processes and provided some performance data for oil removal.

PREVIOUS RELATED RESEARCH

Mukai *et al.*[12] investigated coprecipitation of metal ions with ferric hydroxide with the subsequent elimination of the coprecipitate by flotation; the precipitation and flotation were performed in separate steps. Favorable pH conditions for the removal of the heavy metals were: pH 8.5 to 9.2 for Zn, pH 8.5 to 10.1 for Cu, and pH 8.5 to 11.6 for Cd.

Westra and Rose[18] screened a number of technologies for treatment of alkaline-soak cleaning solutions, including gravimetric separations, membrane processes, electrolytic treatment, carbon adsorption, biological treatment, and chemical emulsion breaking. A mixture of nonpolymeric quaternary ammonium salts proved effective in treatment of oil-in-water emulsions. The use of cationic surfactants suggested the ability of these chemicals to destabilize effectively difficult-to-break emulsions. Some of the heavy metals were also removed with the oil[18]; removals of Cu, Ni, Zn, and total chromium were in the ranges of 15 to 20, 48 to 84, 0 to 17, and 11 to 15 percent, respectively. Effluent oil and grease concentrations ranged from 50 to 400 mg/L.

Using a metal hydroxide precipitation/DAF technique, Ching[5] investigated the simultaneous removal of heavy metals (Cu, Zn, and Pb) and oil using a bench-scale system. The studies were conducted in a batch mode; the experimental variables for the studies included collector type and dosage, pH, applied pressure, and recycle rate. Ching used three types of precipitants (lime, sodium sulfide, and sodium carbonate) in conjunction with four different chemical collectors (sodium lauryl sulfate, Nalco-7734, Atlasep 2A2, and Tretolite TFL-365). The optimal conditions were determined to be: pH ~8.5, an applied pressure of 45 psig, and 30 percent recycle. Removal of the metals was in excess of 70 percent, and oil removal exceeded 90 percent.

A bench-scale metal sulfide precipitation/IAF technique was employed to treat a synthetic wastewater containing Cu, Pb, Zn, and oil[3,16]. Each initial metal concentration was 25.0 mg/L. Optimal metal removals were obtained for pH in the range of 8 to 9 and an air injection rate of about 12,150 cm^3/min (0.429 ft^3/min). Applying the optimal sulfide dosage of 242 mg/L (corresponding to a pH of 8.42) to the particular wastewater employed, the residual Cu, Pb, and Zn concentrations were <0.04, <0.04, and <1.85 mg/L, respectively, for a collector dosage of 80 mg/L. The presence of oil had no noticeable effect on the removal of copper and lead; however, it caused the zinc removal efficiency to decrease by nearly 38 percent. The metal concentrations were consistently in the order of Cu$<$Pb$<$Zn, consistent with the associated metal sulfide solubility products[3]. Because of the low residual metal concentrations achieved using sulfide precipitations, the metal removals were not drastically enhanced by the addition of chemical collectors, resulting in enhancement ratios of about 1.

Gopalratnam *et al.*[8-10] investigated the combined use of hydroxide precipitation/IAF for treatment of an oily metal-laden wastewater. For this particular wastewater, the optimum pH and air injection rate were 9.18 and 12,150 cm^3/min (0.429 ft^3/min), respectively. The optimal collector dosages of Nalco-7182, WOF-67, and Jayfloc-806 were typically in the range of 2 to 5 mg/L. Removals of heavy metals and oils typically exceeded 85 percent, with oil removals as great as 96.2 percent being achieved. The enhancement ratios reached a maximum for each component for collector dosages of 4 to 5 mg/L[9].

Peters and Bennett[15] compared the removals of oil and heavy metals using either metal hydroxide or metal sulfide precipitation in conjunction with air flotation based upon previous research results[5,8,16]. Results were compared for the two precipitation systems using the chemical collectors sodium lauryl sulfate, Nalco-7182, Nalco 77-34, Atlasep 2A2, and Tretolite TFL-365. Removals of metals (Cu, Pb, and Zn) and oil in excess of 93 percent could be achieved using either precipitation process. These precipitation/air flotation treatment systems are capable of producing effluent with Cu, Pb, and Zn concentrations of less than 0.10 mg/L.

RESEARCH OBJECTIVES

This research project focused on the combined chemical precipitation/air flotation process to simultaneously remove oil and heavy metals (Cu,

Ni, Zn, and Pb) from industrial wastewaters. The objectives of our research program are summarized below:

1. To demonstrate the feasibility and applicability of this innovative precipitation/air flotation process by performing a series of experiments in the laboratory
2. To conduct a series of jar tests to determine the optimal pH for heavy metal removal in the flotation trials
3. To investigate the effects of several chemical collectors (including Jayfloc-806 and -824) and their dosages on the simultaneous removal of oil and heavy metals performed at the optimal pH
4. To investigate and compare the performance associated with three types of air flotation systems: dissolved air, induced air, and nozzle air flotation systems.

To our knowledge, few such studies have been performed to date to address the performance of the precipitation/air flotation system for the simultaneous removal of oil and heavy metals. This research project addresses that issue in the technical literature using both a synthetic wastewater and an industrial wastewater to perform these studies.

EXPERIMENTAL EQUIPMENT

To evaluate the removals of dissolved heavy metals and oil from synthetic and industrial wastewaters, three different air flotation units were used: a dissolved air flotation unit, an induced air flotation unit, and a nozzle air flotation unit.

Airflow rates of the equipment used in this study were calibrated using a wet test meter (Precision Scientific, Model 12 ABG). Solution pH was measured using a Corning Model 12 pH meter. Adjustments in pH were made by adding 1.0-M solutions of HCl or NaOH. Uniform mixing of coagulant (chemical collector) in the wastewater was accomplished during jar tests using a Phipps and Bird multi-unit, variable-speed stirred assembly (Model M10). The metal concentrations were determined using a Perkin-Elmer Model 3030 atomic absorption spectrophotometer. A freon Soxhlet extraction was performed in accordance with *Standard Methods*[1] to determine the oil concentrations.

The three air flotation systems are described below.

DISSOLVED AIR FLOTATION UNIT

For the DAF trials, a pressurized vessel connected to a tall column open to the atmosphere was used. The 4.5-L pressurized vessel has a feed hopper for the liquid feed to be poured into the vessel, a valve through which compressed air flows into the liquid being placed in the vessel, a vent for the compressed air to escape from the vessel whenever necessary, a liquid discharge outlet at the bottom of the vessel (into the tall column), and a pressure gauge.

The injection of compressed air into the liquid proceeded until the pressure gauge read 45 psig. Aeration was followed by vigorous shaking of the pressurized vessel to ensure saturation of the liquid with air under pressure. The process of compressed air injection was repeated every time the gauge indicated a significant decrease in pressure from the 45-psig reading following the vigorous shaking of the pressure vessel. This reaeration and shaking process continued until no noticeable pressure decrease was observed upon shaking the vessel.

The pressurized wastewater, saturated with dissolved air under high pressure (45 psig), was allowed to flow into an open-top, rectangular, 2-ft-tall transparent flotation column through a pressure reduction valve connected at the bottom of the pressurized vessel. A clear interface between the float material at the top and the clear liquid at the bottom was immediately observed after termination of the flotation process. The clear liquid was drained from the column for analysis through an outlet at the bottom of the flotation column.

INDUCED AIR FLOTATION UNIT

For the IAF trials, a laboratory model manufactured by Wemco Corporation was used. The cell consists essentially of a batch flotation unit and includes a tachometer to indicate the speed of the machine's rotor and an adjustable, multiribbed rotor dispenser to control the surface turbulence during the flotation runs. A 3.0-L glass, open-top flotation cell with a drain plug at the bottom of the cell (to allow for removal of samples for chemical analyses) was used during the flotation trials. An overflow weir at the upper rim of the flotation cell allowed for removal of wastewater or the froth generated during the experimental run. The unit, containing about 2.5 L of wastewater, was operated at a rotor speed in the range of 850 to 1,500 rpm; typically, the speed was about 1,000 rpm, corresponding to an airflow rate of about 12,150 cm^3/min (0.429 ft^3/min).

NOZZLE AIR FLOTATION UNIT

A recent development in air flotation technology is the NAF unit. The injection device uses a gas aspiration nozzle (an eductor/exhauster) to draw air into recycled treated wastewater and develop a two-phase mixture of air and water. That mixture is discharged into the flotation cell. The fluid flowing into the venturi throat causes a vacuum at the throat. Advantages claimed for the nozzle unit over conventional IAF systems include:

- Lower initial cost and energy use because a single pump provides the mixing and air supply
- Lower maintenance and longer equipment life because the unit has no high-speed moving parts to wear out.

EXPERIMENTAL PROCEDURE

The experiments performed in this research were carried out in three stages: (1) equipment calibration, (2) jar test experiments, and (3) precipitation/air flotation experiments. The first two stages were necessary to determine the operating parameters and conditions for the precipitation/air flotation experiments. This study focused on metal hydroxide precipitation for heavy metal removal, coupled with air flotation for the simultaneous removals of heavy metals and oil. The second and third stages of the research are described below.

JAR TEST EXPERIMENTS

Preliminary jar test experiments were performed to determine optimum conditions for the flotation trials to follow. For the synthetic wastewater portion of the study, a standard solution of Cu, Ni, Zn, and Pb was prepared from ACS-grade cupric acetate, nickel sulfate, zinc sulfate, and lead acetate. The temperature during these jar test experiments and the flotation experiments was held constant at 25.0 $\pm$ 2.0 C. For the jar tests, 300-mL samples of the wastewaters were placed in a series of containers and stirred continuously. Various conditions of pH and collector type and dosage were investigated to determine the optimal conditions. Samples were withdrawn from each of the jar test solutions for chemical analyses.

PRECIPITATION/AIR FLOTATION EXPERIMENTS

Following the jar test experiments, which determined the optimal conditions for the flotation trials, precipitation/air flotation experiments were performed. Two chemical collectors were utilized for these studies: Jayfloc-806 for the synthetic wastewater and Jayfloc-824 for the industrial wastewater. The synthetic wastewater contained 25.0 mg/L each of Cu, Ni, and Zn; 5.0 mg/L of Pb; and 1,000 mg/L of oil (prepared using a 4 percent pine oil solution). The industrial wastewater contained 19.5 mg/L of Cu, 19.63 mg/L of Ni, 11.30 mg/L of Zn, 14.05 mg/L of Pb, and 203 mg/L of oil. After pH adjustment to reach the optimum pH (selected on the basis of the jar test experiments), a known amount of polyelectrolyte (chemical collector) was added, and the flotation tests were run for 4.0 min. The flotation trials were conducted using three different air flotation systems: dissolved air, induced air, and nozzle air. Samples were withdrawn from each unit and filtered through a 0.45-μm filter; the filtrate was analyzed for residual concentrations of Cu, Ni, Zn, Pb, and oil. The variables used in these studies included the type and dosage of the chemical collector(s) used, pH, and type of wastewater. The industrial wastewater used in these tests was obtained from industries located near Toledo, Ohio.

RESULTS AND DISCUSSION

The results for this study are divided into two parts, the first dealing with the synthetic wastewater and the second involving the industrial wastewater. The preliminary jar tests performed on these wastewaters indicated that the optimal pH for both wastewaters was 9.16. In the portions of the study involving the air flotation trials, the wastewaters were subjected to the metal hydroxide precipitation/air flotation treatment described in the Experimental Procedure section. The experiments using each air flotation process were performed under similar conditions. The results from the three flotation systems are presented and compared in terms of their system performance, focusing on the residual concentrations and enhancement ratios. The results from these two portions of the study are described below.

SYNTHETIC WASTEWATER

The synthetic wastewater was prepared using reagent-grade cupric acetate, nickel sulfate, zinc sulfate, and lead acetate. The oil was added in the form of a 4 percent pine oil solution. The initial concentrations of Cu, Ni, and Zn were 25.0 mg/L; the initial concentration of Pb was 5.0 mg/L; and the initial oil concentration was 1,000 mg/L.

Tables 1 and 2 summarize the experimental results for the precipitation/air flotation experiments, using the Jayfloc-806 chemical collector. Table 1 lists the residual concentrations of Cu, Ni, Zn, Pb, and oil as functions of the collector dosage and air flotation system used. Table 2 lists the enhancement ratios for these five species as functions of the collector dosage and the air flotation system used.

The results are presented in graphic form in Figures 1 through 4. Figures 1a, 1b, and 1c show the residual metal concentrations for the three air flotation systems; these figures indicate that the minimum metal concentrations are achieved for collector dosages in the range of 0.125 to 0.25 mg/L. The lowest residual metal concentration is achieved for lead; the residual concentrations of the other metals (Cu, Ni, and Zn) are all comparable. The maximum removals of Cu, Ni, Zn, Pb, and oil are 97.0, 96.5, 96.8, 98.1, and 99.3 percent, respectively.

Table 1. Residual Concentrations for Precipitation/Air Flotation of a Synthetic Wastewater Using Jayfloc-806 Collector

Air Flotation System	Collector Dosage, mg/L	Residual Concentration, mg/L				
		Cu	Ni	Zn	Pb	Oil
DAF	0.00	1.77	1.79	1.87	1.27	127
	0.125	0.75	1.49	0.80	0.65	7
	0.25	0.75	0.87	0.88	0.88	19
	0.50	1.60	1.22	1.65	1.35	23
IAF	0.00	2.45	2.72	2.77	1.27	425
	0.125	1.21	1.49	1.56	0.67	92
	0.25	1.16	1.57	1.62	0.48	115
	0.50	1.31	1.65	1.59	0.57	115
NAF	0.00	4.46	4.19	4.28	1.98	295
	0.125	2.75	2.75	2.80	1.68	45
	0.25	2.77	2.80	2.83	1.72	67
	0.50	2.75	2.40	2.61	1.99	107

Initial concentrations: Cu_i = 25.0 mg/L, Ni_i = 25.0 mg/L, Zn_i = 25.0 mg/L, Pb_i = 5.0 mg/L, Oil_i = 1000 mg/L.

Table 2. Enhancement Ratios for the Precipitation/Air Flotation of a Synthetic Wastewater Using Jayfloc-806 Collector

Air Flotation System	Collector Dosage, mg/L	Enhancement Ratio				
		Cu	Ni	Zn	Pb	Oil
DAF	0.00	1.000	1.000	1.000	1.000	1.000
	0.125	1.044	1.013	1.046	1.166	1.138
	0.25	1.044	1.040	1.043	1.105	1.124
	0.50	1.007	1.025	1.010	0.979	1.119
IAF	0.00	1.000	1.000	1.000	1.000	1.000
	0.125	1.055	1.055	1.054	1.161	1.579
	0.25	1.057	1.052	1.052	1.212	1.539
	0.50	1.050	1.048	1.053	1.188	1.539
NAF	0.00	1.000	1.000	1.000	1.000	1.000
	0.125	1.083	1.069	1.071	1.099	1.355
	0.25	1.082	1.067	1.070	1.086	1.323
	0.50	1.083	1.086	1.081	0.997	1.267

Initial Concentrations: Cu_i = 25.0 mg/L, Ni_i = 25.0 mg/L, Zn_i = 25.0 mg/L, Pb_i = 5.0 mg/L, Oil_i = 1000 mg/L.

Figures 2a through 2e represent cross-plots of the above data, comparing the residual concentrations achieved for the three air flotation systems for Cu, Ni, Zn, Pb, and oil, respectively. With the exception of lead, the DAF system provides the greatest metal and oil removals. For the metal species, the NAF system provides the poorest metal removal. The removal of oil was in the order DAF > NAF > IAF.

Figures 3a through 3c show the enhancement ratios for the three air flotation systems for all five species as functions of the applied collector dosage. The enhancement ratio R is defined as the relative enhancement in the amount of material removed by the addition of the chemical collector as compared with that achieved with no chemical collector present. The enhancement ratio is defined in mathematical terms as[9]:

$$R = \frac{C_i - C_{\text{with collector}}}{C_i - C_{\text{without collector}}} \tag{1}$$

The enhancement ratios are all below 1.6 for this synthetic wastewater. The oil generally had the largest enhancement in terms of its removal with the addition of the chemical collector. The enhancement ratios are all of comparable magnitudes for the metal species.

Figures 4a through 4e represent cross-plots of Figures 3a through 3c, showing the enhancement ratios for the three air flotation systems compared for the cases of Cu, Ni, Zn, Pb, and oil, respectively. The NAF system achieved the largest enhancement ratios for Cu, Ni, and Zn; the IAF system achieved the largest enhancement ratios for Pb and oil.

Comparing Figures 2 and 4, we see that, whereas the NAF system generally has the largest enhancement ratios, use of the DAF system results in the lowest residual concentrations. This can be explained by the fact that the NAF system has the highest residual concentrations in the absence of chemical collectors (as compared with those for the DAF and IAF systems), and hence is the most improved in terms of its removal capability with the addition of the Jayfloc-806 collector; however, use of the NAF system still results in higher residual concentrations than those from the DAF and IAF systems.

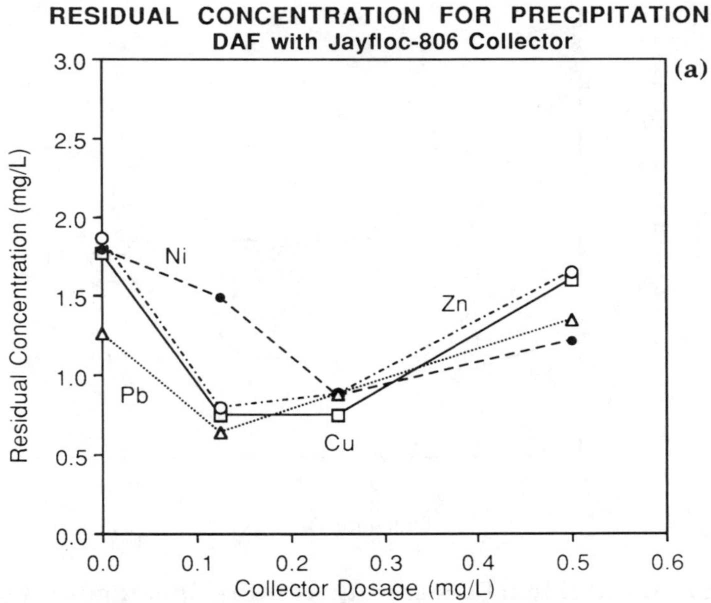

Figure 1. Residual Concentration for Precipitation/Air Flotation of the Synthetic Wastewater with Jayfloc-806, Using (a) Dissolved Air Flotation

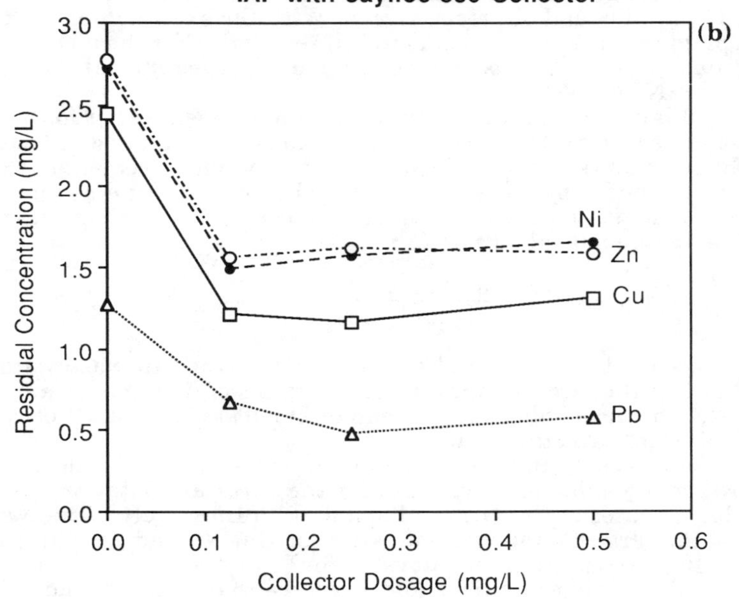

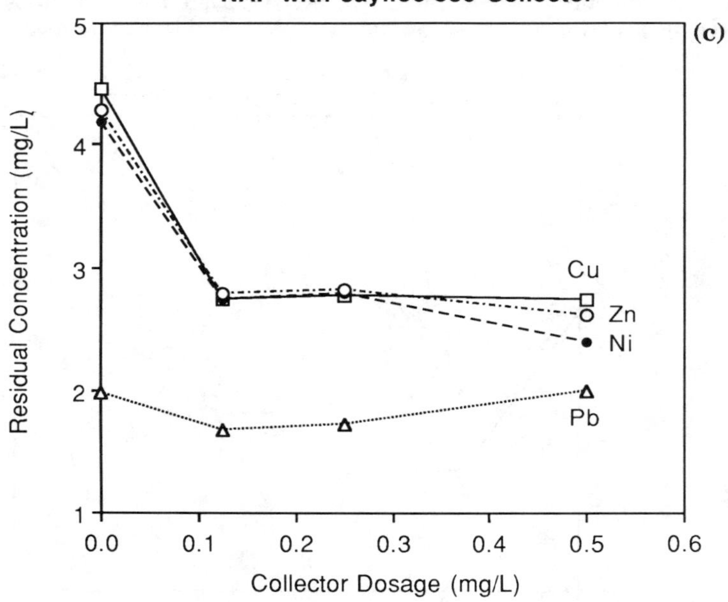

Figure 1. Residual Concentration for Precipitation/Air Flotation of the Synthetic Wastewater with Jayfloc-806, Using (b) Induced Air Flotation, and (c) Nozzle Air Flotation (Continued)

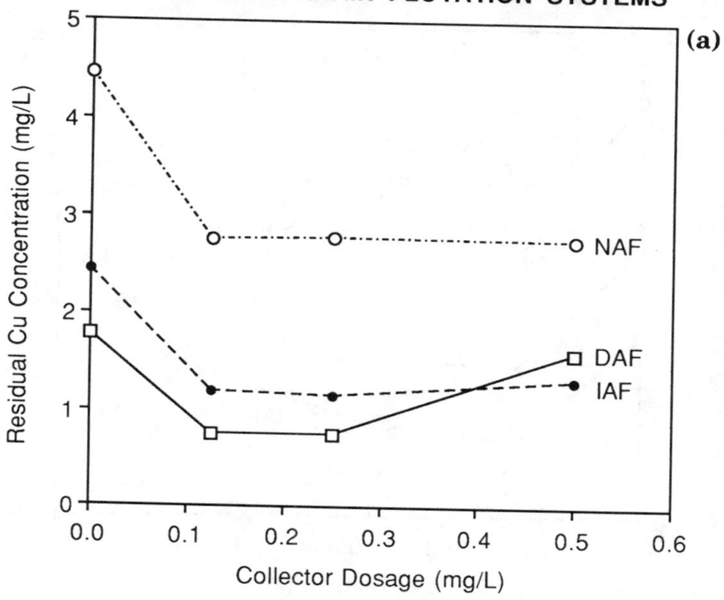

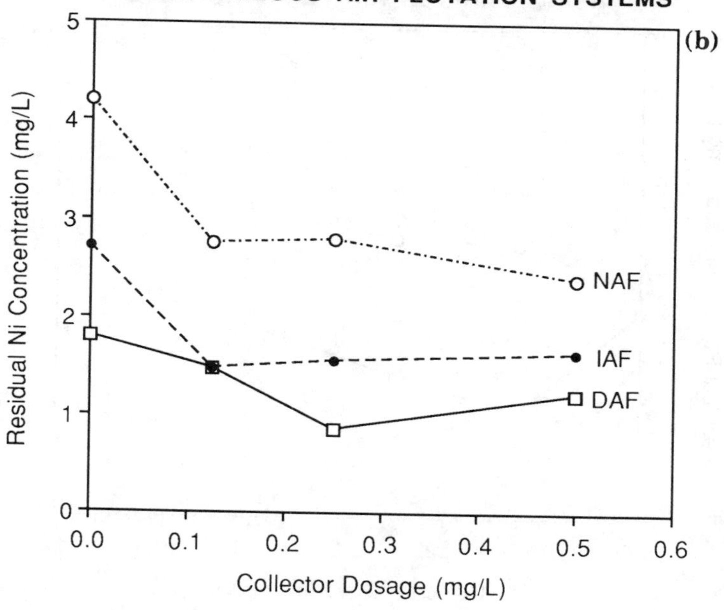

Figure 2. Residual Concentration Comparison Using Various Precipitation/Air Flotation Systems for the Synthetic Wastewater with Jayfloc-806 for the Cases of (a) Cu and (b) Ni

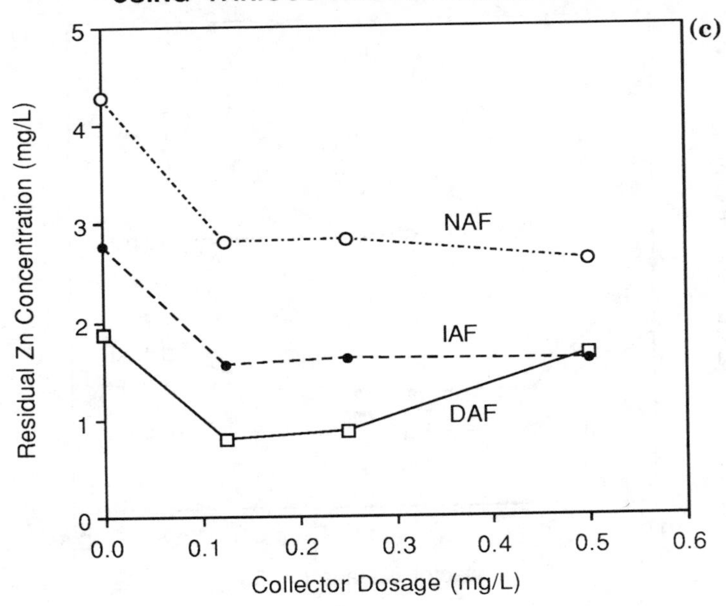

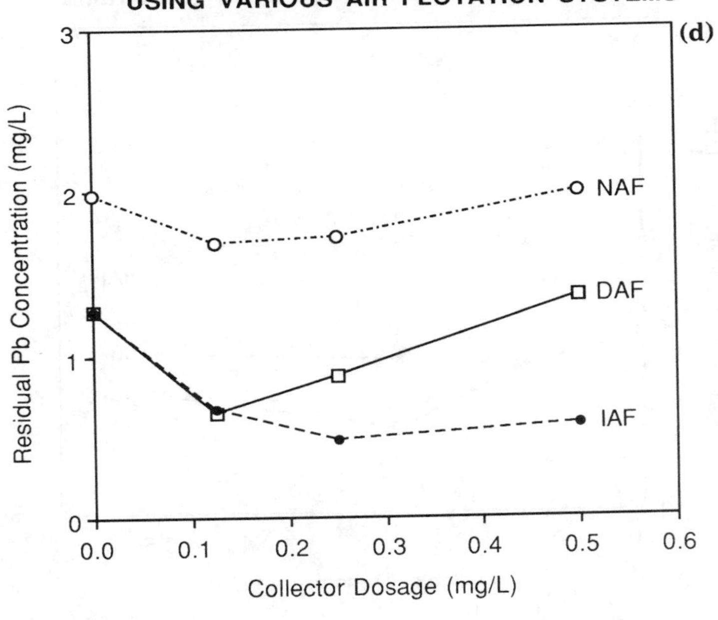

Figure 2. **Residual Concentration Comparison Using Various Precipitation/Air Flotation Systems for the Synthetic Wastewater with Jayfloc-806 for the Cases of (c) Zn and (d) Pb (Continued)**

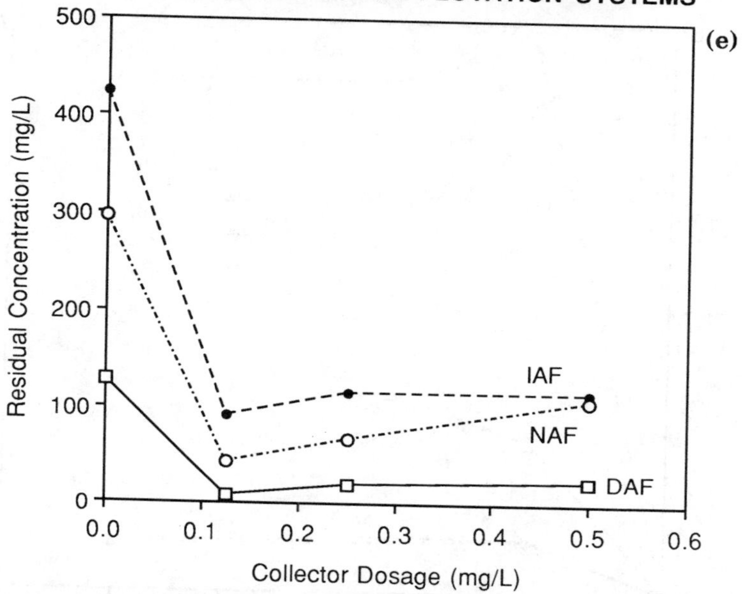

Figure 2. Residual Concentration Comparison Using Various Precipitation/Air Flotation Systems for the Synthetic Wastewater with Jayfloc-806 for the Case of (e) Oil (Continued)

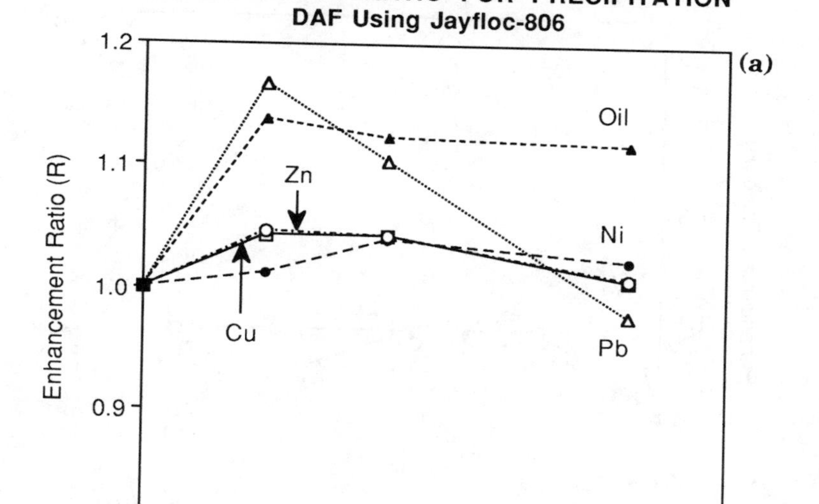

Figure 3. Enhancement Ratios for Precipitation/Air Flotation of the Synthetic Wastewater with Jayfloc-806, Using (a) Dissolved Air Flotation

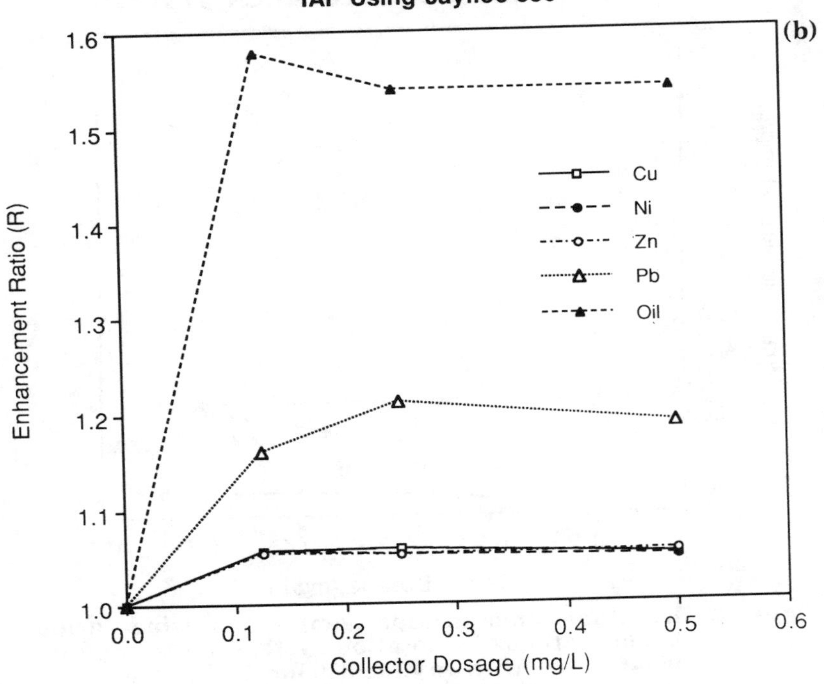

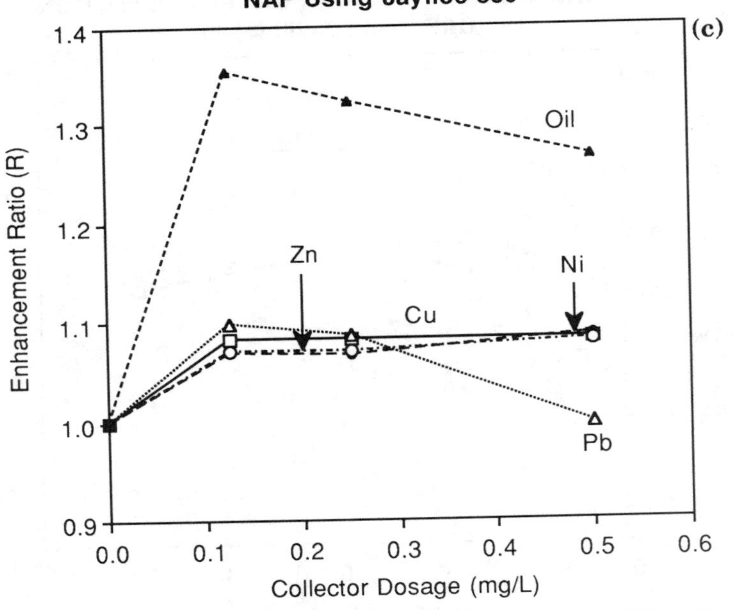

Figure 3. Enhancement Ratios for Precipitation/Air Flotation of the Synthetic Wastewater with Jayfloc-806, Using (b) Induced Air Flotation, and (c) Nozzle Air Flotation (Continued)

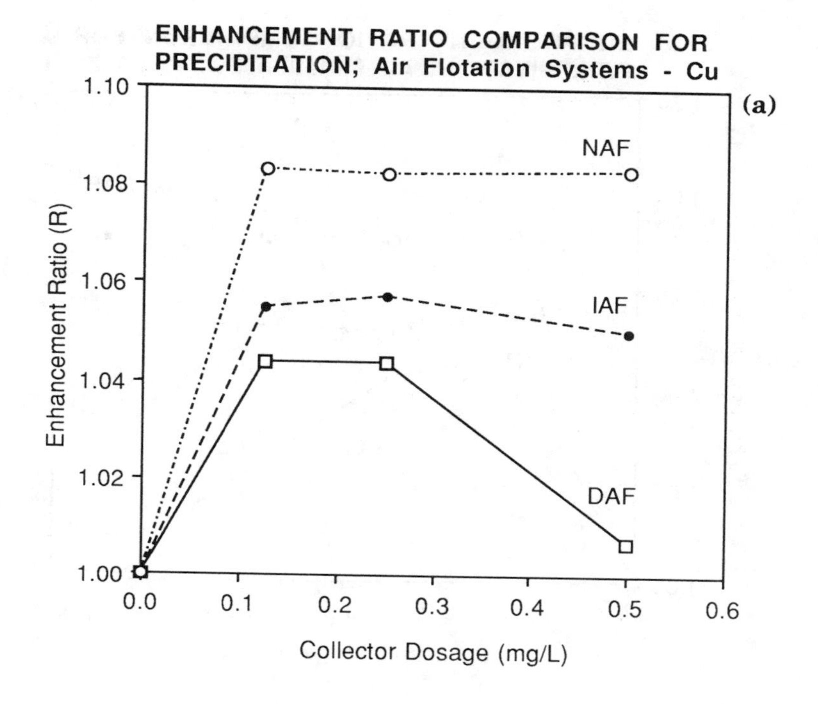

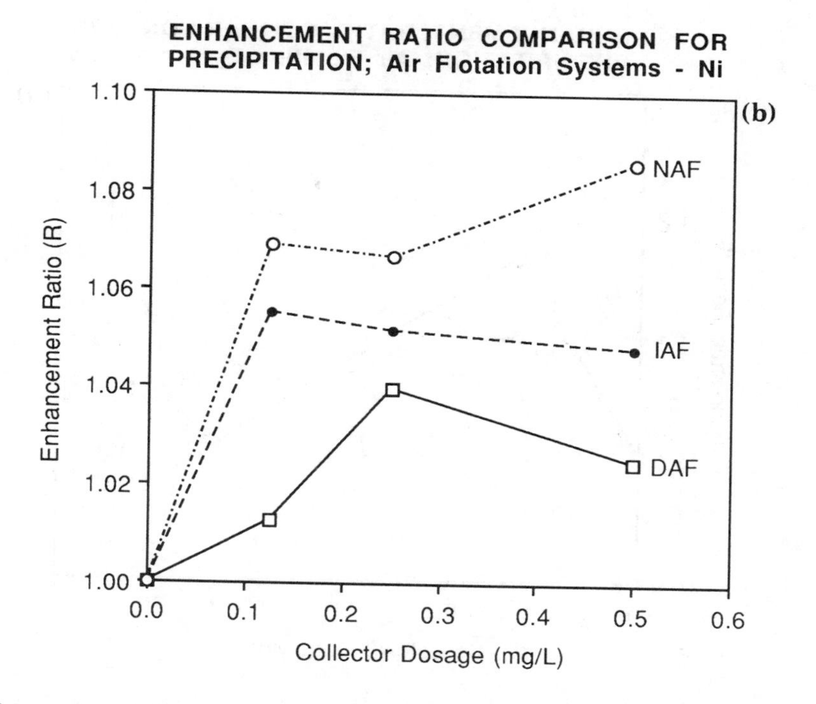

Figure 4. Enhancement Ratio Comparison of the Precipitation/Air Flotation Systems for the Synthetic Wastewater Using Jayfloc-806 for the Cases of (a) Cu and (b) Ni

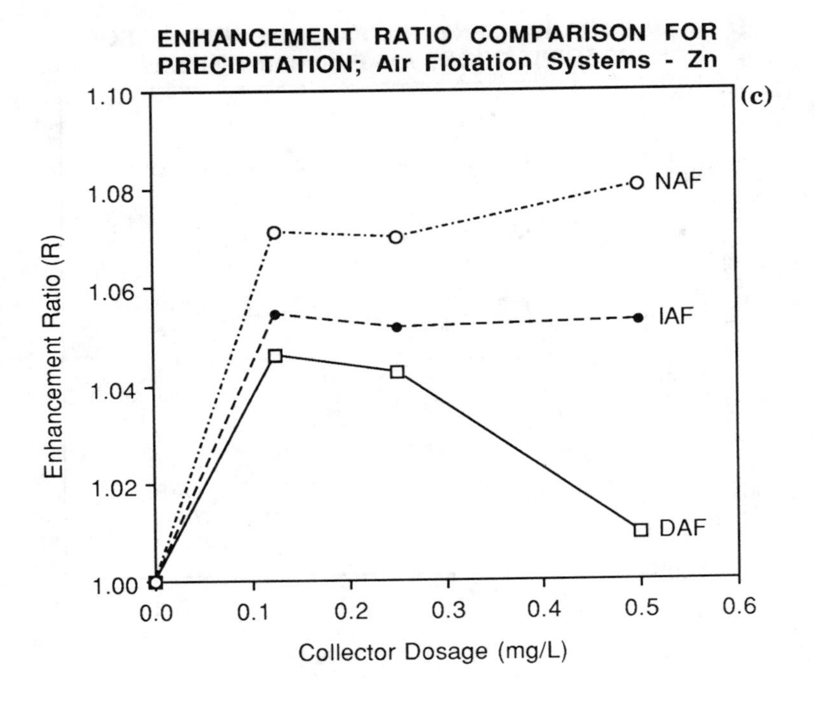

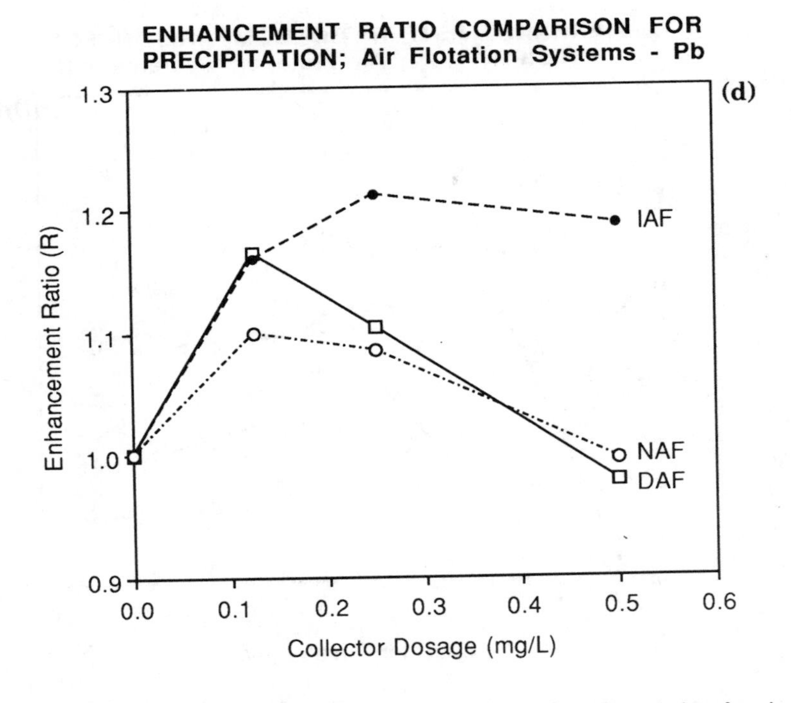

Figure 4. Enhancement Ratio Comparison of the Precipitation/ Air Flotation Systems for the Synthetic Wastewater Using Jayfloc-806 for the Cases of (c) Zn and (d) Pb (Continued)

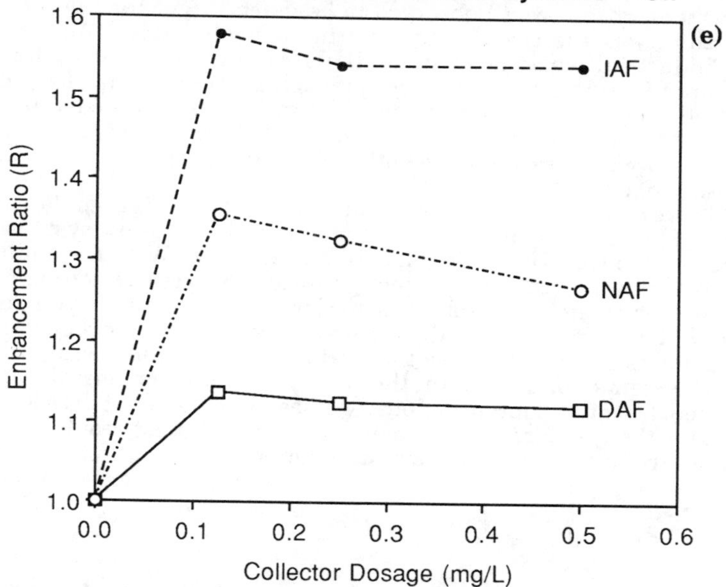

Figure 4. Enhancement Ratio Comparison of the Precipitation/ Air Flotation Systems for the Synthetic Wastewater Using Jayfloc-806 for the Case of (e) Oil (Continued)

INDUSTRIAL WASTEWATER

Tables 3 and 4 summarize the experimental results for the precipitation/air flotation experiments, using the Jayfloc-824 chemical collector. Table 3 lists the residual concentrations of Cu, Ni, Zn, Pb, and oil as functions of the collector dosage and air flotation system used; Table 2 lists the enhancement ratios for these five species as functions of the same two parameters.

The results are presented in graphic form in Figures 5 through 8. Figures 5a, 5b, and 5c show the residual metal concentrations for the three air flotation systems. Zinc had the lowest residual concentration and copper the highest with all three systems. The residual metal concentrations generally decreased with increasing collector (for the limited collector dosage range studied). The lowest residual metal concentrations were achieved for collector dosages in the range of 3 to 5 mg/L. Interestingly, the IAF system consistently yielded the lowest residual concentrations of all five species (Cu, Ni, Zn, Pb, and oil) at a collector dosage of 5.0 mg/L, corresponding to removal efficiencies of 72.8, 77.9, 96.4, 73.5, and 94.1 percent, respectively. The presence of the oil content and other contaminants resulted in the removal efficiencies being reduced from the case involving the synthetic wastewater.

The results from Figures 5a to 5c are cross-plotted in Figures 6a to 6e, showing the residual concentrations for all three air flotation systems presented for the case of Cu, Ni, Zn, Pb, and oil, respectively. For low collector dosages (less than 3 mg/L), the DAF system consistently yielded the lowest residual concentrations, whereas for higher dosages (about 5 mg/L), the IAF system gave the lowest residual concentrations, as previously described. Interestingly, nearly equal removal efficiencies for the cases of copper and lead were achieved with all three air flotation systems.

Figures 7a through 7c show the enhancement ratios for the three air flotation systems for all five species as functions of the applied collector dosage. The enhancement ratios are all observed to be much larger for the industrial wastewater than was the case for the synthetic wastewater. As described earlier, the industrial wastewater most likely contains other contaminants, which are more readily removed via precipitation/air flotation in the presence of chemical collectors. Typically, enhancement ratios are observed to be in the 2 to 14 range. The enhancement ratio is greatest for the case of lead, followed by oil. The enhancement ratios for the other three metals (Cu, Ni, and Zn) are all fairly close to one another.

Figures 8a through 8e represent cross-plots of Figures 7a through 7c, showing the enhancement ratios for the three air flotation systems compared for the cases of Cu, Ni, Zn, Pb, and oil, respectively. With the addition of the Jayfloc-824 collector, the IAF system produced the greatest enhancement over the results with no collector present for the cases involving metals, whereas the NAF system produced the greatest enhancement for removal of oil. Comparing Figures 6 and 8, because the initial concentrations using all three air flotation systems conducted in the absence of any chemical collector resulted in nearly equal concentrations, we see that use of the IAF system resulted in both the greatest enhancement ratios and lowest residual concentrations of the three air flotation systems considered.

Table 3. Residual Concentrations for Precipitation/Air Flotation of an Industrial Wastewater Using Jayfloc-824 Collector

Air Flotation System	Collector Dosage, mg/L	Residual Concentration, mg/L				
		Cu	Ni	Zn	Pb	Oil
DAF	0.00	13.35	13.12	7.83	13.15	177
	0.125	6.63	5.85	1.52	4.76	72
	0.25	7.53	5.13	1.30	5.48	15
	0.50	6.72	5.12	1.50	5.12	27
IAF	0.00	17.37	16.98	9.53	13.22	183
	0.125	7.50	10.87	4.06	5.60	137
	0.25	6.42	5.30	0.67	4.85	53
	0.50	5.21	4.33	0.41	3.73	12
NAF	0.00	17.10	16.33	9.50	12.98	190
	0.125	7.07	7.82	2.43	5.16	125
	0.25	6.74	5.28	0.98	4.99	67
	0.50	5.29	5.50	1.36	4.08	19

Initial concentrations: Cu_i = 19.15 mg/L, Ni_i = 19.63 mg/L, Zn_i = 11.30 mg/L, Pb_i = 14.05 mg/L, Oil_i = 203 mg/L.

Table 4. Enhancement Ratios for the Precipitation/Air Flotation of an Industrial Wastewater Using Jayfloc-824 Collector

Air Flotation System	Collector Dosage, mg/L	Enhancement Ratio				
		Cu	Ni	Zn	Pb	Oil
DAF	0.00	1.00	1.00	1.00	1.00	1.00
	0.125	2.196	2.117	2.818	10.32	5.038
	0.25	2.039	2.227	2.882	9.522	7.231
	0.50	2.181	2.229	2.824	9.922	6.769
IAF	0.00	1.00	1.00	1.00	1.00	1.00
	0.125	6.545	3.306	4.090	10.18	3.300
	0.25	7.152	5.408	6.006	11.08	7.500
	0.50	7.831	5.774	6.153	12.43	9.550
NAF	0.00	1.00	1.00	1.00	1.00	1.00
	0.125	5.893	3.579	4.928	8.308	6.000
	0.25	6.054	4.348	5.733	8.467	10.46
	0.50	6.761	4.282	5.522	9.318	14.15

Initial Concentrations: Cu_i = 19.15 mg/L, Ni_i = 19.63 mg/L, Zn_i = 11.30 mg/L, Pb_i = 14.05 mg/L, Oil_i = 203 mg/L.

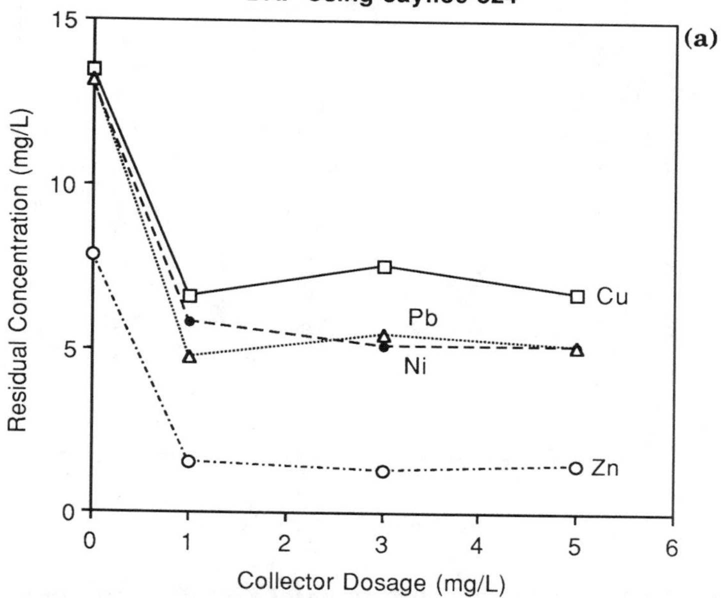

Figure 5. Residual Concentrations for Precipitation/Air Flotation of the Industrial Wastewater with Jayfloc-824, Using (a) Dissolved Air Flotation

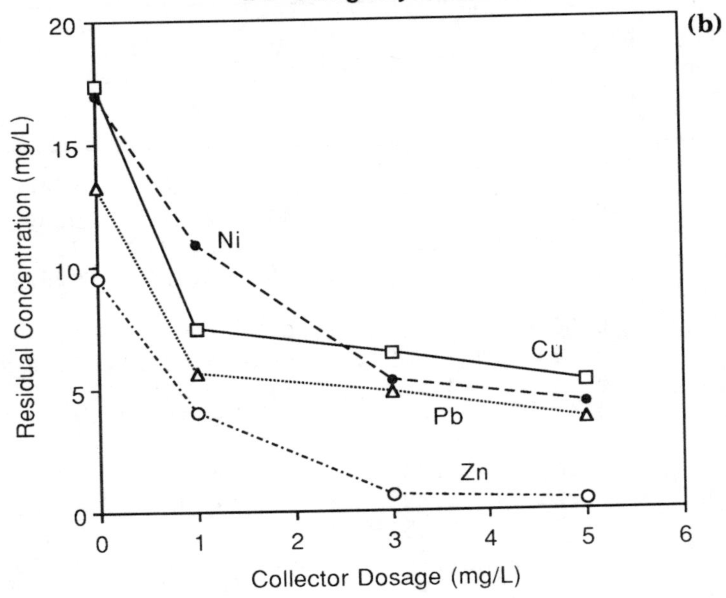

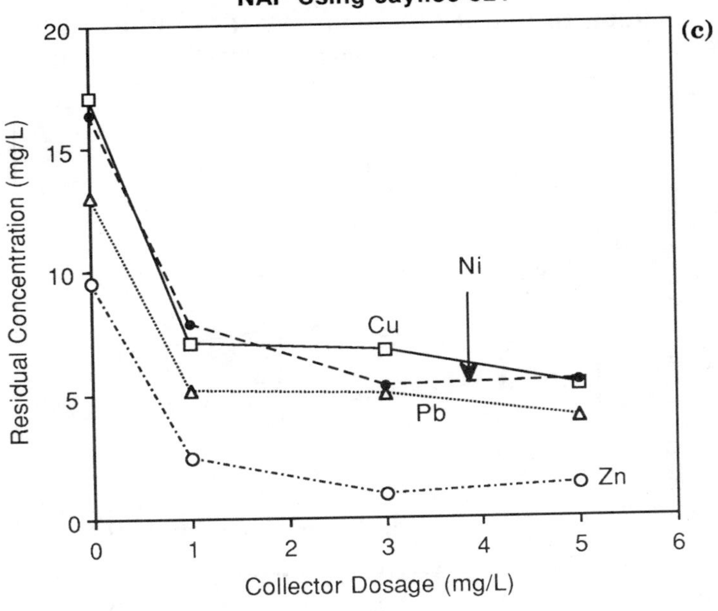

Figure 5. Residual Concentrations for Precipitation/Air Flotation of the Industrial Wastewater with Jayfloc-824, Using (b) Induced Air Flotation, and (c) Nozzle Air Flotation (Continued)

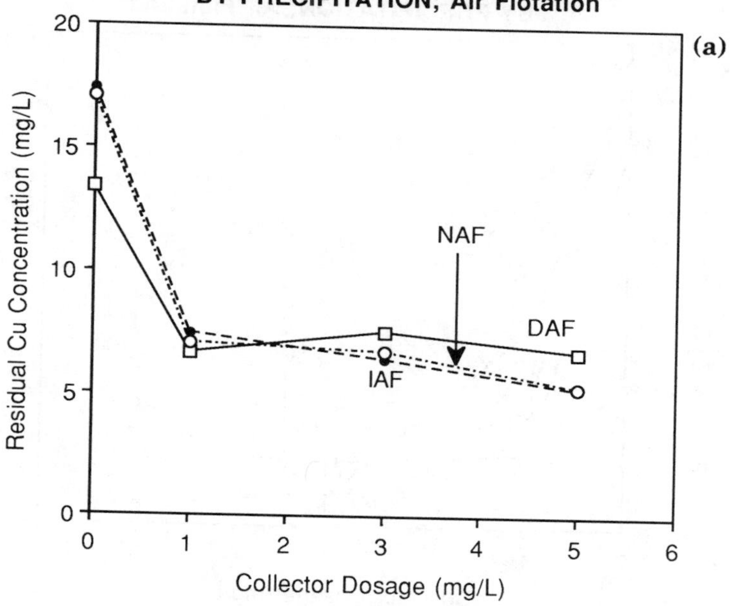

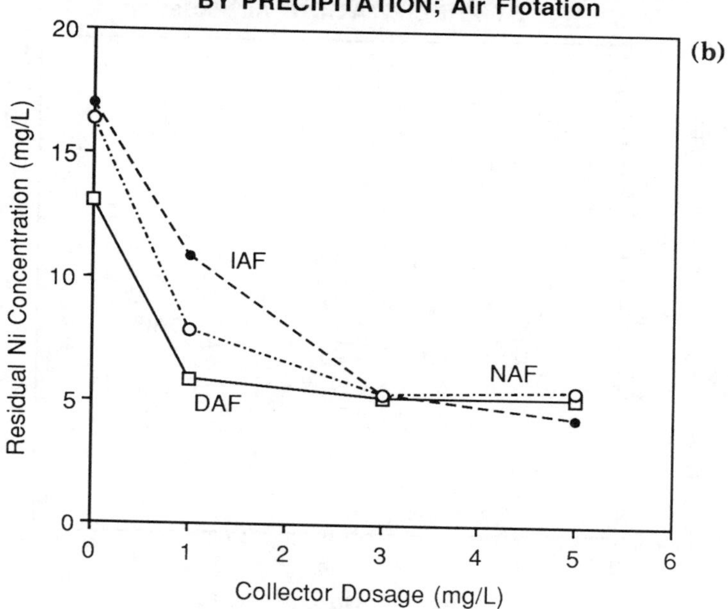

Figure 6. Residual Concentration Comparison Using Various Precipitation/Air Flotation Systems for the Industrial Wastewater with Jayfloc-824 for the Cases of (a) Cu and (b) Ni

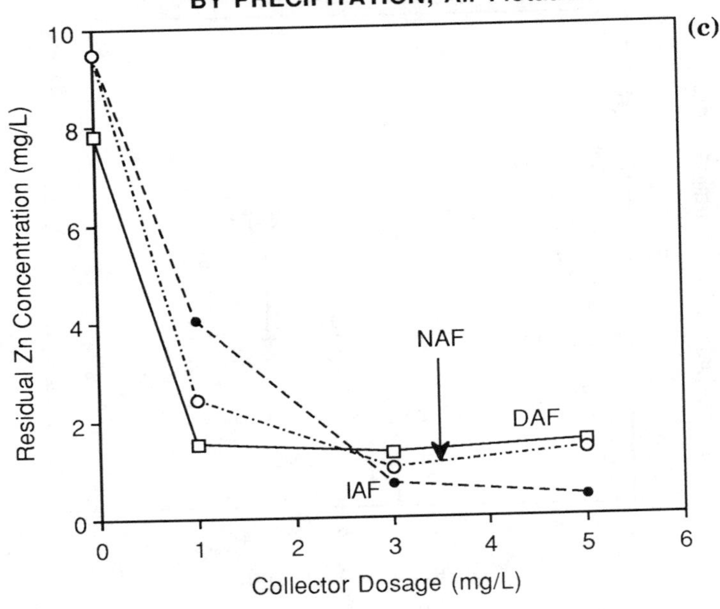

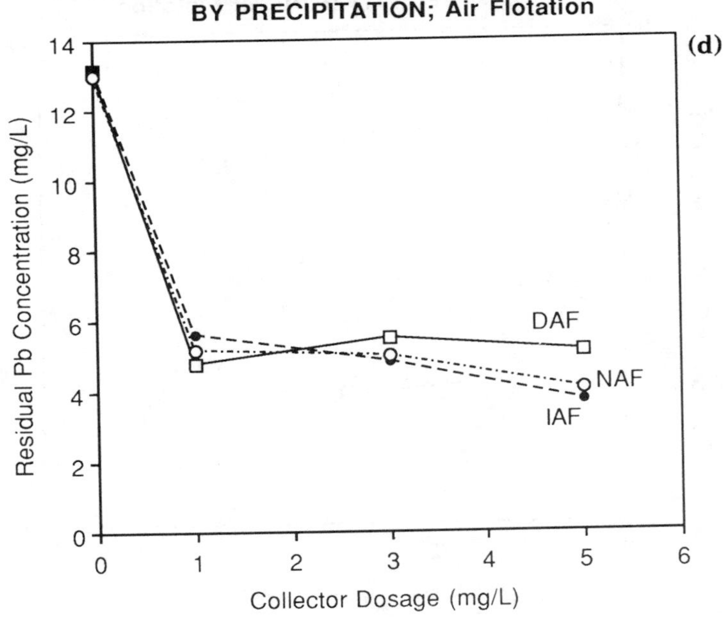

Figure 6. Residual Concentration Comparison Using Various Precipitation/Air Flotation Systems for the Industrial Wastewater with Jayfloc-824 for the Cases of (c) Zn and (d) Pb (Continued)

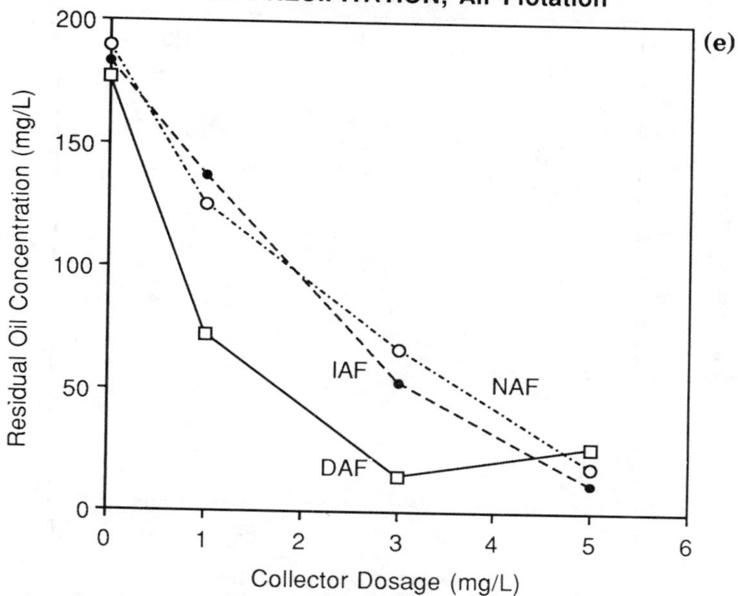

Figure 6. Residual Concentration Comparison Using Various Precipitation/Air Flotation Systems for the Industrial Wastewater with Jayfloc-824 for the Case of (e) Oil (Continued)

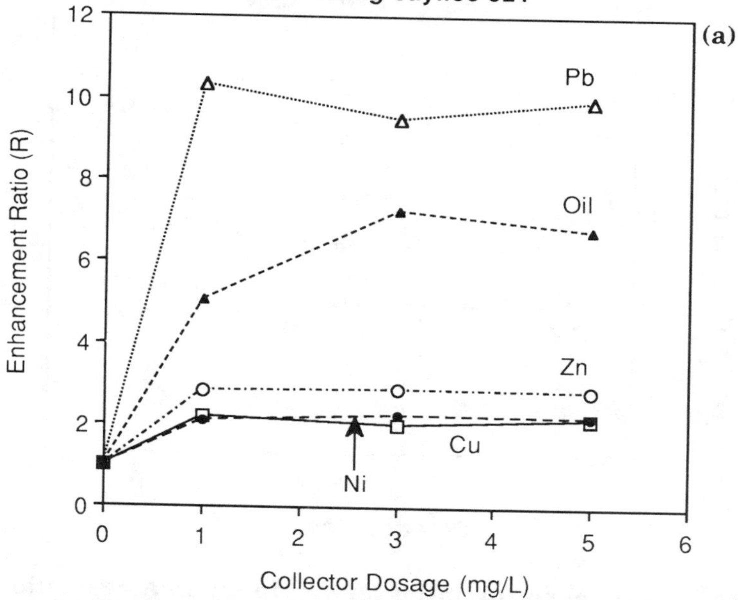

Figure 7. Enhancement Ratios for Precipitation/Air Flotation of the Industrial Wastewater with Jayfloc-824, Using (a) Dissolved Air Flotation

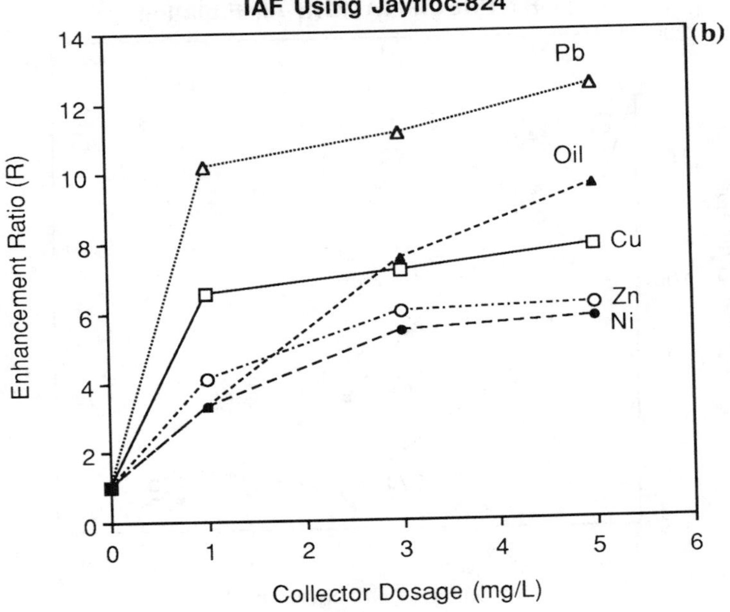

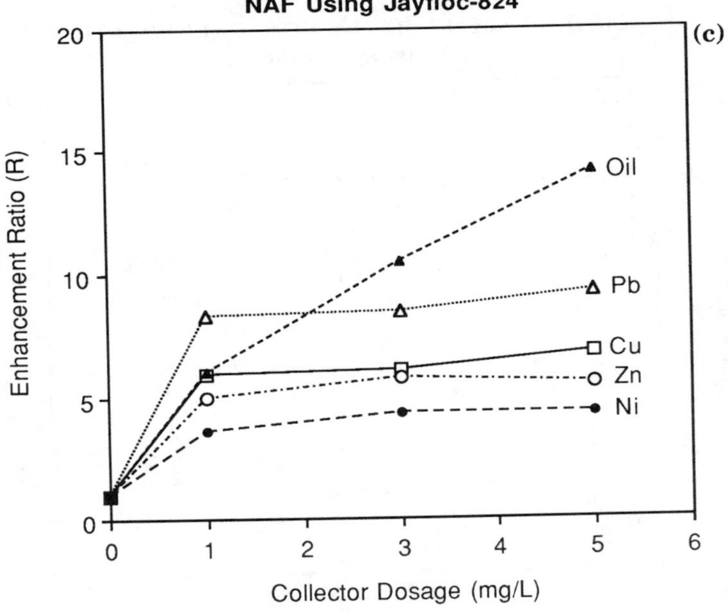

Figure 7. Enhancement Ratios for Precipitation/Air Flotation of the Industrial Wastewater with Jayfloc-824, Using (b) Induced Air Flotation, and (c) Nozzle Air Flotation (Continued)

COMPARISON OF ENHANCEMENT RATIO
BY PRECIPITATION; Air Flotation for Cu

(a)

IAF
NAF
DAF

Enhancement Ratio (R)

Collector Dosage (mg/L)

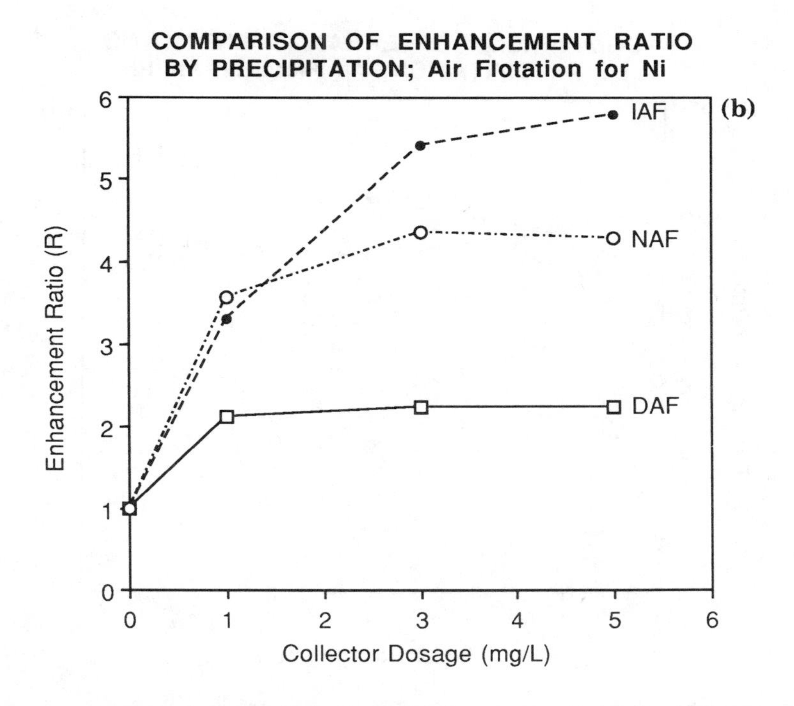

Figure 8. Enhancement Ratio Comparison of the Precipitation/Air Flotation Systems for the Industrial Wastewater Using Jayfloc-806 for the Cases of (a) Cu and (b) Ni

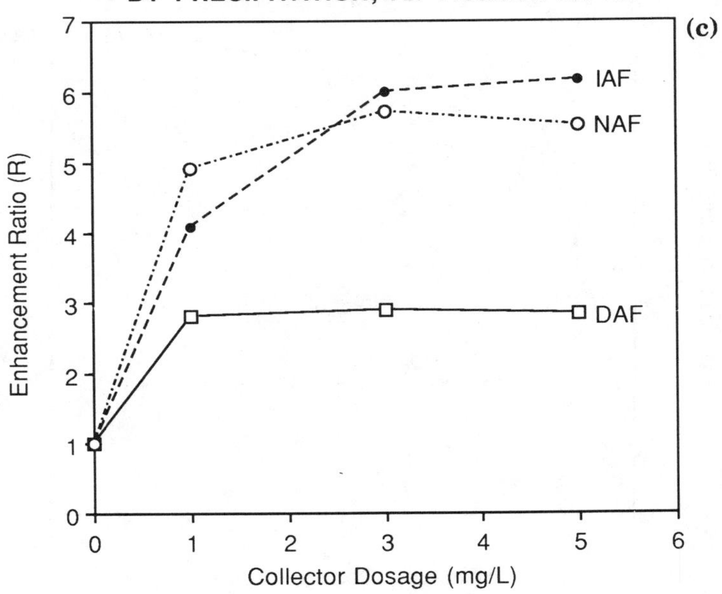

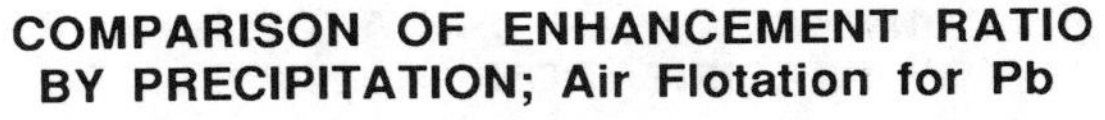

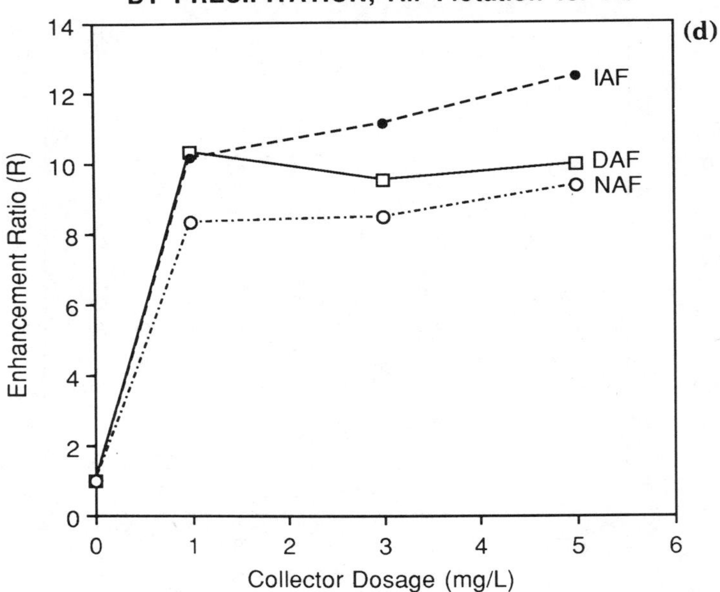

Figure 8. Enhancement Ratio Comparison of the Precipitation/Air Flotation Systems for the Industrial Wastewater Using Jayfloc-806 for the Cases of (c) Zn and (d) Pb (Continued)

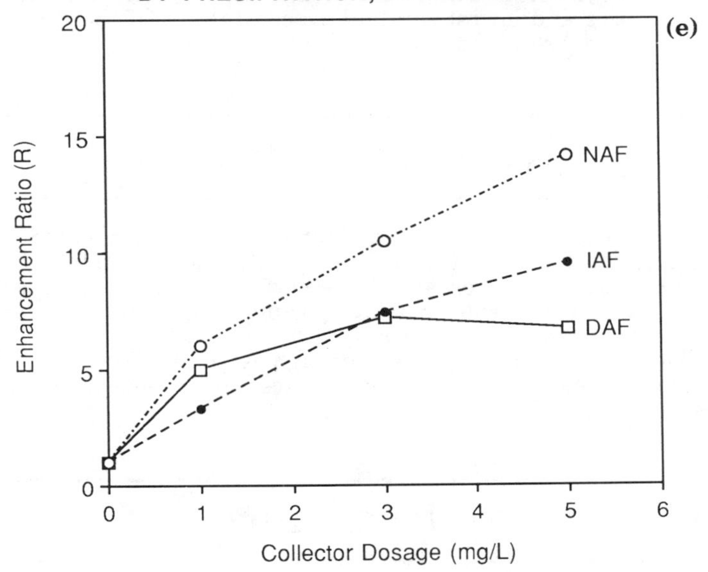

Figure 8. Enhancement Ratio Comparison of the Precipitation/Air Flotation Systems for the Industrial Wastewater Using Jayfloc-806 for the Case of (e) Oil (Continued)

SUMMARY AND CONCLUSIONS

The feasibility of using a combined precipitation/air flotation technique for treatment of oily metal-laden wastewaters has been demonstrated. The performance of this treatment has focused on the residual concentrations and enhancement ratios for three separate air flotation systems, all performed under comparable conditions.

For the synthetic wastewater, removals of the metals and oil can be in excess of 96 percent. The lowest residual metal concentration was achieved for lead. For the metal species, the DAF system produced the best removal, whereas the NAF system produced the poorest metal removal of the three air flotation systems. The removal of oil was in the order DAF > NAF > IAF. The enhancement ratios for the five species were all below a value of 1.6; the oil generally had the largest removal enhancement from the addition of the chemical collector. The NAF system had the largest enhancement ratios for Cu, Ni, and Zn, whereas the IAF system had the largest enhancement ratios for Pb and oil.

For the industrial wastewater, zinc had the lowest residual concentration, and copper the highest, with all three air flotation systems. The residual concentrations generally decreased with increasing collector dosages. The metal removal efficiencies were substantially lower for the industrial wastewater, probably from the presence of other contaminants in the wastewater; removal efficiencies were on the order of 75 percent. For low collector dosages, use of the DAF system consistently resulted in the lowest residual concentration. For higher dosages (of about 5 mg/L), use of the IAF system resulted in the lowest residual concentrations. All three air flotation systems had nearly equal removal efficiencies for the cases of copper and lead. The enhancement ratios for industrial wastewater were typically about 6, 4, 5, 10, and 11 for Cu,

Ni, Zn, Pb, and oil, respectively--much greater than the maximum value of 1.58 observed for the synthetic wastewater. The IAF system had the highest enhancement ratios and lowest residual concentrations of the three air flotation systems studied for the industrial wastewater.

ACKNOWLEDGMENTS

The authors wish to acknowledge the Department of Chemical Engineering at the University of Toledo and Argonne National Laboratory, whose support enabled this research to be performed. This paper will be presented at the Battelle International Symposium on Solid/Liquid Separations, held in Columbus, Ohio, on December 5-7, 1989.

This work was not funded through Argonne National Laboratory, and the viewpoints expressed here are not necessarily those of the Laboratory or its research sponsors.

NOMENCLATURE

API	American Petroleum Institute
C	Metal (or oil) concentration in solution, mg/L
C_i	Initial metal (or oil) concentration in solution, mg/L
C_R	Residual metal (or oil) concentration in solution, mg/L
Cu_i	Initial copper concentration, mg/L
DAF	Dissolved air flotation
IAF	Induced air flotation
NAF	Nozzle air flotation
Ni_i	Initial nickel concentration, mg/L
Oil_i	Initial oil concentration, mg/L
Pb_i	Initial lead concentration, mg/L
pH	- log [H+]
R	Enhancement ratio, defined as:

$$R = \frac{C_i - C_{with\ collector}}{C_i - C_{without\ collector}}$$

Zn_i	Initial zinc concentration, mg/L

REFERENCES

1. American Public Health Association, American Water Works Association, and Water Pollution Control Federation, *Standard Methods for the Examination of Water and Wastewater*, 15th Edition, 1981, American Public Health Association, Washington, DC.

2. Bennett, G. F., *Crit. Rev. in Environ. Control*, 1988, *18* (3), 189-253.

3. Bennett, G. F. and Peters, R. W., submitted for publication in *Environ. Prog*.

4. Burren, D. C., *Atomic Spectrometric Analysis of Heavy-Metal Pollutants in Water*, Ann Arbor Science, Ann Arbor, MI, 1974.

5. Ching, L. A., "A Study of the Simultaneous Removal of Heavy Metals and Oil and Grease by Dissolved Air Flotation," M.S. Thesis, University of Toledo, Toledo, OH, 1981.

6. Churchill, R. J. and Tacchi, K. J., *AIChE Sympos. Ser.*, 1978, *74* (178), 280-299.

7. Dean, J. G., Bosqui, F. L., and Lanouette, K. H., *Environ. Sci. Technol.*, 1972, *6* (6), 518-522.

8. Gopalratnam, V. C., "A Study of the Simultaneous Removal of Oil and Heavy Metals from Industrial Wastewater by Air Flotation," M.S. Thesis, University of Toledo, Toledo, OH, 1986.

9. Gopalratnam, V. C., Bennett, G. F., and Peters, R. W., *Environ. Prog.*, 1988, *7* (2), 84-92.

10. Gopalratnam, V. C., Bennett, G. F., and Peters, R. W., *Proc. 9th AESF/EPA Conference Pollut. Control for the Metal Finishing Industry*, 1988, Orlando, FL, Paper No. P.

11. Lowe, W., *Water Pollut. Control*, 1970, *69*, 270-280.

12. Mukai, S., Wakamatsu, T., and Nakahiro, Y., in *Recent Developments in Separation Science, Volume V*, J. S. Dranoff, J. C. Schultz, and P. Somasundaran, eds., CRC, West Palm Beach, FL, 1979, 67-80.

13. Patterson, J. W., *Wastewater Treatment Technology*, Ann Arbor Science, Ann Arbor, MI, 1975.

14. Patterson, J. W. and Minear, R. A., in *Heavy Metals in the Aquatic Environment*, P. A. Krenkel, ed., Pergamon, Oxford, England, 1975, 261-276.

15. Peters, R. W. and Bennett, G. F., in *Technical Conference Proceedings*, Session L, Environmental I, AESF SUR/FIN '89, 1989, Cleveland, OH, Paper No. L-3.

16. Ubasineke, C. N., "Study of Simultaneous Heavy Metals and Oil Removal by Precipitation and Induced Air Flotation," M.S. Thesis, University of Toledo, Toledo, OH, 1985.

17. U.S. Environmental Protection Agency, *Summary Report: Control and Treatment Technology for the Metal Plating Industry; Sulfide Precipitation*, EPA-625/8-80-003, April 1980.

18. Westra, M. A. and Rose, B. L., in *Technical Conference Proceedings*, Session L, Environmental I, AESF SUR/FIN '89, 1989, Cleveland, OH, Paper No. L-4.

SOLID/LIQUID SEPARATION IN WASTE ACID NEUTRALIZATION: A CASE STUDY

Raja Venkateswar
Principal Chemical Engineer
Ebasco Services Incorporated
111 North Canal Street, Suite 915
Chicago, Illinois 60606-7204
(312) 876-0262

ABSTRACT

Acid neutralization is a part of the treatment process for the management and disposal of waste acids. The amount and the characteristics of the solids generated and their subsequent separation from the liquid phase vary with the type of alkali used for neutralization. This could have a significant impact on ultimate disposal costs. This paper compares the performance of commonly used alkalis for neutralizing a typical waste sulfuric acid. Results from bench-scale experiments are included.

INTRODUCTION

Acid neutralization is frequently a part of the process for treating a waste acid prior to disposal. An alkali is used in the neutralization process, which generates a slurry. The solids from this slurry are separated using conventional dewatering methods. The amount and characteristics of the generated solids, as well as their amenability to dewatering depends upon the type of alkali used for neutralization.

In treating waste acids, the alkali cost comprises only a portion of the total treatment cost. The least expensive alkali is not necessarily the best alkali for neutralization. In many cases, the sludge disposal cost exceeds the alkali cost. This paper presents a case study for neutralizing a typical waste sulfuric acid, compares alkali performance, and discusses the economics of waste acid neutralization.

PREVIOUS WORK

Chung *et al.*[5] performed a field investigation of full-scale processes used at commercial hazardous waste treatment facilities to determine if adequate technologies exist for treating corrosive and metal-bearing wastes. They reported that waste acid neutralization is commonly used in manufacturing and process industries to ensure that waste discharges conform to applicable regulations. Typically, an alkali is added to adjust the pH to 9.5 to 10.5 and to precipitate dissolved metals. A variety of alkalis are used for this purpose. These alkalis perform differently and can significantly affect treatment costs.

Brown[3] discussed different reagents, residence times, and other aspects of industrial waste neutralization. He suggested that sodium hydroxide is the best neutralizing agent for pH control because its compounds are soluble and do not interfere with on-line pH electrodes. Although expensive, sodium hydroxide is extensively used in small effluent treatment plants.

Gray[6] presented a discussion of solutions to pH control problems and concluded that with the proper process control equipment and instrumentation, lime could be the most economical reagent to neutralize large quantities of waste acids.

Brantner and Cichon[2] experimentally compared alternative precipitation processes for heavy metals removal. They conducted a pilot-scale study to evaluate the best available technology (BAT) for removing heavy metals from industrial waste waters. The technologies considered were sulfide, carbonate, and hydroxide precipitation. The metals of concern in their study included cadmium, chromium, copper, lead, and zinc. In general, all of the precipitation technologies produced comparable results for metals removal. However, hydroxide precipitation using lime slurry was the most reliable to operate.

Patterson *et al.*[7] investigated the solubility of metal hydroxides and carbonates over a pH range of 6 to 13. They concluded that carbonate precipitates had better dewatering characteristics.

Bowers *et al.*[1] stated that lime was the most widely used reagent to remove heavy metals from waste waters. They examined the performance of lime as a neutralization agent for various acids and identified information needed to predict treatability theoretically.

Terringo[8,9] examined the performance of magnesium hydroxide as a neutralizing agent for waste acids, and presented field data to show that the resulting sludge had better dewatering characteristics than the sludge generated from using sodium hydroxide. He also presented some cost information to compare treatment costs using different alkalis.

Cappacio and Sarnelli[4] identified some of the factors affecting the cost of waste acid neutralization.

The present work was undertaken to develop data useful for plant design and operation and to dispel some common misconceptions about waste acid neutralization. For example, a waste acid typically requires more alkali, longer reaction time, and produces more sludge than a reagent acid of equivalent strength. Bench-scale experiments were performed with different concentrations of a waste sulfuric acid from pickling operations. The acid was neutralized with different alkalis. In selected cases, reagent sulfuric acid of equivalent strength was also neutralized for comparison purposes.

EXPERIMENTAL PROCEDURE

The initial concentration of the waste acid was 12 percent (measured as sulfuric acid). A portion of this was diluted to 2.5 percent and 5 percent acid. A known volume (500 ml) of the waste acid was treated with the following alkali to a pH of 9.5 to 10.5:

(1) 50 percent NaOH solution
(2) 50 percent $Mg(OH)_2$ slurry
(3) Solid $Ca(OH)_2$
(4) 15 percent $Ca(OH)_2$ slurry
(5) Solid CaO.

In most cases, neutralization to pH 9.5 to 10.5 was achieved within 30 to 60 minutes. In the case of neutralization with $Mg(OH)_2$, a pH greater than 6.8 could not be achieved within a reasonable time period without adding a 50 percent NaOH solution. This is considerably different from Terringo's observation[9] that a 5 percent sulfuric acid solution reaches a pH of 7 in approximately one minute when $Mg(OH)_2$ is used.

The resulting slurry was tested for filterability, the amount of solids generated, and settling.

The filtration test consisted of filtering 100 ml of the neutralized material though a Buchner funnel using ashless Whatman #41 filter paper as the filtering medium. No filter aids were added. The time that elapsed between the start and the appearance of the first crack on the filter cake was noted, and is called filtration time. All filtrations were performed at 480 mm Hg because most vacuum disk filters used in acid neutralization systems operate at 380 to 500 mm Hg. The filtration time obtained under these conditions is not an absolute number, but can be used for determining relative filterability when using different alkalis to neutralize a pickle liquor. In this case the filtrate obtained was analyzed for dissolved metals.

The total solids in the neutralized material was determined by evaporating a weighed sample of material in a drying oven and weighing the dried sample after cooling. Similarly, the dissolved solids in the filtrate were determined by evaporating a known weight of the filtrate. To obtain a random sample for suspended solids, the filter cake was divided into four quarters and any two opposite quadrants taken as samples for drying. All samples were dried overnight at 105 C. Duplicate samples were also dried under the same conditions.

The neutralized material from each acid/alkali pair was poured into a separate 1-L clear plastic cone. No polyelectrolytes or flocculants were added. In each case, the volume of suspended solids was noted at known time intervals. The volume percent clear liquid obtained at each instance was calculated.

DISCUSSION OF RESULTS

Table 1 characterizes the untreated acid samples. Table 2 characterizes the alkalis used in the tests.

Table 1. Characterization of Acid Samples

Type of Acid	Acidity as Sulfuric	Specific Gravity at 70 degrees F	Physical Description
Waste	12.1%	1.288	Very dark green solution with some yellow–brown suspended soils
Waste	5.1%	1.084	Dark green solution
Waste	2.4%	1.029	Light green solution
Reagent	12.0%	1.078	Clear colorless solution
Reagent	5.1%	1.032	Clear colorless solution

Table 2. Source of Alkalis Used for Bench-Scale Tests

Alkali	Source	Form	Type	Concentration
NaOH	RICCA Chemical Company Arlington, Texas	Liquid	Purified Grade	50%
Mg(OH)2	Dow Chemical Company, Midland, Michigan	Slurry	Technical Grade	50–55%
Ca(OH)2	J. T. Baker Chemical Co. Phillipsburg, New Jersey	Solid	Food Grade	95–100%
Ca(OH)2	Prepared in–house by adding de–ionized water to the food grade powder	Slurry		15%
CaO	MCB Manufacturing Chemist Inc., Cincinnati, Ohio	Solid	Technical	95–100%

FILTERABILITY

Table 3 compares the filterability of the resulting material after acid neutralization with various alkalis. The results show that the neutralized waste filters best when solid CaO or $Ca(OH)_2$ is used. Waste acid neutralized with NaOH filters poorly. Filtration is fairly rapid with $Mg(OH)_2$, but in the case of the 5.1 percent acid, filtration was slow because a large amount of NaOH had been used along with $Mg(OH)_2$ to achieve a pH of 9.5 to 10.5. For the 2.4 percent acid waste, the filtration times are similar with solid CaO or $Ca(OH)_2$ slurry, but not with $Ca(OH)_2$ slurry. The reason for this is not clear, and identical results were obtained even upon duplicating the experiment. The 12.1 percent waste acid could not be neutralized beyond a pH of 5.9 to 6.7 because the neutralized material became a semisolid and could not be stirred effectively. However, the 12.1 percent reagent acid could be neutralized to a pH of 9.5 to 10.5 because it did not contain any dissolved metals.

The filtrate analyses in Table 4 show that the choice of alkali does not affect the quality of the filtrate after dewatering. As expected, the solids in

Table 3. Filterability of the Resulting Material After Neutralization with Various Alkalis

Acidity Percent	Type Acid	Alkali Used	Alkali Form	Filtration Time, Sec	g Cake Per 100 g Slurry	Cake Moisture Percent
2.4	Waste	CaO	Solid	10	19.5	71.5
2.4	Waste	Ca(OH)2	Solid	12	17.0	71.4
2.4	Waste	Mg(OH)2	50% Slurry	20	16.6	75.2
2.4	Waste	Ca(OH)2	15% Slurry	40	15.5	69.9
2.4	Waste	NaOH	50% Solution	90	16.3	84.3
5.1	Waste	CaO	Solid	10	47.8	68.9
5.1	Waste	Ca(OH)2	Solid	10	34.2	70.6
5.1	Waste	Ca(OH)2	15% Slurry	15	31.9	75.5
5.1	Waste	Mg(OH)2	50% Slurry	80	35.2	68.7
5.1	Waste	NaOH	50% Solution	160	32.4	79.3
12.1	Waste	Ca(OH)2	Solid	90	55.1	34.4*
12.1	Waste	NaOH	50% Solution	>600	49.9	65.1**
12.0	Reagent	Ca(OH)2	Solid	5	52.7	68.3

* At pH 5.9
** At pH 6.7

Table 4. Filtrate Analysis from Neutralizing Waste Acid

	Metals Concentration, mg/l					
Dissolved Metals	2.4% Waste Acid	50% NaOH Solution	50% Mg(OH)2 Slurry	CaO Solid	Ca(OH)2 Solid	15% Ca(OH)2 Slurry
Cd	0.6	0.1	0.2	0.2	0.2	0.1
Cr	11.0	0.9	0.6	1.1	1.0	0.9
Fe	8846	11.3	9.9	9.5	8.8	10.6
Pb	12.4	2.2	1.3	2.8	2.9	2.4
Ni	5.6	0.5	0.8	0.6	0.7	0.5
Zn	5625	4.0	2.7	4.8	2.2	1.9
Solids, %	4.9	3.5	3.4	0.8	0.7	0.8
pH	1.2	10.1	9.2	10.5	9.5	9.5

	Metals Concentration, mg/l					
Dissolved Metals	5.1% Waste Acid	50% NaOH Solution	50% Mg(OH)2 Slurry	CaO Solid	Ca(OH)2 Solid	15% Ca(OH)2 Slurry
Cd	1.2	0.2	0.3	0.1	0.1	0.2
Cr	22.4	0.8	1.1	1.0	0.3	1.1
Fe	18,400	13.2	11.0	7.3	12.6	13.6
Pb	25.9	2.6	3.2	2.5	0.7	2.7
Ni	11.6	1.0	1.1	0.6	0.5	0.6
Zn	11,700	8.0	4.7	7.2	3.3	4.2
Solids, %	10.3	7.9	7.9	1.1	1.2	1.0
pH	1.0	9.4	9.0	10.5	10.1	9.4

the filtrate obtained from neutralization with limes are lower than those from using 50 percent NaOH. In the case of $Mg(OH)_2$, the higher solids content is due to the solubility of some of the magnesium salts and to the large amount of 50 percent NaOH added to raise the pH beyond 7.0.

SOLIDS GENERATED

Table 5 compares the solids generated from neutralizing different concentrations of waste acid with various alkalis. For example, the table also includes data for neutralizing reagent sulfuric acid. As expected, no solids precipitated upon neutralizing 12 percent reagent acid with 50 percent NaOH solution.

SETTLING CHARACTERISTICS

Table 6 compares the average settling rate (in terms of volume percent clarified liquid) of the generated solids on neutralizing different concentrations of the waste acid with various alkalis. The results show that solids generated through lime neutralization settle faster than those generated from using 50 percent NaOH or 50 percent $Mg(OH)_2$.

Solids produced from using $Mg(OH)_2$ settle slowly, presumably because the addition of 50 percent NaOH made the resulting neutralized material gelatinous and altered its settling characteristics. A greater volume of clarified liquid could be expected from using 15 percent $Ca(OH)_2$ slurry than using solid $Ca(OH)_2$, because the slurry dilutes the waste. However, this has

Table 5. Solids Generated from Neutralization with Various Alkalis

Alkali Used	Alkali Form	*2.4% Waste Acid* Suspended Solids Percent	Dissolved Solids Percent	Total Solids Percent	*5.1% Waste Acid* Suspended Solids Percent	Dissolved Solids Percent	Total Solids Percent
50% NaOH	Solution	2.6	2.9	5.4	6.8	5.3	12.2
50% $Mg(OH)_2$	Slurry	4.1	2.9	7.0	11.0	5.1	16.0
$Ca(OH)_2$	Solid	4.9	0.6	5.5	13.6	0.6	14.3
15% $Ca(OH)_2$	Slurry	5.6	0.5	5.1	10.3	0.5	10.7
CaO	Solid	5.6	0.6	6.2	14.8	0.5	15.2

Alkali Used	Alkali Form	*12% Waste Acid* Suspended Solids Percent	Dissolved Solids Percent	Total Solids Percent	*12% Reagent Acid* Suspended Solids Percent	Dissolved Solids Percent	Total Solids Percent
50% NaOH	Solution	17.7*	13.7	31.5	0	14.7	14.7
50% $Mg(OH)_2$	Slurry	--	--	--	--	--	--
$Ca(OH)_2$	Solid	34.4**	1.1	35.6	16.6	0.2	16.8
15% $Ca(OH)_2$	Slurry	--	--	--	--	--	--
CaO	Solid	--	--	--	--	--	--

* At pH 5.9
** At pH 6.7

not been the case with either concentration of the waste acids studied. The volume of clarified liquid produced after 1 hour of settling suggests that a clarifier is not suitable for dewatering this sludge.

As expected, no suspended solids are generated when 5.1 percent reagent sulfuric acid is neutralized with 50 percent NaOH solution. However, a considerable amount of suspended solids is generated when a waste acid of equivalent concentration is neutralized to the same pH. Although the sludge generated from neutralizing reagent sulfuric acid with lime settles quickly, the same is not true for the sludge produced from neutralizing waste acid of equivalent strength. This shows how the presence of other constituents, besides acid and water, can affect the settling characteristics of the generated solids.

Table 6. Comparison of the Average Settling Rate of Solids in the Neutralized Waste After 1 Hour

Acidity Percent	Type Acid	Alkali Used	Alkali Form	Vol. % Clarified
2.4	Waste	Mg(OH)2	50% Slurry	3.9
2.4	Waste	NaOH	50% Solution	13.7
2.4	Waste	CaO	Solid	47.7
2.4	Waste	Ca(OH)2	15% Slurry	58.3
2.4	Waste	Ca(OH)2	Solid	66.0
5.1	Waste	NaOH	50% Solution	0.7
5.1	Waste	Mg(OH)2	50% Slurry	2.6
5.1	Waste	CaO	Solid	5.4
5.1	Waste	Ca(OH)2	15% Slurry	7.6
5.1	Waste	Ca(OH)2	Solid	18.0
5.1	Reagent	CaO	Solid	23.6
5.1	Reagent	Ca(OH)2	Solid	38.9
5.1	Reagent	NaOH	50% Solution	100.0

THE ECONOMICS OF ACID NEUTRALIZATION

Figure 1 shows the block diagram for a typical acid neutralization system used in a hazardous waste treatment, storage, and disposal (TSD) facility. The total treatment cost consists of the following:

- Delivered alkali cost
- Filter cake disposal cost
- Filtrate disposal cost
- Capital equipment cost
- Operating cost.

Each of these cost elements will vary from one location to another. Table 7 compares the relative treatment costs for the different alkalis used in this study.

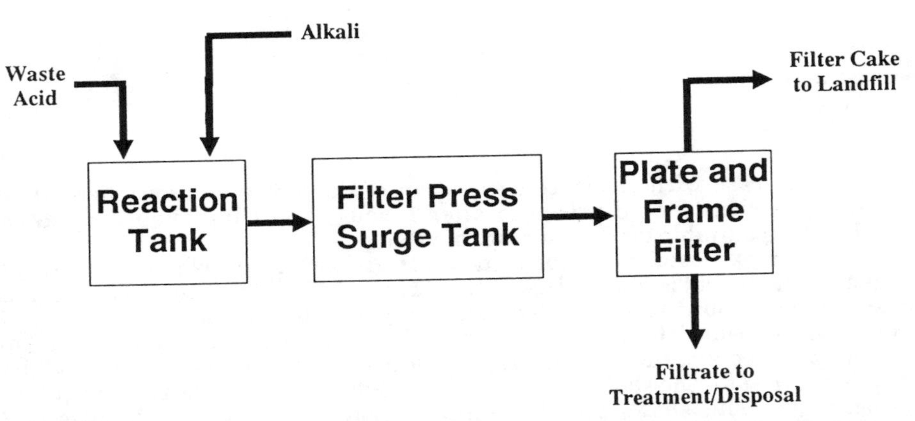

Figure 1. Block Diagram of a Typical Acid Neutralization System

Table 7. **An Example of Relative Treatment Cost for Neutralizing a Waste Acid with Various Alkalis**

BASIS: 30,000 gallons of 5% waste sulfuric acid treated to pH 9.5 – 10.5

Alkali Used	Alkali Form	Delivered* Cost/Ton as Received	Alkali Cost	Filter Cake Volume (yd3)	Filtrate Volume (Gallons)	Capital Equipment
$Ca(OH)_2$	Solid	$ 82	$ 599	48.5	21,524	Base Case
CaO	Solid	$ 91	$ 619	69.4	16,628	Same
NaOH	50% Solution	$115	$1,437	49.4	22,030	Increases Slightly
$Mg(OH)_2$	58% Slurry	$157	$2,808	56.2	22,209	Increases
$Ca(OH)_2$	15% Slurry	$ 70	$3,136	60.0	27,342	Increases

* Delivered to Chicago, IL.

The delivered costs were obtained through local vendor quotations. The alkali cost for 15 percent $Ca(OH)_2$ is the highest because of the additional cost for transporting water.

The filter cake and filtrate disposal costs are not included in Table 7 because they will vary depending upon the treatment facility. Furthermore, the volume of solids generated will vary depending upon the concentration of dissolved metals in the untreated waste acid. Therefore, the filtrate and cake disposal volumes presented in Table 7 can be used only as an example to estimate these costs for a given facility. Since sodium salts are generally water soluble, a lower volume of filtered solids would be expected for disposal (as compared with $Ca(OH)_2$) when waste acid is neutralized with 50 percent NaOH solution. Experimental results show that this is not always true. In the case shown in Table 7, the cake resulting from 50 percent NaOH addition had a higher moisture content and a lower density than the cake from $Ca(OH)_2$.

Table 7 presents a qualitative comparison of capital costs in each case, considering a system using $Ca(OH)_2$ as the base case. Since filtration is considerably slower with 50 percent NaOH, additional filters may be required to maintain the same throughput. At the same time, a smaller reaction tank may be needed because NaOH reacts faster than other alkalis. A system using $Mg(OH)_2$ requires an additional alkali feed system to raise the pH above 7.0. In addition, a larger reaction tank may be needed to provide longer retention time for the slow reaction. A system using 15 percent $Ca(OH)_2$ slurry would require a larger reaction tank to accommodate the volume increase from neutralization. A greater surface area may also be needed to filter the solids.

CONCLUSIONS

These results clearly show that a waste acid behaves differently from a reagent acid of equal strength. The maximum concentration of waste acid that can be neutralized with an alkali is primarily limited by the concentration of dissolved metals in the untreated waste. In this case, the acid concentration in the waste as received was 12 percent (measured as sulfuric acid), and contained a total of about 10 percent dissolved metals. The maximum concentration that could be neutralized practically was 5 to 7 percent. At higher acid concentrations, the neutralized material became a paste.

Gravity settling without the addition of flocculants or other settling aids is not suitable for dewatering the sludge produced. Filtration is an applicable dewatering technique, but the moisture content in the cake produced varies with the type of alkali used. However, the quality of the filtrate in terms of dissolved metals is not affected by the choice of alkali.

Alkali selection for acid neutralization should be based on several factors such as the volume and filterability of solids generated, reaction time, and delivered alkali cost. Bench-scale experiments should be performed to evaluate waste treatability, alkali requirements, alkali performance, and disposal options for the dewatered solids and the liquid before designing a waste acid neutralization system.

ACKNOWLEDGMENT

The author wishes to thank Chemical Waste Management Inc., Geneva Research Center, Geneva, Illinois, for providing the financial and technical support in performing this work.

REFERENCES

1. Bowers, A. R., Ching, G., and Huang, C. P., *Predicting the Performance of a Lime-Neutralization/Precipitation Process for the Treatment of Some Heavy Metal-Laden Industrial Wastewaters*, 1981, 51-62, Industrial waste, Proceedings of the Thirteenth Mid-Atlantic Conference sponsored by the Department of Civil Engineering, University of Delaware, Newark, Delaware, June 29-30, 1981.

2. Brantner, K. A. and Cichon, E. J., *Heavy Metals Removal: Comparison of Alternative Precipitation Processes*, 1981, 43-50, Industrial waste, Proceedings of the Thirteenth Mid-Atlantic Conference sponsored by the Department of Civil Engineering, University of Delaware, Newark, Delaware, June 29-30, 1981.

3. Brown, M., *Process Engineering*, 1983, *64*, 53-55.

4. Capaccio, R. S. and Srnelli, R. J., *Plating and Surface Finishing*, 1986, *73* (9), 18-19.

5. Chung, N. K., Crawford, M. A., and Shively, W. E., *Field Evaluation of Treatment Process for Corrosives and Metal Bearing Wastes*, 1986, 329-337, U.S. Environmental Protection Agency, Proceedings of the Twelfth Annual Research symposium entitled "Land Disposal, Remedial Action, Incineration and Treatment of Hazardous Waste," Cincinnati, Ohio, April 21-23, 1986; EPA/600/9-022.

6. Gray, D. M., *Pollution Engineering*, 1984, *16*, 54-56.

7. Patterson, J. W., Scala, J. J., and Herbert, A. E., *Heavy Metal Treatment by Carbonate Precipitation*, 1975, 132-150, Proceedings of the 30th Industrial Waste Conference sponsored by Purdue University, Lafayette, Indiana, May 6-8, 1975.

8. Terringo III, J., *Plating and Surface Finishing*, 1986, *73* (10), 36-39.

9. Terringo III, J., *Pollution Engineering*, 1987, *19* (4), 78-83.

FILTRATION AND DEWATERING OF FINE COAL AND REFUSE

Y. S. Cheng and S. H. Chiang
School of Engineering
University of Pittsburgh
1249 Benedum Hall
Pittsburgh, Pennsylvania 15261
(412) 624-9630

ABSTRACT

Filtration and dewatering characteristics of fine coal and refuse were experimentally studied. The dewatering was enhanced by the change of driving force (vacuum/air pressure) and the use of flocculants/coagulants as additives. Among the various treatments, the flocculant addition was found to be most effective for increasing the filterability of the coal and refuse samples. The permeability of the filter cake was increased by as much as an order of magnitude when a proper dosage of flocculant was used. The final moisture content of the filter cake was effectively reduced by using air pressure filtration in conjunction with flocculant addition. The results showed that the performance of the dewatering operation was strongly influenced by the molecular weight and flocculant/coagulant dosage. The optimum values of these variables, at which the filterability of the samples was maximum, were observed.

INTRODUCTION

Fine coal and refuse dewatering has been a challenging task in the coal processing industry due to the difficulties caused by a large portion of ultrafine particles and high mineral content present in coal refuse. The problems, such as low filtration rate, high final moisture content, and serious medium plugging, have not been solved especially for dewatering of fine coal refuse.

Filtration, by definition, is an operation separating solid particles from a fluid by passing the slurry through a filter medium. At the end of filtration, where a 100 percent saturated cake is formed, the moisture content is usually much too high to satisfy the requirements for most solid/liquid separation processes. Dewatering is therefore necessary. During dewatering, air penetrates into the filter cake and displaces the water within the void spaces. As the dewatering proceeds, the moisture content of the cake continuously decreases until the equilibrium value is reached. The efficiency of filtration and dewatering depends on many factors, such as the particle size distribution of a sample, chemical additives used, and operating conditions applied. Among various factors, filtration rate and final moisture content of filter cakes are commonly used to evaluate filtration and dewatering efficiency(5).

The purpose of this work is to investigate the filtration and dewatering characteristics of a fine coal and a refuse using two different types of laboratory filters, a vacuum filter and an air pressure filter. The experiments were designed to determine quantitatively the most important factors on enhanced dewatering of the fine coal and refuse and to find improved methods.

EXPERIMENTAL

The coal sample used in this work was a Pittsburgh seam raw coal containing 6 percent ash with particle size of -28 mesh. The refuse sample tested was typical of that obtained from an operating coal preparation plant in southwestern Pennsylvania when cleaning Pittsburgh seam coal. The chemical reagents originally present in the refuse sample were American Cyanamid anionic 1202, Dow M813 cationic polymer, and residual frother Dow M222. The particle size distribution, the ash content, as well as the density of the sample were experimentally measured and are listed in Table 1.

The chemical reagents used include one inorganic coagulant obtained from Fisher Scientific and four polymeric flocculants supplied by American Cyanamid. The key properties of these additives are listed in Table 2.

Table 1. **Characteristics of Coal Refuse Sample**

Size (mesh)	Wt. Fraction (%)	Ash Content (wt.%)	Density
+28	2.94	5.49	1.338
-28 + 200	30.31	8.16	1.354
-200	66.75	42.16	1.789
Composite	100.00	31.10	1.647

Table 2. **Key Properties of Chemical Additives Used**

Name	Type	Mol. Wt.
Aluminum Sulfate	Inorganic Coagulant	342.1
Accoal-Floc 550	Flocculant, Anionic	250,000
Accoal-Floc 204	Flocculant, Anionic	$4\text{-}6 \times 10^6$
Accoal-Floc 218	Flocculant, Anionic	$18\text{-}20 \times 10^6$
Accoal-Floc 16	Flocculant, Nonionic	$2\text{-}4 \times 10^6$

The experimental apparatus employed for filtration and dewatering, as shown in Figure 1, has been successfully used in previous work(3,4,9). The vacuum filtration cell consists of two Plexiglas cylinders and a load cell transducer system. Filter cakes were formed in the upper Plexiglas cylinder of 0.05 m I.D. The filtrate was collected and weighted by the load cell and associated recording system. The air pressure filter, as shown in Figure 2, is similar to the vacuum filtration cell. It is equipped with a pressure distributor and a

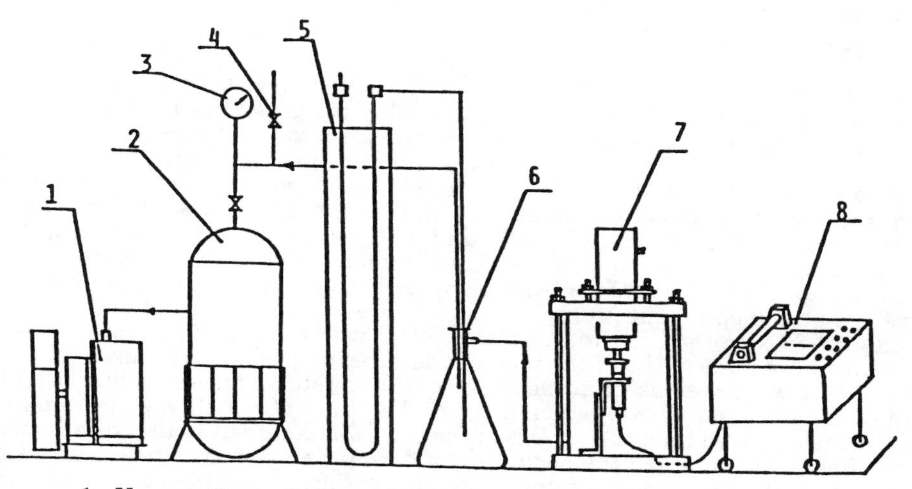

1 – Vacuum Pump, 2 – Surge Tank, 3 – Vacuum Gauge, 4 – Bleed Valve
5 – Hg Monometer, 6 – Vacuum Trap, 7 – Filtration Unit, 8 – Recorder

Figure 1. Schematic Diagram of Experimental Apparatus

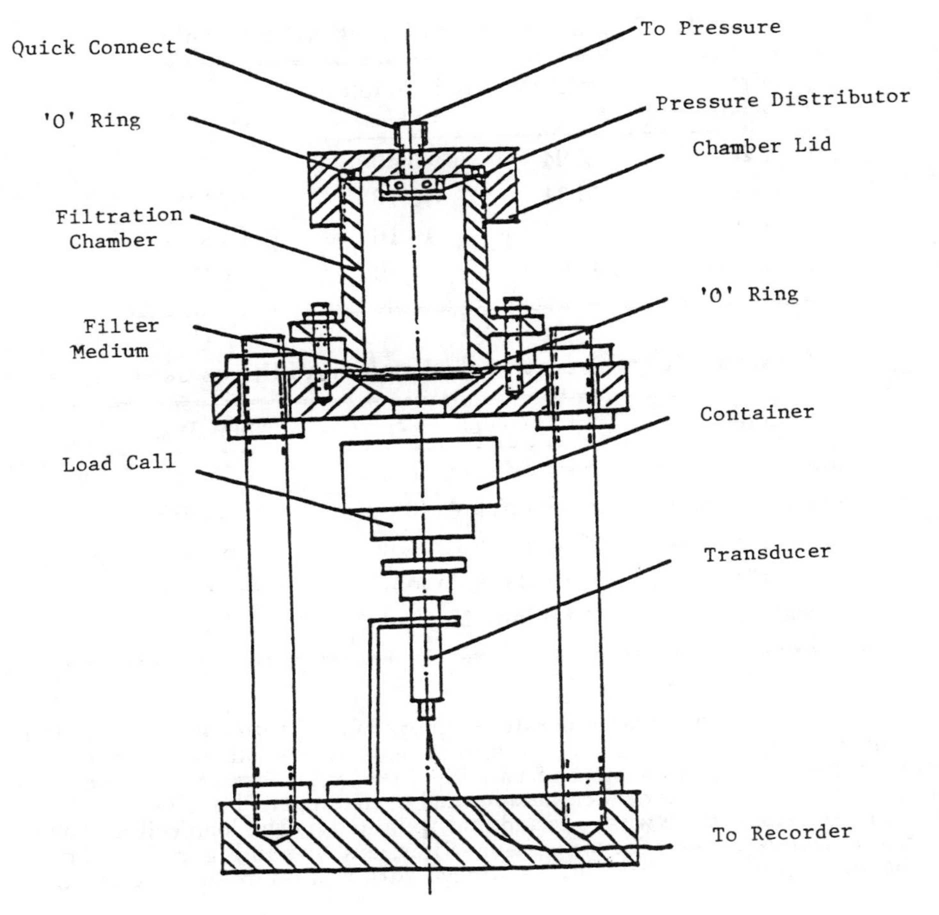

Figure 2. Air Pressure Filtration Cell

quick connect to ensure an instant and uniform pressure at the start of the filtration experiment. Filter cakes were formed in the pressure chamber, and filtrate was collected in a container placed on the top of the load cell.

For each experimental run, a slurry containing 25 g of coal and refuse was prepared with distilled water. The solid concentration of the slurry was maintained at 25 weight percent for all runs including those with chemical additives. After premixing with and without inorganic coagulant for 30 minutes at 380 rpm, a predetermined amount of flocculant solution was added into the slurry and mixed for another 30 seconds (coal) or 60 seconds (refuse) at 380 ppm. The slurry was then poured into the filtration cell, and a vacuum (67 kPa) or air pressure (345 kPa) was immediately applied and remained constant in both filtration and dewatering period. Whatman #1 and #42 paper were used as filter media for coal and refuse, respectively. The filtrate was recorded as a function of time. The time at which the last drop of liquid disappeared from the surface of the cake was recorded as the cake formation time, which marked the beginning of dewatering period. The experiment was terminated at 10 minutes total time for coal and 5 minutes of dewatering time for refuse. The filter cake was weighed, dried at 50 C for 48 hours, and weighed again to determine the final moisture content of the cake. The single-phase permeability of the cake was calculated using the Ruth equation[8] and Darcy's equation[2].

RESULTS AND DISCUSSION

In this work, the effectiveness of dewatering enhancements was evaluated by two factors, the ease of filtration and the moisture content attainable. The first one, the filterability of the coal and refuse, was measured by the single-phase permeability or the formation time of filter cakes. The second one, the suitability of the filter cakes for use (coal) or for disposal (refuse), was determined by the final moisture content of the cakes.

EFFECT OF POLYMERIC FLOCCULANT ON VACUUM FILTRATION

The effectiveness of flocculant depends on the interaction between the polymer and the particles, which in turn is governed by the charge characteristics, the concentration, and the molecular weight of the polymer. The effects of the ionic nature of flocculant on dewatering fine coal and refuse have been studied in our early work(3). It was found that among various flocculants tested, the anionic flocculant was most effective. For this reason, three anionic flocculants with different molecular weights were selected for this work.

Figure 3 shows the effects of the three flocculants, Accoal-Floc 550 (Mol. Wt. = 250,000), Accoal-Floc 204 (Mol. Wt. = 4-6 $\times$ 10^6), and Accoal-Floc 218 (Mol. Wt. = 18-20 $\times$ 10^6), on the filterability of the refuse samples. As can be seen, the addition of Accoal-Floc 204 increases the single-phase permeability of refuse filter cake by as much as a factor of 11 at low concentrations. The increase by Accoal-Floc 204 with molecular weight of 4 to 6 million is higher than those by the flocculants having either higher or lower molecular weight. The final moisture content shown in Figure 4 is about the same for each flocculant at concentrations less than 60 ppm.

It is clear that both molecular weight and concentration of flocculant are important for dewatering enhancement. The fact that an optimum mole-

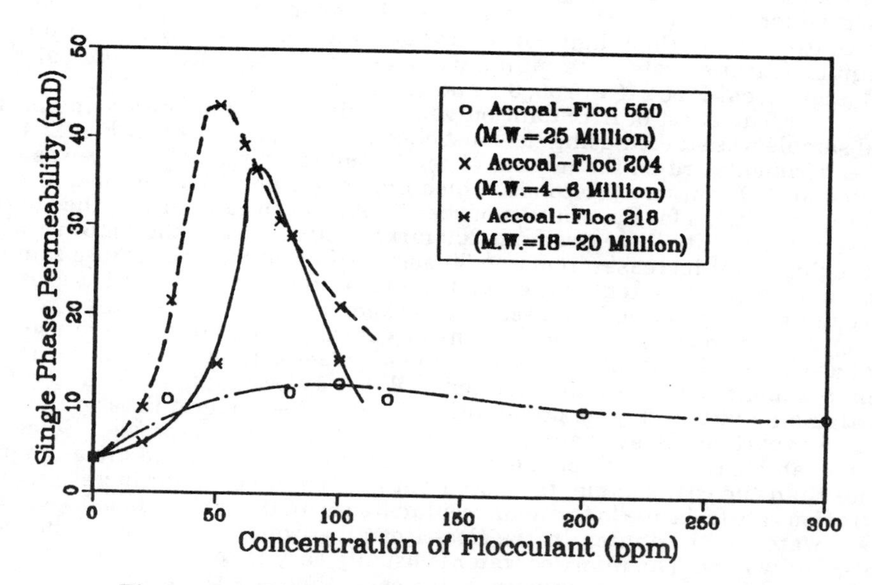

Figure 3. Effects of Flocculants on Permeability of Refuse Cakes (Vacuum Filtration)

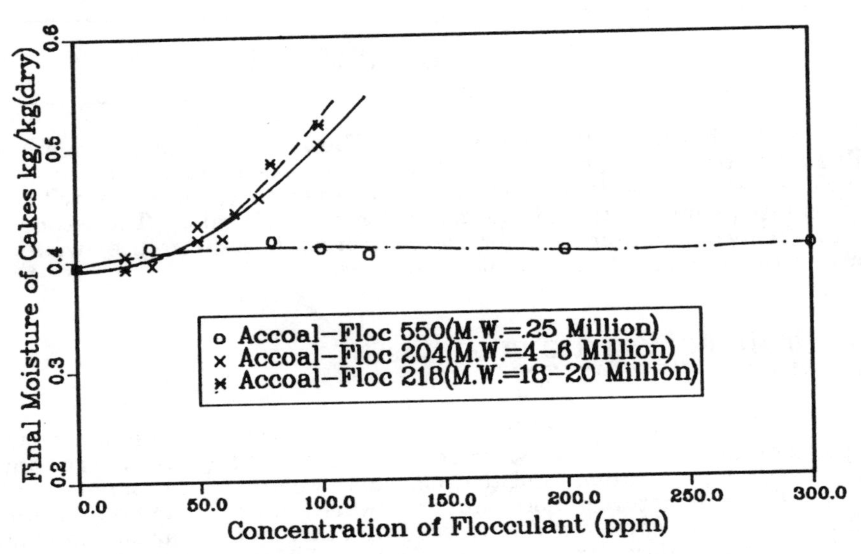

Figure 4. Effects of Flocculants on Final Moisture Content of Refuse Cakes (Vacuum Filtration)

cular weight exists results from changes in the shape of dissolved polymers. As the molecular weight increases from 5,000 to 15 million, the shape of the dissolved polyelectrolete molecules changes from extended long chains to strongly swelled and highly inflated balls[6]. The swelled and inflated molecules knit suspended particles to form flocs. With higher molecular weights, the flocs formed become larger and looser, which is less favorable for filtration and dewatering.

It is also noted from Figure 3 that the dosage of flocculant is another main factor in the slurry pretreatment. An optimum concentration was observed for each of the three flocculants. As the molecular weight increases, the control of the flocculant concentration becomes more critical to obtain optimum filtration rate. The optimum dosages for Accoal-Floc 550, Accoal-Floc 204, and Accoal-Floc 218 are 50, 65, and 100 ppm, respectively.

The effect of flocculant on vacuum filtration and dewatering of the coal sample was studied using the most effective flocculant, Accoal-Floc 204. In all experimental runs, the mixing of the flocculant with coal slurry was controlled at 380 rpm for 30 seconds, which previously proved to be the optimum mixing conditions for flocculation of coal[3]. As shown in Figure 5, the single-phase permeability increases from 200 mD to 350 mD when the concentration of the flocculant increases from 0 to 50 ppm. Different from the refuse sample, the final moisture content of the coal filter cakes continuously decreases as the flocculant concentration increases up to 50 ppm.

The differences between the dewatering behavior of the refuse and coal are believed to be attributable to the differences in mineral matter (clay) contents and particle size distributions. When a flocculant is added, the fine particulate matter (clay) is preferentially aggregated to form flocs so that the apparent particle sizes are enlarged and the filterability is improved. Since the refuse sample contains much more mineral materials (clay) and ultrafine particles than the coal sample, the flocculant is more effective for increasing the filtration rate (relative to the nonflocculated case) of the refuse. In a flocculated cake, water is distributed in the flocs (intrafloc water) and among the flocs (interfloc water). The interfloc water is usually easy to remove, whereas much of the intrafloc water may remain in the cake. Because there are fewer ultrafine particles in the coal sample, only a small amount of water is entrapped in the flocs, which in turn results in a lower final moisture content.

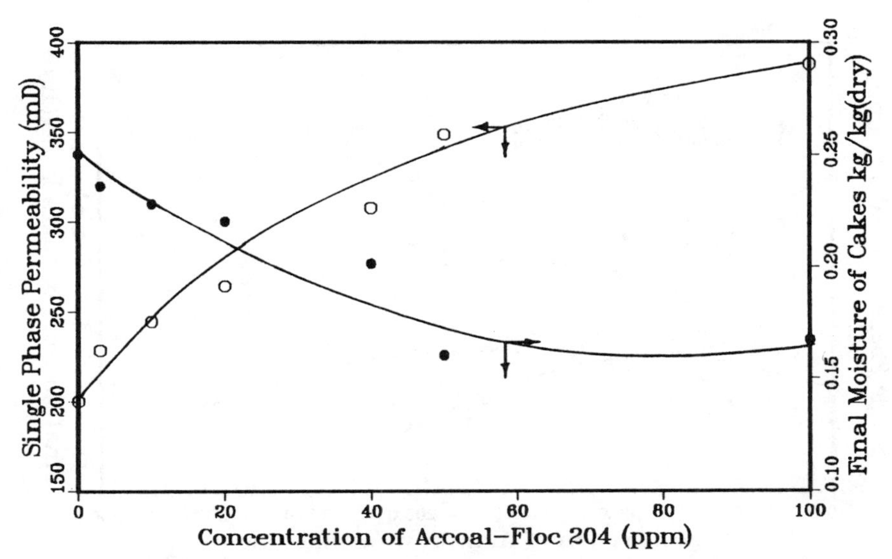

Figure 5. Effect of Coagulant on Vacuum Filtration and Dewatering of Refuse in Conjunction with Flocculant

EFFECT OF INORGANIC COAGULANT

In addition to the properties of flocculant, the particle characteristics, especially the surface charge, are also believed to influence the flocculation of the particles. In aqueous solutions containing finely divided clay, a double layer is formed with a negative zeta potential(7,10). An inorganic electrolyte can serve as a coagulant by providing cations that reduce the zeta potential. Since the surface charge properties can be altered by an inorganic coagulant, flocculant-aided filtration and dewatering may be enhanced through the addition of a coagulant.

A commonly used coagulant, aluminum sulfate, was tested in conjunction with two flocculants, Accoal-Floc 204 (anionic) and Accoal-Floc 16 (nonionic), at various concentrations. The coagulant was added into refuse slurry first and mixed with the slurry for 30 minutes. The concentrations of the flocculants were kept as constants of 20 ppm and 50 ppm for Accoal-Floc 204 and Accoal-Floc 16, respectively. Figure 6 shows the variations of single-phase permeability and final moisture content of filter cakes with the concentration of the coagulant. As the concentration of aluminum sulfate increases, the final moisture content first marginally decreases and then increases, whereas the permeability increases to a maximum value and then decreases. The existence of maximum points may indicate that a certain amount of an inorganic electrolyte is needed to modify the zeta potential of the particles so that the best flocculation can be obtained using polymeric flocculant. Also noteworthy is the increase in permeability brought about by the coagulant; it is much smaller than that resulting from the increase in concentration of the flocculants.

EFFECT OF APPLIED AIR PRESSURE

The effects of applied air pressure on dewatering of the fine coal and refuse were studied at a pressure range of 69 to 379 kPa (10 to 55 psi). Figure 7 shows that the refuse filter cake formation time decreases from 10.8 to 4.1 minutes, and the final moisture content of filter cakes is reduced by about 47 percent (from 0.379 to 0.200 kg/kg dry basis) as the applied air pressure increases from 69 to 379 kPa (10 to 50 psi). It is also noted that both improvements in cake formation time and final moisture content occur sharply at the

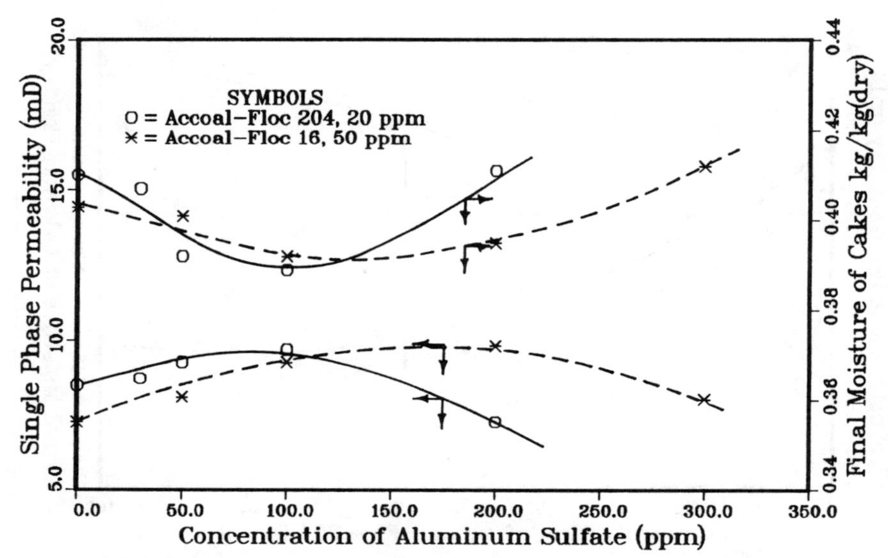

Figure 6. Effect of Flocculant on Vacuum Filtration and Dewatering of -28 Mesh Pittsburgh Coal

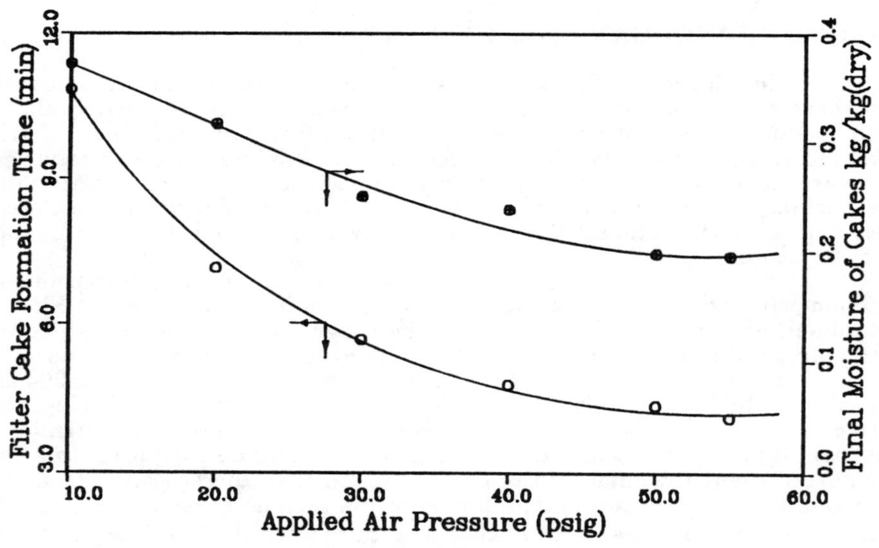

Figure 7. Effect of Applied Air Pressure on Filtration and Dewatering of Refuse

beginning and become less significant as the applied pressure approaches 345 kPa (50 psi).

Similar to what was observed in refuse dewatering, both the formation time and the final moisture content of coal filter cakes decrease with an increase in the applied air pressure. Figure 8 shows a similar trend in the improvements obtained in the filtration rate and the extent of dewatering, i.e., a very sharp decrease in both the cake formation time and the final moisture content at the beginning, and then a gradual approach to an asymptotic value as the applied pressure increases to 345 kPa (50 psi).

Compared with vacuum filtration, the major advantage of the air filtration seems to be the significantly lower final moisture content of filter cake.

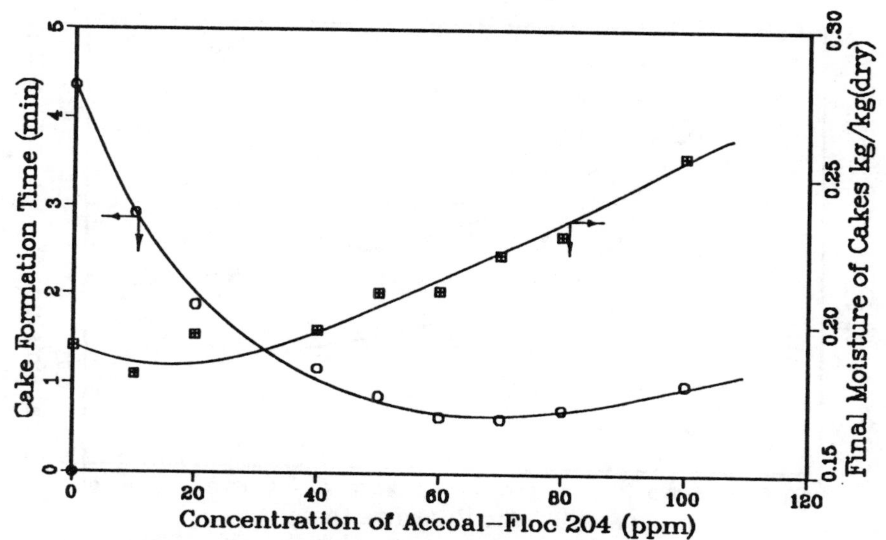

Figure 8. Effect of Flocculant on Air Pressure Filtration and Dewatering of Refuse

This indicates that higher air pressure effectively opens a large number of small pores in the filter cake; the water in these small pores could not be displaced by air under vacuum conditions.

EFFECT OF FLOCCULANT ON AIR PRESSURE FILTRATION

The effects of flocculant on air pressure filtration and dewatering of the coal and refuse were determined using Accoal-Floc 204, the most effective flocculant found in the vacuum filtration, at a constant applied pressure of 345 kPa (50 psi). The mixing conditions of the flocculant with slurry were the same as those used in the vacuum filtration. Since the filter cake formed may be compressible under the applied pressure, the cake formation time (instead of cake permeability) was used to evaluate the filterability of the samples.

In the case of refuse, it was found that the cake formation time was greatly shortened from 4.4 minutes, corresponding to the case without flocculant, to 0.6 minute at 60 ppm of the flocculant (see Figure 9). A further increase in the flocculant concentration beyond 60 ppm resulted in a longer cake formation time. The final moisture content, however, remained substantially the same at low flocculant concentrations and increased as the concentration increases beyond about 40 ppm. These results show a similar trend as that found in the vacuum filtration. The addition of the flocculant can dramatically improve the filtration rate of the refuse sample in the low concentration range (<60 ppm). A lower final moisture content may be achieved by changing the mixing conditions of the flocculant[1,3].

Air pressure filtration and dewatering of the coal sample were also carried out with the addition of Accoal-Floc 204. As shown in Figure 10, the cake formation time decreases sharply (from 8.5 to 3.1 seconds) when the flocculant concentration increases from 0 to 50 ppm. A further increase in flocculant dosage up to 150 ppm only results in a slight decrease in the cake formation time. At a still higher concentration (>150 ppm), the cake formation time becomes longer. In contrast to the results obtained for the refuse, the final moisture content of the coal cakes decreases sharply as the flocculant concentration increases up to 50 ppm, then it remains unchanged with a further increase in flocculant dosage. This result also agrees with the finding in the vacuum filtration.

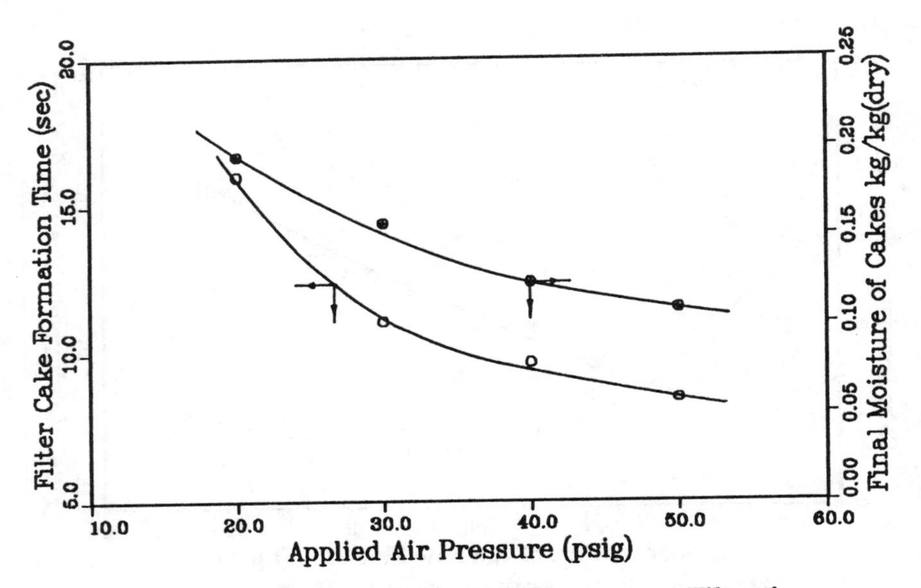

Figure 9. Effect of Applied Air Pressure on Filtration and Dewatering of -28 Mesh Pittsburgh Coal

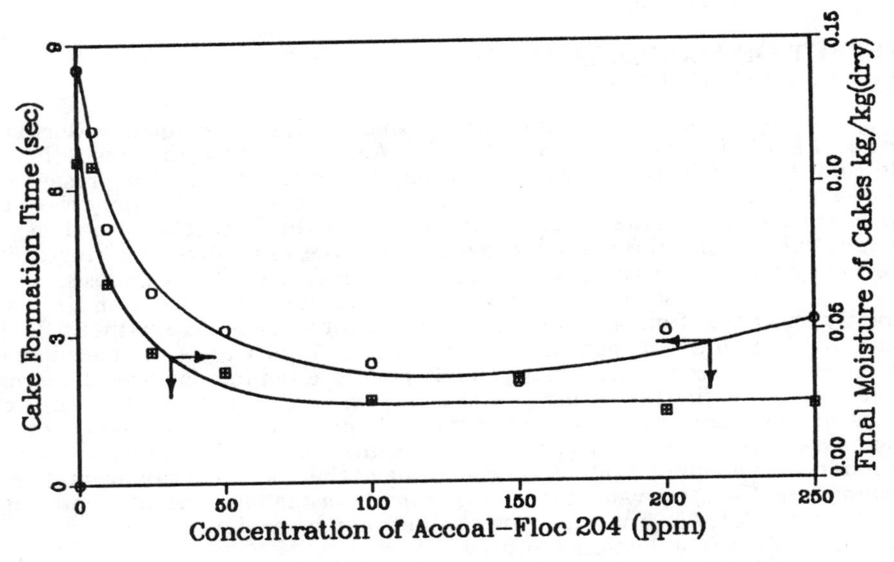

Figure 10. Effect of Flocculant on Air Pressure Filtration and Dewatering of -28 Mesh Pittsburgh Coal

CONCLUSIONS

The following conclusions can be drawn from the results obtained in this work:

1. The addition of flocculant is a very effective treatment in filtration and dewatering of fine coal and refuse. In selecting a proper flocculant, the molecular weight is an important factor. Among the flocculants tested, the one with molecular weight of $4\text{-}6 \times 10^6$ is found to be most effective for the refuse.
2. Dosage of flocculant is another critical factor in slurry pretreatment. Optimum concentration exists for each flocculant. Overdosing may lead to deterioration of the performance.
3. Modifying the surface charge properties of particles using coagulant in conjunction with flocculant may result in better filterability and a reduction of final moisture content of filter cakes. The improvements, however, are not large. The concentration of coagulant should also be properly controlled.
4. An increase in applied air pressure up to 345 kPa (50 psi) can effectively increase the filtration rate and dramatically decrease the final moisture content of fine coal and refuse.
5. The addition of flocculant can significantly improve the air pressure filtration rate of both fine coal and refuse. It can also result in a great reduction of final moisture content of fine coal cakes.

ACKNOWLEDGMENT

The generous support provided by the U.S. Department of Energy under Contract No. DE-AC2285PC81582 is gratefully acknowledged.

REFERENCES

1. Cheng, Y. S., Ph.D. Thesis, University of Pittsburgh, 1988.
2. Darcy, H.P.G., *Les Fontaines Publiques de Ville de Dijon*, Dalrmont, 1856.
3. Fang, S. R., Cheng, Y. S., and Chiang, S. H., *Interfacial Phenomena in Biotechnology and Materials Processing*, Elsevier Science Publishers B. V., Amsterdam, 1988, 515-529.
4. Gala, H. B., Ph.D. Thesis, University of Pittsburgh, 1984.
5. Gala, H. B. and Chiang, S. H., *Filtration and Dewatering--Review of Literature*, U.S. Department of Energy, No. 00EIT/14291-1, September 1980.
6. Reuter, J. M. and Hartan, H. G., *World Congress of Particle Technology*, IV, Nurenberg, Germany, April 1986, 269-287.
7. Riddick, T. M., *Control of Colloid Stability Through Zeta Potential*, Zeta-Meter, Inc., 1968.

8. Ruth, B. F., Montillon, G. H., and Montonna, R. E., *Industrial Engineering Chemistry*, 1933, *25* (2), 153.

9. Venkatadri, R., Klinzing, G. E., and Chiang, S. H., *Filtration and Separation*, 1985, May/June, 172-177.

10. Wakeman, R. J., Mehrotra, V. P., and Sastry, K. V., *Hydraulic Conveying*, 1980, *1* (2), 281-293.

MECHANICAL AND THERMAL TREATMENT OF DREDGED MATERIAL TO PRODUCE CERAMIC PELLETS

H. J. Hampel and P. Mehrling
Battelle-Institut e.V.
Am Römerhof 35
Postfach 90 01 60
D-6000 Frankfurt am Main 90
West Germany

and

H. Kröning
Strom- und Hafenbau
Hamburg, West Germany

SUMMARY

The maintenance work required to provide for the necessary water depth in the harbor of Hamburg yields about 2 million m^3 dredged material annually, the disposal of which becomes increasingly difficult because of polluted river sediments.

As an alternative to dumping the dredged material, the Freie und Hansestadt Hamburg, through its dredging authority, the Strom- und Hafenbau division, has built a large-scale pilot plant (1,200 m^3/h throughput) for separating the almost pollutant-free sand and dewatering the remaining fine-grained fraction containing the pollutants. Large-scale experiments are carried out to establish design data for the final industrial-scale plant.

With a view to the further treatment of the separated fine-grained dredged material, the investigations conducted at the laboratories of Lurgi GmbH and Battelle Frankfurt resulted in the development of a process permitting an agglomerate to be produced that satisfies existing standards for building materials.

INTRODUCTION

Navigable waterways, in particular harbor basins, must be dredged at regular intervals to maintain the necessary water depth. Each year, around 2 million m^3 of sand and fine-grained material have to be dredged from the Harbor of Hamburg. Hamburg Harbor is open to the tide of the North Sea and has a mean tide lift of 3.3 m; it is located in the estuary of the River Elbe[4]. Due to the tidal influence, thick deposits of sandy materials are found, especially near the harbor entrance where eddies occur. By contrast, fine-grained material is deposited in the weakly agitated rear harbor basins[1]. The organic and inorganic pollutants contained in this fine-grained fraction of the dredged material, the silt, are environmentally unacceptable for storage in containment areas or for land application to agricultural soil[10]. In this situation, Freie und Hansestadt Hamburg, through its dredging authority, the Strom- und Hafenbau division, established a Dredged Material Investigation Program in the late 1970s. The main objective of this investigation program was to find reasonable, long-term solutions to the problem of how to dispose of or use the dredged material produced each year.

In 1981, Battelle in Frankfurt was commissioned to identify and develop new methods of treating the material. Freie Hansestadt Bremen also supported this project. The very complex problem required Battelle's collaboration with harbor authorities, university institutes, consulting companies, and industrial experts[2,3]. This joint effort was to ensure that ecologically safe and economically acceptable techniques were developed for treating, utilizing, and disposing the dredged material.

DEVELOPMENT OF NEW TREATMENT METHODS FOR DREDGED MATERIAL

The research project carried out by Battelle Frankfurt was divided into two phases:

- System Analysis: All promising approaches to a solution were identified and a preselection was made.
- Development: The selected solutions were investigated and assessed in detail. Selected methods were optimized for industrial-scale use.

SYSTEM ANALYSIS

To begin with, all candidate methods for disposing of or utilizing the dredged material were considered; this was done in two steps. First, possible approaches to a solution were identified. Second, the most promising of these solutions were pinpointed.

In a first screening, 100 possible solutions were examined for their technical feasibility, ecological consequences, costs, etc. This resulted in a shorter list of 25 methods, which were then subjected to closer scrutiny by various experts who prepared brief evaluation studies. The second screening step resulted in the six possible solutions to be developed further:

- Use the fine-grained fraction (the silt) as an additive in brick production
- Use the silt as an additive in cement production
- Use the silt as a substitute raw material in mineral fiber production
- Substitute silt for gravel as an additive to concrete
- Solidify the silt for deposition or utilization
- Deposit the silt in pits, or pile it up as hills.

DEVELOPMENT

The second project phase, the development phase, was aimed at investigating the selected solutions experimentally and judging them by technical, ecological, and economic criteria. This assessment yielded those methods which were the most practical and ecologically most beneficial.

Strom- und Hafenbau carried out investigations concerning the hill-shaped deposition and the optimization of mechanical separation of dredged material. Battelle's efforts, on the other hand, were concentrated on the thermal treatment, solidification, and dewatering of the dredged material. This involved a series of laboratory-scale, pilot-plant-scale, and field experiments.

TREATMENT CONCEPT

It was found that the best possible alternative to depositing the dredged material in containment areas involves separating the unpolluted coarse-grained fraction of the dredged material (sand) and dewatering the fine-grained fraction (silt). Afterwards, the dewatered silt would be piled up in hills or subjected to thermal treatment and used as a construction material aggregate.

The basic principle of dredged material treatment according to this strategy is shown in Figure 1 and explained below.

MECHANICAL TREATMENT (METHA)

The dredged material is passed first through the mechanical separation units where the uncontaminated sand fraction is recovered. The residual fine-grained fraction is dewatered. The wastewater is partly recirculated and partly passed into a sedimentation and biological nitrification unit. The purified wastewater is returned then into rivers or lakes.

HILL-SHAPED LAND DEPOSITION

According to the short- to medium-term concept, the partially dewatered, fine-grained material is to be deposited in stable hill-shaped land

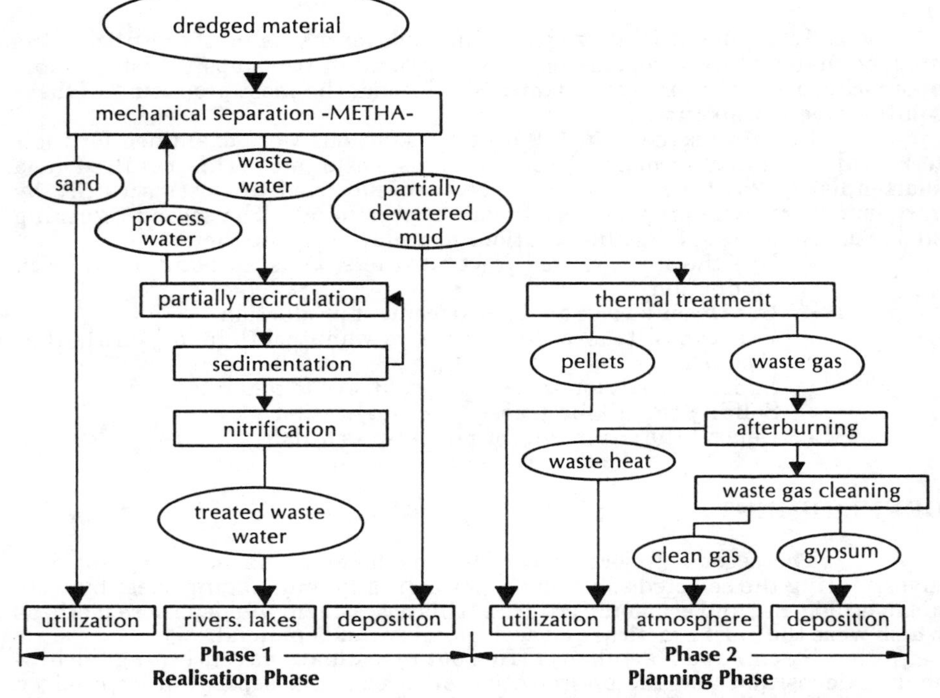

Figure 1. Principle of the Treatment of Dredged Material

deposits. With a view to safety, hill-shaped land deposits will be designed in the same way as hazardous waste deposits with a multiple sealing system. The sand-free and partially dewatered fine-grained material is piled up in slightly inclined layers, with sand layers in between. The sand strata serve as drains for leakage and interstitial waters. The hills are provided with surface covers (with drainage) on which layers of normal soil will be spread eventually, where trees and plants can be grown.

THERMAL TREATMENT

As an alternative to hill-shaped land disposal, the partially dewatered fine-grained material is treated continuously on a pelletizing disk. The resultant wet pellets are subjected to thermal treatment to produce an agglomerate that satisfies the standards for building materials.

After choosing this concept of treatment, disposal, and utilization of dredged material, the three processing steps--mechanical treatment, hill-shaped land deposition, and thermal treatment--were further developed and optimized.

Following is a description of how the mechanical and thermal treatment processes--the major elements of the new concept--have been optimized.

MECHANICAL TREATMENT OF DREDGED MATERIAL (METHA)

Optimizing the METHA treatment was important particularly because it is at this stage that the material is prepared for subsequent hill-shaped deposition or thermal treatment. The basic principle of the mechanical treatment is shown in Figure 2.

The industrial-scale experimental plant, METHA II, will be used to separate and dewater dredged material to examine and demonstrate the feasibility of the treatment concept. Individual plant components such as hydrocyclones, centrifuges, belt presses, thickeners, etc., will be tested in their maximum final size; the entire treatment process will be optimized; and suitable measuring and control strategies will be developed under practical operating conditions. These large-scale experiments are considered the only way to establish well-founded design data for the final industrial-scale plant, which will process the total volume of dredged material produced.

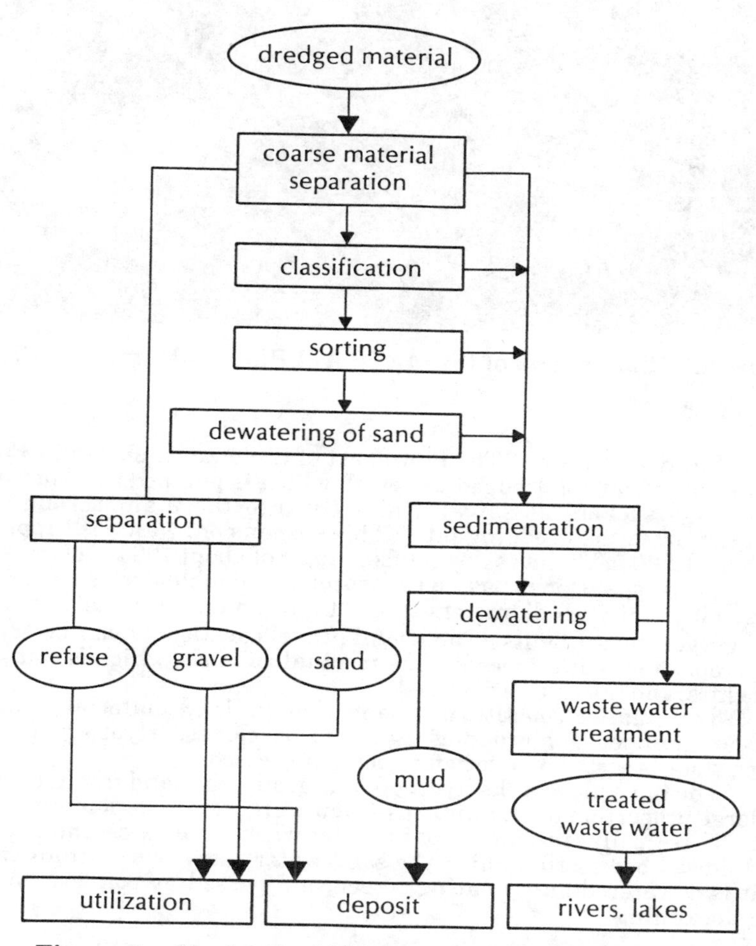

Figure 2. Mechanical Treatment of Dredged Material

DESCRIPTION OF THE PROCESS

On the basis of various laboratory and pilot-plant experiments(7,8,11), the METHA II plant was constructed and put in operation in 1987 (Figure 3). The plant and process concept is outlined below.

Figure 3. Photograph of the METHA II Plant in Hamburg, FRG

Sand Recovery. The flow sheet of the plant is shown in Figure 4. Part of the suspension of dredged material, which is pumped into containment areas by barge suckers, is supplied directly into the plant through a line branching off from the dredging pipe. This suspension, which is supplied at a rate of about 1,200 m^3/h, has a dry solids content of about 100 to 300 t.

The suspension is passed first through a double-deck screen to separate coarse material with diameters larger than 4 mm, essentially, woody and mineral (gravel) constituents. The useful gravel constituent can be separated by an appropriate sorting process. The residual rubbish, which is composed of wood, plastics, and the like, is disposed.

The smaller constituents--sand, fine-grained material, and water mixture--are pumped to a group of eight hydrocyclones. Hydrocyclones work best when stock is sorted by size before being processed.

The underflow of the hydrocyclone group, the sand fraction, still contains a large proportion of organic constituents and small agglomerates of clay and silt. These pollutants are separated by washing in a second separation unit, a fluidized-bed sorting unit. The sand-water suspension withdrawn from this unit is dewatered on a floating screen and passed by conveyor for intermediate stockpiling.

Dewatering. The overflow of the hydrocyclone group and the fluidized-bed sorting unit are combined. The design of the various plant units is based on the performance data of their individual components, and the throughput of the dewatering unit is about 200 m^3/h suspension with a dry solids content of 5 to 15 t. This roughly corresponds to the maximum performance of individual units that can be used for dewatering dredged material.

Various units, such as centrifuges, belt presses, and high-intensity pressure filters, were tested to determine the optimum dewatering process.

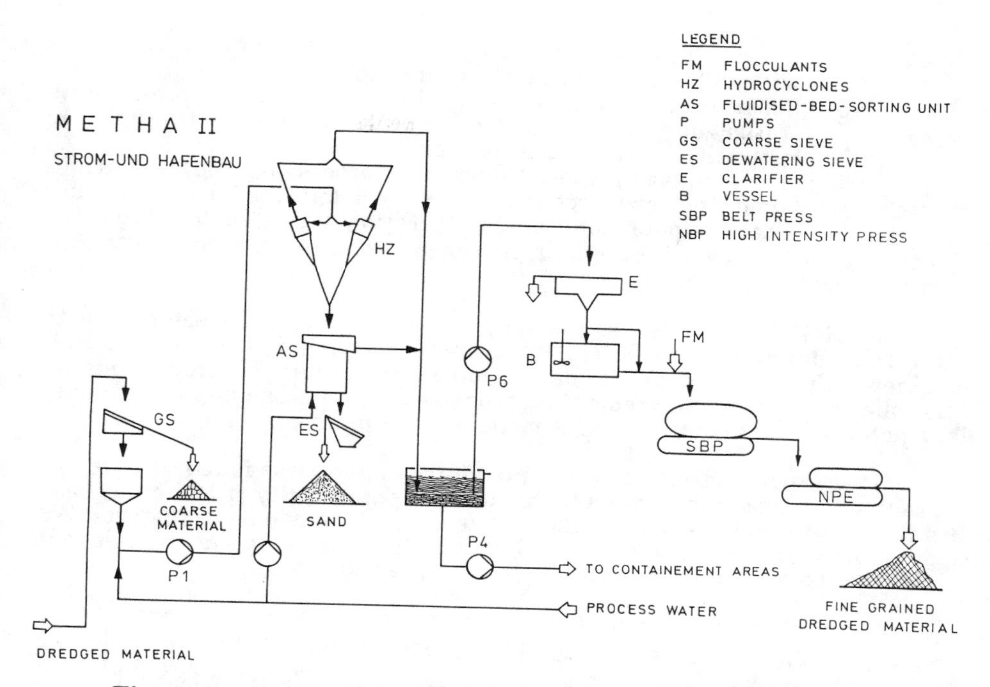

Figure 4. Process Flow Sheet of the Mechanical Treatment of Dredged Material

EXPERIMENTAL PROGRAM

Since the first experiments have shown that classification and sorting yields a sand of standard quality, efforts now concentrate on optimizing the dewatering stage. First of all, quality requirements had to be defined to yield a material suitable for hill-shaped deposition or thermal treatment.

Material for Hill-Shaped Deposition. To optimize construction concept of the hill-shaped deposits(9), the fine-grained, dewatered material must have a specific minimum shear strength and consistency. At a solids content of more than 50 weight percent, this requirement is satisfied in most cases.

Material for Thermal Treatment. The pelletizing and thermal treatment experiments conducted so far have shown that the physical as well as the ecological properties of the pellets can be controlled in the firing process and are less dependent on the composition or contamination of the dredged material. The main requirement in this respect is that the dewatered material must have a solids content of at least 50 weight percent.

It follows that any dewatering method that guarantees a dry solids content of 50 weight percent or better will be suitable for both process routes, hill-shaped deposition and pelletizing. The best possible way of achieving this has been determined in an extensive experimental program. The first dewatering experiments were carried out at the pilot-plant scale, over a limited period of time; continuous large-scale experiments are under way at present. Following are the most important dewatering methods and systems that have been investigated so far:

1. Concentration and centrifugation
2. Concentration, centrifugation, and high-intensity pressing
3. Concentration and single-stage dewatering in a belt press

4. Concentration and two-stage dewatering, using a belt press with additional high-intensity pressing
5. Separation of the superfine-grained sand fraction, pressure filtration, and parallel dewatering of the silt fraction through concentration and two-stage pressing.

When investigating these alternative processing routes, various characteristic data of the dewatered material were determined (e.g., dry solids content, soil-mechanical data, pollutants). Operational data, such as the material throughputs to be achieved and the consumption of auxiliary materials, were also identified.

Results. Figure 5 shows major experimental results achieved with the various dewatering methods. When comparing the dewatered materials obtained with the different methods, it is found that, in terms of product quality (dry solids content, shear strength, and consistency) and consumption of auxiliary materials, the fourth method is particularly favorable.

Future Activities. Since the fourth method was found to have significant advantages, a belt press with a throughput capacity of 10 t/h has been installed. The press is being subjected to a continuous-operation test. An appropriate high-intensity press is being installed at present, and the relevant trials are to begin within the current operating period.

PROCESSES DEVELOPED	SOLID CONTENT % wt	STRENGTH, CONSISTENCY
→ Z →	38-44	-
→ Z → NPE →	44-50	+
→ SBP →	40-46	0
→ SBP → NPE →	44-49	+
→ (SBP → NPE) ∥ DF →	> 50	++

Figure 5. Dewatering Methods and Results--Optimization of the Dewatering Stage

THERMAL TREATMENT OF DREDGED MATERIAL

Based on the positive results of previous lab-scale experiments for pellet manufacture from dredged material[5], a combined effort of the Central Research and Development Department of Lurgi GmbH and of Battelle

Frankfurt, in cooperation with Strom- und Hafenbau, Hamburg, resulted in the development of a process using fine-grained dredged material to produce an agglomerate that can be used as building material[6]. The objectives and results of this research work are discussed below.

OBJECTIVES

The investigations and experiments concerning a thermal treatment process for fine-grained and dewatered dredged material were carried out with the following objectives:

- To manufacture a lumpy product resembling gravel
- Product to be used as high-grade building material (concrete aggregate or road-building material)
- Almost complete destruction of the organic pollutants (particularly PCB), necessitating that the temperature of the solid constituents exceed 1,000 C (with sufficient retention time)
- Environmentally safe binding of heavy metals in the ceramic structure
- No use of aggregates (e.g., clay) to avoid any artificial increase in product quantity
- Integration of an off-gas cleaning process in the sediment pelletizing plant with a minimum production of effluent and residue
- Exclusive use of proven processes and units.

SELECTION OF PROCESS

The following alternatives were investigated with a view to their suitability for pellet production:

- Fluidized-bed process for incinerating sewage sludge
- Rotary kiln process for thermal treatment of lumpy to sandy materials
- Sintering process for melt agglomeration of sinter ores (<6 mm, for instance)
- Traveling grate process for firing pellets of fine-grained mineral substances.

Of these four alternatives only the traveling grate process satisfies all the requirements stated above. It permits even complex temperature-time profiles to be accurately kept, leads to a lumpy ceramic product, and permits high retention times just below the softening point of the dry sediment (1,200 C). The process permits the complete destruction of the organic pollutants and the environmentally safe binding of the remaining heavy metals in the ceramic lattice structure of the clay minerals.

EXPERIMENTS ON THE LURGI TRAVELING GRATE FOR PELLET PRODUCTION

Extensive experiments and investigations have been conducted according to the following strategy (see Figure 6) to determine the process data required for planning the industrial unit, to define the specific properties of the dredged material pellets, and to analyze the ecological effects of the manufacture and use of the pellets.

EXPERIMENTAL RESULTS

The resulting ceramic pellets were analyzed with regard both to the binding and mobility of the pollutants and to their usefulness as building material.

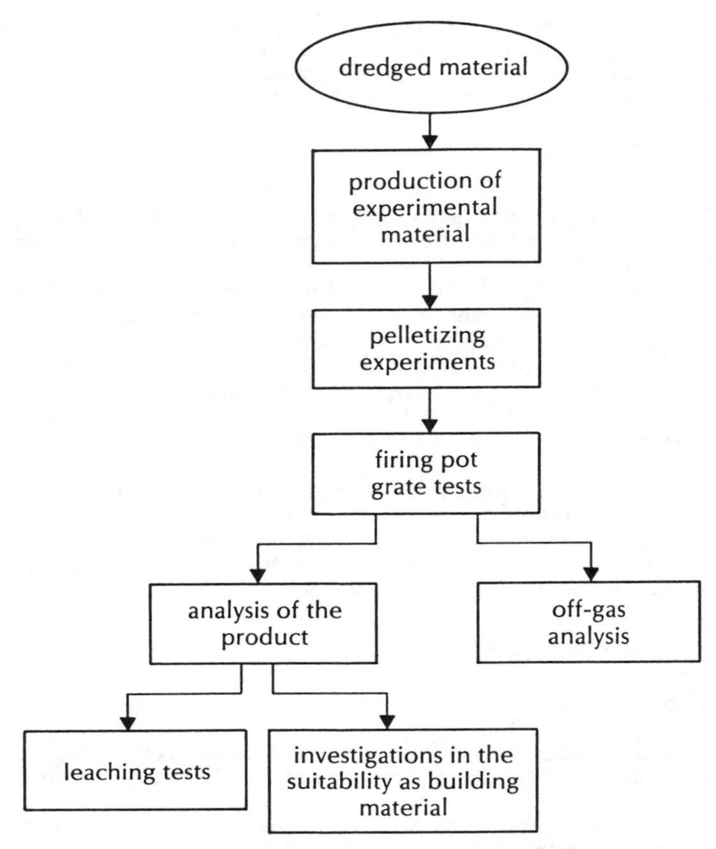

Figure 6. Flow Sheet for the Preparation and Examination of Dredged Material Pellets

Binding and Mobility of the Pollutants. To determine the pollutant quantities released in the pellet firing process, the pollutant content of unburned (green) pellets was compared with that of burned pellets. The investigations show that mercury is released almost completely, while the major part of all other heavy metals is bound in the ceramic structure of the pellets. Thermal treatment of the pellets in the firing process (at about 1,150 C) leads to an almost complete combustion of the organic pollutants. No PCDD or PCDF was found in the burned pellets.

To determine the mobility of the pollutants contained in the dredged material and bound in the burned pellets, leaching experiments were conducted under different conditions to simulate, among other things, the long-term behavior:

- According to the German Standard Regulation S4 (DIN 38414, Part 4) with distilled water over a period of 24 hours
- Following S4 with sulphuric acid for 5 to 14 days, adjusting the pH value every 24 hours to 3.7 to simulate acid rain, until the pH no longer increases substantially
- According to the methods stated above, with ground pellets (<0.1 mm) to simulate the leaching behavior after mechanical destruction of the pellets.

The results of the leaching experiments (Figure 7) show that substantially smaller amounts of pollutants are elutriated from burned pellets (bulky and ground) than from the equivalent amount of untreated dredged material.

CONTAMINANTS \ SAMPLE	CRUDE DREDGING MUD mg/kg	DREDGING MUD CLINKER PELLETS mg/kg
LEACHING CONDITIONS	7 days at $p_H = 3.7$ (H_2SO_4)	5 days at $p_H = 3.7$ (H_2SO_4)
Cd	2.9	0.02
Hg	0.004	<0.002
As	0.13	0.24
Ni	20	0.08
Pb	0.11	<0.01
Cr	0.74	0.01
Cu	11	0.92
Zn	580	1.5
NH_4^+	1100	3.0
Cl^-	340	3.5
F^-	15	4.0
NO_3^-	32	0.1

Figure 7. Immobilization of Contaminants by Thermal Treatment of Dredged Material Pellets by the Traveling Grate Process

Furthermore, the investigations yielded more favorable leaching values for burned pellets than for comparable industrial by-products, such as iron silicate stones (NA stones), foamclay produced from river (Neckar) sediment, and waste incineration slag, which are used as building and road-building materials.

Suitability of the Pellets as Building Material. To determine the general suitability of the dredged material pellets as concrete agglomerate and as road-building material (surface, binder, base, and frost-protection layers), various experiments were conducted according to the relevant German DIN standards and according to TL-Min-StB 83. The results of the investigations (Figure 8) may be evaluated as follows.

Suitability as Concrete Agglomerate. With a bulk density of 1.98 kg/dm^3, the pellet material under investigation belongs to the group of lightweight aggregates. Lightweight aggregates have a bulk density of ≤2.2 kg/dm^3. The chipping quantities of 0.02 weight percent found in the freeze-thaw alternating experiments are far below the permitted values of ≤4.0 and ≤2.0.

Investigations into the suitability of the material when used in PZ 25 F concrete have shown that an amount of 55 volume percent of the gravel aggregate with a grain-size of 4 to 16 mm (larger pellets were not available) can be substituted with pellets without decreasing strength. After a setting time of 28 days, the compressive strength of the pellet-containing concrete amounted to 40 N/mm^2, 2 N/mm^2 higher than that of gravel concrete. Consequently, the pellet material under investigation is considered to be a suitable aggregate for PZ 35 F concrete.

Suitability as Road-Building Material. The chipping quantities of 0.02 weight percent, a measure of frost resistance, lie far below the values permitted for twice-crushed and screened chippings. Thus, pellets with a sufficient inherent strength (impact shattering values) can be used in frost-protection layers.

Product pellets		
* diameter	14 - 17	mm
* bulk density	0.98	kg/dm^3
* pressure strength	1000 - 1500	N/P
* 1/5 ISO-tumble-test.		
>6.3 mm	95.3	M-%
<0.5 mm	3.0	M-%
* apparent pellet density (DIN 52102)	1.98	g/cm^3
* water absorption capacity (DIN 52103)	4 - 5	M-%
* pressure strength of pellet/gravel concrete after 28 days	40/38	N/mm^2
* fraction of fragments by alternating frost and thaw (DIN 4226, 3)	0.02 - 0.08	M-%
* impact destructions before/after freezing (TP Min-StB. 5.2.1.4)	21.7/21.8 - 27.8/28.4	M-%

Figure 8. Properties of the Pellets

Considering the impact shattering values (inherent strength) determined in the experiments, pellets may be used in bituminous surface layers for road classifications* IV and V, and in binder layers for roads classed II, IV, and V, if the affinity of the pellet material to bituminous binders can be improved. For base layers, dredged material pellets may be grouped under the mineral aggregate blast-furnace lump slag. In accordance with the RG Min-StB 83 standards, the impact shattering values $SZ_{8/12}$ = 21.8 and 21.7 weight percent permit the pellet material to be used for base layers of all classes.

Summarizing, it may be stated that the pellet material under investigation is suitable for frost-protection layers and base layers of all classes as a qualified mineral according to TL Min-StB 83, but can be used for other applications only as a nonqualified mineral.

CONCEPT OF A PLANT FOR THE MANUFACTURE OF DREDGED MATERIAL PELLETS

According to the present state of the investigations, an industrial pellet plant may operate as described below (see Figure 9).

---

* Classifications range from I, highest traffic volume and applied load, to V, lowest.

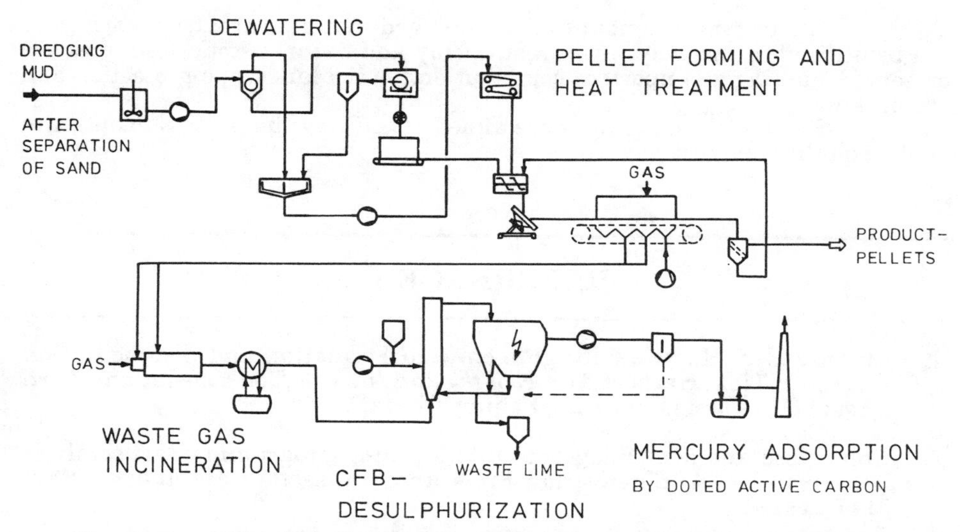

Figure 9. Flow Sheet of the Dredged Material Pelletizing Process

The fine-grained and dewatered (50 weight percent water content) dredged material coming from a mechanical separation plant (METHA) is mixed with returns from the pellet plant. The admixing of returns is necessary for the agglomeration of the dredged material into green pellets (wet pellets). The returns consist of soft-burned material. In addition, the use of returns dilutes the organic carbon in the mixture to be pelletized, so that uncontrollable temperature peaks can be avoided during the firing process.

The pellets are formed continuously on pelletizing disks, yielding pellets with diameters ranging between about 10 and 15 mm. The pellets are subjected to thermal treatment on a Lurgi traveling grate, which permits accurate adjustment of the process parameters. During the drying and heating process, humidity and volatile constituents are expelled from the dredged material pellets by sucking hot gases through the pellet packing from above. In the firing zone the hot gas reaches temperatures of 1,100 to 1,150 C. As a result, the heavy metals, with the exception of mercury, are bound in the ceramic matrix.

The burned pellets are cooled on the traveling grate before being separated by screening into ceramic material and soft-burned material. The ceramic pellets--that is, the pellets hardened in the firing process--undergo intermediate storage, whereas the soft-burned material is subjected to dry grinding and, mixed with dewatered dredged material, is formed again into green pellets.

The off-gases from the firing process are conducted to a purification unit essentially consisting of afterburning, off-gas desulfurization, and dust and mercury separation.

OUTLOOK

The investigations described in this paper have shown that mechanical separation of dredged material and subsequent manufacture of pellets from the fine-grained dredged material are technically feasible and ecologically safe. Furthermore, this process can also be used for treating similar sludges and contaminated soils.

A substantial amount of the dredged material produced each year can be utilized in this way. As a result, clay and natural stone resources are preserved, making an important contribution toward landscape protection and preservation.

Further investigations are aimed at refining the process to optimize product quality and cost.

REFERENCES

1. Christiansen, H., "New Insights on Mud Formation and Sedimentation Process in Tidal Harbors," *Proc. of the 2nd Int. Conf. on Coastal and Port. Eng. in Developing Countries*, Beijing, China, 1987.

2. Ergebnisse aus dem Baggergutuntersuchungsprogramm, Fachseminar Baggergut, 1984, Eigenverlag Freie und Hansestadt Hamburg, ISSN 0177-1191.

3. Frey, M., *et al.*, "Durchführbarkeitsstudie Stufe I: Sammlung, Vorauswahl und Beschreibung der aussichtsreichsten Lösungsmöglichkeiten," Report of Battelle-Institut e.V., August 1982, September 1983.

4. Göhren, H., Probleme der Baggergutunterbringung des Hamburger Hafens, *Z. f. Kulturtechnik und Flurbereinigung*, 1982, *23*, 95-104.

5. Hampel, H. J., Hankel, D., *et al.*, "Halbtechnische Versuche zur Herstellung von gebrannten Schlickpellets," Report of Battelle-Institut e.V. and Lurgi, November 1986.

6. Hampel, H. J., Hankel, D., *et al.*, "Thermal Treatment of Dredged Material," *Proc. of the Second International TNO/BMFT-Congress on the Clean-Up of Old Dumping Sites*, Hamburg, April 11-15, 1988.

7. Hilligardt, R., "Zum Einsatz von Hydrozyklonen für die mechanische Aufbereitung organikhaltiger Baggerschlämme," Doctoral thesis, 1986, TU Hamburg-Harburg.

8. Kröning, H. and Bracker, U., "Mechanische Aufbereitung von kontaminiertem Hafenschlick mit dem Ziel der Minimierung von Ablagerungsflächen unter Verwertung unbelasteter Sedimentanteile, Bau und Betrieb der Anlage," Final Report of the BMFT Project No. 1430327, Hamburg 1986.

9. Tamminga, P. G. and Hirschberger, H., "Einlagerungskonzeption, Baubetrieb und Überwachungsmaßnahmen für die Schlicklagerstätten im Hamburger Hafen," *Proc. of the International Environmental Congress, The Harbor--An Ecological Challenge*, Hamburg, September 11-15, 1989.

10. Tent, L. and Wild, S., Ergebnisse aus dem Baggergutuntersuchungsprogramm No. 2, Sedimentuntersuchungen im Hamburger Hafen, 1986, Eigenverlag Freie und Hansestadt Hamburg, ISSN 0177-1191.

11. Werther, J., *et al.*, "Mechanische Aufbereitung von kontaminiertem Hafenschlick mit dem Ziel der Minimierung von Ablagerungsflächen unter Verwertung unbelasteter Sedimentanteile," Final Report of the BMFT Project No. 1430325/9, Hamburg 1986.

TOTAL RECYCLING OF WASTEWATER IN WET PROCESS PHOSPHORIC ACID PRODUCTION

R. R. Stana
IMC Fertilizer, Inc.
New Wales Operations
P.O. Box 1035
Mulberry, Florida 33860
(813)428-2531

ABSTRACT

The wet process phosphoric acid industry historically treated their acid-containing wastewaters with lime and then discharged the treated water. The IMC Fertilizer plant at New Wales was designed to avoid all treatments and discharges by controlling waste generation and the use of recycling. This plant has been successfully operating without any liquid discharge since its start-up in 1975. The benefits and unique problems associated with this operation will be discussed.

INTRODUCTION

In 1972, the predecessor of IMC Fertilizer, Inc., International Minerals and Chemicals Corporation, decided to build a large facility for the production of wet process phosphoric acid and associated fertilizer products. The site chosen was in central Florida near existing phosphate rock mining operations.

The plant would employ the wet process for making phosphoric acid. This process consists of reacting phosphate rock (see Table 1 for typical composition) with sulfuric acid to precipitate a gypsum waste product that leaves a crude phosphoric acid solution. The phosphoric acid would then be treated to remove some of the impurities, evaporated to remove some of the water, and then reacted with ammonia or phosphate rock to produce dry granular fertilizers. The waste gypsum would be stacked in a confined area, since there has been and still is no significant use for the large quantities of crude gypsum generated.

The plant would require a cooling pond to reject the heat associated with the various exothermic processes, or the heat would have to be removed by the condensers. As was the practice at that time, the pond would also be the destination for all plant process waste streams as well as for the drainage from the gypsum stack.

At that time, it was decided that the facility would be designed for total recycle of all pond water to meet the 1977 and 1983 requirements of the Federal Water Pollution Control Act Amendments of 1972. This would avoid the requirement for the time-consuming and costly application for a NUDES permit. This was a bold step, since until then, all wet process phosphoric acid plants continuously or at least periodically treated and discharged 500 to 10,000 gpm of treated wastewater. Operating without a discharge meant that two balances had to be considered, a water balance and a salt or impurity balance.

WATER BALANCE

Extensive water balance calculations were made by Davy McKee, the engineering firm hired by IMC to design the plant. They showed that a properly designed plant and cooling pond could be operated with no discharge from the cooling pond, provided that the pond had enough surge capacity to accommodate the seasonal rainfall pattern.

When one thinks of Florida, one thinks of the year-round hot days. Evaporation must be large and a water balance would be no problem! With long-term lake evaporation averaging about 49 in. per year in central Florida, this is certainly true. However, it also rains a lot in Florida, especially in the summer. Typically, rain exceeds evaporation by about 5 in. per year. Worse,

Table 1. Typical Composition of Central Florida Pebble(6)

Major Components	Weight %	Trace impurities	ppm
BPL	68.24	Arsenic (As_2O_3)	5-30
P_2O_5	31.26	Beryllium (Be_2O_3)	ca. 10^2
CaO	45.96	Boron (B_2O_3)	20-100
CaO/P_2O_5	1.47	Cadmium (CdO)	ca. 10
Fe_2O3	1.38	Cesium (Cs_2O)	ca. 10
Al_2O_3	1.30	Chromium (Cr_2O_3)	1-130
CO_2	3.67	Copper (CuO)	<5-30
SiO_2	9.55	Iodine (I)	5-15
F	3.70	Lead (PbO)	ca. 10
Na_2O	0.22	Lithium (Li_2O)	ca. 10
K_2O	0.15	Manganese (MnO)	20-500
MgO	0.37	Mercury (HgO)	ca. 10^2
SO_3	0.82	Molybdenum (MoO_3)	20-50
Cl	0.012	Nickel (NiO)	20-50
Organic Carbon	0.60	Nitrogen (N)	60-150
Total Carbon	1.60	Rare Earths (Re_2O_3)	300-700
		Rubidium (Rb_2O)	ca. 10^2
		Selinium (SeO_2)	0-15
		Silver (Ag_2O)	ca. 10^2
		Strontium (SrO)	ca. 10^3
		Tin (SnO_2)	10-50
		Titanium (TiO_2)	300-7007
		Uranium (U_3O_8)	100-200
		Vanadium (V_2O_3)	10-200
		Zinc (ZnO)	<5
		Zirconium (ZrO_2)	ca. 10

most of the rain comes in the summer, commonly associated with a near-miss hurricane that can dump 10 in. or more of water in a single day.

Not only would our cooling pond collect any rain falling on it, but also drainage from the adjacent gypsum stack and runoff from the plant itself. Overall, the pond would have to dispose of the rainfall from about 2.8 times its planned area. Additionally, there were many other water inputs from the plant itself.

Production of phosphoric acid and phosphate products requires the evaporation of the acid and drying of the products. All the evaporated water ends up in the pond, typically 300 gal/ton of P_2O_5 produced. Due to strict air pollution regulations, the water driven off by drying operations also ends up in the pond. This occurs because all the dryer gasses must be scrubbed with cooling water to remove any fluorine or particulate matter. In the process the water is also condensed from the gas.

However, the phosphoric acid process also helps the net water balance in the pond. Almost all the heat generated by the burning of sulfur to make sulfuric acid ends up in the cooling pond, either directly or indirectly. This is because most of this heat is turned into steam, which is used throughout the plant. Although some of the energy is removed from the steam to make electricity, ultimately it is condensed with the heat being rejected to the cooling pond. This heat load on the pond is very significant. It causes evaporation to increase by a factor of 2 to 3, depending on the production rate.

The phosphoric acid process also consumes some water from the pond (see Table 2). About 600 gal/ton of P_2O_5 is added in the digestor to produce the 26 to 30 percent P_2O_5 acid and carry away the process heat by evaporation. Pond water is or can be added in other parts of the process, such as in wet rock grinding.

Table 2. Water Balance for Wet Phosphoric Acid

PROCESS INPUTS (Kg of Water/Kg of P205)	
3.47	As Pond water to filter
1.65	With wet rock
0.06	Free water with H2SO4
0.04	Combined water with rock
0.46	Combined water with H2SO4
TOTAL	**5.68**
PROCESS OUTPUTS (Kg of Water/Kg of P205)	
2.08	With acid
0.99	Evaporated to remove heat in flash cooler
1.30	Free water with gypsum
0.96	Combined water with gypsum
0.35	Combined water with acid
TOTAL	**5.68**

Even with the above uses of pond water in the plant, it would be impossible to keep the pond from overflowing. Fortunately, there is another place where pond water is "used." As stated earlier, the major waste product from wet process phosphoric acid is gypsum dihydrate. Not only is water required to hydrate it, but the gypsum also retains a significant quantity of moisture. Typically, the gypsum on our stack contains 25 percent free water. The combination of free and crystallized water represents a consumption of about 550 gal/ton of P_2O_5 produced.

Fortunately, when all the inputs and outputs are added together, there is a net deficit in the water balance. The overall pond-gypsum-stack plant complex will consume or evaporate more water than it receives.

Controlling the water balance in the plant is much more complex. To control the pond level between the two extremes of dry and overflowing (a difference of only 5 feet in pond height), we needed to control the plant inputs and outputs of water to balance the expected net effect of rainfall and evaporation. The pond freeboard is also large enough to accommodate any short-term differences between predicted and actual weather. The real problem is controlling the pond level for seasonal weather differences.

To help us manage our pond water balance, a computer program was written by our engineering department in 1982. This program was rewritten and updated in 1988 to take into account an expected expansion of the gypsum stack. Both programs use long-term weather patterns (rainfall, evaporation, wind, and humidity records) and the expected production rates for New Wales. They are used to predict the pond level 12 months ahead. (Figure 1 is a typical output of the program.)

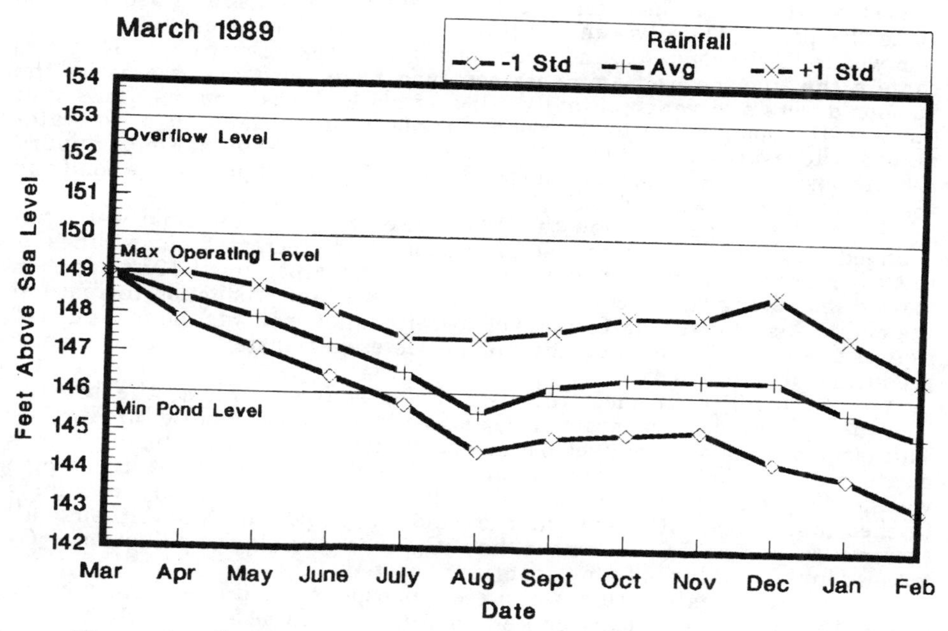

Figure 1. Cooling Pond Level August 22, 1988--Forecast

If the program predicts the pond will overflow, then steps are taken in the plant to reduce the addition of fresh water into the pond. This has occurred twice during the 15 years of New Wales operation. Projects implemented during those times included diversion of parking lot runoff away from the pond, use of pond water in wet rock grinding, collection of fresh water used in pump seals, use of mechanical pump seals instead of packed seals (which require fresh water for flushing), and the use of pond water instead of fresh water in several areas of the plant.

If the model predicts that the pond will go dry (the more likely case most of the time), then water is simply added to the pond, but not until it is at the lowest level that it can be operated.

Occasionally, wet years bring a crisis in the pond water management. In September, 1988, the pond was only 18 in. from overflowing after a very wet summer. A hurricane was also approaching, with all indications of one that would give us at least 10 in. of rain. Fortunately, the hurricane did not reach us and there was very little rain. The weather that followed was very dry, allowing the pond to drop quickly to a low level. If the weather had not been so cooperative, an emergency overflow pond can hold the equivalent of about two back-to-back hurricanes. If the pond does overflow, the excess water can go to the emergency pond where it will be held until it can be fed back into the cooling pond system.

We are now very comfortable with the management of the water balance in our pond and gypsum stack. We now have the scientific understanding necessary to predict accurately the effect of any plant or operating changes and have published two papers on the subject(1,2).

IMPURITY BALANCE

When the plant was designed, little consideration was given to the impact of impurities. The operation of the plant requires that all waste streams go to the pond. These waste streams contain all the scale-forming solids removed from pipes, pumps, and tanks during the weekly scrub days. They also contain the fluorine and ammonia removed from the stack gasses by the various dryers and vents. Finally, all acid spills, leaks, and overflows that occur in the complex go to the pond. Table 3 shows detailed pond water analyses at several times during the plant's operation. Note that most impurities in phosphate rock end up in the pond water. Typical inputs to the pond and their source are shown in Table 4.

Figure 2 shows how the P_2O_5 concentration in the pond water has changed with time. As can be seen the concentration increased almost linearly until it reached 1.8 percent in 1981, where it remained until 1988. It then started increasing again, coinciding with a significant production increase at the plant. Also during the 15 years of operation, not only was the plant debottlenecked and expanded, but additional processes were installed to produce new products and by-products.

This P_2O_5 comes from two major sources. First, the routine operation of the plant generates P_2O_5 losses of about 5 percent of production. This is not as high as it may seem considering that crude wet phosphoric acid is a saturated solution with many impurities that will precipitate readily on pipe walls, heat exchangers, valves, tank walls, etc. All systems are regularly flushed or scrubbed with pond water to remove the P_2O_5 buildup. Although all systems are purged first to remove the P_2O_5, there are always pockets of P_2O_5 trapped in the buildup or in low spots in the system.

Additionally, crude wet process phosphoric acid is both very corrosive and erosive. The typical composition is given in Table 5. The nature of the acid causes leaks in the many miles of pipe lines, tens of thousands of valves, and over 1,000 pumps in the complex. These occur in spite of the use of exotic metals and linings, conductivity alarms on key ditches, and continuous sampling and bihourly and hourly analyses of samples from all ditches in the plant.

However, P_2O_5 is added to the pond from another very significant source that we have discovered only in the last few years. We have always known that the gypsum contains 3 to 7 percent of the P_2O_5 contained in the incoming rock. This P_2O_5 can be in the gypsum as unreacted rock coated with gypsum or as P_2O_5 trapped or coprecipitated in the gypsum. These losses were always thought to be permanent, with recovery requiring the dissolution of the gypsum. However, both P_2O_5 balances around the pond system and laboratory tests show that at least 20 percent of this P_2O_5 is recovered by contact with pond water. The reaction appears to involve the exchange of fluoride or fluosilicic acid for P_2O_5 in the gypsum.

Although the higher P_2O_5 in the pond water means that P_2O_5 losses are high, it also means that P_2O_5 recovery from the pond is high. As seen in Table 2, there is over 2 kg of pond water (net = pond water to filter minus free water associated with the gypsum) per kg of P_2O_5 consumed by the phosphoric

Table 3. Typical Pond Water Analysis

Sample Date	6/9/79	8/5/80	7/1/82	4/3/87	9/18/87
Analysis					
Turbidity NTU	10			29.5	
Color	320			1000	
Specific Conductance	31200			26700	32000
COD mg/1	1200	6800			
pH	1.6	1.35	1.6	1.4	1.6
N-org (ppm)	26			77	
N-amon.(ppm)	460	440	800	1421	1108
P	4500	5590	5280	6500	6900
Carbon, org.	150	200			
Ca	70	1466	1300	1248	1076
Mg	87	217	240	279	300
Na	1700	1692	1560	2020	2210
K	180		209	292	271
SO_4	5300	6500	6350	5728	4216
F	8500	9250	7745	9400	9000
Si	4700	2630	2100	4452	5318
Al				228	273
As				0.12	0.33
Be				0.10	
Cd				0.41	
Cu				0.28	
Cr				1.52	
Fe				190	252
Pb				0.15	
Mn				11.4	
Hg				<0.0002	
Ni				1.40	
Se				0.006	
Ag				0.06	0.06
Zn				4.79	

acid process. This means that at the present pond P_2O_5 concentration, about 0.05 ton of P_2O_5 is recovered from the pond per ton of P_2O_5 produced.

Figure 3 shows how the fluorine concentration varied with time. Unlike the P_2O_5 concentration, fluorine rapidly rose to a concentration of about 1.2 percent. Except for fluctuations caused by extreme wet spells, the fluorine value has remained relatively steady since then. Since 40 to 70 percent of the fluorides originally present in the rock are continually sent to the pond or gypsum stack, it is clear that there must be some sort of sink for the fluorine.

Table 4. Typical Inputs to a Phosphoric Acid Cooling Pond

Impurity	Input Kg/Kg P205	Typical Input tons/year for 1,000,000 ton/yr P205 Plant
F	.04-.07	50,000
P205	.04-.07	50,000
NH3	.004-.01	7,000

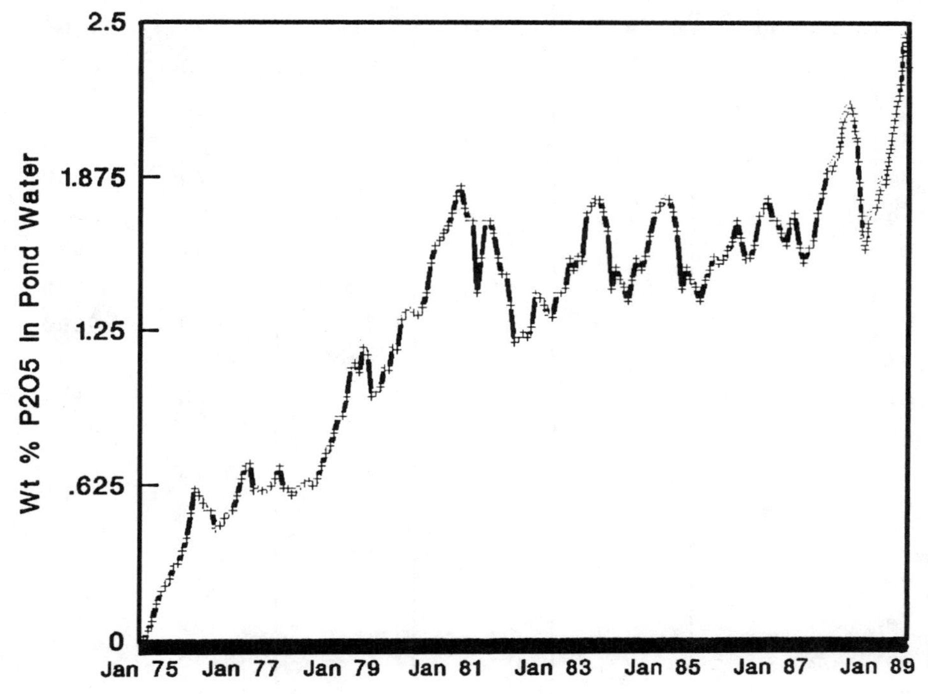

Figure 2. P_2O_5 in Pond Water Versus Time

The vapor pressure over pond water has been measured by the U.S. EPA and others and has been found so low that loss of fluorine to the atmosphere is negligible[5,7].

The fluorine in pond water is typically present both as fluosilicates and fluorides. These are in equilibrium with all the cations contained in the water. The high ionic strength of the pond water will affect equilibrium constants of all of these species. The present knowledge of pond water chemistry is not sufficient to fully understand all the reactions that will occur. In general, however, it is known that the high ionic strength of the pond water will oppose reactions that contribute multivalent ions to the solution. Equilibria will tend to be more balanced toward the uncharged or low-charged species[4].

Work by Lehr[4] identified 12 fluoride compounds present in gypsum and pond water, all of which have solubilities that are affected by the cations present and the ionic strength of the pond water. Recent work conducted by IMC has suggested that the bulk of our fluorine is precipitated as the compound

Table 5. Typical Phosphoric Acid Composition, 30 Acid

NOMINAL % P_2O_5	30
P_2O_5	28.85
MgO	0.43
Fe_2O_3	0.92
Al_2O_3	0.80
H_2SO_4 (total)	1.85
Solids	0.25
F	2.18
CaO	0.35
H_2O	53.39
Sp. Gr. (at 25^{0}C)	1.340

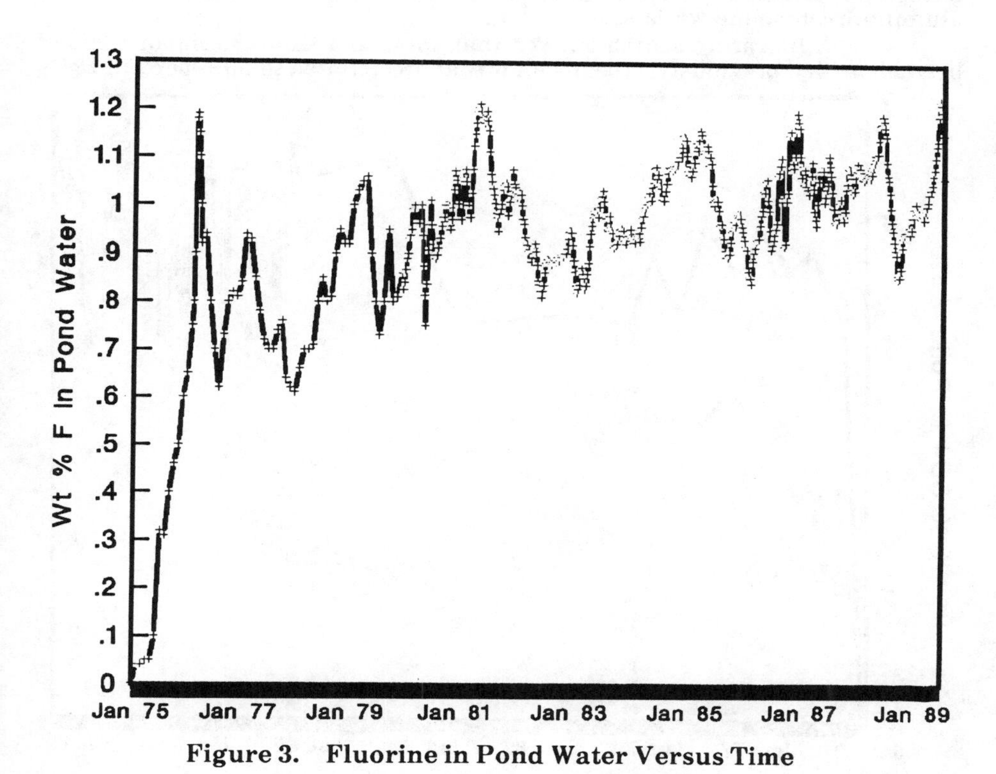

Figure 3. Fluorine in Pond Water Versus Time

chukhrovite ($Ca_4AlSiF_{13}SO_4 \cdot 10H_2O$). The fluorine compounds form by reactions between gypsum and the ions contained in the pond water. As explained earlier, at least some of these reactions liberate the P_2O_5 contained in the gypsum.

Figure 4 shows how the sulfate concentration has varied with time. As can be seen, it is similar to the fluoride graph in that it rapidly rose to an equilibrium value. Clearly, the sulfate is in equilibrium with the calcium in the water. The seasonal variations in the concentration are due to the change in solubility of gypsum with changing temperature.

Figure 5 shows how the sodium content of the pond water changed with time. Again the sodium content rapidly rose to an equilibrium value. Although considerable sodium is sent to the pond from ion exchange and carbon columns regenerants, the bulk of the sodium comes from the phosphate rock itself. The sodium precipitates primarily as sodium fluosilicate, maintaining the equilibrium values of sodium seen in Figure 5.

Figure 6 shows how the nitrogen content of the pond water has varied with time. It also increased very rapidly and seemed to equilibrate at about 0.04 percent during the first 4 years of operation. In fact, the daily measurement of nitrogen was terminated at that time. However, occasional spot checks showed that the nitrogen concentration started to increase in 1982. At that time we increased production of ammoniated phosphates. Additionally, initial operation of some of these units gave high ammonia losses to the pond.

Since previous work had not identified any ammonia-containing compounds that precipitate in phosphoric acid at low ammonia concentrations, there was no concern about the increase in nitrogen concentration. However, in 1985 we began to experience a very high rate of scale formation both on our phosphoric acid filters and in the carbon columns used to pretreat the acid prior to uranium recovery. The compound was identified as an iron ammonia phosphate, principally $Fe_3NH_4H_{14}(PO_4)_8 \cdot 4H_2O$. A small amount of a similar aluminum compound was also identified.

A literature search showed that these and similar compounds had been identified previously as compounds that precipitated in 50 + percent P_2O_5

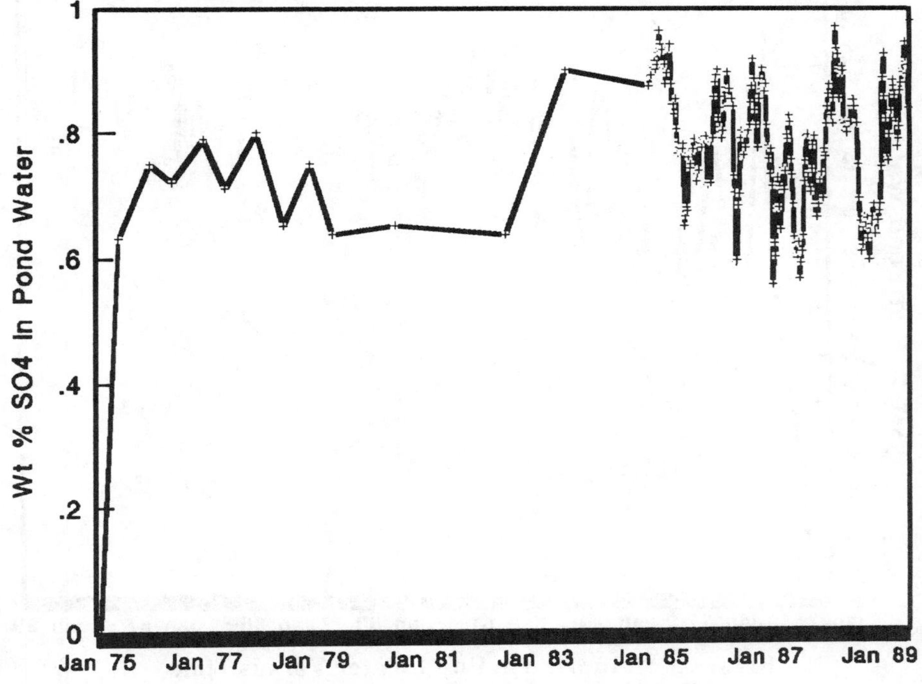

Figure 4. SO_4 in Pond Water Versus Time

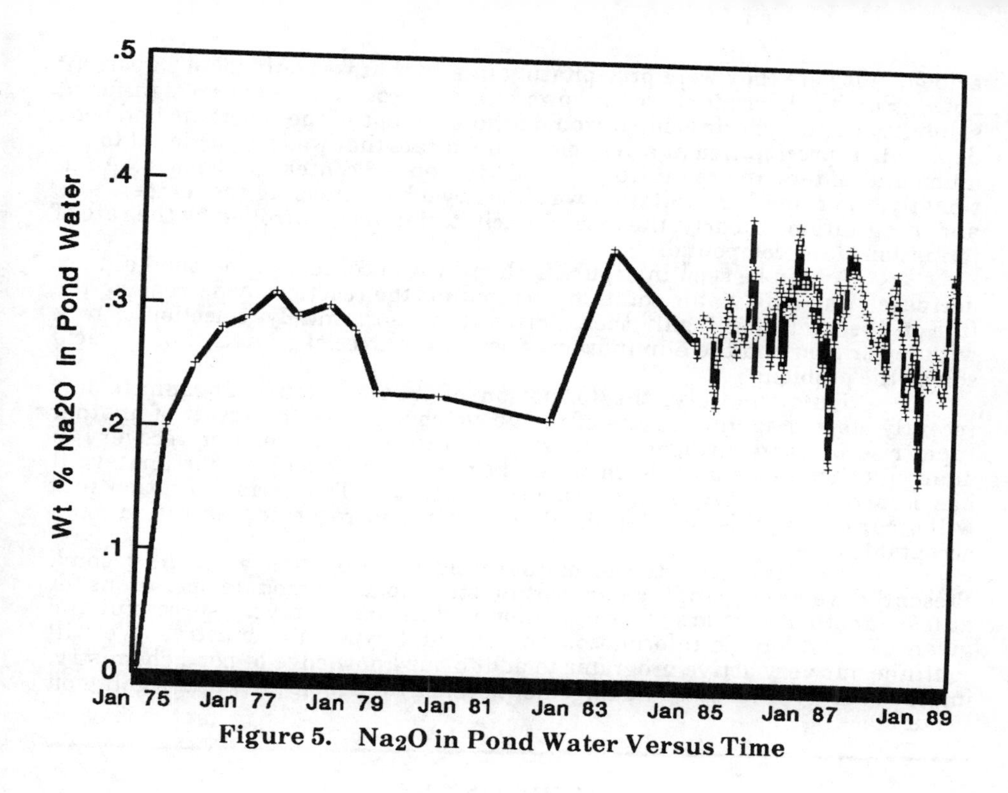

Figure 5. Na_2O in Pond Water Versus Time

Figure 6. N_2 in Pond Water Versus Time

acid[3]. But here they were precipitating in acid that was only 26 to 28 percent P_2O_5! Further laboratory work showed that the compound had no significant solubility in phosphoric acid. It would form until one of the reactants had been depleted. It precipitated at a very slow rate, a rate that was proportional to the ammonia content and the third power of the iron +3 content of the acid. Additionally, the rate of precipitation was increased by increasing the temperature and by agitation. Clearly, the rate of precipitation was controlled by the rate of formation of the compound.

As can be seen in Figure 6, the nitrogen content of the pond did not decrease. The only viable solution was to limit the reaction by decreasing the iron in the acid, specifically the oxidized iron. Fortunately, selection of rock with lower iron and the elimination of several sources of oxidation of the acid solved the problem.

Unfortunately, the formation of iron-ammonia phosphate has recently started again, in spite of the above changes, as the result of another rapid rise in pond nitrogen content. A faulty pre-scrubber (for recovery of ammonia back to the process) in one of the plants was to blame. Fortunately, it has occurred at the beginning of our rainy season. The normal dilution that will occur during this summer should reduce the nitrogen concentration to an acceptable level.

IMC is committed to maintaining a "zero release" cooling pond. Presently, we are expanding our gypsum stack to accommodate operations for at least another 20 years. Our operation to date has been very successful and given us considerable information on pond and gypsum chemistry. We will continue our very active programs to add to our knowledge of pond chemistry, improve our gypsum stack management system, and monitor what is going on in our stack.

REFERENCES

1. Cameron, J. E., Project Specialist, IMC Fertilizer, Inc., "Land Planning for Phosphogypsum Stacks in Central Florida," presented at American Society of Civil Engineers Florida Section/South Florida Section 1988 Joint Annual Meeting, Sheraton Inn, Sand Key, FL, September 29-October 1, 1988.

2. Cameron, J. E., Project Specialist, and O'Connor, J. J., Project Engineering Manager, "Design and Operating for a Gypsum Stack/Cooling Pond from an Owner/Operator's Point of View," presented at Second International Symposium on Phosphogypsum, University of Miami and J. L. Knight Conference Center, December 10-12, 1986.

3. *Crystallographic Properties of Fertilizer Compounds*, Chemical Engineering Bulletin No. 6, TVA, 16.

4. Lehr, J. R., *Fluorine Chemical Distribution Interrelation to Gypsum Storage Pond Systems*, TVA, Muscle Shoals, AL, undated.

5. Preece, J. W., "Derivation of Emission Factors for Fluorides from Gypsum Ponds," Memo to File, TRW Environmental Engineering Division, Research Triangle Park, NC, August 27, 1980.

6. Sepehri-Nik, E., *Fertilizer Technical Data Book*, FCI Chemical Engineers, 1989.

7. U.S. Environmental Protection Agency, "Measurement of Fluoride Emissions from Gypsum Ponds," Contract 69-01-4145, Task No. 10 U.S. EPA, Division of Stationary Source Enforcement, Washington, DC, September, 1978.

AN OVERVIEW OF SEPARATION TECHNOLOGIES DEMONSTRATED BY THE EPA'S SUPERFUND INNOVATIVE TECHNOLOGY EVALUATION (SITE) PROGRAM

Stephen C. James
SITE Demonstration Evaluation Branch
U.S. Environmental Protection Agency
26 West Martin Luther King Drive
Cincinnati, Ohio 45268

INTRODUCTION

The U.S. EPA's Superfund Innovative Technology Evaluation (SITE) Program's major focus is the technology demonstration program, which is designed to field demonstrate/evaluate cleanup technologies for the purpose of providing engineering reliability and economic data on operational aspects. This joint program between the EPA and technology developers will provide extensive data that validate technology capabilities and assess performance, reliability, and cost of new and innovative technologies for the cleanup of hazardous waste/Superfund sites.

The program has completed over 15 field demonstrations in the areas of thermal, chemical, solidification/stabilization, biological, and physical processes that treat, destroy, recycle, or separate contaminants from the waste material. Currently the program's major emphasis is on evaluating technologies applicable to sludges and soils. This presentation will focus on the results from completed demonstration projects and will include lessons learned. Specific technologies that will be discussed are

- Terra Vac's In-Situ Vacuum Extraction Technology
- CF System's Solvent Extraction Technology.

The presentation will also discuss various aspects of conducting field demonstration projects. This includes sampling and analysis, waste feed preparation, site preparation, equipment decontamination, permit issues, and community relations.

TERRA VAC (VACUUM EXTRACTION TECHNOLOGY)

The vacuum extraction process is technology for removal and venting of volatile organics compounds (VOCs) from the unsaturated zone of soils. Once a contaminated area is completely defined, an extraction well or series of wells, depending on the extent of contamination, can be installed. A vacuum system induces airflow through the soil, stripping and volatilizing the VOCs from the soil matrix into the air stream. Liquid water is generally extracted along with the volatile contamination flow. This two-phase flow of contaminated air and water flows to a vapor/liquid separator where contaminated water is removed. The contaminated airstream then flows through activated carbon units arranged in series. Primary adsorbing canisters are followed by a secondary or backup adsorber to ensure that no contamination reaches the atmosphere.

Terra Vac Inc.'s vacuum extraction system was demonstrated at the Valley Manufactured Products Company, Inc., site in Groveland, Massachusetts, under the U.S. EPA's Superfund Innovative Technology Evaluation (SITE) Program. The property is part of the Groveland Wells Superfund site and is contaminated mainly by trichloroethylene (TCE).

This technology uses readily available components such as extraction and monitoring wells, manifold piping, vapor/liquid separator, vacuum pump, and emission control equipment such as activated carbon canisters. Once a contaminated area is completely defined, an extraction well is installed and connected by piping to a vapor/liquid separator device. A vacuum pump draws the subsurface contaminants through the well, separator device, and activated carbon system before the airstream is discharged to the atmosphere. Subsurface vacuum and soil vapor concentration are monitored via vadose zone monitoring wells. Figure 1 provides a schematic drawing.

The in-situ vacuum extraction demonstration was conducted at the Groveland Wells Superfund site from December 1, 1987, to April 26, 1988. Four extraction wells were used to pump contaminants. Four monitoring wells

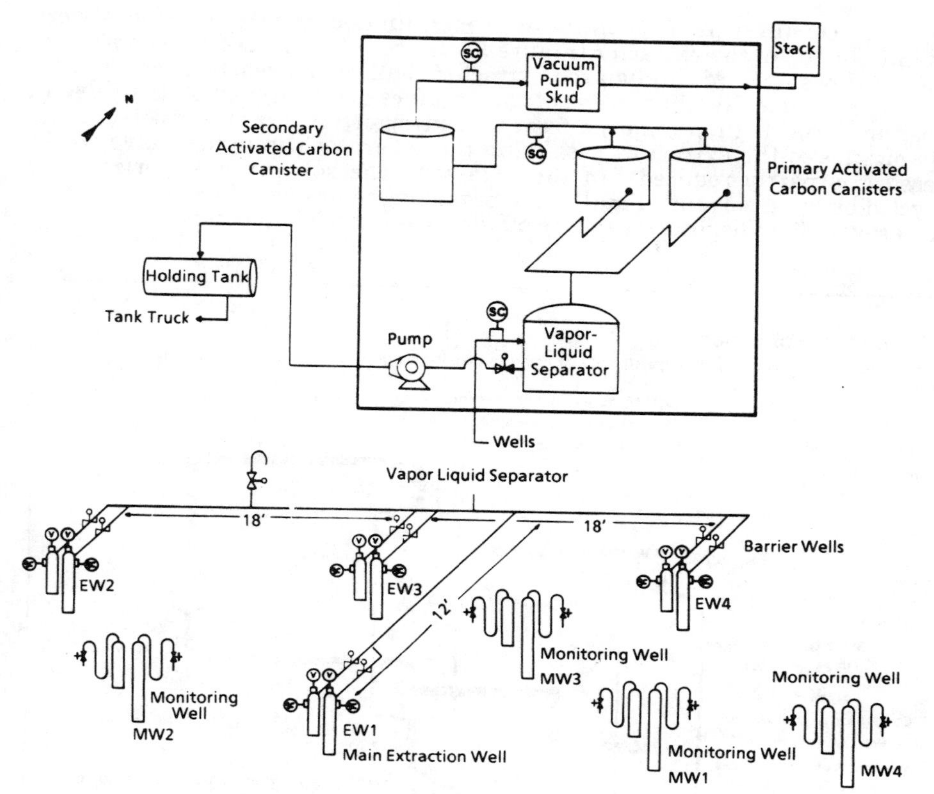

Figure 1. Schematic Diagram of Equipment Layout

were utilized to determine the effectiveness of the process. The 8-week demonstration produced the following results:

- Extraction of 1,300 lb of VOCs
- A steady decline in the VOC recovery rate with time
- A marked reduction in soil VOC concentration in the demonstration area
- An indication that the process can remove VOCs from clay strata--VOCs removed from both highly permeable strata and low-permeability clays.

CF SYSTEMS (ORGANIC EXTRACTION UTILIZING SOLVENTS)

This technology utilizes liquified gases as the extracting solvent to remove organics, such as hydrocarbons, oil, and grease, from wastewaters or contaminated sludges or soils. Carbon dioxide is generally used for aqueous solutions, and propane is used for sediments, sludges, and soils.

The SITE demonstration of CFS solvent extraction technology was conducted during September 1988 on PCB-contaminated sediments from the New Bedford, Massachusetts, harbor. The pilot-scale commercial unit, rated at 20 barrels/day, was used during the demonstration. The unit was a trailer-mounted organic extraction system for nonaqueous materials and used a propane/butane mixture as the extraction solvent.

During this demonstration, contaminated harbor sediments were fed into the top of the extractor (Figure 2). Solvent (condensed by compression at 70 F) flows upward through the extractor, making nonreactive contact with the waste. Typically, 99 percent of the organics are dissolved by the solvent. Clean material is then removed from the extractor. A mixture of solvent and organics leave the extractor, passing to the separator through a valve where pressure is partly reduced. In the separator, the solvent is vaporized and recycled as fresh solvent. Finally, the organics are drawn off from the separator, recovered for disposal, or reused off-site in industrial processes.

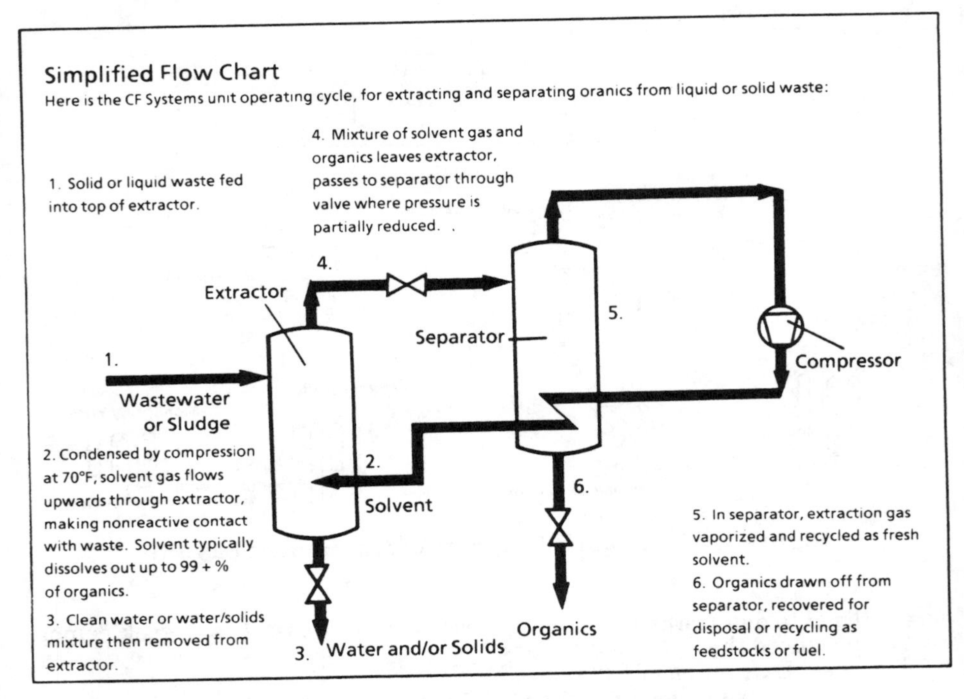

Figure 2. Simplified Flow Chart

During the demonstration, PCB concentrations (ranging from 300 to 5,000 ppm) and the number of passes through the unit were varied for each of the four separate test runs. About one-half drum (30 gallons) of sediment, with additional water added to obtain the appropriate consistency, was processed for each run. Results from the demonstration are:

- PCB removal efficiencies of 90 percent were achieved for sediments containing PCBs ranging from 350 to 2,575 ppm. A high removal efficiency was achieved after several passes, or recycles, of treated sediments through the unit.
- Extraction efficiencies greater than 60 percent were achieved on the first pass of each test. Later passes of treated sediments through the unit resulted in efficiencies that ranged from zero to 84 percent.
- Metals were not removed from the sediments, as expected.

FLOCCULATION

Flocculation plays a vital role in the solid/liquid separation industry. Floc is generally defined as a porous agglomerate consisting of tiny particles that form as a result of interparticle collisions. Flocculation strongly affects the dewaterability of sludges. Good flocculation improves both the extent of solids dewatering and the quality of the filtrate sent for further treatment. Flocculants should possess certain important size, density, and strength characteristics, and toxicity and other detrimental effects should be considered before their application.

The desired floc properties vary depending on the application. For example, a strong, porous, but lower density floc is desired for a filtration process, but a thickening process will work better with a large, dense floc of minimum porosity. Recent advances in the design of sedimentation devices owe their success to the recent development of synthetic flocculants. Selection of a polymer as a flocculant is complex, requiring a thorough optimization of a number of parameters.

In this session a number of interesting papers are being presented. Dr. Moudgil is a well-known colloid scientist and has worked extensively in the area of flocculant selection for the mineral and allied industries. Professor Somasundaran of Columbia University is the chief editor of the *Journal of Colloid and Interface Science* and has been pursuing very interesting research in this area for several years. Dr. Attia from The Ohio State University has done some original work in selective flocculants, and Mr. Kim of Stranco will be presenting some real-world experience in flocculation mixing. Professor Rubin of The Ohio State University, Dr. Wickramanayake of Battelle, and Mr. Bowen of Metcalf & Eddy will also discuss interesting work in the field.

CHARACTERIZATION OF FLOCS FOR SOLID/LIQUID SEPARATION PROCESSES

Brij M. Moudgil and Mark E. Springgate
Mineral Resources Research Center
Department of Materials Science
and Engineering
University of Florida
161 Rhines Hall
Gainesville, Florida 32511
(904) 392-6670

and

T. V. Vasudevan
Department of Mineral Engineering
Columbia University
New York, New York 10027

ABSTRACT

Flocs of specific size, density, and shear strength are required to achieve maximum solid/liquid separation efficiency by sedimentation or filtration. Large dense flocs are desired in sedimentation, whereas highly porous flocs are expected to yield optimum filtration performance. To generate flocs of given characteristics, however, is difficult if not impossible with the current knowledge of polymer properties and floc structure correlations. In the present study, the nature of such correlations is being developed so that flocs of desired properties can be generated. In this presentation, techniques developed to characterize flocs with respect to size, density, and shear strength are discussed.

INTRODUCTION

Aggregation of fine particles by polymer flocculation is important in many applications such as wastewater treatment and clay and coal slurry dewatering. The effectiveness of this technique depends upon the ability to form flocs of the proper structure for the specific solid/liquid separation process. For example, if filtration is being employed, highly porous flocs are desired; however, sedimentation requires large, high-density flocs. Another important factor is shear strength: the flocs formed should be strong enough to withstand the turbulent forces encountered during handling and further processing.

The formation of flocs depends upon the physical variables of agitation intensity, rate of polymer addition, length of time of flocculation, and pulp density of the slurries being employed. The chemical parameters of the polymer that are important are charge, molecular weight, and dosage. The suspension properties such as pH, ionic strength, and temperature of the slurry are also known to affect the flocculation process.

A systematic study was undertaken to understand the role of the above parameters in the flocculation process. The specific goal of this effort is to establish techniques that will yield flocs of desired properties for a given polymer/particle system.

EXPERIMENT

MATERIALS

- Dolomite--from Agrico Chemical Company (-400 mesh)
- Kaolinite--from Ward's Natural Sciences, Inc.
- Alumina--from Alcoa (A-14, -325 mesh)
- Montmorillonite--from Occidental Chemical Co.
- Polyethylene oxide (PEO)--from Polysciences, Inc., 5 million mol. wt.
- SUPERFLOC 127, a nonionic polyacrylamide; SUPERFLOC 214 and SUPERFLOC 204, anionic polyacrylamides--from American Cyanamid Co.
- PAA-15, a nonionic polyacrylamide--from National Chemical Lab, Pune, India

PROCEDURE

Flocculation tests were conducted using a fully baffled (0.85-cm width), 7.6-cm i.d., 14-cm tall, 500-cm^3 standard reaction vessel obtained from the Fisher Scientific Company. It was fitted with a 2.85-cm diameter, six-bladed stainless steel turbine impeller for all the flocculation experiments. The power input was initially determined by connecting a voltmeter across an ammeter in series with the controller. Later the same impeller was attached to a HST-10 stir controller (G. K. Heller Corp.) with torque and velocity readouts. Results by both methods were found to be comparable.

For all experiments, a desired amount of polymer solution was added to 500 cm^3 of a well-agitated suspension. The flocculation was carried out for 2 minutes, followed by increasing the shear rate to a predetermined speed for another 2 minutes to break the aggregates. The resulting suspension was then poured into a large tray to prevent reaggregation of flocs. Some experiments were stopped after the initial 2-minute flocculation period to obtain an original floc size.

FLOC SIZE

The measurement of size is critical to all other floc characterization tests. In these experiments the desired flocs are transferred from the large holding tray to glass slides. The size of the flocs is then determined using a Bausch and Lomb microscope with a measuring grid.

FLOC DENSITY

The density of flocs has traditionally been obtained by measuring the buoyant density of a floc while settling in a quiescent column[4]. This method is difficult to utilize because of the need for a correlation between shape, porosity, and flow conditions of the flocs in question.

A simpler method to measure floc density was developed. In this technique a single floc is removed from the holding tray and placed on a glass slide. The exterior water is removed and the floc is weighed using a precision analytical balance. The floc is dried for 4 hours at 50 C to remove the entrapped water. The volume fraction of solids and water in the flocs and thus the floc density (ρ_f) is then determined using the following equations:

$$(1-\varepsilon) = \frac{M_{df}/\rho_s}{M_{df}/\rho_s + (M_{wf} - M_{df})/\rho_w}$$

$$\rho_f = (1-\varepsilon)\rho_s + \varepsilon\rho_w$$

$$V_f = \frac{M_{wf}}{\rho_f}$$

where $1-\varepsilon$ = volume fraction of solids in the floc, ε = volume fraction of liquid, M_{df} = weight of dry floc, M_{wf} = weight of wet floc, ρ_s = density of the solid, ρ_w = density of the liquid, and V_f = volume of floc.

Floc density experiments were carried out by Moudgil and Vasudevan[7], using kaolinite and montmorillonite clays flocculated with PEO. The effect of three variables--solids loading, flocculant dosage, and chemical nature of the flocculant--on floc density was investigated. No significant effect of solids loading on floc density was found. These results were in agreement with those of Klimpel and Hogg[3] for flocculation of quartz with nonionic polyacrylamide, and computer simulation results of Meakin[6], which also predicted no dependency of the density of aggregates formed on the frequency of collisions between the particles and the aggregates.

Additionally, it was determined in the above study that within the range of polymer dosages examined, the amount of flocculant had no significant effect on floc density. This result also agrees with Klimpel and Hogg[3]. However, optimization of surface coverage leading to the maximum amount of polymer/particle attachments for flocculation should result in a lower floc density[8,11]. As flocculant dosage is increased one would expect to form flocs of lower density since more polymer attachments should result in a stronger floc, thus less floc breakage would occur at a given shear speed. This phenomenon was reported by Tambo and Watanabe[10] for the coagulation of kaolinite with alum. Beyond optimum surface coverage, steric stabilization would occur, thus increasing the density of the flocs formed. It is possible that the range of flocculant dosages used in the above study was not broad enough to allow these observations.

The results of the effect of the chemical nature of the flocculant upon characteristics of kaolinite flocs in the above study showed that smaller flocs formed by the nonionic polyacrylamide (15 million mol. wt.) are less dense than the flocs formed by the PEO (5 million mol. wt.). For larger floc sizes, there was no significant difference in the floc densities. This could result from the fact that the flocs formed with the polyacrylamide were produced at a higher agitation intensity, and thus breakage of larger aggregates could have yielded flocs of higher density.

FLOCS AS FRACTALS

It is known that aggregation of colloidal particles exhibits fractal characteristics. A cluster-cluster aggregation model was developed by Meakin[5,6] to describe the aggregation process. Two variations of the model, diffusion-limited and reaction-controlled aggregation, have been proposed[11].

It is known that flocculation occurs mostly through a process known as bridging. In bridging, a polymer chain initially adsorbs onto an active site of a particle. The same polymer chain also adsorbs onto a bare active site on another particle, thus binding the two particles together[8]. The collision probability/efficiency (E_c) between two similar particles is dependent upon the fraction of active sites on the surface (Φ) and the fractional surface coverage (Θ) of the polymer. The following equation describes this collision probability/efficiency for a given mineral/polymer system[8].

$$E_c = 2\Theta\Phi^2(1-\Theta)$$

In systems with high values of Φ and low values of Θ, or vice versa, many collisions between the particles with adsorbed polymers would have to occur before flocculation would result. Under these conditions the probability of flocculation (E_c) is low, resulting in the need for many collisions before irreversible attachments of polymer occurs. The higher number of collisions results in a greater possibility of penetration between clusters and thus leads to the formation of flocs of higher density. This phenomenon is analogous to the cluster-cluster aggregation processes under reaction-controlled conditions[11]. At moderate values of both Φ and Θ, the probability of flocculation of the particles would be high, and the agglomeration would follow cluster-cluster aggregation under the diffusion-limited model.

Values of Φ, Θ, and the collision probability (E_c) for dolomite and kaolinite flocculated with PEO were calculated. It was found that the collision probability (E_c) of dolomite was higher than that of kaolinite because of the larger value of Θ for dolomite. From this information one would expect dolomite to aggregate as described by the diffusion-limited model. It can be seen from density measurements in Figure 1 that dolomite yields a fractal dimension of 1.68, which indeed represents the diffusion-controlled aggregation model. For

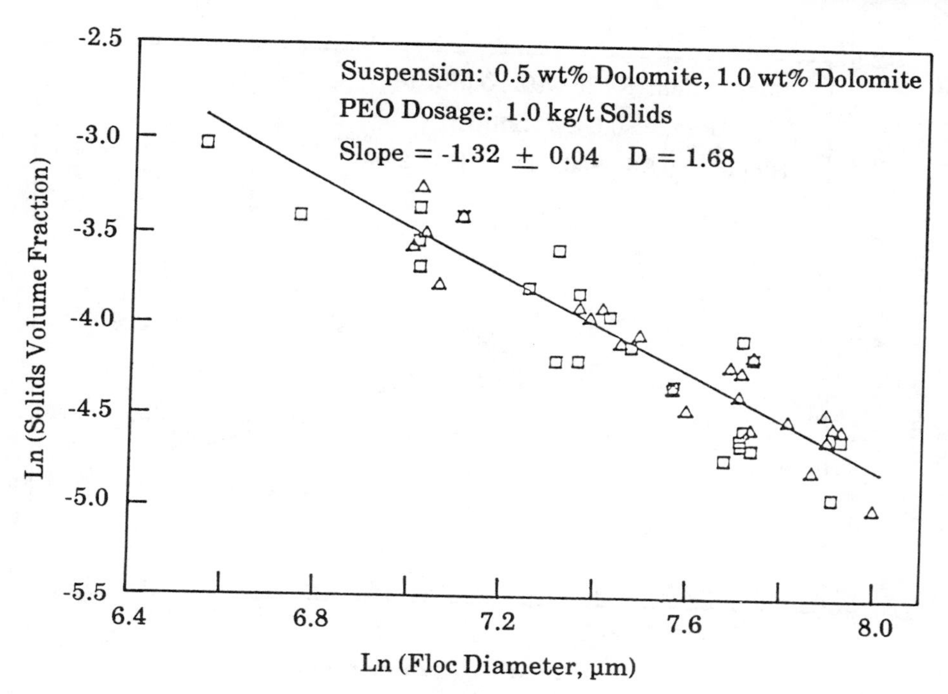

Figure 1. Relationship Between Floc Diameter and Solids Volume Fraction for a Dolomite Suspension; Effect of Solids Loading on Fractal Dimension

kaolinite the number of collisions between flocs would have to be higher before irreversible attachment between the colliding aggregates could occur, since the collision/sticking probability (E_c) is low. From Figure 2, the fractal dimension for kaolinite is seen to be 2.04. Thus aggregation of kaolinite with PEO can be considered reaction controlled. Figure 1 shows that increasing the pulp density of the slurry of dolomite did not change the density of the flocs formed, and therefore the fractal dimension remained unchanged. These results are in agreement with Meakin's hypothesis[6] that the fractal dimension is independent of the frequency of collisions, which would be larger at higher solids loading.

FLOC STRENGTH

The strength of kaolinite, dolomite, and Al_2O_3 flocs was evaluated with various polymers. Previous studies of floc strength[1,2,9] employed methods based on measuring the size and number of flocs that remain after previously formed flocs are subjected to a known shear stress. In most of the past studies the strength of the flocs was measured under viscous flow conditions. In the present investigation, however, the strength of flocs is determined under turbulent conditions, which are more realistic of filtering and pumping flocculated suspensions.

According to Tambo and Hozumi[9], the binding force (S_f) of a floc can be expressed by the following relationship:

$$S_f = \frac{\rho_w}{2}\left[\frac{\varepsilon_0 D_f}{\rho_w}\right]^{2/3} a_f$$

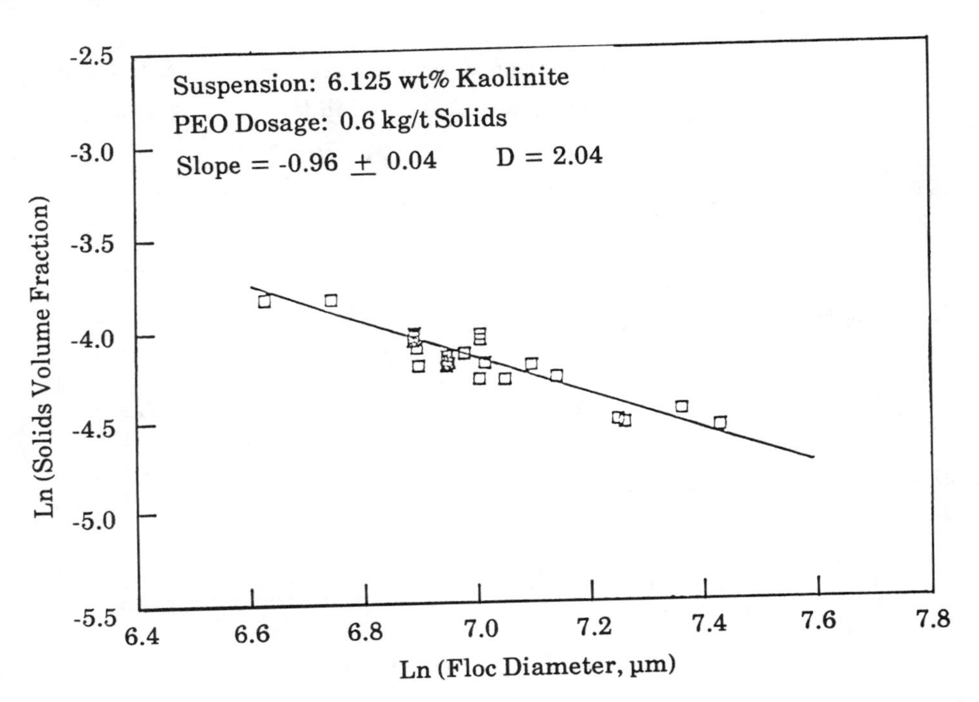

Figure 2. Relationship Between Floc Diameter and Solids Volume Fraction in Floc for a Kaolinite Suspension

where: ε_0 = energy dissipation rate per unit volume of suspension, erg/cm^{3}sec
D_f = mean floc diameter obtained after shearing, cm
ρ_w = density of suspension medium, g/cm^3
a_f = cross sectional area of floc, cm^2.

Normalization with respect to cross-sectional area results in the binding force per unit area termed as binding strength (σ),

$$\sigma = \frac{S_f}{a_f} = \frac{\rho_w}{2}\left[\frac{\varepsilon_0 D_f}{\rho_w}\right]^{2/3}$$

Using dolomite as a substrate, binding strength experiments were carried out by Moudgil and Vasudevan(7) using polymers of different types, molecular weights, and dosages. It can be seen from the results presented in Table 1 that the binding strength reaches a maximum with an increase in polymer dosage and then decreases as the dosage is increased further. This phenomenon probably results from optimum surface coverage for flocculation at a given polymer amount and onset of steric stabilization beyond it. Furthermore, it can be seen that the binding strength obtained using the high molecular weight polyacrylamide was almost the same as that obtained with the low molecular weight polyacrylamide. However, the optimum polymer dosage is 2.5 times greater for the lower molecular weight polymer.

The effect of the polymer's chemical nature on binding strength is presented in Table 2. The binding strength is similar for both the nonionic and the highly anionic polymers. This can be attributed to polymer adsorption characteristics being dominated by entropic rather than enthalpic effects because of the high molecular weight of the polymers used in the study.

The effect of pH upon floc binding strength is shown in Figure 3. In these experiments Al_2O_3 was used as a substrate with a high molecular weight

Table 1. Effect of Polymer Molecular Weight and Dosage on Binding Strength

Suspension: 500 ml of 4 wt% dolomite

Speed of Agitation during Flocculation = 1000 rpm

Polymer*	Polymer Dosage	Shearing Conditions		Binding Strength, σ
	kg/t solids	Time, min.	Speed, rpm	dynes/cm^2
PAA-15 (3.6 million m.w.)	1.00	0.5	1100	34
	1.25	0.5	1100	46
	2.50	0.5	1100	25
SF-127 (15 million m.w.)	0.25	2.0	1200	25
	0.50	2.0	1200	44
	1.00	2.0	1200	28

* Nonionic polyacrylamides

Table 2. Effect of Polymer Type on Binding Strength of Flocs

Suspension: 500 ml of 4 wt% dolomite

Polymer Type	Polymer Dosage	Speed of Agitation, rpm		Binding Strength
	kg/t solids	Flocculation	Shearing	σ, dynes/cm^2
Nonionic, 15 million m.w. (SF 127)	0.5	1000	1200	44
Anionic 12-15 million m.w. (SF 214)	0.5	1000	1200	45

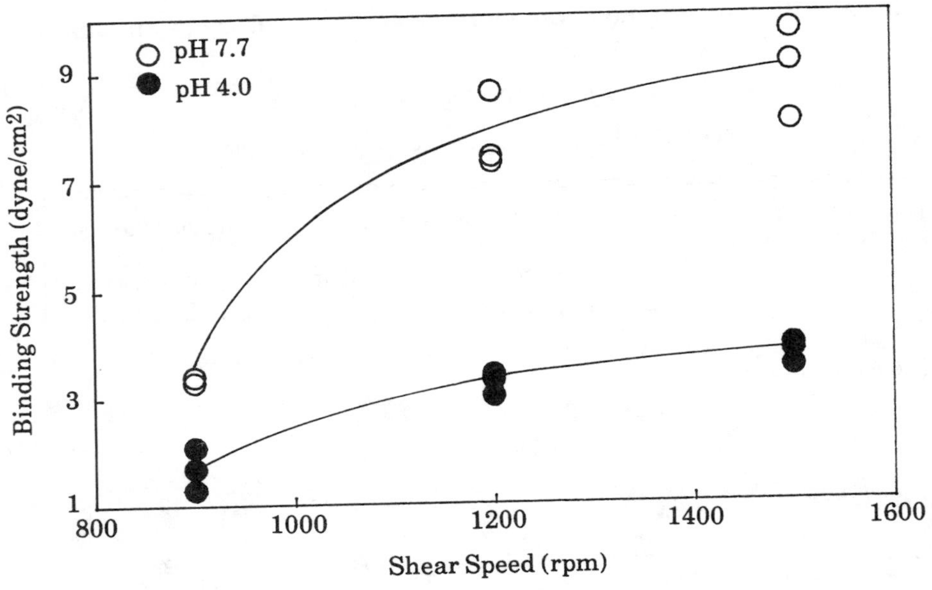

Figure 3. Effect of pH upon Floc Binding Strength

anionic polyacrylamide. For a given polymer dosage, the binding strength increases as the shearing speed is increased, but the flocs at pH 4.0 have a substantially lower binding strength than at the natural pH (7.7). This is because the zeta potential of the Al_2O_3 particles is higher at pH 4.0 (+22.0 mV) than at pH 7.7 (-2.4 mV). At the higher zeta potential the particles repel each other, and a loose network of particles constitutes the aggregates. On the other hand, at pH 7.7 the zeta potential is low, and more dense aggregates with higher shear strength are formed.

SUMMARY

Techniques to characterize floc size, density, and shear strength have been established. Effect of chemical nature of the polymer charge characteristics, and molecular weight on floc density was examined.

Binding strength of flocs attained an optimum at a given polymer dosage. This was attributed to less than adequate polymer adsorption at low dosages and the onset of steric stabilization at very high dosages. The optimum binding strength was found to be independent of molecular weight of the polymer; the dosage, however, was determined to be 2.5 times lower for the higher molecular weight polyacrylamide. Again, this can result from different amounts of polymers required for achieving the optimum surface coverage for maximum flocculation. Anionic and nonionic polyacrylamides yielded similar maximum binding strength. It appears that the major mode of polyacrylamide adsorption on dolomite is predominantly enthalpic in nature.

Binding strength of alumina with polyacrylamide at different pH values indicated that interparticle distance plays a major role in determining the strength of the aggregates formed.

It should be realized that the conclusions reached in the present study are system specific. It is hoped that, as more information is generated and correlation between polymer properties and floc structures are established, general guidelines to produce flocs of desired properties can be formulated.

ACKNOWLEDGMENTS

The authors acknowledge partial financial assistance of the NSF-PYI award (MSM #8352125) and the U.S. Geological Survey, Department of Interior (Grant #14-08-0001-G1555). The contents of this publication do not necessarily reflect the views and policies of the Department of the Interior, nor does mention of trade names or commercial products constitute their endorsement by the U.S. Government.

REFERENCES

1. Glasgow, L. A. and Hsu, J. P., "An Experimental Study of Floc Strength," *AIChE Journal*, 1982, *28*(5), 779-785.

2. Hannah, S. A., Cohen, J. M., and Robeck, G. G., "Measurements of Floc Strength by Particle Counting," *J. of American Water Works Association*, 1967, *59*, 843-855.

3. Klimpel, R. C. and Hogg, R., "Effects of Flocculation on Agglomerate Structure," *J. of Colloid Interface Sci.*, 1986, *113*(1), 121-131.

4. Matsumoto, K. and Mori, Y., "Settling Velocity of Flocs - New Measurement Method of Floc Density," *J. of Chemical Engg. of Japan*, 1975, *8*(2), 143-147.

5. Meakin, P., "Formation of Fractal Clusters and Networks by Irreversible Diffusion-Limited Aggregation," *Phys. Rev. Lett.*, 1983, *51*, 1119-1122.

6. Meakin, P., "Diffusion-Limited Aggregation in Three Dimensions: Results from a New Cluster-Cluster Aggregation Model," *J. of Colloid Interface Sci.*, 1984, *102*(2), 491-504.

7. Moudgil, B. M. and Vasudevan, T. V., "Evaluation of Floc Properties for Dewatering of Fine Particle Suspensions," *Minerals and Metallurgical Processing*, 1989, in press.

8. Moudgil, B. M., Shah, B. D., and Soto, H. S., "Collision Efficiency Factors in Selective Flocculation," *J. of Colloid Interface Sci.*, 1987, *119*(2), 466-473.

9. Tambo, N. and Hozumi, H., "Physical Characteristics of Flocs - II: Strength of Floc," *Water Research*, 1979, *13*, 421-427.

10. Tambo, N. and Watanabe, Y., "Physical Characteristics of Flocs - I: The Floc Density Function and Aluminum Floc," *Water Research*, 1979, *13*, 409-419.

11. Weitz, D. A., Huang, J. S., Lin, M. Y., and Sung, J., "Limits of Fractal Dimension for Irreversible Kinetic Aggregation of Gold Colloids," *Phy. Rev. Lett.*, 1985, *54*(13), 1416-1419.

SURFACE CHARGE ON WASTEWATER SLUDGES

Paul T. Bowen
Associate/Staff Consultant
Metcalf & Eddy, Inc.
10 Harvard Mill Square
Wakefield, Massachusetts 01880

and

Upendra N. Tyagi
ERM, Inc.
855 Springdale Drive
Exton, Pennsylvania 19341

ABSTRACT

Charged sites on the surface of three types of wastewater sludge were identified. Sludge samples were treated with cationized ferritin (CF) to label anionic charged sites on the surface of sludge particles. Sludge surfaces were examined using transmission electron microscopy. The presence of cationized ferritin granules indicated charged sites. Waste-activated, primary sedimentation, and anaerobic-digested sludge responded differently to CF staining. Each sludge had different attachment patterns and varied in the amount of CF attracted. Location of charge was associated with the presence of exocellular polysaccharide material. Waste-activated sludge attracted the most CF granules. Anaerobic-digested sludge was stained the least by CF.

INTRODUCTION

Many factors are known to influence sludge conditioning. This process is influenced by properties of the conditioning agent, process variables in the mixing operation, and sludge characteristics. Sludge biochemical properties have been shown to influence conditioning[2].

Biological sludges are comprised of viable and nonviable microorganisms, cellular debris, biopolymer and abiotic substances[1]. Each type of material has different surface characteristics and properties. Surface characteristics of sludge particles are significantly altered by the presence of biopolymers and exocellular material. Protein, carbohydrate, and lipid are components of cellular surfaces, as well as constituents of biopolymers[7,8]. Both exocellular biopolymer and cell wall structure determine anionic surface charge. A coating of exocellular biopolymer may provide a new surface to sludge particles. Changes in surface charge of sludge particles may be due to variations of the biochemicals on the surface and/or the environmental conditions of the sludge.

This study sought to identify the distribution of anionic sites on sludge surfaces. By using various stains to identify cell components and structures, the relationships between these elements and surface charge were identified.

METHODOLOGY

Three types of organic sludges were selected for this study: waste activated, anaerobically digested and primary sedimentation sludges. These sludges had different biochemical properties and surface characteristics, which led to variations in the distribution of anionic charge. Sludge samples were collected from fully operating domestic wastewater treatment facilities. The waste-activated and primary sludge were collected from Oklahoma City's Chisolm Creek treatment facility. Anaerobic sludge was collected from the Del City treatment plant. Sludge samples were characterized by measuring total solids; lipid, carbohydrate, and protein concentration; and capillary suction time (CST). Standard analytical procedures were used for this characterization. Results of these analyses were used to examine sludges of similar composition.

Anionic sites on biological surfaces have been successfully labelled with cationized ferritin (CF)[4,5,6,9]. CF (purchased from Polyscience) was prepared by coupling electron microscope grade horse spleen ferritin with N,N-dimethyl-1,3-propanediamine via carbodiimide activation of the ferritin carboxyl groups. CF has distinct size and shape that can be easily identified using transmission electron microscopy (TEM). To obtain true representation of anionic sites, sludges were labelled with CF prior to preparation for electron microscope viewing.

Two drops of sludge sample to be labelled were placed in a 4-mL vial. Approximately 2 mL of CF (concentration 250 µg/mL) in 0.067 M cacodylate buffer (CB) were added to the vial and mixed with sludge. This suspension was allowed to stand for 15 minutes, then it was washed twice with 0.067 M CB to remove excess CF from solution. The stained sample was prepared for TEM study using 3 percent glutaraldehyde and 1 percent osmium tetroxide for fixation. Following a graded series of ethanol dehydrations, the samples were embedded in Spurr's resin. The embedded specimens were thin-sectioned with glass knives, mounted on 400-mesh grid, and examined with a Zeiss 10 TEM. Numerous sections were examined and photographed to ensure representative sampling. The magnification of the TEM was maintained constant throughout the study at 20,000 X.

Ruthenium Red (RR) has been used to stain exocellular polymers(3). RR replaced CF in the procedure outline above for evaluation of sludge biopolymers. Unstained samples, samples with either RR or CF, and samples with both RR and CF also were prepared.

Micrographs of sludges were grouped according to sludge type. Changes in the CF attachment among sludge types were identified. Comparison of the micrographs revealed differences in the quantity of CF attached and the pattern of attachment for the different sludges.

RESULTS AND DISCUSSION

The results of this study showed that there are variations in the surface charge on wastewater sludges. These differences are due to the treatment processes that generated the sludge. Sludge biochemical properties and micrographs of RR-stained sludge were used to explain these variations.

Results are based on visual observation of TEM scans and micrographs. Often detail in photomicrographs does not reproduce well in print. To enable a better understanding of the observations, portions of the photomicrograph in Figure 3 were enhanced for publication. Other photomicrographs were printed unretouched. In all micrographs CF appears as small dark granules or dots.

SLUDGE BIOCHEMICAL PROPERTIES

The amount of protein, lipid, and carbohydrate in sludge changes with its type. These differences are due to intrinsic differences in the source of the sludge. Biochemical characteristics for the three types of sludge that were investigated showed distinguishable differences (Table 1). Also shown in Table 1 are the total solids and CST. These variables provided additional support for obtaining sludge samples with similar properties.

The trend seen for protein concentration for the three sludge types reflected the active nature of the biomass. Primary sludge had the lowest protein concentration, indicating this sludge is low in active biomass. Waste-activated sludge shows an increase in protein concentration, which was expected due to the active biological conversions brought about by the microorganisms utilizing organic substrate and in an active growth phase. Anaerobic sludge had the highest protein concentrations. Although an indication of high biological activity, residual protein from cell destruction added to active level increased this value.

Lipid concentration was highest for primary sludge and minimal for waste-activated sludge. Domestic sewage contains lipids and grease as two of its main constituents. The sludge derived from primary clarification would

Table 1. Sludge Biochemical and Physical Characteristics

Parameter	Waste Activated Sludge	Primary Sludge	Anaerobic Sludge
CST, sec	32.08	93.13	718.15
total solids, g/L	8.8649	12.5472	20.0624
protein, mg/g	445.63	359.32	701.69
carbohydrate, mg/g	119.97	460.82	76.04
lipid, mg/g	72.15	186.34	104.37

have high concentrations of lipids and grease because of limited biological activity in this process. Waste-activated sludge had lower lipid concentrations due to degradation of these materials. Lipid concentrations in anaerobic sludge were high for several reasons. Scum and grease skimmings from the clarifiers were added to the sludge en route to digestion. Secondly, lipid stored in cells is released upon lysis of the cells during digestion. This lipid withstood further anaerobic degradation and accumulated in the digester.

Carbohydrates in the three sludges showed maximum concentration in the primary sludge and minimum in the aerobic sludge. This level of carbohydrate concentration in anaerobic sludge was due to limited substrate availability in anaerobic digesters. Microorganisms in these reactors did not produce exocellular polysaccharides or other storage products. Due to limited biological activity, the carbohydrate level in primary sludge was high. Waste-activated sludge contains lower amounts of carbohydrate. During aerobic metabolism, carbohydrates were utilized, as well as generated, as exocellular biopolymer. Production slightly offset the decrease due to utilization.

LOCATION OF SURFACE-CHARGED SITES

Unstained and stained sludge samples were studied using transmission electron microscopy. Particles of unstained (no RR or CF) waste-activated sludge (Figure 1) were stained by the glutaraldehyde and osmium tetroxide used in the general preparation of samples for TEM analysis. There was a wide variety of particles in this figure. Cell A was surrounded by a clear space (S) and a layer of exocellular material. Other particles (B) were not easily distinguishable. Many of these particles could not be identified, but appeared in many of the sections from waste-activated sludge samples. Single cells as well as clusters of material appeared in the waste-activated sludge samples.

When stained with RR to highlight exocellular material (Figure 2), the boundaries of the cells were more distinct. Various biopolymers adsorb the stain differently and produce an uneven distribution of stain around the cells. The staining pattern or dark spots indicated the presence of exocellular material. The structure of cells was easier to evaluate in these micrographs than in unstained samples. Cells A and B contained a central mass, surrounded by a clear zone (S) and enclosed by a darker band of exocellular material. The dark spots in the outer layer are biopolymers that accept RR in different fashion. The cluster of four cells (floc) in this figure was "held together" by stained material (C) that was a biopolymer. This type of clumping showed one of the functions of exocellular material.

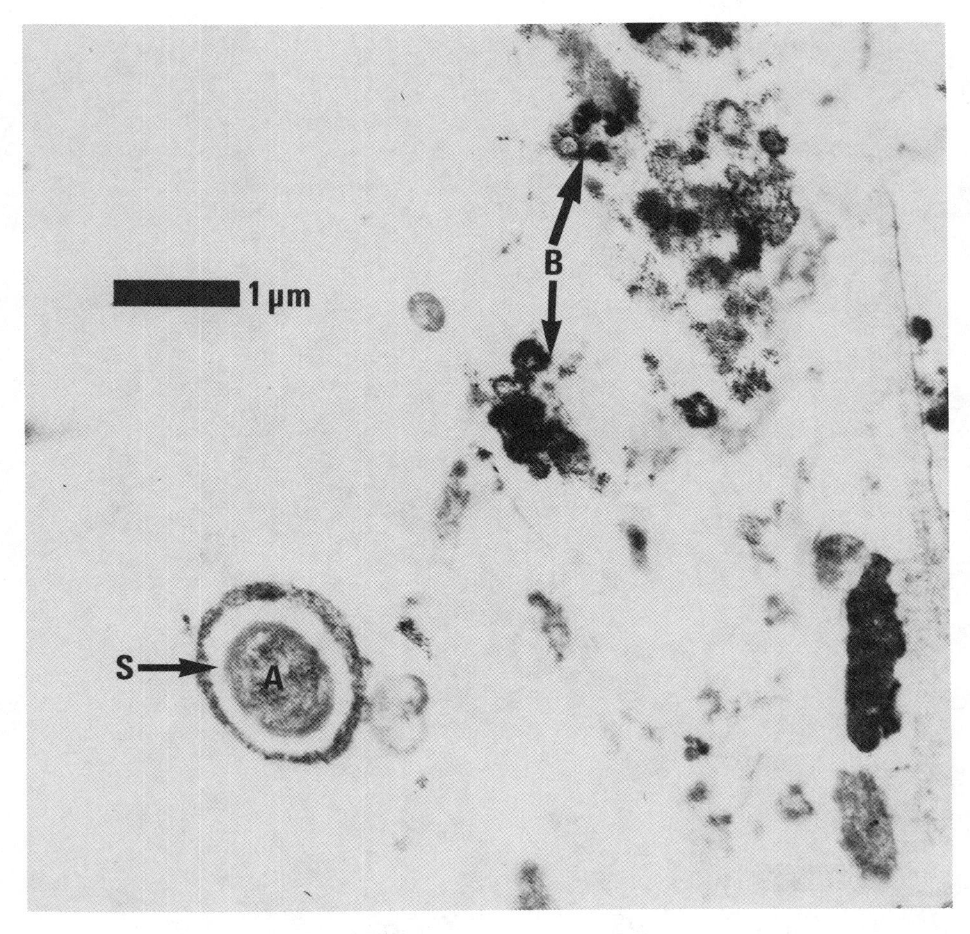

Figure 1. Micrograph of Unstained, Waste-Activated Sludge

Figure 2 was representative of waste-activated sludge. When primary sludge was examined, less exocellular material was observed. Anaerobic sludge contained extremely small amounts of RR-stainable material. Based on the microbial growth stage and activity of the sludges, these differences were expected.

A sample of waste-activated sludge was stained with both RR and CF (Figure 3). RR staining resulted in dark stained cell perimeters. Within the stained exocellular material, small dark granules of CF were visible. Penetration of CF into the cluster of cells was dependent on the amount of biopolymer present. In areas where little RR staining occurred (A), CF appeared around the cell. Where biopolymer held cells together (B), little CF staining was visible.

RR-stained material also appeared away from cells surfaces (C). In these locations, anionic-charged sites were evident by the CF staining. Note the pattern was uniform coverage, but not complete coverage of the exocellular mass. Distinct charged sites existed on the surface of this mass.

Figure 3 was typical of waste-activated sludge. The interaction of CF with the material stained by RR was apparent. For primary and anaerobically digested sludge, where less RR-stained material was present, less CF staining occurred. Anionic sites were associated with exocellular polysaccharide material in the sludges and with cell walls.

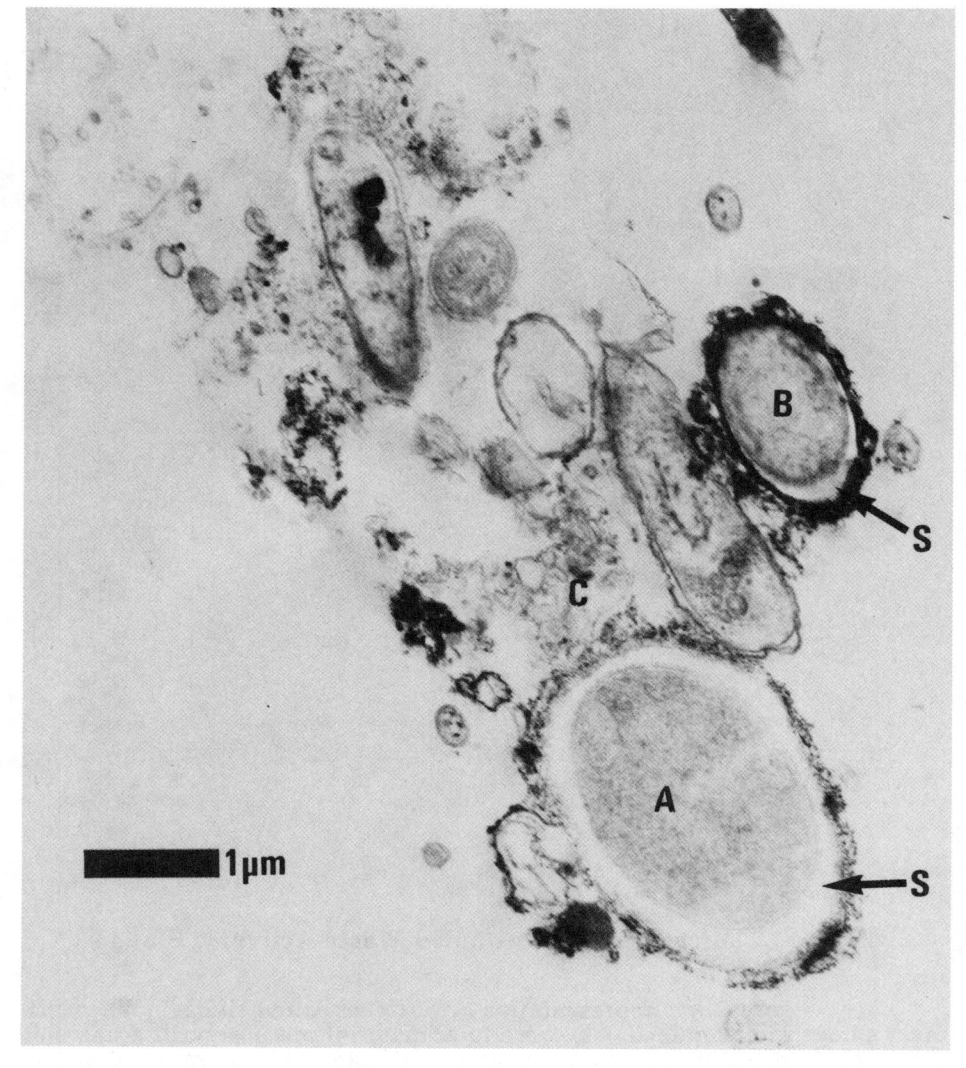

Figure 2. Micrograph of Waste-Activated Sludge Stained with Ruthenium Red

CHARGED SITES OF DIFFERENT SLUDGES

Trends in CF attachment for different types of sludges were identified by comparing numerous micrographs.

For waste-activated sludge (Figure 4), a distinct pattern of CF attachment around the sludge particles was observed. CF was uniformly attached within a band around the individual microbial sludge cells. The distribution of anionic sites was not at the cell surface, but within this band of material. A clear space (S), similar to ones observed in Figures 1 and 2, surrounded the central cell mass. A band of CF enclosed this space and the cell. This band contained biopolymer and exocellular material generated by the microorganisms. Band thickness varied for individual particles. Variation was due to the angle at which the section was sliced or the growth stage of the organism. Surface characteristics such as the amount of biopolymer on the individual particles also accounted for differences in CF distribution.

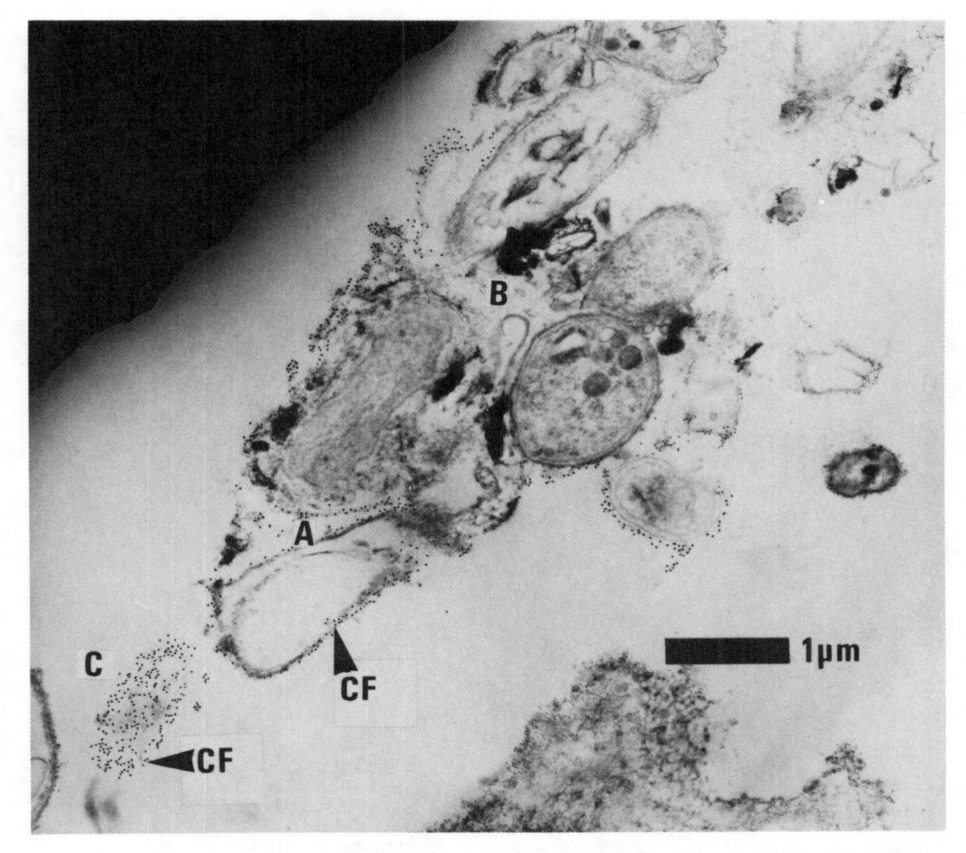

Figure 3. Micrograph of Waste-Activated Sludge Stained with Ruthenium Red and Cationized Ferritin. CF Staining Enhanced

When observing a cluster of waste-activated sludge cells (floc), a band of CF surrounded the entire cluster with only slight penetration of CF into the floc. Individual cells within the floc were void of any CF labelling. The floc was covered with biopolymers that accepted CF stain. In comparing micrographs, it was important to observe not only the similarity in microorganisms, but also their positions with respect to floc boundaries.

In the case of primary sludge (Figure 5), patches of CF were seen around the individual particles, instead of the uniform band seen on waste-activated sludge. Gaps, identified by G, were present in the band of material surrounding the cells. The amount (density) of CF attached was similar for primary and waste-activated sludge, with primary sludge showing slightly less attachment overall. Two micrographs, Figures 4 and 5, show microorganisms (cells A) of similar size and similar CF densities. The variation in the pattern of CF attachment observed between the sludge types was due to uneven distribution of exocellular material.

The clear space (S) that surrounds cell B was different from the spaces seen in Figures 1 and 4. In this micrograph, the clear space does not have a boundary separating CF from the space. CF appeared to migrate into this clear zone.

The pattern of CF attachment on anaerobic sludge (Figure 6) was uniformly distributed around the cells. However, the density of attached CF was less than the other two sludges. The band width around the cells was narrower than for waste activated and primary sludge. The CF attachment pattern within the band and the size of the band suggested anionic sites were

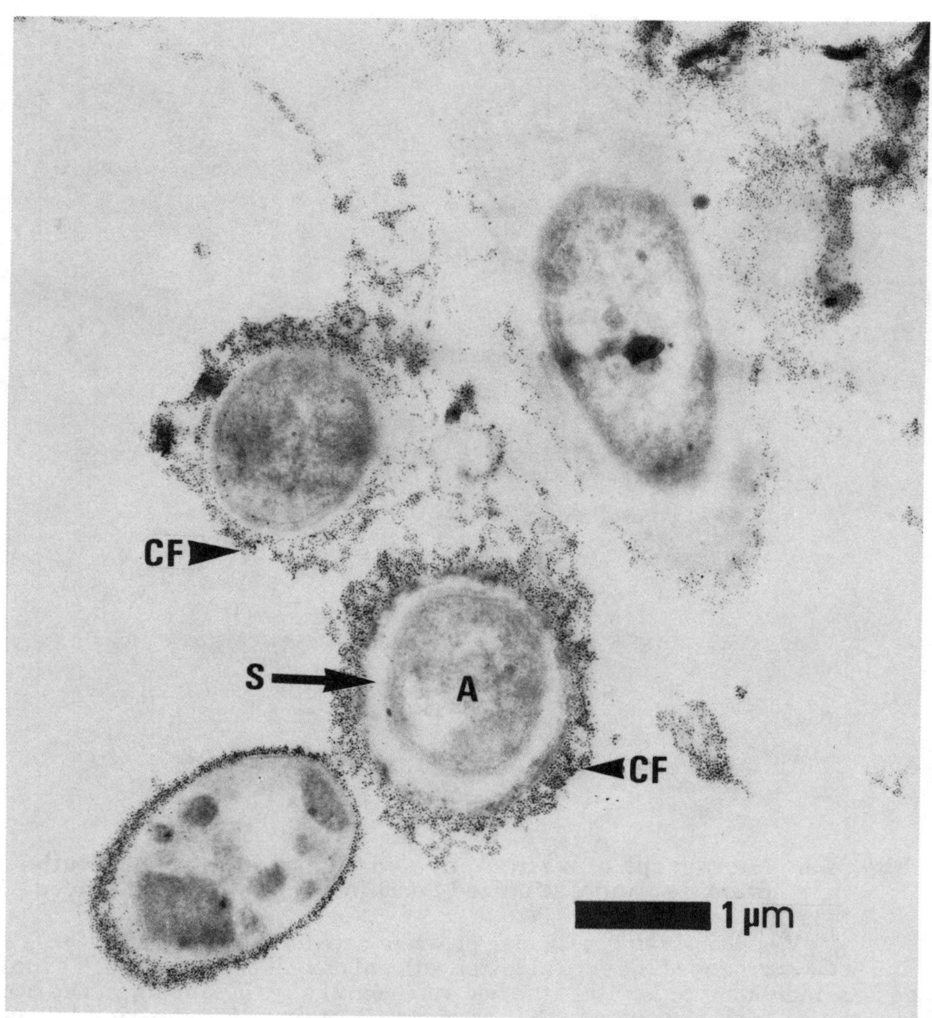

Figure 4. Micrograph of Waste-Activated Sludge Stained with Cationized Ferritin

practically on the cell wall. Closeness to the cell wall corresponded to the small RR pattern and low carbohydrate concentration observed for anaerobic sludge.

Individual cells were much smaller than those in the other sludges. This size accounted for less surface area and fewer anionic sites.

CONCLUSIONS

The three sludges showed differences in the magnitude and distribution of anionic charge sites. These differences were due to surface characteristics of the sludge particles. Variations in the amount of exocellular polymers and biochemicals in the sludge matrix affected the location of anionic charge. These concentrations were affected by the treatment processes that generated the sludge.

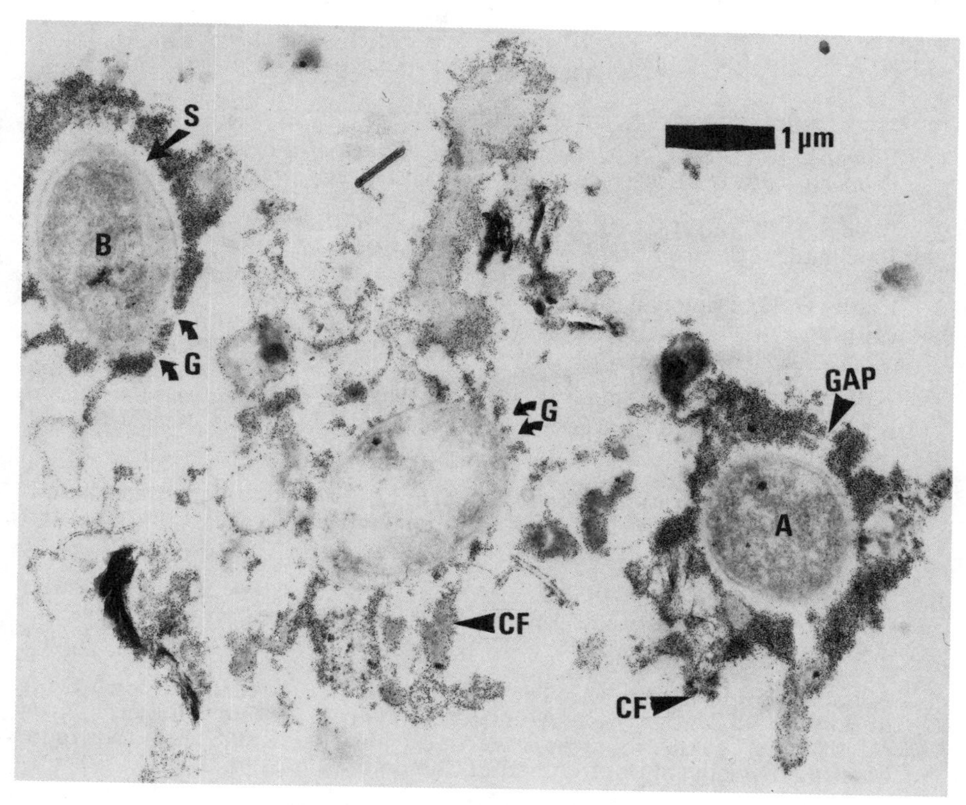

Figure 5. Micrograph of Primary Sludge Stained with Cationized Ferritin

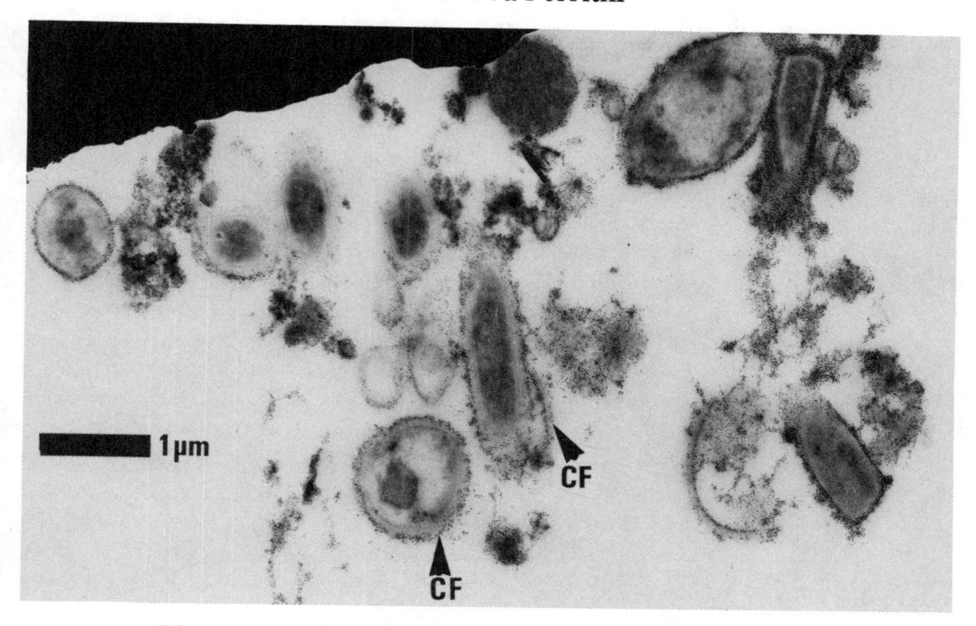

Figure 6. Micrograph of Anaerobic Sludge Stained with Cationized Ferritin

REFERENCES

1. Bowen, P. T., *Sludge Conditioning: Effect of Polymer and Sludge Properties*, Ph.D. Dissertation, Clemson University, December 1982.

2. Bowen, P. T. and Keinath, T. M., "Sludge Conditioning: Effects of Sludge Biochemical Composition," *Wat. Sci. Tech.*, 1984, *17*, 505-515.

3. Cagle, G. D., Pfizter, R. M., and Vela, G. R., "Improved Staining of Extracellular Polymer for Electron Microscopy: Examination of Azabactu, Leuconostoc, and Bacillus," *Appl. Microbio.*, 1972, *24*, 477-487.

4. Danon, D., Goldstein, I., and Marikovski, Y., "Use of Cationized Ferritin as a Label of Negative Charges on Cell Surfaces," *J. Ultrast. Res.*, 1972, *38*, 500-510.

5. Eighmy, T. T., Marates, D., and Bishop, P. L., "Electron Microscope Examination of Wastewater Biofilm Formation and Structural Components," *App. Environ. Microbio.*, 1983, *45*, 1921-1931.

6. Johansson, B. R., "Distribution of Anionic Binding Sites in Extravascular Space of Skeletal Muscle Demonstrated with Polycationized Ferritin," *J. Ultrast. Res.*, 1983, *83*, 176-183.

7. Roberts, K. and Olsson, O., "The Role of Colloidal Particles in Dewatering of Activated Sludge with Aluminum Hydroxide and Cationic Polyelectrolytes," *Preprints of Papers*, American Chemical Society National Meeting, Division of Environmental Chemistry, Chicago, IL, 1972, 5-8.

8. Steiner, A. E., McLareen, D. A., and Foster, C. F., "The Nature of Activated Sludge Flocs," *Water Res.*, 1976, *10*, 25-30.

9. Thurauf, N., Dermietzel, R., and Kalweit, P., "Surface Charges Associated with Fenestrated Brain Capillaries - I. In Vitro Labeling of Anionic Sites," *J. Ultrast. Res.*, 1983, *84*, 103-110.

AN INVESTIGATION OF SOL PROTECTION IN CHEMICAL CLARIFICATION

Alan J. Rubin and Paul R. Schroeder
Water Resources Center
The Ohio State University
590 Woody Hayes Drive
Columbus, Ohio 43210
(614) 292-7336

ABSTRACT

Examined was the clarification of colloidal dispersions of coal, soluble starch, and a mixture of the two that served as a model protected sol. Turbidity changes during settling were used as the criteria for aggregation and stability. The entire pH-aluminum concentration stability limit diagram was established from these data for each of the sol systems. Sol protection, as demonstrated in these studies, had several significant effects on settling rates and the boundaries of the stability diagrams. The lower boundaries were raised; that is, the smallest amount of aluminum salt that would result in removal of coal at any pH was significantly increased. Boundaries above these minimum coagulation values were noticeably changed in alkaline solutions, but were relatively unaffected on the acid side.

INTRODUCTION

BACKGROUND AND PURPOSE

The chemical clarification process, also known as coagulation or flocculation, is used in the purification of various types of process and drinking waters. Removed are finely divided particulates that are often colloidal in nature and have relatively high surface areas. In perfectly clean systems these particles are hydrophobic hydrosols that are easily destabilized and thus separated without much difficulty. In many cases, however, as is common with surface waters, organic species are present that are capable of being adsorbed by the particulates and enhancing their colloidal stability. This phenomenon is known as sol protection. The result is a more difficult separation often with increased chemical costs and less perfectly clarified water. An understanding of sol protection thus has very practical aspects as well as being of theoretical interest.

The phenomenon of sol protection is known to occur in natural colloidal systems and wastewaters; however, only a few studies have examined its role in the practical application of coagulation in water and wastewater treatment. In these studies clay in the presence of soluble humic substances, being either natural or prepared systems, was aggregated by alum or cationic polymers to determine the effects that soluble organics have on turbidity removal[3,7,18,26]. In general, it was found that the concentration of humic materials controls the dose of alum needed for effective clarification. Aqueous organic compounds raise the required coagulant dose and lower the optimum pH for turbidity removal to near pH 5. This is about two pH units lower than is usual in the absence of organics. The environmental engineering literature contains virtually no other works on the effects of soluble organics on turbidity removal. Several relevant studies have been reported in the chemistry literature on the aggregation of colloidal suspensions stabilized by adsorbed polymers, surface active agents, or other species[8,17,19,28].

The sol systems examined in this study were suspensions of oxidized and unoxidized colloidal coal, potato starch (amylose), and a mixture of the two that served as an example of a protected sol. The experimental techniques employed were the same as used in previous studies in this laboratory[4,20,21,25]. The general approach was to obtain sample turbidities during periods of quiescent settling as a function of either pH or coagulant concentration. In properly designed experiments the sides of the settling curves are quite sharp and, in many instances, independent of time. Extrapolation of the steep portions of the settling curves yields "critical values," which are characteristic of a given pH-coagulant concentration pair. The critical values obtained over broad ranges of pH and coagulant concentration, when plotted, form a stability limit diagram (domain of stability) that is characteristic of the sol-coagulant system. These diagrams are similar in construction to phase diagrams. They are convenient for summarizing large amounts of data, and their boundaries, at least with aluminum(III) salts in organic-free systems, can be related to the hydrolytic chemistry of the coagulant.

COLLOIDAL SYSTEMS AND SOL STABILITY

Sol protection is the increase in stability of a lyophobic particulate sol resulting from the addition of a suitable component, usually a soluble lyophilic sol. This protective agent adsorbs to the particle surface. The mechanism of protection varies with its nature. For example, macromolecules may sterically and entropically stabilize a hydrosol. The flexible polymeric chains extend from the particle surface into the solvent and, upon approach, interact with macromolecules on other particles. Interpenetration of the chains decreases the configurational entropy of the polymer loops and tails. Mixing of the polymer clouds also produces osmotic repulsion. Upon closer approach, elastic forces deriving from compression of the macromolecules further contribute to the stability[5,16].

Many methods have been proposed for classifying colloids. Differentiation has been made between inorganic and organic sols, lyophobic and lyophilic colloids, linear and globular particles, macromolecular solutions, micellar suspensions and particulates, and emulsions and suspensions[6,27]. A proposed classification scheme that includes protected sols is shown in Figure 1. Three classes of simple colloids are identified: macrosolutes (macromolecules), association colloids (micelles), and colloidal particulates (hydrosols). Each class may contain sols that vary considerably from other sols within its class. For example, hydrosols include amorphous solids, gels, microorganisms, simple emulsions (oil hydrosols without emulsifiers), and crystalline solids. In this scheme, protected sols represent a fourth class of colloidal systems: these sols are particulates that are stabilized by macromolecules, micelles, or other species. Included in this class are emulsions chemically stabilized with emulsifiers such as surface active agents.

The classes in the schematic are arranged as a function of hydration or hydrophilicity. In general, the hydration decreases from left to right; that is, macromolecules are more hydrated than particulates. Also, in general, pro-

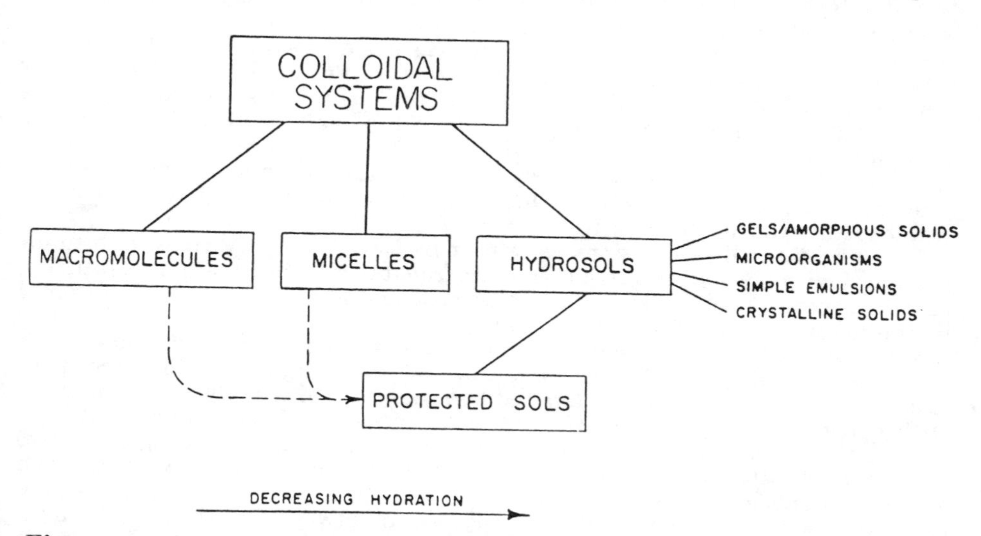

Figure 1. Classification of Colloidal Systems, Including Protected Sols, as a Function of Solvation

tected sols are more hydrophilic than unprotected particulates. However, the hydration of the classes overlaps, and therefore, some particulates may be more hydrated than some macromolecules. It is quite likely that the hydration of the sol is the single most important factor that determines its stability and coagulation behavior.

The destabilization of colloidal suspensions can be accomplished by several different mechanisms, depending on the nature of the coagulant and of the sol. Lyophobic sols are aggregated primarily by reducing the electrical double-layer repulsion (coagulation), by flocculation (chemical bridging), and by enmeshing (gathering). Methods for destabilizing lyophilic sols include desolvation (salting out) as well as precipitation, flocculation, and enmeshment. Many sols possess both lyophobic and lyophilic properties, and therefore, the aggregation mechanism is not always readily apparent.

Enmeshing is a destabilization process whereby colloidal particles are "gathered" in the unstable precipitate of a metal salt. The sol is then swept out of suspension as the precipitate settles. This mechanism is particularly important for aggregation by hydrolyzable salts such as iron(III) and aluminum(III). With these aggregates, enmeshing most frequently occurs in the neutral pH range at concentrations of metal salts above the critical supersaturation point of the precipitate[20,21,25].

EXPERIMENTAL METHODS AND MATERIALS

PREPARATION OF SOLUTIONS AND SUSPENSIONS

Carbonate-free, demineralized double-distilled water was used in the preparation of all solutions and suspensions. Reagent grade chemicals were used, and solutions were stored in polyethylene bottles. Fresh solutions of sodium hydroxide were prepared periodically to prevent contamination from absorption of atmospheric carbon dioxide. Dilutions of the aluminum stock solution were prepared just prior to their use to eliminate aging effects.

Soluble potato starch (amylose; Fisher Scientific) used for iodometry was employed in the preparation of starch solutions. Amylose is a linear macromolecule with a molecular weight of approximately 10^6 g/mol. The starch was prepared in 1-L batches by mixing the powder with 200 mL of room-temperature distilled water. The solution was then diluted to 1 L with boiling distilled water to completely hydrate the starch. Amylose swells and hydrates at about 60 C. In suspensions of greater than 1 g/L, all of the water is weakly bonded with the starch and the amylose slowly retrogrades to the crystalline form. Fresh sols were prepared every 3 days to minimize retrogradation and to prevent bacterial contamination. Stock starch solution was diluted with boiling water to form the dilute working sol.

The coal suspensions were prepared with a highly volatile, bituminous coal with low sulfur and ash content from the Elkhorn seam in Kentucky, obtained from Battelle Memorial Institute. The coal was ground and prepared in the manner described earlier[22] to produce stable, reproducible sols. The ground and milled coal was sieved through a standard U.S. Series 400 mesh sieve and the -400 mesh fraction was stored in an airtight container. Suspensions of unoxidized coal were prepared by mixing 5 g of the -400 mesh material with 1 L of 3×10^{-4} $\underline{M}$ NaOH. The coal was mixed in a Waring blender at high speed until it was thoroughly wetted and dispersed. To ensure homogeneity two additional minutes of blending were employed. The suspensions were then allowed to settle and age in glass columns to form sols of the most stable particles. After 1 week the pH typically fell to between 9.5 and 10.0 from its initial pH value of 10.5. Only 1 to 2 percent of the coal remained in suspension after aging.

Stable suspensions of oxidized coal were prepared as follows. Two-gram batches of the -400 mesh coal were oxidized with 10 mL of 30 percent hydrogen peroxide and 200 mL of distilled water in 500-mL Erlenmeyer flasks. The flasks were shaken for 4 hours in an Eberbach constant shaker bath at 80 C. Following oxidation the coal slurries were filtered through a 0.3-μ pore size, 47-mm diameter type pH Millipore membrane filter, suspended and washed in 200 mL of distilled water, and refiltered. Next, the oxidized coal was dispersed in distilled water by shaking in glass columns at concentrations ranging from 2 to 10 g/L and then allowed to settle and age. About 3 percent of the coal remained in suspension after aging. The particles in the aged suspensions were spherically shaped and uniformly about 0.5 μ in diameter, although initially, the suspension contained many nonuniform, irregularly shaped particles. Typically, the pH of the sols ranged from 5.5 to 6.0. Sols having a pH below 5.0 were discarded. Before each study the suspension was carefully siphoned from the top of the columns to avoid resuspending settled coal particles, and then diluted with distilled water to produce a working suspension with the desired turbidity.

The coal sols protected by adsorbed starch were prepared from highly aged oxidized coal sols and dilute soluble potato starch solutions just described. Starch was mixed with an oxidized coal suspension to form a sol with an absorbance of 1.2 and twice the starch concentration desired in the aggregation experiments. The sol was diluted by half during these tests. The starch-oxidized coal mixture was shaken for 1 hour and then allowed to age for at least 24 hours before use. The quantity of starch adsorbed was measured for correlation with the degree of protection.

PROCEDURES AND TREATMENT OF DATA

Changes in sample turbidity during settling were used to indicate aggregation. Preliminary investigations showed that turbidity was a linear function of sol concentration. The aggregation experiments were performed directly in 19×105 mm round cuvettes using the "nonmixing" technique[20,21,25] at room temperature under nearly constant conditions. Two separate series of solutions were prepared for each experimental run. In one series, 5-mL aliquots of the suspension were added to small glass vials with NaOH solution for pH adjustment (when required). A second series of solutions consisting of the coagulant, dilution water, and HNO_3 for pH adjustment (when needed) were prepared in cuvettes. In some experiments with starch, 0.5 mL of 0.02 $\underline{M}$ iodine solution with potassium iodide was also added to the cuvettes. Each 10-mL test sample was formed by transferring the contents of a single vial to a cuvette containing the coagulant. Both acid and base were never added to the same sample. The cuvette was then stoppered, shaken for 10 seconds, and allowed to settle without further mixing. The "initial turbidity" of the sample was measured within 1 minute following its preparation. Subsequent readings were taken as required to observe the extent of settling and removal. The pH of each sample was measured at the end of each experiment. The reported concentrations of the coagulants and suspensions are those present in the test sample after the final preparation and dilution.

The turbidity and removal of starch sols were analyzed by several methods. Starch complexes produce either a dark blue color at high starch-to-iodine ratios or a dark brown color at low starch-to-iodine ratios. A wavelength scan on a Perkin-Elmer model 124 double-beam grating spectrophotometer indicated that a strong broad absorption peak occurred with its center at 590 nm. Absorbance was linear up to 80 mg/L of starch, and above 5×10^{-4} $\underline{M}$ I_2, it was nearly independent of iodine concentration. The iodine solution was prepared by dissolving iodine crystals and potassium iodine in distilled water at the molar ratio of 1 to 2.5, respectively. The 0.1 $\underline{M}$ I_2 solution was stored in amber glass bottles wrapped in aluminum foil to prevent photochemical degradation of the iodine.

Aggregation of concentrated starch sols was examined by light scattering measurements on a Coleman model 9 nephelometer. The starch suspension without coagulant was used as the standard solution. Following the settling period, the samples were centrifuged for 15 minutes at 2,500 rpm (890 g) and light scattering was remeasured. For experiments with the 1-g/L starch sols, 1 mL of the supernatant was diluted to 15 mL with distilled water and iodine solution. With the 100-mg/L starch samples, 5 mL of the supernatant was diluted to 10 mL with iodine solution. The absorbance of the starch in the presence of 10^{-3} M I_2 was measured at 590 nm. Zero absorbance was established using a 10^{-3} M iodine solution as the blank. Standard solutions were prepared from the same starch suspension without coagulant and treated in the same manner as the samples to be analyzed. Aggregation tests with 25-g/L starch sols were conducted in the presence of 10^{-3} M iodine.

Sample turbidities of coal and coal-starch mixtures were estimated by absorbance readings at 400 nm using a Coleman model 14 spectrophotometer. The concentration of the coal suspensions was determined by measuring the suspended solids concentration. A known volume of the coal suspension was filtered through a 0.3-μ membrane filter, and the retained coal was then dried and weighed with the tared filter. Measurements of pH were taken with Sargent-Welch model LS or NX pH meters. Electrophoretic mobilities were measured with a device manufactured by Zeta Meter of New York using procedures described in the instrument manual.

Two experimental approaches were used to study aggregation. In one, the series of test samples was prepared at constant pH while systematically varying the concentration of coagulant. The resultant turbidities of the test series were plotted as a function of their log molar coagulant concentration. The intersection of the extrapolated steep samples yielded the critical concentrations for coagulation and stabilization. The critical coagulation concentration (c.c.c.) is defined as the lowest solute concentration that merely initiates aggregation. The critical stabilization concentration (c.s.c.) is the lowest concentration at which stabilization is just completed.

The second experimental approach involved several series of samples, each prepared at a different given aluminum concentration while the pH of the samples in each series was varied systematically. The turbidities of the samples at each aluminum(III) concentration were plotted as a function of their pH. The critical pH values for coagulation and stabilization were determined by extrapolation in the same manner as described for the critical concentrations. The definitions of the critical pH of coagulation (pH_c) and stabilization (pH_s) are analogous to those of the corresponding critical concentrations in that a given sol is stable in the pH range greater than or equal to a pH_s and less than a pH_c.

EXPERIMENTAL RESULTS

EFFECT OF pH ON STABILITY

The first studies were conducted to examine the effect of pH on the stability of the oxidized and unoxidized coal sols in the absence of coagulants other than hydrogen ion. The electrophoretic mobility of sols aged for 8 days was measured, and their aggregation was indicated by turbidity changes during quiescent settling. The results are summarized in Figure 2. Data for oxidized coal are represented by open circles, and those for unoxidized colloidal coal by blackened circles. In the top portion of the figure their electrophoretic mobilities are plotted as a function of pH. Negative mobility values indicate negatively charged particles. The pH at which the sol has zero mobility is the isoelectric point (pH_i). The bottom part of the figure shows the sample turbidities following 24 hours of settling. The settling data are plotted as relative absorbance, which is the value measured at the end of the settling period divided by the initial absorbance. The critical pH for stabilization (pH_s) for both sols was also obtained from these data.

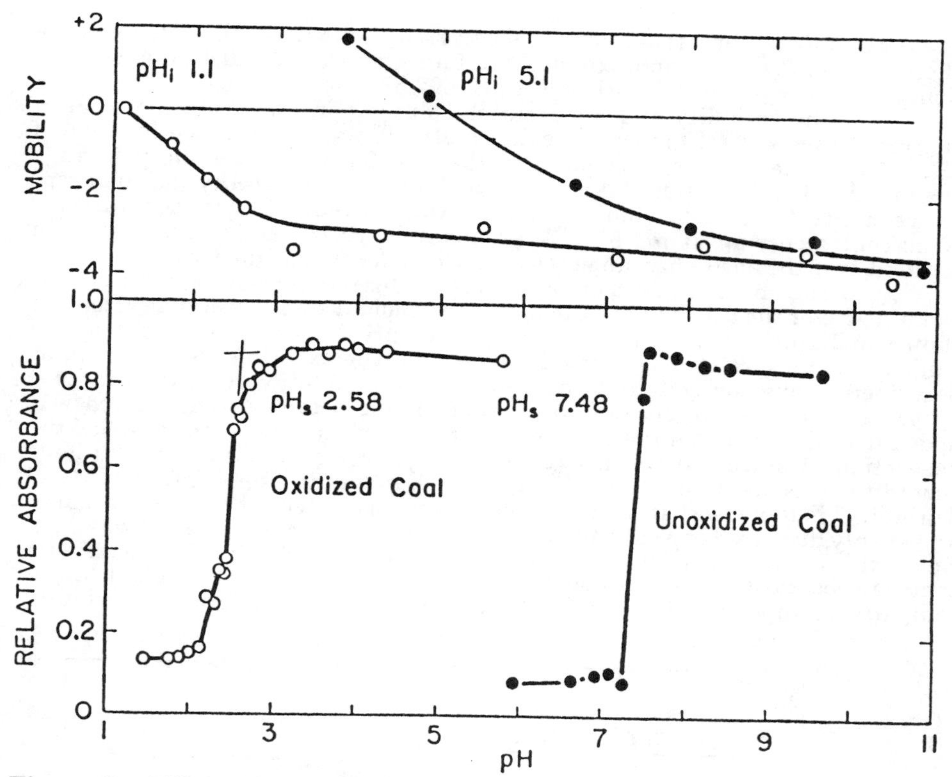

Figure 2. Effect of pH on the Stability and Electrophoretic Mobility of Colloidal Coals. 24-hr settling data(25).

The two coal sols exhibited similar negative mobilities throughout the alkaline region above pH 8, but their mobilities differed drastically in the acid range. Unoxidized coal had an isoelectric point at pH 5.1, whereas the pH_i of the oxidized coal was at pH 1.1. The mobility of the unoxidized coal in the vicinity of the pH_i changed most rapidly between pH 3.9 and pH 8. At higher pH values the mobility was nearly independent of pH. The maximum negative mobility measured was approximately 3.4 µm-cm/v-sec at pH 10.8. Positive mobilities were not observed for oxidized coal since measurements could not be made below pH 1 due to limitations of the electrophoresis apparatus. Between pH 1.1 and pH 3, the negative mobility increased sharply to approximately -2.8 µm-cm/v-sec. Above pH 3, the mobility changed gradually to -3.8 µm-cm/v-sec at pH 11.

The initial absorbances of the oxidized and unoxidized coal suspensions were 0.67 and 0.53, respectively, for coal concentrations in the range of 25 to 30 mg/L. Below the pH_S, which was established from these data, both suspensions settled at similar rates, with 85 to 90 percent removal of their turbidity occurring within 24 hours. The coal with minimum oxidation, due solely to exposure to air, was coagulated by hydrogen ion below pH 7.48, whereas the highly oxidized coal was stable above pH 2.95. These pH_S values were obtained by extrapolating the steepest portion of the 24-hour settling curves up to intersect a line through the turbidity values of the stable samples at higher pH. The data at earlier times are not shown, but the resulting curves were very similar except that the removals were less. The pH_S values obtained from the curves at these time periods were virtually the same, indicating the time independence of the critical values.

Vigorous oxidation of the coal sol with hydrogen peroxide lowered the pH_S by 4.5 pH units. This is similar to the reduction in the isoelectric point of 4.0 pH units. The pH_s of both sols occurred at virtually the same electrophoretic mobility value, approximately -2.4 μm-cm/v-sec. Oxidized coal suspensions aged 22 days had a slightly lower pH_s of 2.58.

The turbidity of 1.0 g/L starch solutions in the absence of coal at different times was estimated by nephelometric measurements as a function of pH. Very little settling took place in the first 96 hours over the pH range examined of 1 to 13. After 200 hours of settling, some instability was apparent between pH 4.3 and pH 6.3, with a maximum reduction in turbidity of 55 percent occurring at pH 5.0. Approximately 10 percent of the starch in solution over the entire pH range retrograded to the crystalline form.

The effect of starch on protecting oxidized coal from coagulation was examined over the pH range of 1 to 3.5. The colloidal stability of the system, as shown in Figure 3, was not examined above pH 3.5 since both starch and oxidized coal are relatively stable above pH 3. The experiments were run, as described above for colloidal coal, on 32-mg/L, 38-day-old oxidized coal suspensions in the presence of 25 mg/L of soluble potato starch. Starch was added to coal suspension the day before the start of an experiment to allow the adsorption of starch to reach equilibrium. The figure shows the 70-hour turbidity data as a function of pH. The horizontal line labeled "Initial Turbidity" is the absorbance of a stable sample just after it had been prepared and evenly dispersed by shaking. The pH_s determined from these data was 2.58. Below this critical pH of stability, the coal settled very slowly. The maximum removal observed in the 70 hours of settling was 55 percent at pH 1.0. In comparison, unprotected oxidized coal had an 85 percent reduction in turbidity

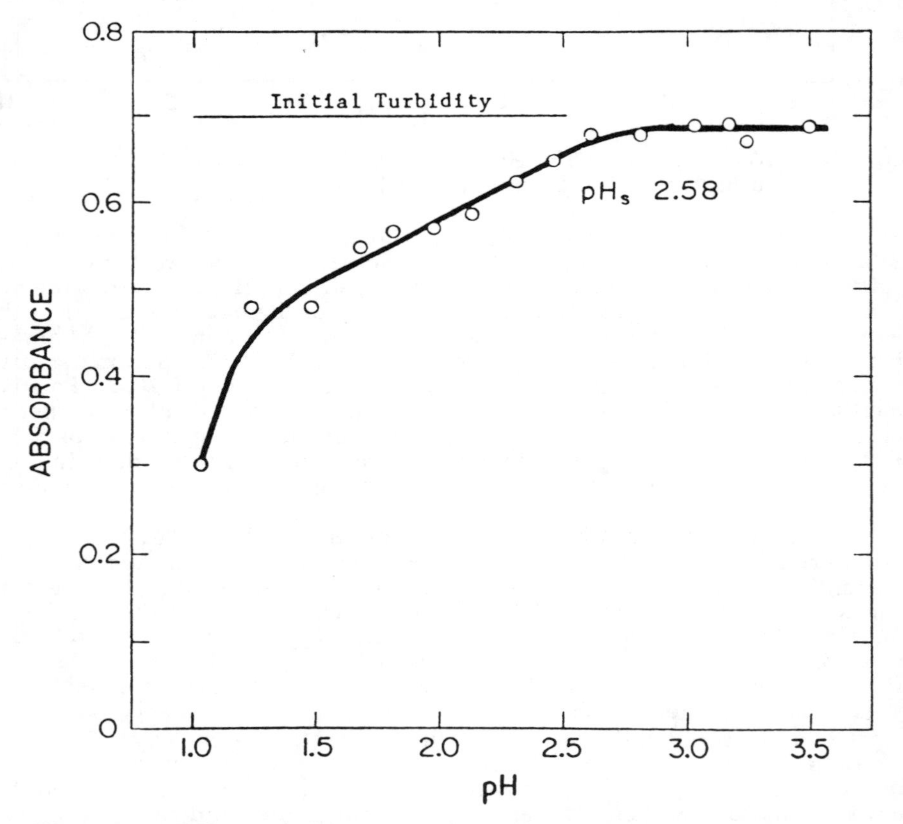

Figure 3. Effect of pH on the Stability of Oxidized Coal Protected by 25 mg/L of Soluble Potato Starch (Amylose). 70-hr settling data.

in just 24 hours. Above pH 2.85, the coal was stabilized by the starch as the turbidities of these samples remained nearly constant during the 70 hours of testing. In comparison, without starch, the stable oxidized coal samples at pH values above the pH_s lost 10 to 15 percent of their initial turbidity in just 24 hours.

AGGREGATION WITH ALUMINUM SALTS

The aluminum(III)-pH stability limit diagrams for the systems' colloidal coal (without starch) and coal with starch were established by determining the critical values for coagulation and stabilization. These pairs of critical pH (pH_c and pH_s) and critical concentration (c.c.c. and c.s.c.) values are the coordinates of points on boundaries between regions of stability and destabilization. The experiments were run by systematically varying either the pH or the Al(III) concentration in a series of samples, while holding the other constant. Typically, aggregation was examined from pH 2 to 12 and from 10^{-7} M to 10^{-1} M Al(III). In these tests the turbidity of the samples was measured immediately after preparation and several more times over the next 48 hours to obtain sharp limits between a region of stability and clarification by settling.

Typical turbidity data for a given Al(III) concentration as a function of pH and for a given pH at different coagulant concentrations are shown in Figures 4 and 5, respectively. Sample absorbances from several time periods are plotted. Additional examples are shown elsewhere[25]. Experiments were also performed at several other Al(III) concentrations and pH values, and the resulting critical concentrations and pH values are listed in Table 1.

Figure 4 shows results at 1.0×10^{-4} M $Al(SO_4)_{3/2}$ (corresponding to 33 mg/L of alum as $Al_2(SO_4)_3 \cdot 18H_2O$). The rate of settling was very slow, and

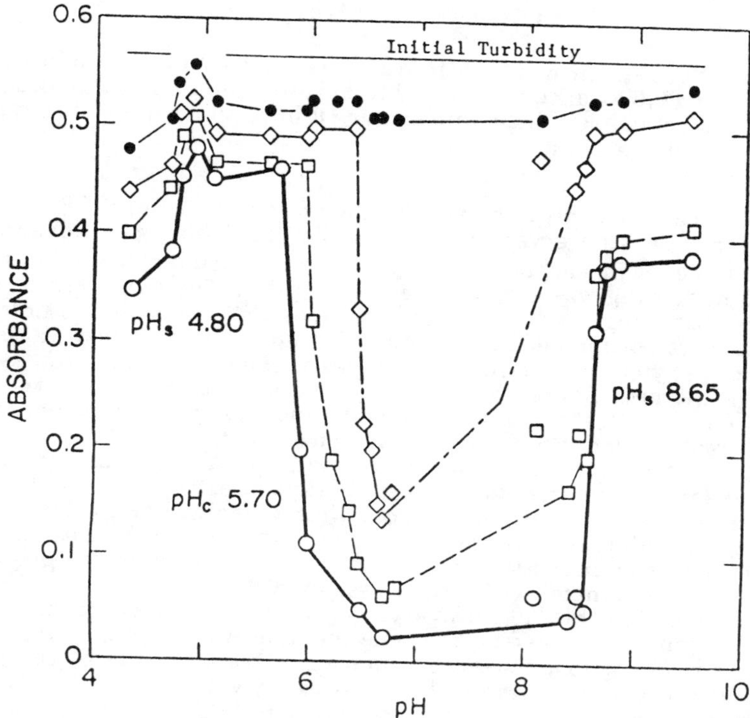

Figure 4. **Aggregation of Colloidal Coal with 1.0×10^{-4} M $Al(SO_4)_{3/2}$ as a Function of pH. Blackened circles are 3-hr, diamonds are 6-hr, squares are 12-hr, and open circles are 20-hr settling data.**

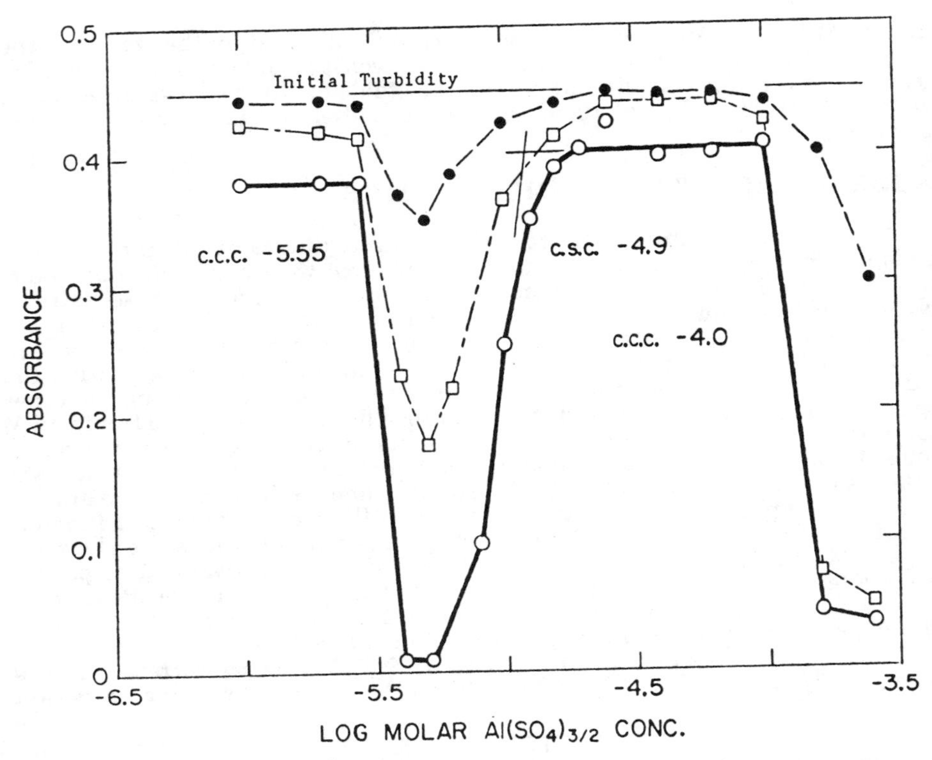

Figure 5. Aggregation of Colloidal Coal with Aluminum Sulfate at pH 6.0 as a Function of Aluminum Concentration. Blackened circles are 3-hr, squares are 10-hr, and open circles are 20-hr settling data[25].

clarification occurred only in a narrow region between about pH 5.7 and pH 8.7. Little change was observed in the turbidity in 3 hours; however, nearly complete removal was achieved near pH 7 after 24 hours of quiescent settling. Some slight settling was also observed below pH 5. The figure can be divided into four distinct regions or zones separated by steep boundaries. Extrapolation of the steepest portions of these boundaries up to the turbidity value of the stable samples yielded the critical pH values as shown on the figure. As will be evident in the next figure, this $Al(SO_4)_{3/2}$ concentration was also the critical coagulation concentration for this pH range and was the highest concentration at which restabilization was observed. Consequently, the pH_c changed with time because it was moving along the horizontal restabilization boundary. The right boundary, however, is independent of time as extrapolation of the turbidity curves at other time periods produced almost identical pH_s values. Increasing the applied aluminum concentration, as illustrated by the data in Table 1, results in a broadening of the area (the sweep zone) between the pH_c and the right-most pH_s and increases the rate of settling[20,21,25].

Results at pH 6.0 with varying aluminum sulfate concentrations are shown in Figure 5. As before, the zones of stabilization and aggregation are distinct and separated by steep boundaries. Extrapolation of the steepest portions of the settling curve to the absorbance value of the stable samples yielded the log c.c.c. and log c.s.c. values. An $Al(SO_4)_{3/2}$ concentration of 2.8×10^{-6} $\underline{M}$ (about 1 mg/L alum) was barely sufficient to induce aggregation. Increasing the dose to 1.3×10^{-5} $\underline{M}$ resulted in a stable coal dispersion. The sol remained dispersed until sufficient sulfate ion had been added to coagulate once again the restabilized oxidized coal. This occurred at $Al(SO_4)_{3/2}$ concentrations

Table 1. **Summary of Critical pH Values for the Aggregation of Coal Sols by Aluminum(III) Salts**

Molar Conc.	log M	pH_s	pH_c	pH_s
28 mg/L Oxidized Coal and Aluminum Nitrate				
2.0×10^{-5}	-4.70	-	6.94	7.24
3.1×10^{-4}	-3.51	-	-	9.46
1.0×10^{-2}	-2.00	-	-	11.73
2.3 mg/L Oxidized Coal and Aluminum Sulfate				
1.3×10^{-4}	-3.89	-	4.63	8.85
1.8×10^{-3}	-2.74	-	4.13	10.66
28 mg/L Unoxidized Coal and Aluminum Sulfate				
2.0×10^{-4}	-3.70	4.58	4.70	9.34
1.0×10^{-2}	-2.00	-	-	11.65
28 mg/L Oxidized Coal and Aluminum Sulfate				
0	-	2.58	-	-
6.3×10^{-6}	-5.20	-	-	6.80
1.0×10^{-5}	-5.00	5.25	6.15	6.52
1.6×10^{-5}	-4.80	5.20	6.25	6.91
3.0×10^{-5}	-4.50	5.05	6.60	7.60
5.0×10^{-5}	-4.30	-	6.50	7.90
8.0×10^{-5}	-4.10	4.81	6.45	-
1.0×10^{-4}	-4.00	4.80	5.70	8.65
1.3×10^{-4}	-3.89	4.55	4.95	8.50
2.5×10^{-4}	-3.60	-	4.65	9.00
4.0×10^{-4}	-3.40	-	4.45	9.05
5.0×10^{-4}	-3.30	-	4.40	9.15
1.0×10^{-3}	-3.00	-	4.35	9.88
3.0×10^{-3}	-2.50	-	4.10	10.70
1.0×10^{-2}	-2.00	-	4.00	11.75
2.0×10^{-2}	-1.70	-	3.68	12.08
1.0×10^{-1}	-1.00	-	3.70	12.62

greater than 1.0×10^{-4} M. The results at this pH and similar experiments are summarized in Table 2.

Critical pH values obtained with aluminum nitrate using these same coal suspensions, which are also summarized in Table 1, were very similar to those for aluminum sulfate. Values of pH_s that form the alkaline boundary of the rapid coagulation zone were unchanged. Apparently, neither sulfate nor nitrate ions affected the aggregation of colloidal coal by aluminum(III), except in the restabilization region where anions coagulate the restabilized positive sol.

Dilute suspensions, containing 2.3 mg/L of oxidized coal, were aggregated by $Al(SO_4)_{3/2}$ at 1.26×10^{-4} M and at 1.8×10^{-3} M over a wide pH range to

Table 2. **Summary of Critical Concentrations for 28 mg/L Oxidized Coal with Aluminum Sulfate**

pH	Log c.c.c.	Log c.s.c.	Log c.c.c.
2.50	-6.50	-	-
3.00	-5.80	-	-
3.50	-5.54	-	-
3.95	-5.55	-	-
4.25	-5.54	-	-
4.60	-5.55	-	-
5.00	-5.55	-4.60	-3.90
5.50	-5.55	-4.91	-3.95
6.00	-5.55	-4.94	-4.05

study the effects of sol concentration on the boundaries of the rapid coagulation zone. The turbidity was measured by nephelometry because greater sensitivity was needed for these dilute suspensions. The distilled water blank had a turbidity or nephelos value of four standard units. The initial turbidity of the coal sol corresponded to a nephelos value of 63. The clarification was very similar to that for oxidized coal at 1.0×10^{-4} M $Al(SO_4)_{3/2}$, except that restabilization was not observed. Only a very narrow region of stability ("spike") occurred at pH 4.60. The pH_s values for the dilute suspensions were slightly, though possibly insignificantly, higher than obtained for the undiluted sol. The pH_c values forming the acid boundary of the rapid coagulation zone were virtually identical for both sols.

Experiments were also conducted on unoxidized coal suspensions aged 14 days at $Al(SO_4)_{3/2}$ concentrations of 1.0×10^{-2} M and 2.0×10^{-4} M to compare the aggregation of oxidized and unoxidized coals. Clarification was rapid and complete within 12 hours in both the rapid coagulation zone and the normally slower ionic coagulation zone. A narrow but sharp and distinct stability spike occurred between these zones. The critical pH values for unoxidized coal (see Table 1) were essentially identical to those for oxidized coal.

The aggregation of soluble potato starch was investigated over the aluminum nitrate concentration range of 10^{-6} to 10^{-1} M and the pH range of 3 to 12. Two starch concentrations were examined: 0.10 g/L and 1.0 g/L. The turbidity of the samples was measured by nephelometry, and the starch concentration was determined by absorbance measurements in the presence of iodine. The flocs produced by aggregation were often fluffy and settled poorly, and therefore, the samples were centrifuged to separate the stable starch and precipitate from the settleable flocs. Typical results are shown in the next figure.

Figure 6 shows the results for the aggregation of 0.10 g/L starch by 5.6×10^{-4} M $Al(NO_3)_3$. Turbidity data are represented by circles joined by solid lines, whereas the starch concentrations in relative absorbance units are given as squares connected by dashed lines. The absorbance data are presented in units relative to the initial absorbance, which was set to equal 1.0. The initial turbidity of the samples corresponded to a nephelos value of about 10 units.

Four distinct changes in turbidity were observed as a function of pH. The turbidity increased from 10 to about 43 nephelos units between pH 4.4 and 6.85. A stable precipitate, presumably $Al(OH)_3(s)$, was formed in this pH region. The critical pH for precipitation, 4.4, was determined by extrapolating the turbidity curve to the initial turbidity of the starch solution. The starch was also stably dispersed below pH 6.85 as shown by the absorbance curve. The turbidity and starch concentration decreased sharply at the pH_c. Virtually all

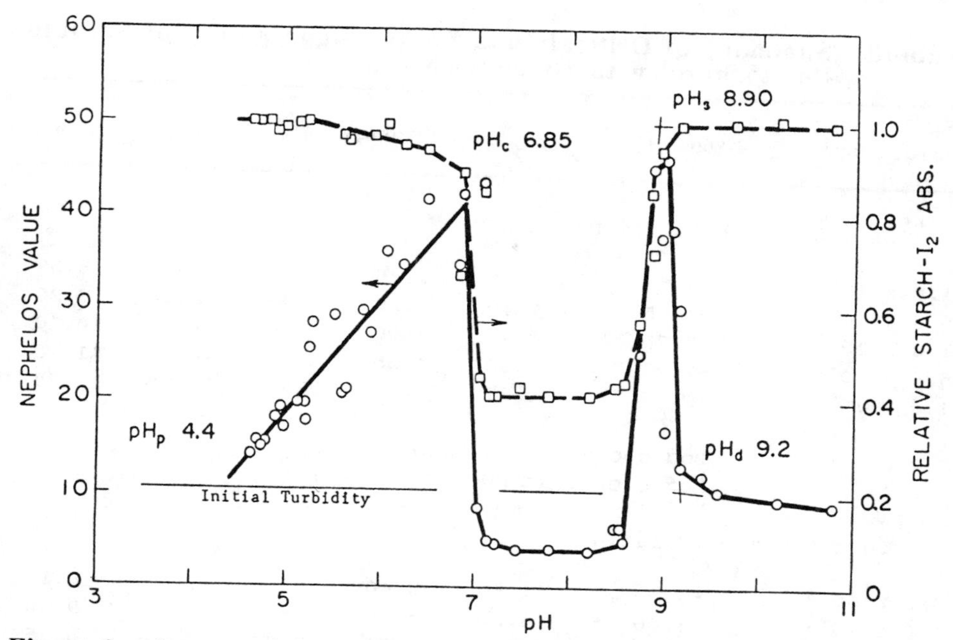

Figure 6. Aggregation of 0.10 g/L Starch with 5.6×10^{-4} M $Al(NO_3)_3$ as a Function of pH. Data following centrifugation at 24 hr.

of the turbidity and precipitate settled between this pH_c at 6.85 and the pH_s at 8.90, with about 60 percent of the starch being removed. At pH 9.2, the turbidity dropped sharply to the initial turbidity level as the precipitate dissolved. Starch was removed only in the sweep zone where significant quantities of precipitate were produced and settled.

Similar results were obtained at higher Al(III) doses. For example, increasing the concentration by about a factor of three to 1.8×10^{-3} M $Al(NO_3)$ again yielded two precipitate restabilization zones and a sweep zone of rapid selling. The sweep zone was wider as the pH_s increased nearly one pH unit to 9.78. Ninety-five percent of the starch was removed in the sweep zone at this Al(III) concentration. Starch removal was not observed outside the precipitation region. The precipitate was stable between pH 4.4 and 6.9, and the concentration of dispersed starch decreased in this pH range as the concentration of precipitate increased. The precipitate was also stable in a narrow pH range just below pH 10.05 where the precipitate dissolved.

Starch removal was not observed at 1.8×10^{-4} M $Al(NO_3)_3$, a level about three times lower than that in Figure 6. A stable precipitate formed between the pH_p at 4.7 and the pH_d at 8.59 (rapidly above pH 7.26). Starch inhibited precipitation below pH 7.26 and at higher Al(III) concentrations, but the difference between the critical pH values for precipitation and for rapid precipitation decreased. The minimum $Al(NO_3)_3$ concentration at which starch removal was observed in the sweep zone was 4.5×10^{-4} M.

The restabilization zones at any given aluminum nitrate concentration were wider for 1.0 g/L starch than for 0.10 g/L starch. As with 0.10 g/L starch, aggregation of 1.0 g/L starch by higher aluminum nitrate concentrations was more rapid and yielded greater removals. The sweep zone and precipitation regions were wider, whereas the restabilization zones were narrower. Table 3 summarizes the critical values derived from the studies with the starch-aluminum nitrate system.

The aggregation of 1.0 g/L starch by aluminum sulfate and by aluminum nitrate in the presence of sodium sulfate was examined at an Al(III) concentration of 6.0×10^{-3} M, a concentration sufficient to aggregate starch in the absence of sulfate ions. The sulfate ions prevented clarification across the

Table 3. **Summary of Critical Data for the Aggregation of Soluble Potato Starch with Aluminum Nitrate**

Molar Conc.	log M	pH_p	pH_p*	pH_c	pH_s	pH_d
0.1 g/L Starch Solutions						
1.8×10^{-4}	-3.75	4.70	7.26	-	-	8.59
5.6×10^{-4}	-3.25	4.40	5.20	6.85	8.90	9.20
1.8×10^{-3}	-2.75	4.40	5.00	7.00	9.75	10.10
5.6×10^{-3}	-2.25	4.30	4.55	5.90	10.90	11.10
1.8×10^{-2}	-1.75	4.25	4.50	5.40	11.55	11.75
	Log c.c.c. at pH 5.10: -3.15					
	Log c.c.c. at pH 7.80: -3.35					
1.0 g/L Starch Solutions						
3.0×10^{-4}	-3.50	4.35	5.40	-		8.15
1.0×10^{-3}	-3.00	4.30	5.05	-		9.30
3.0×10^{-3}	-2.50	4.20	4.70	7.00	9.15	10.25
1.0×10^{-2}	-2.00	4.15	4.35	5.95	10.30	10.60
3.0×10^{-2}	-1.50	4.10	4.30	4.93	11.25	11.46
	Log c.c.c. at pH 7.15: -3.00					

*pH of rapid precipitation without starch inhibition

entire pH range from 3 to 11. A stable, milky, gel-like, peptized precipitate of starch, aluminum hydroxide, and sulfate was formed. Experiments with alum were also run on 25 mg/L and 0.1 g/L starch in the presence of iodine. In these tests the sulfate did not form a noticeably stable gel, and the starch and precipitate settled in the sweep zone.

The aggregation of oxidized coal sols in the presence of 25 mg/L of soluble potato starch was studied to determine the protective effects of the nonionic macromolecule. Starch at this concentration required an alum concentration of 1.3×10^{-4} M to initiate removal. This alum dose was 45 times higher than the minimum concentration required to aggregate the oxidized coal sol. Therefore, starch at 25 mg/L offered significant protection for the oxidized coal sol and, consequently, was studied further. Aggregation of protected oxidized coal sol was examined over the pH range of 2 to 13 and the $Al(SO_4)_{3/2}$ concentration range of 1.0×10^{-5} to 3.1×10^{-2} M. Starch removal was not examined directly, but it should have coincided with the clarification of the coal.

Typical results are shown in Figure 7 for aggregation by 3.1×10^{-4} M $Al(SO_4)_{3/2}$. The results were very similar to those obtained for oxidized coal in the absence of starch except that the protected sol settled somewhat more slowly. Distinct regions of stability and settling were present, and clarification was rapid and complete in the sweep zone between pH 4.95 and pH 8.75. This zone was narrower than for oxidized coal (between pH 4.55 and pH 9.05). Starch stabilized the coal in the pH ranges between 4.55 and 4.95 and between 8.75 and 9.05. Apparently, starch stabilized the small quantity of aluminum hydroxide precipitate formed in these pH ranges, and thereby the coal. This was shown by the results for aggregation of starch sols in the absence of coal. The same trends were observed at higher alum concentrations. Ionic coagulation occurred at pH

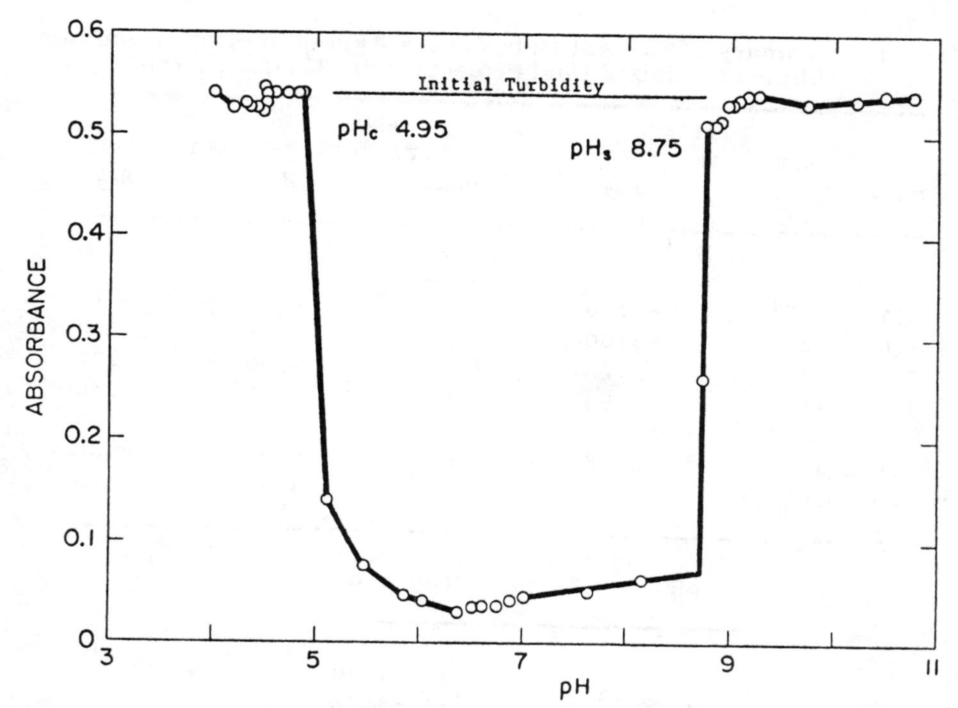

Figure 7. Aggregation of Oxidized Coal Protected by 25 mg/L of Starch with 3.1×10^{-4} $\underline{M}$ $Al(SO_4)_{3/2}$ as a Function of pH. 24-hr settling data.

values below the critical pH for rapid coagulation. Restabilization was not observed in the acid pH range, however. Apparently a narrow zone of sol protection persisted along the alkaline boundary of the sweep zone. The width of the sweep zone and the rate of clarification were greater at the higher alum concentrations. Several tests were also run to determine the minimum aluminum(III) concentration for aggregation. The critical coagulation concentration in these tests varied with pH; the lowest value observed was 1.0×10^{-4} $\underline{M}$ Al(III) at pH 7.05.

The critical concentrations and pH values obtained from the studies with oxidized coal protected with starch are summarized in Table 4. The critical values were virtually identical to the critical pH values found with alum for 0.1 mg/L starch sols and the c.c.c. values determined for 25 mg/L starch sols.

DISCUSSION AND CONCLUSIONS

COLLOIDAL COAL

The colloidal stability and electrophoretic mobility of the oxidized and unoxidized coal suspensions were examined as a function of pH. Oxidation lowered the isoelectric point four pH units by producing more easily ionized, acidic groups on the surface of the coal particles. More importantly, oxidation widened the pH range in which the coal sol was stable. The pH_s was lowered from 7.5 to below 3 by increasing the ionization and, therefore, the surface potential in the acid pH range. With both oxidized and unoxidized coal, the pH_s was approximately 2 pH units above the pH_i. The pH_s value for both sols

Table 4. Summary of Critical Data for the Aggregation by Aluminum Sulfate of Oxidized Coal Protected with 25 mg/L Starch

Aluminum Molar Conc.	log M	Critical pH Data pH_s	pH_c	pH_s
0	-	2.58	-	-
3.1×10^{-4}	-3.50	4.47	4.95	8.75
1.0×10^{-3}	-3.00	-	4.29	9.60
5.6×10^{-3}	-2.25	-	4.06	11.30
3.1×10^{-2}	-1.50	-	3.79	12.10
1.8×10^{-4}*	-3.75	-	6.95	8.08
1.0×10^{-2}**	-2.00	-	3.96	11.10

pH	log c.c.c.
3.20	-4.30
4.00	-3.35
5.30	-3.65
6.00	-3.77
6.80	-3.81
7.05	-4.00
7.55	-3.90
5.00#	-3.76
7.60#	-3.90

*Aluminum Nitrate, 25 mg/L Starch
**Aluminum Sulfate, 0.1 g/L Starch without coal
#Aluminum Sulfate, 25 mg/l Starch without coal

occurred at nearly the same electrophoretic mobility, presumably representing the minimum potential required for stabilization by double-layer repulsion.

Sol aggregation with aluminum salts can be examined systematically with the aid of a diagram of the critical coagulation and stabilization values plotted against their associated pH or Al(III) concentration. These experimentally determined critical values for coal sols have been summarized in Tables 1 and 2 and are plotted in Figure 8, the log aluminum sulfate concentration-pH stability limit diagram (domain). Each critical value is a point on a boundary separating regions of settling and of stability. The open symbols are critical coagulation values, and the blackened symbols are critical stabilization values; the critical concentrations are represented by diamonds. All other symbols are critical pH values. The squares represent values obtained for a dilute (2.3 mg/L), oxidized coal suspension. The dels are critical pH values for the unoxidized coal sol. The triangles are for experiments on oxidized coal sols with $Al(NO_3)_3$ instead of $Al_2(SO_4)_3$.

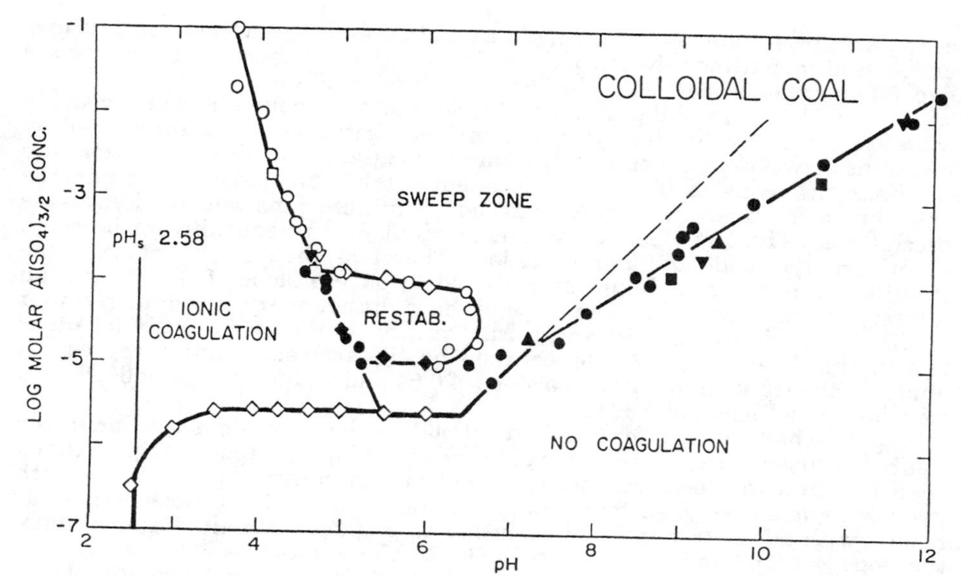

Figure 8. Aluminum Sulfate Concentration-pH Stability Limit Diagram for Colloidal Coal. Oxidized -400 mesh coal aged 25 days. Blackened symbols are stabilization values, and open symbols are coagulation values. Diamonds are turbidity-concentration data. All other symbols are turbidity-pH data[25].

The stability limit diagram has four distinct zones. The main zone of aggregation with aluminum is the central region labeled the sweep zone. This region of rapid clarification lies between the boundaries formed by the pH_c values on the acid side and by the pH_s values in alkaline solutions. The bottom boundary of the zone is delineated by c.c.c. values. In this region aluminum hydroxide was precipitated and enmeshed the oxidized coal. The floc settled rapidly, clarifying the suspension by "sweeping" the coal out of suspension.

Aggregation also occurred in the area to the left of the sweep zone. In this region the coal sol settled slowly, destabilization resulting from coagulation by ions, both soluble aluminum species and hydrogen ions. Below pH 2.48, the coal suspension aggregated in the absence of aluminum because of coagulation (double-layer compaction) by hydrogen ions. Above pH 3.5, the minimum Al(III) concentration required to induce aggregation was 2.8×10^{-6} $\underline{M}$ or about 1 mg/L, as $Al_2(SO_4)_3 \cdot 18H_2O$. This concentration is quite low for simple ionic coagulation with Al(III) species between pH 3.5 and pH 5.5. Most probably, the low c.c.c. value is due to adsorptive coagulation[13]. Soluble humates whose hydroxycarboxylic functional groups are very similar to those on the oxidized coal surface are known to precipitate with aluminum in this pH range[3,11]. Presumably, the ionic aluminum species react specifically with the coal surface, neutralizing its surface charge and thereby lowering the c.c.c. value in this pH range.

Oxidized coal suspensions were stable over two regions of the diagram. The principal zone of no coagulation was located to the right of the sweep zone and below the horizontal line formed by the lowest c.c.c. values. To the right of the sweep zone, stability occurred because at high pH, aluminum hydroxide hydrolyzes to form the soluble aluminate ion, $Al(OH)_4^-$[4]. This negatively charged ion is incapable of coagulating the negatively charged coal particles. The other stability region was the restabilization zone in the center of the diagram. Here, highly charged aluminum ions, i.e., $Al_8(OH)_{20}^{4+}$, are sufficiently adsorbed onto the coal and aluminum hydroxide precipitate to reverse the charge[14,15,21]. At concentrations above the c.s.c., the restabilized sol was coagulated by sulfate ions. The set of c.c.c. values for coagulation by sulfate ions added as alum outlined the top of the zone. The size and location of the zone are dependent on sol properties, including its surface area[21].

The shape and position of the right and left boundaries of the sweep zone are controlled by the hydrolysis and precipitation of aluminum and by reactions between the sol and aluminum. Typically, significant interactions are minimal with hydrophobic sols. Consequently, the left boundary represents the equilibrium between positively charged aluminum ions and its hydroxide precipitate. The right boundary is dependent on the equilibrium between aluminum hydroxide and aluminate ion. Therefore, the right boundary of the stability limit diagram would be expected to have a slope of +1.00 and an intercept of log K_4[4,20,21]. The slope of the right boundary of the diagram for oxidized coal differed significantly from results obtained previously for other sols in this laboratory. As determined by linear least squares regression analysis, the right boundary had a slope of 0.64 and an intercept of -9.35, with a correlation coefficient of 0.991.

The boundary for colloidal TiO_2[21] is shown as a dashed line in the figure to illustrate the significance of the change. This difference suggests that a specific interaction occurred between coal and aluminum, presumably with hydroxyl and carboxyl surface groups. Mangravite *et al.*[11] obtained similar results with a 5-mg/L humic acid solution aggregated with aluminum sulfate; the slope was 0.67 and the intercept -9.60. At higher humic acid concentrations the slope was smaller and the intercept increased. Humic acid and oxidized coal are closely related, both possessing similar ionic groups; vigorous oxidation of coal produces humic acids[1,12]. Additionally, there was no discernible difference between the pH_s values for the oxidized coal-aluminum sulfate system and those obtained for $Al(NO_3)_3$ (triangles), or a dilute suspension (squares), or with the unoxidized coal (dels).

The left boundary was very similar in shape to the aggregation boundary found with other hydrophobic sols and with the $Al(OH)_3$ precipitation boundary obtained with aluminum sulfate[4]. Its slope and intercept for the coal system were -3.25 and 10.94, respectively. A simple analysis of this boundary cannot be performed since at least two ionic species, $AlOH^{2+}$ and $Al_8(OH)_{20}{}^{4+}$, are in equilibrium with $Al(OH)_3(s)$ in significant relative concentrations. However, comparisons with other sols indicate that the left boundary for colloidal coal is controlled by the hydrolysis of aluminum. Coal does not alter the hydrolysis; that is, it does not interact specifically with Al(III) at the acid boundary.

COAL AND STARCH

The colloidal stability of a 1.0 g/L starch solution (in the absence of coal) was examined as a function of pH. The solution pH had little or no effect on the starch except for a slight reduction in the turbidity between pH 4.3 and pH 6.3. Similar observations were reported by Samec and co-workers[23,24]. They concluded that amylose (starch) is electrically neutral, but that a weak charge may arise from phosphorus and nitrogen impurities.

The critical values for the aggregation of 0.1 g/L and 1.0 g/L potato starch solutions with aluminum nitrate are plotted in Figure 9. Also shown in these stability limit diagrams are the critical limits of the $Al(OH)_3$ precipitation region as determined by nephelometric measurements. The blackened squares and dels are critical precipitation pH values, and open squares are critical dissolution pH values. Diamonds represent critical coagulation concentrations. The open and blackened circles are critical pH values for coagulation and for stabilization, respectively.

Starch changed the precipitation and dissolution boundaries of $Al(OH)_3$. Starch lowered the critical precipitation pH values (shown as dels), but inhibited precipitation. The dashed line represents the precipitation boundary in the absence of starch. The squares show the minimum pH values at which $Al(OH)_3$ precipitated rapidly and in large quantities. The dissolution

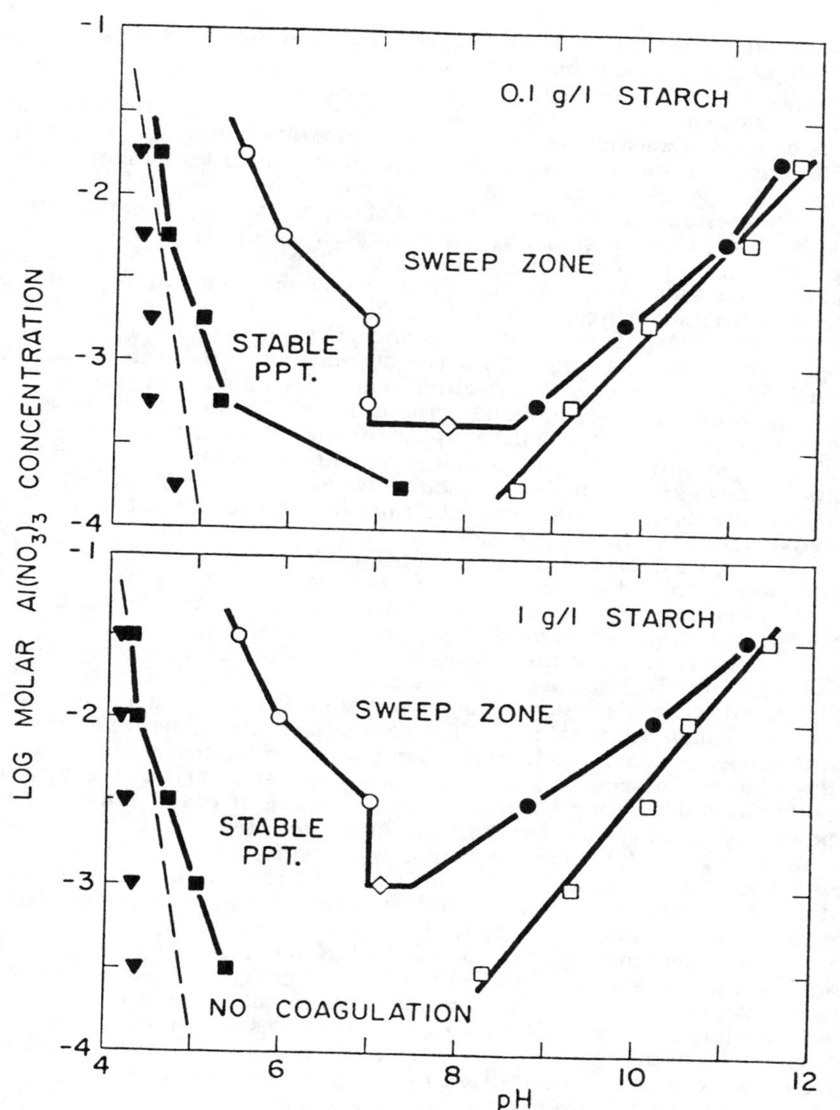

Figure 9. Aluminum Nitrate Concentration-pH Stability Limit Diagrams for Soluble Potato Starch. Blackened squares and dels are precipitation pH data, and open squares are dissolution pH data. Blackened circles are stabilization pH data, and open circles are coagulation pH data. Open diamonds are critical coagulation concentrations.

boundary contrasts significantly in that its slope and intercept are, respectively, 0.160 and -8.88 for 1.0 g/L starch, and 0.615 and -8.62 for 0.1 g/L starch. In general, the precipitation range was widened, particularly in the alkaline pH region at Al(III) concentrations above 1×10^{-3} $\underline{M}$. The precipitation boundary for the 1 g/L starch system was about 0.1 pH units lower than for the 0.1 g/L starch system, and the dissolution boundary was 0.6 pH units lower.

Aggregation in the sweep zone was by enmeshment in the flocs of aluminum hydroxide precipitate. The starch was removed only in the region where large quantities of precipitate were formed; clarification was not achieved outside the precipitation region. That is, starch was not aggregated by ionic coagulation. This behavior is consistent with that of hydrophilic colloids, such as silica and Teflon®, with adsorbed nonionic surfactant[8-10].

Starch stabilized the $Al(OH)_3$ precipitate at high ratios of starch/Al. This is evident from comparisons between the two diagrams. Increasing the starch concentration increased the minimum concentration of $Al(NO_3)_3$ required for aggregation and enlarged the restabilization zone in the acid pH range. This restabilization zone is typical of aggregation by aluminum nitrate. It results from charge reversal by the adsorption of highly charged aluminum complexes onto the sol and the precipitate. Finally, starch created a second restabilization zone along the dissolution boundary. The width of this zone increased with increasing starch concentration.

In all Al(III) concentrations and pH regions where the quantity of precipitate formed was small, both the starch and precipitate were stable. Starch is known to act as a protecting agent; Freudlich[2] measured its gold number according to the Zsigmondy procedure. He found that 20 mg of starch were required to prevent 10 mL of a gold sol from changing color. This corresponds to a gold number of 20, which represents weak protecting power. The results for the aggregation of starch showed weak protection of $Al(OH)_3$. Starch at concentrations of 25, 100, and 1,000 mg/L stabilized 11, 24, and 78 mg/L, respectively, of aluminum hydroxide.

Figure 3 showed the effect of pH on the colloidal stability of oxidized coal in the presence of 25 mg/L soluble potato starch. The critical pH_s for the coal was unchanged by starch, although the rate of clarification decreased markedly, providing partial protection from aggregation.

The results for the aggregation by Al(III) of oxidized coal in the presence of 25 mg/L starch are summarized in the stability limit diagram shown in Figure 10. The blackened symbols are stabilization values, and the open symbols are coagulation values. Circles are critical pH values, and squares are critical concentrations. The triangles are critical pH values for 100 mg/L starch in the absence of coal. The diamonds represent critical coagulation concentrations for 25 mg/L starch in the absence of coal. For comparison purposes, the boundaries of the diagram for unprotected coal are shown as light lines in the figure. The main region of aggregation was the sweep zone, as it was for both sols alone. Clarification was both rapid and complete, and the location of its boundaries was controlled by the starch, as is evident from the agreement with the data obtained in the absence of coal.

Comparison with the SLD for oxidized coal in the absence of starch shows the effect of starch to protect coal and $Al(OH)_3(s)$ at low Al(III) concentrations. Starch also shifted the alkaline boundary of the sweep zone to a lower pH where a larger mass of $Al(OH)_3$ precipitate is present. In the region between the boundaries of the sweep zones for oxidized coal with and without starch, the $Al(OH)_3$ and coal were restabilized in the same manner as for starch alone. Apparently, starch was absorbed at the solid surfaces, preventing floc formation and subsequent clarification. In the sweep zone, 25 mg/L starch raised the critical coagulation concentration of Al(III) by as much as a factor of 60. Therefore, it is apparent that starch increases the stability of coal with respect to aggregation by enmeshment. Coal was effectively clarified by enmeshment above the c.c.c. The boundaries of the sweep zone correspond to the location where the quantity of $Al(OH)_3$ precipitate is sufficiently large to overcome the protecting power of starch.

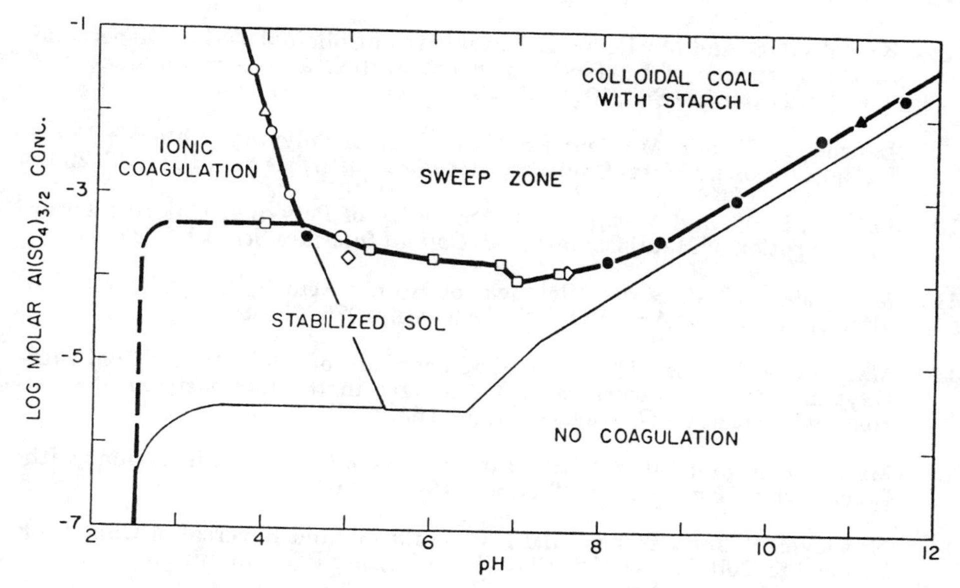

Figure 10. Aluminum Sulfate Concentration-pH Stability Limit Diagram for Oxidized Coal Protected by 25 mg/L of Soluble Potato Starch. Blackened symbols are stabilization values, and open symbols are coagulation values. Squares and diamonds are turbidity-concentration data. Circles and triangles are turbidity-pH data. The triangles and diamonds are critical values for starch in the absence of coal. The light lines are the SLD boundaries for unprotected coal (see Figure 8).

REFERENCES

1. Czuchajowski, L., "Infra-red Spectra of Coals Oxidized with Hydrogen Peroxide and Nitric Acid," *Fuel*, 1960, *39*, 377.

2. Freundlich, H., *Kapillarchemie*, Vol. II, Leipzig, 1932, as cited in *Colloid Science*, Vol. 1, H. R. Kruyt, ed., Elsevier, New York, NY, 1952.

3. Hall, E. S., and Packham, R. F., "Coagulation of Organic Color with Hydrolyzing Coagulants," *J. Am. Water Works Assoc.*, 1965, *57*, 1149.

4. Hayden, P. L., and Rubin, A. J., "Systematic Investigation of the Hydrolysis and Precipitation of Aluminum(III)," in *Aqueous Environmental Chemistry of Metals*, A. J. Rubin, ed., Ann Arbor Science, Ann Arbor, MI, 1974.

5. Hesselink, F. Th., Vrij, A., and Overbeek, J.Th.G., "On the Theory of the Stabilization of Dispersions by Adsorbed Macromolecules II. Interactions between two flat particles," *J. Phys. Chem.*, 1971, *75*, 2094.

6. Jirgensons, B. and Straumanis, M. E., *A Short Textbook of Colloid Chemistry*, Elsevier, New York, NY, 1962.

7. Kavanaugh, M. C., "Modified Coagulation for Improved Removal of Trihalomethane Precursors," *J. Am. Water Works Assoc.*, 1978, *70*, 613.

8. Kratohvil, S. and Matijevic, E., "Stability of Colloidal Teflon Dispersions in the Presence of Surfactants, Electrolytes, and Macromolecules," *J. Colloid Interface Sci.*, 1976, *57*, 104.

9. Lauzon, R. V. and Matijevic, E., "Stability of Polyvinyl Chloride Latex I. Coagulation by Metal Chelates," *J. Colloid Interface Sci.*, 1971, *37*, 296.

10. Lauzon, R. V. and Matijevic, E., "Stability of Polyvinyl Chloride Latex II. Adsorption of Metal Chelates," *J. Colloid Interface Sci.*, 1972, *38*, 440.

11. Mangravite, F. J. *et al.*, "Removal of Humic Acid by Coagulation and Microflotation," *J. Am. Water Works Assoc.*, 1975, *67*, 88.

12. Marinov, V. N., "Self-ignition and Mechanisms of Interaction of Coal with Oxygen at Low Temperatures. 3. Changes in the Composition of Coal Heated in Air at 60°C," *Fuel*, 1977, *56*, 165.

13. Matijevic, E. and Allen, L. H., "Interactions of Colloidal Dispersions with Electrolytes," *Environ. Sci. Technol.*, 1969, *3*, 264.

14. Matijevic, E. and Stryker, L. J., "Coagulation and Reversal of Charge of Lyophobic Colloids by Hydrolyzed Metal Ions III. Aluminum Sulfate," *J. Colloid Interface Sci.*, 1966, *22*, 68.

15. Matijevic, E., Janauer, G. E., and Kerker, M., "Reversal of Charge of Lyophobic Colloids by Hydrolyzed Metal Ions. I. Aluminum Nitrate," *J. Colloid Interface Sci.*, 1964, *19*, 333.

16. Napper, D. H., "Steric Stabilization," *J. Colloid Interface Sci.*, 1977, *58*, 390.

17. Napper, D. H. and Netschez, A., "Studies of the Steric Stabilization of Colloidal Particles," *J. Colloid Interface Sci.*, 1971, *37*, 528.

18. Narkis, N. and Rebhun, M., "The Mechanism of Flocculation Processes in the Presence of Humic Substances," *J. Am. Water Works Assoc.*, 1975, *67*, 101.

19. Packter, A., "Coagulation of Peptised Sols by Simple Electrolytes," *Kolloid-Z.*, 1960, *170*, 48.

20. Rubin, A. J. and Hanna, G. P., "Coagulation of the Bacterium Escherichia coli by Aluminum Nitrate," *Environ. Sci. Technol.*, 1968, *2*, 358.

21. Rubin, A. J. and Kovac, T. W., "Aggregation and Restabilization of Colloidal Titania by Aluminum Sulfate," *Separ. Sci. Technol.*, 1986, *21*, 439.

22. Rubin, A. J. and Kramer, R. J., "Recovery of Fine-Particle Coal by Colloid Flotation," *Separ. Sci. Technol.*, 1982, *17*, 535.

23. Samec, M. and Haerdtl, H., *Kolloidchem. Beihefte*, 1920, *12*, 281, as cited in *Starch and its Derivatives*, J. A. Radley, ed., Chapman and Hall, London, England, 1968.

24. Samec, M. and Mayer, A., *Kolloidchem. Beihefte*, 1921, *13*, 272, as cited in *Starch and its Derivatives*, J. A. Radley, ed., Chapman and Hall, London, England, 1968.

25. Schroeder, P. R. and Rubin, A. J., "Aggregation and Colloidal Stability of Fine-Particle Coal Suspensions," *Environ. Sci. Technol.*, 1984, *18*, 264. Figures 2, 5, and 8 are reprinted with permission.
Copyright 1984, American Chemical Society.

26. Semmens, M. J. and Field, T. K., "Coagulation: Experiences in Organics Removal," *J. Am. Water Works Assoc.*, 1980, *72*, 476.

27. Shaw, D. J., *Introduction to Colloid and Surface Chemistry*, 2nd Edition, Butterworths, London, England, 1970.

28. Willims, D.J.A. and Ottewill, R. H., "The Stability of Silver Iodide in the Presence of Polyacrylic Acids of Various Molecular Weights," *Kolloid-Z.u.Z. Polymere*, 1971, *243*, 141.

IMPORTANCE OF FLOCCULANT PREPARATION FOR USE IN SOLID/LIQUID SEPARATION

Yong H. Kim
Stranco, Inc.
Bradley, Illinois 60915
(815)932-6022

ABSTRACT

To achieve the optimum performance of water and wastewater treatment processes using polymeric flocculants, it is essential to manipulate the polymers adequately. Much information is needed to achieve the desirable state of polymer solution, such as characteristics of polymer, nature of dilution water, and hydrodynamic environment involved in the process. The first part of the paper deals with the effect of water chemistry on the polymer dissolution. Various mixing schemes are considered based upon the role of hydrodynamic force in the polymer dissolution mechanism. Two-stage mixing of polymeric flocculant is found to be very effective by reducing the necessary aging time, remarkably without damaging polymer structure.

INTRODUCTION

Water-soluble polymers of high molecular weight have been of growing importance as flocculants in water and wastewater treatment. Polymers based upon acrylic acid and its derivatives, such as acrylamide and various cationic derivatives, are an important class of materials in this respect. Their effectiveness stems, among other things, from the extended and flexible nature of the polymer molecules when in solution in water. This extension enables them to adsorb onto the surface of many suspended particles, thus forming flocs. A system thereby flocculated is much more amenable to separation of the solid and liquid constituents by one of the numerous dewatering devices available today.

To obtain optimum effectiveness in this function, polymeric flocculants must be dissolved in water to prepare solutions. In these solutions the polymer molecules, originally in highly entangled form, are allowed to absorb water in such a manner as to permit a substantial degree of disentanglement. This dissolving and disentangling process, which is referred to as "activation" in this paper, improves their ability for multiple particle interaction during the floc formation by an interparticle bridging effect. The principal objective of the paper includes the investigation of the effects of dilution water chemistry and the hydrodynamic environment on the activation of polymeric flocculants.

POLYMERS IN FLOCCULATION

Many polymeric flocculants are available commercially for a wide range of applications. Some products are based upon natural materials such as gelatine and starch; however, the use of synthetic polymers as flocculants in water and wastewater treatment is of growing technological importance. The important characteristics of polymeric flocculants include molecular weight and the nature of functional groups. In general, the flocculating efficiency of a polymer (assessed in terms of floc size, settling rate, filterability) increases along with molecular weight for a homogeneous series of the polymer. On the other hand, with increasing molecular weight, the solubility of a polymer generally tends to decrease, causing the technical difficulty involved in the synthesis of water-soluble, high molecular weight polymers. To be effective in destabilizing the particle suspension, the polymer molecule must contain functional groups that can interact with sites on the surface of suspended particles. The adsorption of polymer segments at the particle surface can be due to electrostatic attraction, hydrogen bonding, and covalent bonding. Polymers with high ionicity can adsorb at the oppositely charged particle surface. Hydrogen bonding plays a role when the particle surface and the polymer molecules have suitable hydrogen bonding sites[8]. Polymers can also adsorb at the lightly charged particle surface due to covalent interactions. If the charge density at the particle surface is not very high, the electrostatic repulsion between the charged polymer segments and the particle surface is not strong enough to prevent polymer adsorption through covalent and/or hydrogen bonding.

The repulsion between charged particles extends over a range whose breadth depends upon the effective thickness of diffuse layer and hence upon the ionic strength of the suspension. For polymer-bridging flocculation to occur, the adsorbed polymer chains must be able to span the gap over which double-layer repulsion operates. Depending upon the molecular weight of the polymer (2×10^4 to 2×10^6), the contour length of the linear polymers in solution ranges from 0.1 µm to 1.5 µm[21], and the effective thickness of the diffuse double layer is about 0.03 µm in water containing electrolytes at an ionic strength of 10^{-4} M. When a polymer molecule comes in contact with a suspended particle, some of the functional groups of polymer adsorb at the particle surface, leaving the remainder of the molecule extending out into the solution. If a second particle with some vacant sites contacts these extended segments, attachment can occur. A particle-polymer-particle complex is thus formed in which the polymer serves as a bridge. This is the bridging mechanism first proposed by La Mer[11] and there is no doubt that the bridging mechanism operates with nonionic polymers and with polymers charged the same to the particles of the colloidal suspension.

When the polymers with opposite charge to the suspended particles are used to destabilize the suspension, flocculation is accomplished by charge neutralization, bridging, or a combination of these two mechanisms. The polymer may serve first as a coagulant to reduce the particle surface charge density such that particles may approach each other sufficiently closely that the attractive forces become effective. Then bridging may occur via extended segments of adsorbed polymer molecules[1]. On the other hand, Mabire *et al.*[13] claimed that the bridging mechanism becomes less important in the case of a strong attraction leading to charge neutralization. Because most polymer molecules may be so strongly adsorbed on the particle surface, there may not be enough extended polymer chains in the solution to bring about bridging. Gregory[7] proposed the electrostatic patch model, in which a one-to-one correspondence between opposite charges on the particle surface and the adsorbed polymer is not necessary. If the charge density of polymer in its adsorbed state is higher than that of the particle surface, the adsorbed polymers will distribute unevenly, resulting in patches of positive and negative charge. Thus, although the particle as a whole may be electrically neutral, particles having this patch type of charge distribution may interact in such a way that positive and negative patches come into contact, resulting in attachment.

POLYMER DISSOLUTION

THEORY

Theoretical treatment of the solution behavior of linear polymer would involve both the complicated statistics of elongated chains beyond the range of usefulness of normal approximations and the even more difficult problem of their electrostatic interactions when highly charged. Stochastic approach[9] may be the most appropriate way to describe such a system, but its inherent complexity is beyond the scope of this paper.

The study on the swelling of polymeric network was greatly advanced by Flory[4], who described the swelling process as the diffusion of ions and solvent from an external solution into a polymer network. He suggested two kinds of forces are involved: elastic retractive force and swelling force induced by the localization of charges on the polymer structure. The swelling force resulting from the presence of the charged sites is identified with the swelling pressure or the osmotic pressure which, assuming the solution to be dilute, is given by

$$\pi_i = RT\Sigma\,(c_i - c_i^*) \tag{1}$$

where R is the gas constant, T the absolute temperature, and c_i and c_i* the mobile ion concentration inside and outside the polymer gel, respectively. From the equation, we notice the favorable effect of temperature on polymer swelling and the reason why the polymers with higher charge density are easy to activate in a polymer dissolution process. When the concentration difference of the mobile ions between inside and outside of the gel is comparable in magnitude to the concentration of fixed charge in the gel (ic), equation (1) can be written

$$\pi_i = \frac{RT(ic)^2}{4I} \tag{2}$$

where I is the ionic strength of external solution. In the limit of large ratio of ionic strength compared with the fixed charge concentration, the swelling behavior of the ionic polymer network reverts to that for the nonionic polymers.

More intensive research was performed by Tanaka and coworkers(18,19) on the swelling and even shrinking of polymers in various external conditions such as pH, temperature, composition of solvent, and electrical field. They proposed that these phenomena in polymer gels are influenced by three forces: polymer elasticity, polymer-solvent affinity, and mobile ion effects. The total osmotic pressure is a consequence of the net sum of these three components acting either positively or negatively. In the swelling of linear polymer, however, the strength of polymer elasticity may be very small due to its least degree of crosslinkage. Consequently, linear flocculant can dissolve into homogeneous solution even below its overlap concentration with a sufficient amount of water. The polymer-solvent affinity can be either attractive or repulsive, depending mainly upon the electrical properties of the molecules. In an acrylamide polymer gel and where the gel is placed in a solvent that is a mixture of water and acetone, a polymer chain has a greater affinity for other polymer chains, so the polymer tends to shrink. However, as the amount of acetone is decreased in the solvent, the polymer-solvent affinity is increased and the polymer chain tends to expand in pure water.

The kinetics of polymer swelling in a solvent have been explored by Tanaka and Fillmore(18) using the equation of motion of the polymer gel network. Although their derivation was based on the network structure of polyacrylamide and some modification may be needed to apply it to the linear polymer system, it is worth examining the results. They showed that the radius of the polymer gel during the swelling process can be predicted by introducing the characteristic swelling time (t_s), which is used to scale the actual swelling time,

$$t_s = \frac{r^2}{D} \tag{3}$$

That is, the swelling time is proportional to the square of the gel radius and inversely proportional to the diffusion coefficient of the gel network. This argument was confirmed by the experimental data obtained from the swelling of polyacrylamide gels(15).

EXPERIMENTS

Three acrylamide-based polymers with different ionicity were used in the study. They are high molecular weight emulsion polymers in which active contents are about 30 percent and dispersed in hydrocarbon oil. The properties of these polymers are illustrated in Table 1. Sodium hydroxide and nitric acid were used to adjust the pH of the solution to the desired value.

Polymers were dissolved in the water by injecting through a syringe while agitation was provided with a stirrer (Jiffy mixer LM) in a 600-mL beaker. The concentration of polymer was 0.15 percent by weight unless otherwise noted. The stirrer employed was known to be very effective in bench-scale experiments, since its unique design (large diameter and prolonged height of impeller blade) helps to provide more uniform mixing within the vessel than other kinds.

To measure the degree of polymer activation, viscosity measurements were performed with a Brookfield viscometer (Model LVTDV-II) at the spindle speed of 12 rpm. For the same purpose other techniques can be employed such as zeta potential, streaming current, and colloid titration. They are based on the degree of exposure of charged sites on the chain into the solution and have been used to detect the effectiveness of the flocculation process. However, they may not be as good an indicator of polymer activation as viscosity, since the physical state (length of the chain) of the polymer structure is as important as its chemical state (charge exposure) in functioning as a flocculant.

Table 1. Properties of Polymers Used in Study

Polymer	A	B	C
Commercial Name	Praestol A3035L	Praestol K226FL	Praestol N3100L
Charge	Anionic	Cationic	Nonionic
Backbone monomer	Acrylamide	Acrylamide	Acrylamide
Functional group	Acrylic acid	Quaternary amine	---
Charge density	25%	30%	---
Active content	36%	28%	30%
Molecular weight	$>10*10^6$	$>6*10^6$	$>16*10^6$
Product viscosity* (cp)	305	1870	1530

* Brookfield viscometer, LVTDV-II, spindle 2, 12 rpm.

Furthermore, the colloid titration technique has a serious interference with the presence of other electrolytes[20]. It is known that electrolytes of higher charge, divalent or trivalent ions, seriously interfere with the titration unless their presence is kept below 50 ppm.

RESULTS

To prove the validity of using viscosity as a measure of polymer activation, both viscosity and conductivity measurements were performed for 0.1 percent solution of polymer A. Polymer A was mixed with distilled water for 7 minutes with a Jiffy mixer at 600 rpm. After the solution was aged for 4 hours, it was agitated at very high speed of 1200 rpm for the next 30 minutes. As presented in Figure 1, solution viscosity showed a sharp decrease with employing excessive agitation, indicating the severe damage on the polymer chain. However, conductivity of the polymer solution continued to increase with the excessive agitation. The importance of polymer extension has been thoroughly demonstrated in the work of Sakaguchi and Nagase[17], who found a good correlation between viscosities of various solutions of polyacrylamide and the settling rates of the flocculated kaolin suspensions.

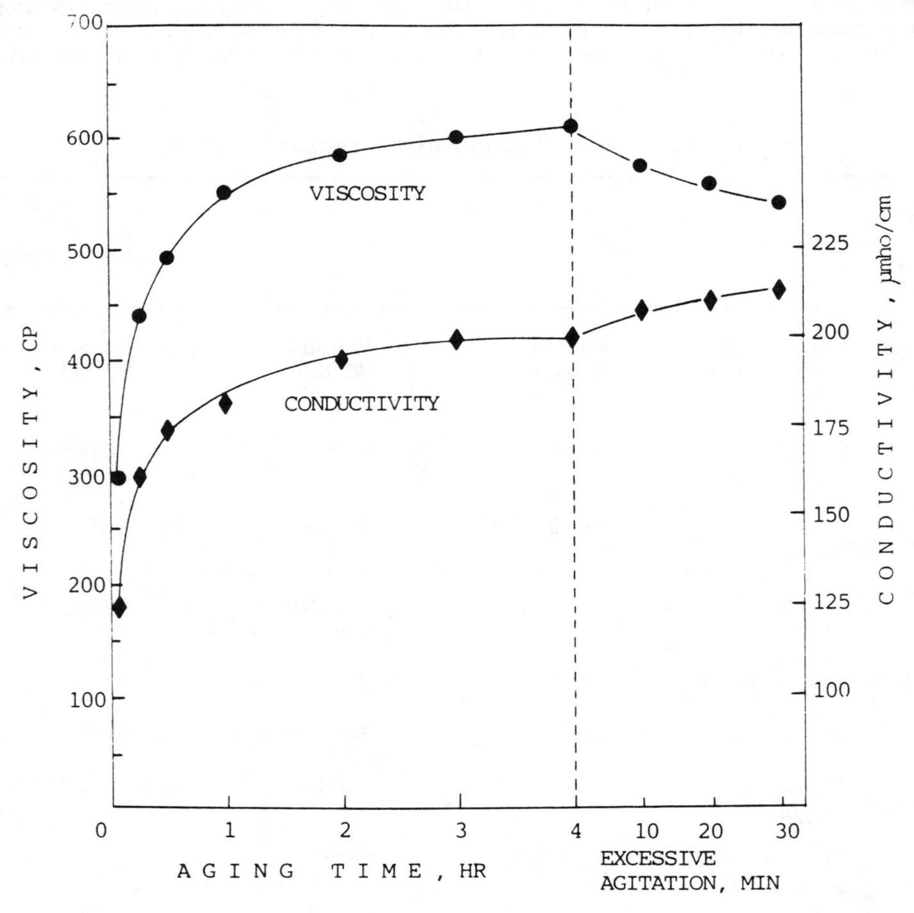

Figure 1. Development of Viscosity and Conductivity of Polymer A Solution After Preparation and Subsequent Excessive Agitation at 1200 rpm; 23 C, 0.10%, Brookfield Viscometer, 12 rpm

To investigate the effect of dilution water hardness on polymer activation, water hardness was varied from 0 (distilled water) to 420 mg/L. Polymers were mixed with water for 10 minutes with a Jiffy mixer at 600 rpm. The experimental data shown in Figure 2 presents the viscosity development of polymer solution as a function of aging time. Anionic polymer A showed sharp viscosity increase during the first 1 hour after mixing, indicating the rapid extension or disintegration of polymer molecules into solution. It is striking to note the effect of hardness on the expansion of polymer chains. Figure 3 illustrates the ratio of the solution viscosity of polymer prepared with water of various hardnesses to that prepared with distilled water. Within the range of hardness studied, the solution viscosities vary from the 20-fold increase for an anionic polymer A, 14-fold for a cationic polymer B, and 7-fold for a nonionic

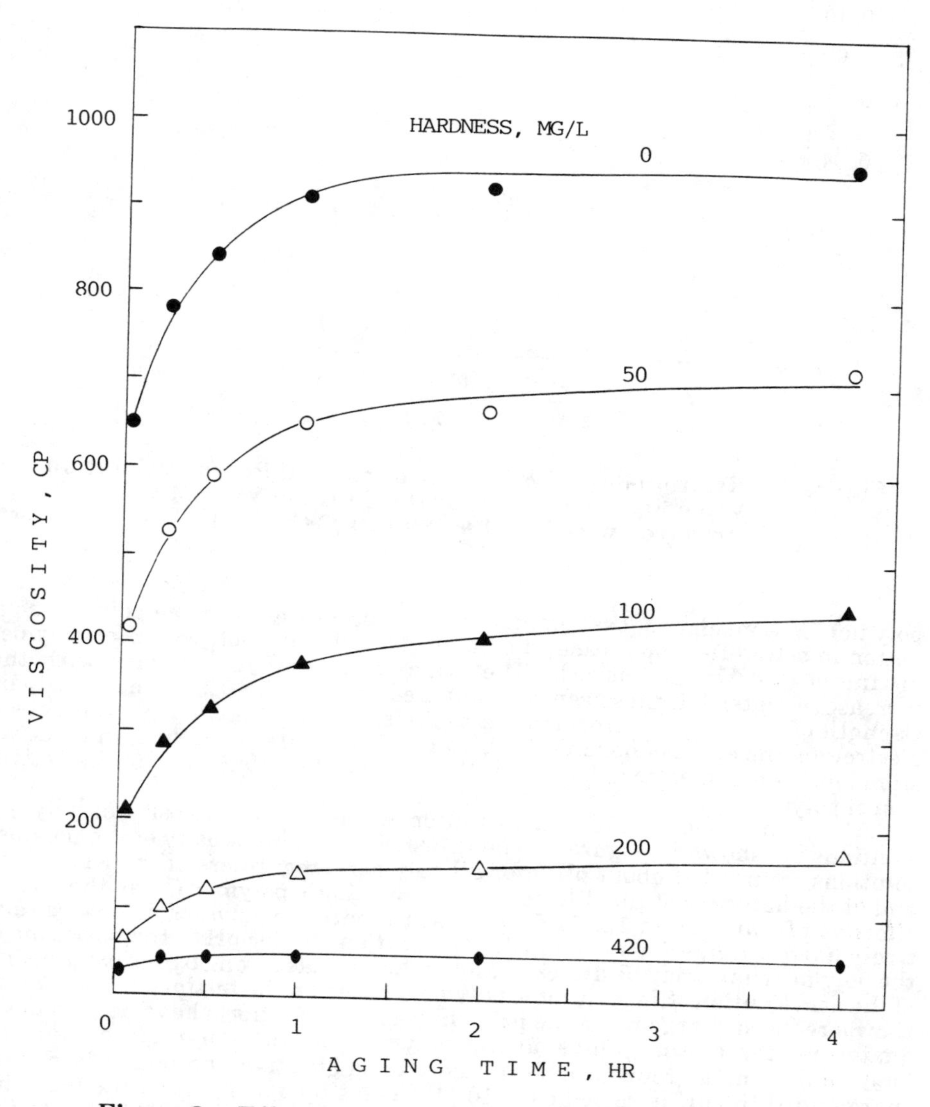

Figure 2. Effect of Water Hardness on the Viscosity of Polymer A Solution; 23 C, 0.15%

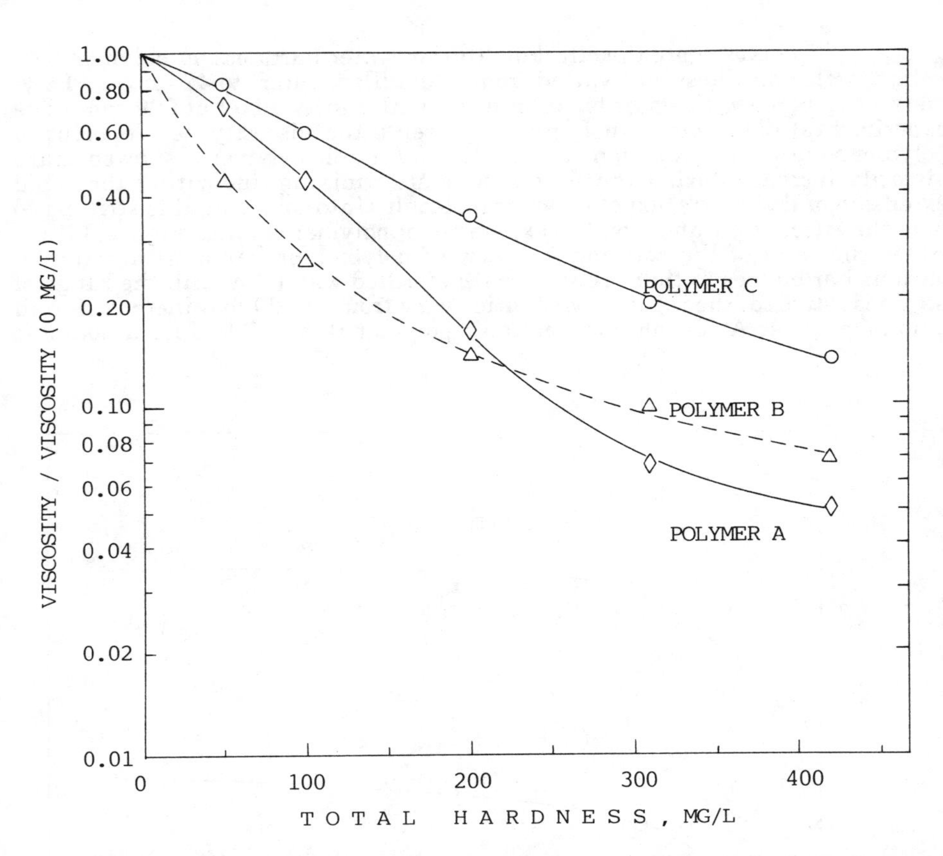

Figure 3. Relationship of Water Hardness and Polymer Solution Viscosity Normalized by the Solution Viscosity Prepared with Distilled Water; 23 C, 0.15%

polymer C. Considering the common use of process water for the polymer activation in actual field operation, the importance of using soft water for polymer mixing should be emphasized. The dramatic fall-off in viscosity with the increase of water hardness can be explained by equation (2). Since the ionic strength of a solution is proportional to the square of the valency of dissolved electrolyte, the effect of hardness, which is the measure of calcium and magnesium ions, is much more detrimental than the single valency ion to the activation of polymers.

The effect of pH of the dilution water on the viscosity of polymer solutions are shown in Figure 4. The threshold of high viscosity of the polymer solutions occurred at about pH 3 for all polymers, regardless of their ionicities and of the hardness of the dilution water. Nonionic polymer C has the widest effective pH range from 3 to 11.5, whereas the cationic polymer B shows a relatively narrow range from pH 3 to 9. For solutions under pH 2, the backbone of the polymer chain may be attacked and broken by the strong oxidative property of the acid solution. Since cationic polymers based on the tertiary amine groups show a reduced charge when the pH is higher than 8, efforts have been made to quaternize the amine groups on the polymer chain. Polymers containing quaternary amine groups are more likely to be effective and do not lose their charge until the pH is raised over 10[14]. The slight increase in the viscosity with the increase of pH for both polymers A and C may be attributed to the increased ionization of the carboxyl groups.

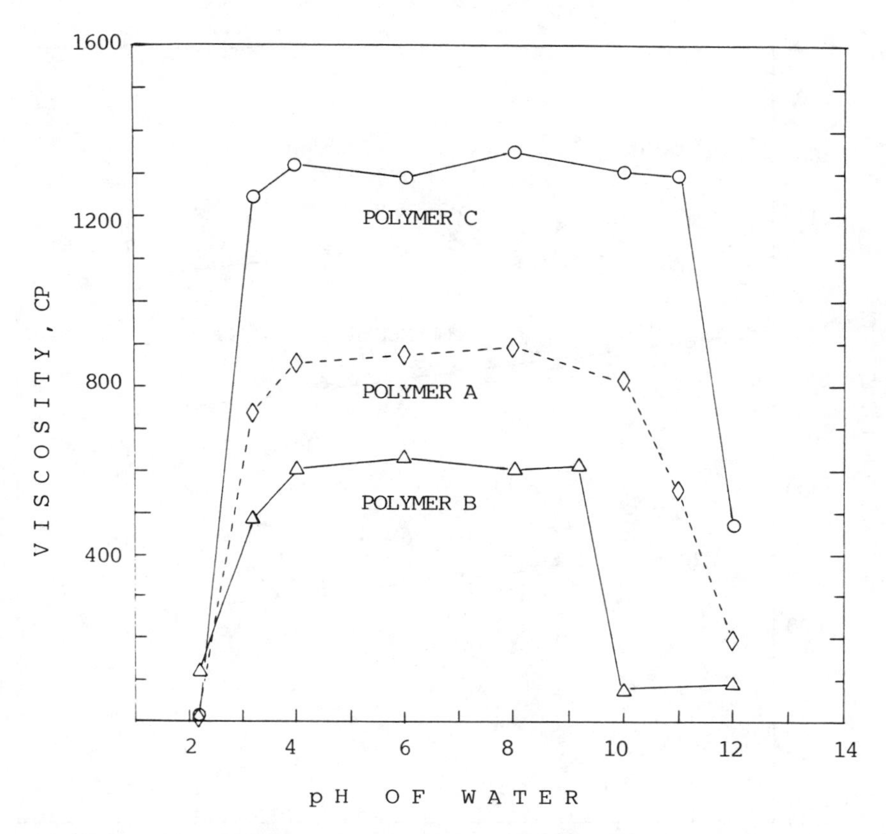

Figure 4. pH Effect on the Viscosity of Polymer Solution Prepared with Distilled Water; 23 C, 0.15%

To correlate the viscosity of polymer solution to temperature, polymer solutions were prepared and aged for 24 hours at room temperature. Then, viscosity was measured while polymer solutions were cooled to 5 C. The simple logarithmic equation below, as well as the Arrhenius equation, has been used to relate the viscosity-temperature data,

$$\mu = a \exp(-bT) \tag{4}$$

where a and b are empirical constants. Equation (4) applied to polymer solutions fits the experimental data well, as shown in Figure 5. The data regression yields to the following relations for each polymer solution:

$$\begin{aligned} \mu_A &= 10{,}027 \exp(-0.0080\,T) \\ \mu_B &= 5{,}049 \exp(-0.0071\,T) \\ \mu_c &= 8{,}283 \exp(-0.0064\,T) \end{aligned} \tag{5}$$

Rheological behavior of nonionic polymer C is less sensitive to the change in temperature than that of ionic polymers A and B.

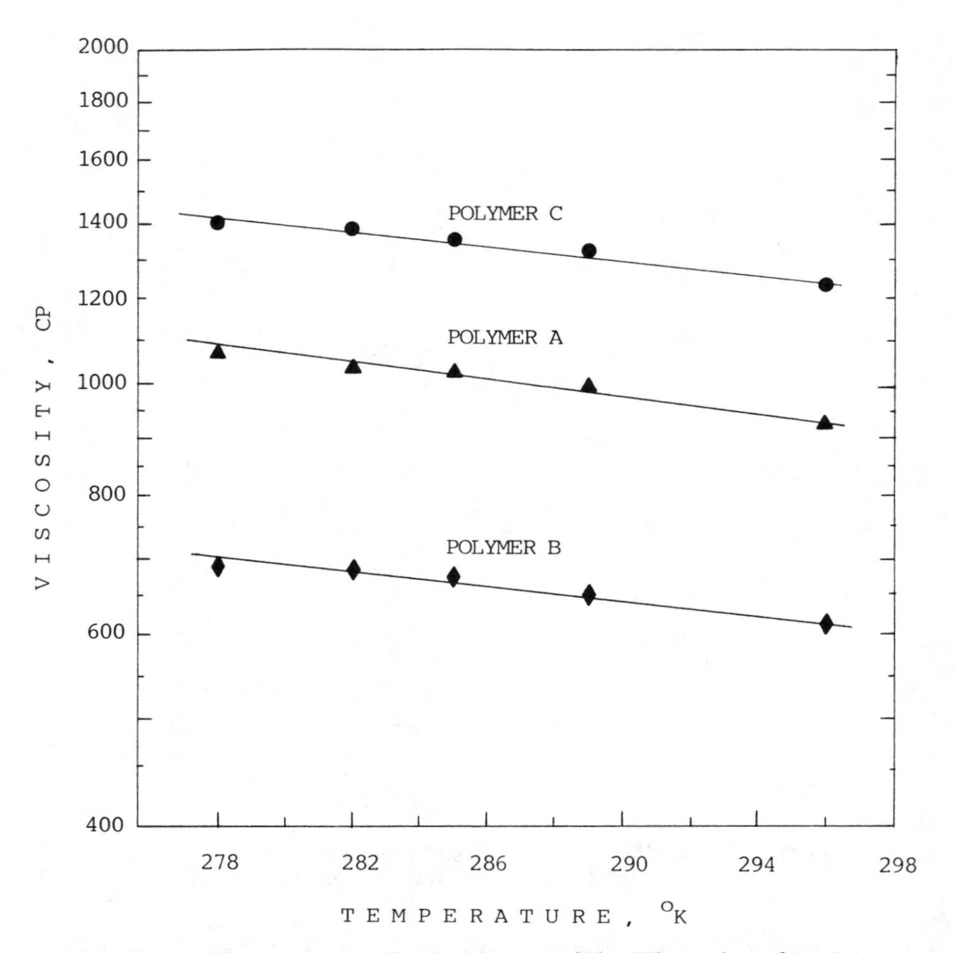

Figure 5. Temperature Dependence of the Viscosity of 0.15% Polymer Solutions Prepared and Aged for 24 Hours Before Cooling

HYDRODYNAMICS IN POLYMER DISSOLUTION

Agitation must be applied in the polymer activation process to disperse dry polymer or polymer gel particles in emulsion polymer into the dilution water. Hydrodynamic force also helps to reduce the size of partially swollen gel particles to increase the rate of dissolution. It is well accepted that vigorous agitation is required, and conventionally the agitation is applied for a considerable time with the intention of improving dissolution. Once the polymer gels are dispersed in the dilution water and begin to swell, however, excessive hydrodynamic force should be avoided lest the swollen or extended polymer chains be disrupted.

At high shear conditions there exists the possibility that the polymer chain can be ruptured. The force required to break a carbon-carbon bond in a polymer chain is of the order of 1 to 10 nN[12], and the maximum hydrodynamic force exerted on the spherical particle of 1 µm under the shear rate of 1000/sec would become over 30 nN according to Goren's expression[6]. Because of this fragility of the polymer chain, a considerable number of polymer chains may be easily broken during the activation process, when the polymer molecules stay long under the high shear environment with the extended conformation.

Nonuniformity of the agitation intensity within the mixing vessel is another factor to be considered in the design of polymer activation system. Okamoto *et al.*[16] measured the agitation intensity distribution in a stirred tank with turbine impeller and found that the energy dissipation rate near the impeller blade tips is nearly 60 times greater than that away from the impeller. In other reports[3,5], the difference in the state of turbulence is on the order of tens or hundreds of times within the mixing vessel. In practice, agitation by use of a large paddle type or close-clearance impeller has been effective to achieve the optimum activation of polymeric flocculants. These findings conclude that the larger the impeller diameter and blade width by comparison with the tank diameter, the more uniform the agitation intensity within the tank[16].

CONCEPT OF TWO-STAGE MIXING

The activation process of emulsion of dry polymers is known to be difficult because once the polymer contacts water, a film of concentrated solution of the polymer is built up around each of the polymer gels, and they become so gluey that they easily form aggregates. It is at this stage that most problems occur. The activation of emulsion polymer was postulated as two separate processes in this study: inversion of emulsion and swelling of polymer gel. First the water (containing polymer gels)/oil emulsion must be inverted to the oil/water emulsion by introducing the sufficient amount of dilution water. High mixing energy in a short time period is needed at this stage to prevent the polymer gels from forming agglomerates. Without proper mixing energy, polymer gel agglomerates form that increase polymer gel size and require much longer swelling time, as indicated in equation (3). After this initial wetting, polymer gels begin to swell and gradually become disintegrated into a solution. This second step may be well described by de Gennes' rotation model. The basic idea is that the entanglement of the polymer chains prevents any large scale motions and the chains can only rotate in a snakelike motion. It is obvious that high mixing energy does not help, but rather deteriorates the polymer chains at this stage.

Based upon this inversion-swelling concept, experiments to explore the effectiveness of two-stage mixing were performed by employing very high shear mixing in the first short period of time, followed by the gentle agitation. The results are shown in Figure 6 with the single batch mixings at various impeller speeds. Anionic polymer A was mixed with soft water (100 mg/L) in this part of the study, and the mixing equipment was the same as that in the previous sections. When impeller speeds lower than 600 rpm were applied to the system, more than 3 hours of aging time were required to attain about 90 percent of the maximum viscosity. A high impeller speed of 1000 rpm, though, resulted in serious damage to the polymer structures, as indicated by the significant decrease in the maximum viscosity. The optimum impeller speed was found to be 800 rpm with the aging time of 1.5 hours to approach the maximum viscosity. Two-stage mixing was carried out by applying a very high impeller speed of 1350 rpm for the first 3 minutes and then reducing it to 250 rpm for the next 10 minutes. The result was remarkably improved both in instantaneous solution viscosity right after mixing and in shortened aging time to obtain the maximum viscosity. The resultant polymer solution approached the maximum viscosity within 15 minutes after mixing.

IN-LINE TWO-STAGE MIXING

Acknowledging the benefit of the two-stage mixing in polymer activation, the original continuous polymer mixing chamber was redesigned to

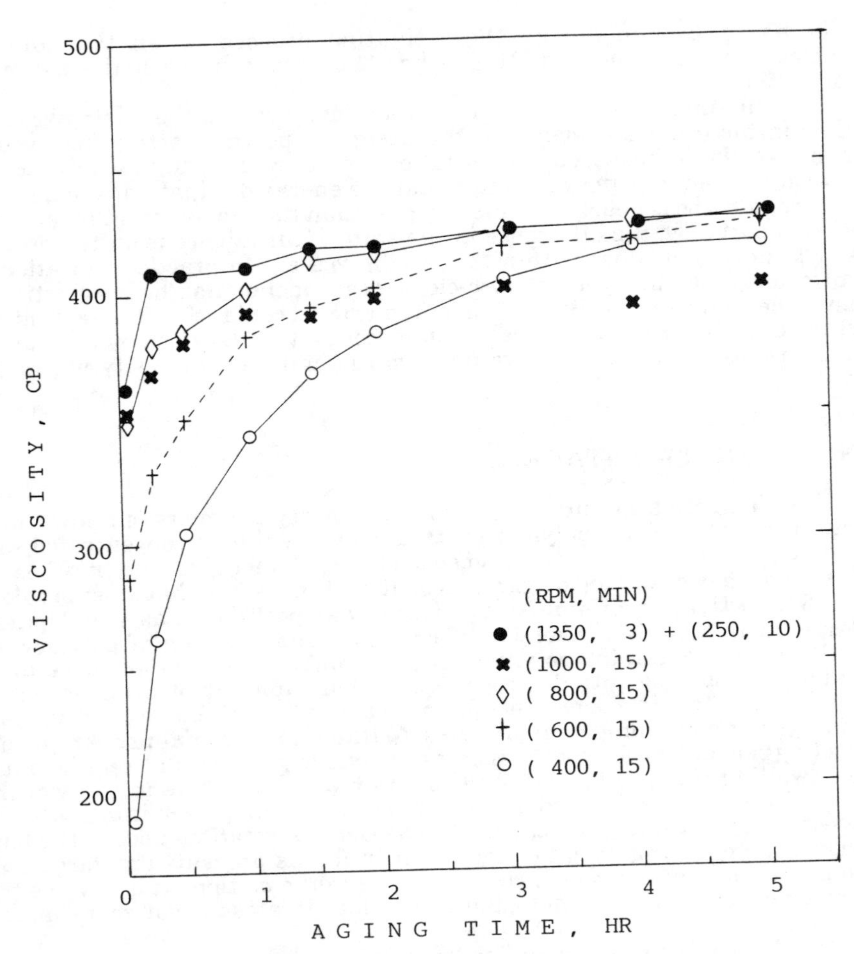

Figure 6. Effect of Different Mixing Schemes on the Viscosity of Polymer A Solution; 23 C, 0.15%, Water Hardness, 100 mg/L

accommodate the idea of two-stage mixing. Schematic diagrams of a single-stage mixing chamber (PB600, Stranco, Inc.) and a two-stage mixing chamber (MZ600, Stranco, Inc.) are presented in Figure 7. In the new design, inlets and outlets of polymer and dilution water are the same, but the mixing chamber (19 cm ID x 37 cm H) is partitioned to yield two different shear fields. First, polymer contacts with water in very high shear environment produced by perforated turbine impeller. At this step, the water/oil emulsion can be quickly inverted to an oil/water emulsion, while minimizing the agglomeration of polymer gels exposed to the water and keeping the size of polymer gels small. Then, partially swollen polymer gels enter the low shear zone, which is agitated by a disk-type impeller that is designed to accelerate the swelling of polymer gels without degrading the polymer structure.

Although there have been arguments[2,10] about the validity of equating the G value with the root-mean-square velocity gradient within the mixing vessel, the familiar Camp-Stein equation is widely used in the design of rapid mixing and flocculation facilities. The oft-cited relation used for determination of G is

$$G = \left(\frac{P}{\mu V} \right)^{1/2} \tag{6}$$

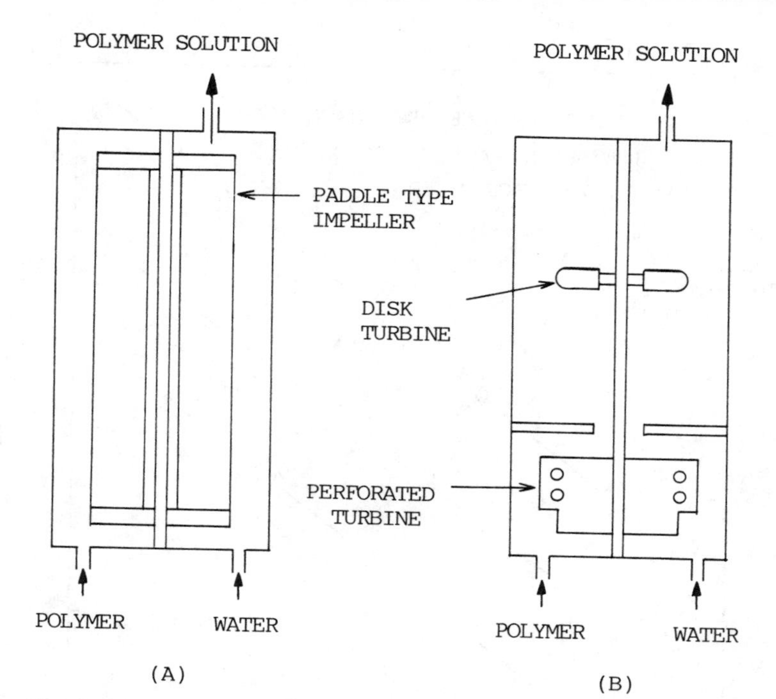

Figure 7. Schematic Diagrams of In-Line Polymer Mixing Chambers; (A) Single-Stage Chamber, (B) Two-Stage Chamber

where P is the power transmitted to the fluid, μ the dynamic viscosity of the fluid, and V the fluid volume in the mixing vessel. By measuring the torque imposed on the fluid by impeller motion, equation (6) can be rewritten as

$$G = \left(\frac{2\pi NT}{\mu V} \right)^{1/2} \tag{7}$$

where N is the rotational speed of impeller and T is the applied torque. In the turbulence literature, of course, one would expect to find the dissipation rate per unit mass of fluid, since the understanding of the energy cascade is based upon dissipation at small scales and production of turbulent energy at large scales. By employing the mean dissipation rate per unit of mass of fluid (ε), equation (6) can be rewritten

$$G = \left(\frac{\varepsilon}{\nu} \right)^{1/2} \tag{8}$$

where ν is the kinematic viscosity of the fluid.

The power delivered by the impeller motion was determined by directly measuring the torque on the impeller shaft using a torque monitor (TDS-DS-TM1, General Thermodynamics Corp.). Correction was made on the measured torque by subtracting the small torque needed to overcome the friction, owing to the bearings and the couplings in the system. This correction was obtained by measuring the torque at various impeller speeds with the chamber empty. Figure 8 shows the turbulence characteristics of both single- and two-stage mixing chambers. Very high mixing intensities were successfully produced with the perforated turbine impeller blades in the first stage of the two-stage mixing chamber; mean shear rate of 6300/sec or mean dissipation rate of 4×10^5 cm^2/sec^3 at 725 rpm, and 4000/sec or 1.6×10^5 cm^2/sec^3 at

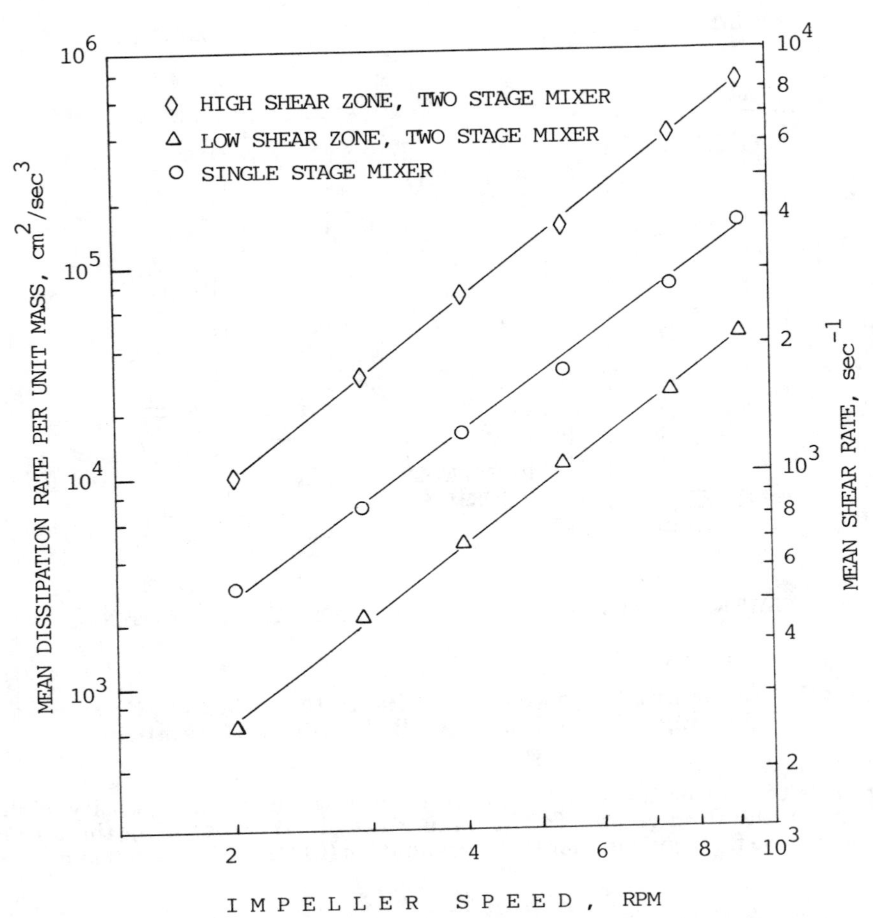

Figure 8. Mean Dissipation Rate per Unit Mass of Fluid (ε) and Mean Shear Rate (G) as a Function of Impeller Rotational Speed for Continuous Mixing Chambers

520 rpm. The ratio of G values in the high shear zone to those in the low shear zone in the two-stage mixing chamber ranges from 2.5 to 4.0, depending upon the impeller speed.

In order to investigate the performance of the two-stage mixer, polymer A was chosen and dissolved in tap water (hardness, 170 mg/L). Bench-scale batch mixings were performed as well as continuous mixings with in-line mixers. For the two-stage mixer, mean residence times of polymer solution in the high and low shear zones were about 25 seconds and 50 seconds, respectively. Figure 9 shows the performance comparison between batch mixing and continuous mixing. Performance data for the single-stage continuous mixer are presented; they show the merit of continuous mixing compared with the batch mixing in polymer activation process. It is clear that polymer structure was deteriorated significantly by employing 800 rpm for 15 minutes in batch mixing. When the impeller speed of 600 rpm was applied to the batch mixing system, about an hour of aging time was required to obtain the maximum viscosity. However, 90 percent of the maximum viscosity was achieved within 15 minutes after preparation using the in-line two-stage mixer. Considering the extremely high molecular weight of polymer A, this is a remarkable achievement in shortening the aging time normally recommended by polymer manufacturers. In fact, it has been proven that most of the cationic polymers

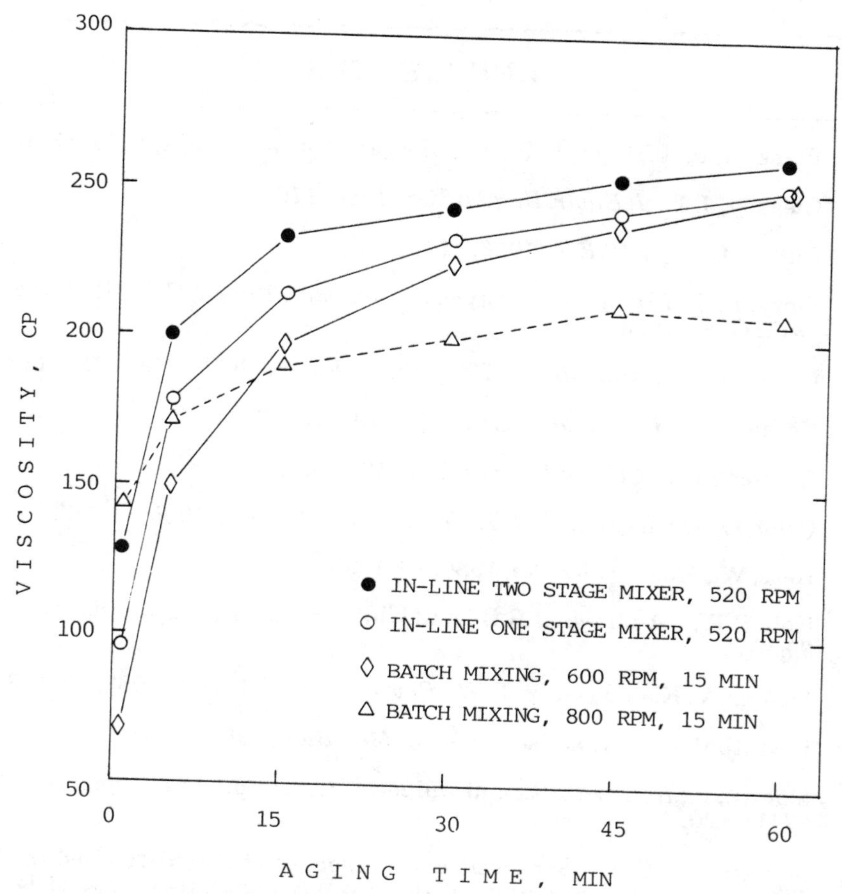

Figure 9. Viscosity Developments of 0.15% Polymer A Solution with Different Mixing Schemes; 21 C, Water Hardness, 170 mg/L

with medium-high molecular weight do not need any additional aging to obtain the maximum solution viscosity when the two-stage mixer is utilized. It is also observed experimentally that, for most of the anionic polymers, 3 to 5 minutes of aging is enough to attain about 90 percent of maximum viscosity of polymer solution with the use of the two-stage mixer.

SUMMARY

The effects of dilution water chemistry and hydrodynamic environment on the dissolution of polymeric flocculants have been discussed in this study. The hardness of dilution water showed the very serious effect on the polymer dissolution process among other factors. Ionic polymers were more sensitive to the water hardness than nonionic polymers. As generally known, polymeric flocculants were successfully dissolved in the water over a wide pH range. A two-stage activation mechanism was proposed for the dissolution of emulsion polymer, and the two-stage mixing utilizing this mechanism was found to be very effective in polymer activation by reducing the aging time remarkably without damaging polymer structure.

REFERENCES

1. Black, A. P., Birkner, F. B., and Morgan, J. J., *J. AWWA*, 1965, *57*, 1547.
2. Cleasby, J. L., *J. Envir. Eng.*, ASCE, 1984, *110*, 875.
3. Cutter, L. A., *AIChE J.*, 1966, *12* (1), 35.
4. Flory, P. J., *Principles of Polymer Chemistry*, Ithaca, Cornell University, 1971, Chapter 13.
5. Glasgow, L. A. and Kim, Y. H., *J. Envir. Eng.*, ASCE, 1986, *112*, 1158.
6. Goren, S. L., *J. Colloid Interface Sci.*, 1971, *36*, 94.
7. Gregory, J., *J. Colloid Interface Sci.*, 1973, *42*, 448.
8. Griot, O. and Kitchener, J. A., *Trans. Faraday Soc.*, 1965, *61*, 1026.
9. Hess, W., *Macromolecules*, 1986, *19*, 1395.
10. Koh, P.T.L., Andrews, J.R.G., and Uhlherr, P.H.T., *Chem. Eng. Sci.*, 1984, *39*, 975.
11. La Mer, V. K. and Healy, T. W., *Rev. Pure Appl. Chem.*, 1963, *13*, 112.
12. Lavinthal, D. and Davison, P. F., *J. Mol. Biol.*, 1961, *3*, 674.
13. Mabire, F., Audebert, R., and Quivoron, C., *J. Colloid Interface Sci.*, 1984, *97* (1), 120.
14. Mangravite, F. J., "Synthesis and Properties of Polymers Used in Water Treatment," in *Proceedings* of the AWWA Seminar on Use of Organic Polyelectrolytes in Water Treatment, Las Vegas, NV, 1983.
15. Nakano, Y., Naruoka, H., and Murase, M., *J. Chem. Eng. Japan*, 1986, *19*, 274.
16. Okamoto, Y., Nishikawa, M., and Hashimoto, K., *Int. Chem. Eng.*, 1981, *21*, 88.
17. Sakaguchi, K. and Nagase, K., *Bull. Chem. Soc. Japan*, 1966, *39*, 88.
18. Tanaka, T. and Fillmore, D. J., *J. Chem. Phys.*, 1979, *70* (3), 1214.
19. Tanaka, T., Fillmore, D. J., Sun, S. T., Nishio, I., Swislow, G., and Shah, A., *Phys. Rev. Letters*, 1980, *45* (20), 1636.
20. Ueno, K. and Kina, K., *J. Chem. Ed.*, 1985, *62*, 627.
21. Ying, Q. and Chu, B., *Macromolecules*, 1987, *20*, 362.

ADEQUACY OF ANALYTICAL TECHNIQUES TO DETERMINE THE FATE OF POLYELECTROLYTES IN WASTE TREATMENT

G. B. Wickramanayake and B. W. Vigon
Research Scientists
Battelle
505 King Avenue
Columbus, Ohio 43201-2693
(614)424-5758

ABSTRACT

Some of the polyelectrolytes used in wastewater treatment and industrial applications are known to be toxic to aquatic life if released above certain concentrations. However, very little is known about the fate of these compounds during treatment processes and in the environment. The purpose of this paper is to identify the advantages and disadvantages of available methods to determine the fate of the polyelectrolytes.

The available analytical methods for polyelectrolytes include nephelometry/turbidimetry, spectrofluorometry, spectrophotometry, interferometry, viscometry, colloid titration, luminescence titration, gel permeation chromatography, pyrolysis gas chromatography, bromine oxidation, and radioimmunoassay. The major limitations associated with most of these analytical methods include lack of specificity and sensitivity. In order to determine the distribution of polyelectrolytes in complex waste matrices at low concentrations, it may be necessary to develop a hybrid technique combining the advantages of several of the above techniques. Radiochemical methods (e.g., using C-14 labeled polyelectrolytes and scintillation counting procedures) may be employed to validate and confirm such techniques before recommending them for routine analytical applications.

INTRODUCTION

Polyelectrolytes are used widely in wastewater treatment processes as agents that facilitate solid/liquid separations. Applications of nonionic, cationic, or anionic polyelectrolytes in wastewater treatment have been studied by numerous researchers[3,4,13,20,31,32]. These polymers are applied at different stages of municipal waste treatment processes to improve suspended solids and phosphate removal. Polyelectrolytes also have useful applications in sludge dewatering[7,24,29].

Other than wastewater treatment, some industries such as the pulp and paper industry use polyelectrolytes at different stages in the manufacturing process. The currently extensive use of these materials also may be attributed to their effectiveness and economy when compared with other chemicals and processes. In the case of waste treatment, the ease of handling and storage, no added solids volume, and lower sludge ash content make them more attractive when compared with the inorganic coagulants[29].

Very little is known about the fate of the polyelectrolyes in industrial applications or in wastewater treatment. This is of concern because of the potential toxicity posed by these materials if released to receiving waters[5,6]. Prior to determining the fate and toxicity associated with the released polyelectrolytes, it is necessary to establish a suitable set of techniques to analyze and quantify these materials in wastewater effluents and in the ambient aqueous environment. It is the purpose of this paper to present feasible analytical methods identified from the literature for the quantification of polyelectrolytes in the effluents from wastewater treatment systems where polyelectrolytes are used in the thickening and clarification unit operations.

Although the research described in this article has been funded by the U.S. Environmental Protection Agency (Contract No. 68-03-3248) to Battelle Memorial Institute, it has not been subject to the Agency's review and therefore does not necessarily reflect the views of the Agency, and no official endorsement should be inferred.

ANALYTICAL METHODS

The literature search resulted in a total of 23 analytical procedures, which were assigned to one of 11 method categories. These categories and the percentage of methods in each are shown in Figure 1. The majority of polyelectrolyte analytical techniques have been developed for potable water. Most of these techniques, however, appear to have a moderate to low potential for the analysis of polyelectrolytes in wastewater. Wickramanayake *et al.*(35) gives a detailed description of these techniques. The following discussion provides applicability of those methods, along with the advantages and disadvantages, in wastewater analysis.

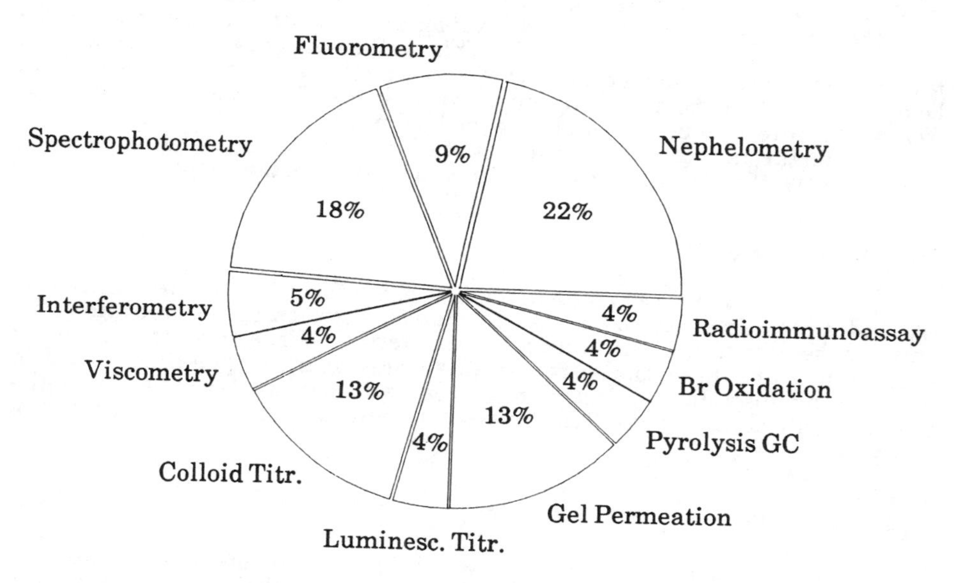

Figure 1. Procedures by Method Category

NEPHELOMETRY/TURBIDIMETRY

The nephelometric/turbidimetric method is based on the measurement of association complexes formed from interaction of polyelectrolytes with macromolecules bearing opposite charges(2,8,14,17). This method has been reported to be applicable for the determination of cationic as well as nonionic polyelectrolytes in water. The addition of EDTA may reduce the interferences caused by heavy metals, but only iron interference has been tested. Further work is necessary to investigate the effect of organic and inorganic substances that can interfere with this method.

SPECTROFLUOROMETRY

There are two spectrofluorometric techniques for the quantification of polyacrylamide present in aqueous solutions. One is based on conversion of polymers to its amine derivatives followed by reacting them with fluorogenic reagents and measuring the fluorescence intensity(16). Applicability of this technique may be limited because of the interferences caused by organic amine and amides in wastewater. In the second method, Nagata and Okamoto(26), using the fluorescence probe technique with the terbium ion Tb^{3+}, noted an enhancement in the fluorescence intensity in the presence of polyelectrolytes. Interferences resulting from NaCl, NaOAc, EDTA, and biopolymers make this method less attractive for wastewater analyses.

SPECTROPHOTOMETRY

Analysis of cationic polyelectrolytes in alcoholic beverages and potable water by spectrophotometry has been investigated[9,21]. Available information is limited with respect to interfering substances. Since this method was successful in polyelectrolyte analysis of wine and brandy containing moderate levels of salt and low levels of some organic compounds, there may be a potential for this technique to be used in wastewater analysis. Another spectrophotometric method involves incorporation of a chromophobe into the polyacrylamide of interest[12]. This method appears to be promising if the selected colored monomer has a specific absorption spectrum that would not be affected by the substances present in wastewater. This method is different from all other techniques in that the tagged polymers have to be synthesized by the manufacturers to meet technical criteria for performance and deductibility. Consequently, the manufacturing cost may increase substantially.

INTERFEROMETRY

Interferometry is based on the fact that small differences in refractive index resulting from concentration changes between the sample and reference can be determined with an interferometer[28]. Since this method can be used to measure these concentration differences, it may have limited use in determining absolute polyelectrolyte concentrations in wastewaters. Also, the effects of interfering agents can have significant impact on the specificity of analysis.

VISCOMETRY

The viscometric technique, which is based on the determination of viscosity of a polyelectrolyte solution by measuring the laminar flow rate through a membrane filter, has been used to analyze the levels of polymer flocculants in mine water[18]. It was shown that the laminar flow rates of dilute polyacrylamide solutions through membrane filters under vacuum are inversely proportional to their concentration. The viscometric technique, however, has not been evaluated for wastewater. Numerous high molecular weight dissolved organic compounds present in wastewater are likely to affect the method's accuracy.

COLLOID TITRATION

The colloid titration method involves quantitative determination of charged polyelectrolytes in solutions. Here, the charge on the polyelectrolyte is neutralized by titrating with a known quantity of an oppositely charged polyelectrolyte. It is assumed that charge neutralization complex is formed stoichiometrically. Although colloid titration is commonly used for cationic and anionic polyelectrolyte analyses in potable waters[33,34], this technique is expected to have limited applicability for wastewater analysis because of the potential severe interferences resulting from dissolved organic compounds. However, colloid titration was successful in quantification of both anionic and cationic polyelectrolytes in filtered pulp suspensions[27]. The sensitivity of the analysis was improved by employing electrochemical endpoint detection methods.

LUMINESCENCE TITRATION

Luminescence titration, another polyelectrolyte determination method for potable waters, has not been evaluated for wastewaters, but appears to have a moderate potential in this application. This method is based on the observation that binding of a luminescent polyelectrolyte to an oppositely charged polyelectrolyte produces a light output-intensity that proportionately increases up to the equivalent weight of the polyelectrolyte of interest(1).

GEL PERMEATION CHROMATOGRAPHY

Gel permeation chromatography is identified as a suitable technique for isolation and quantification of polyelectrolytes present in wastewater. Its principal attractiveness is as a separation and concentration step prior to another detection method. The gel permeation chromatography method is based on the principles of liquid chromatography. The minimum equipment for a high-pressure system capable of providing the necessary resolution consists of a liquid chromatograph with a suitable column, a mobile phase delivery system, and a detector. A variety of cationic polymers as well as anionic polymers have been analyzed by the gel permeation chromatography method(11,19).

PYROLYSIS GAS CHROMATOGRAPHY

Pyrolysis gas chromatography, which is used to analyze some nonvolatile organic compounds, also has been used to determine polyelectrolytes present in water and soils. Hatanaka and Yamaguchi(15) pyrolyzed a sample of polyacrylamide at 700 C before injecting into the GC column. The high temperature thermally degrades the polymer into smaller fragments that are sufficiently volatile for GC analyses. A potential disadvantage of the technique with respect to the analysis of complex wastewater samples is the lack of qualitative identification of the composition of each fraction eluted from the chromatograph. This disadvantage may be overcome by combining the chromatograph with a mass spectrometer, albeit at a much higher cost per analysis.

BROMINE OXIDATION OF PRIMARY AMIDES

In this method, primary amide groups present in the water-soluble polymers are oxidized by bromine. The amide oxidation product is used to oxidize iodide ion to iodine, which is then detected spectrophotometrically as the starch-trioxide complex(30). This method was applied to oil field brine, but cannot be used for municipal wastewaters without process modification to reduce interferences. Some trivalent cations, free amides, and amines are potential interfering chemical components.

RADIOIMMUNOASSAY

The radioimmunoassay technique is based on the positive antibody response observed in rabbits immunized with a polyacrylamide bovine serum albumin conjugate(10). This technique involves mixing of known concentrations of polyacrylamide with its radiolabeled compounds and deriving an appropriate curve from the percent of counts added minus the percent of counts bound nonspecifically by the control serum. The concentration range over which the technique has been demonstrated may not be low enough to use in some wastewater analyses. However, the specificity of method is far superior to the other procedures. Since radioimmunoassay is expensive and time consuming, it may not be recommended for routine analysis, but can be used to validate or confirm other methods.

DISCUSSION

Some of the polyelectrolytes used in wastewater treatment and industrial applications are known to be toxic to aquatic life if released into the environment above certain concentrations during treatment processes. In this paper, some of the available methods to determine the fate of the polyelectrolytes were identified. A total of 11 categories of methods was presented. These include nephelometry/turbidimetry, spectrofluorometry, spectrophotometry, interferometry, viscometry, colloid titration, luminescence titration, gel permeation chromatography, pyrolysis gas chromatography, bromine oxidation, and radioimmunoassay. Five of the categories--nephelometry/turbidimetry, spectrofluorometry, spectrophotometry, luminescence titration, and pyrolysis/GC--approach some of the desired criteria for method performance. A sixth, gel permeation chromatography, is well suited to sample pretreatment, and a seventh, radioimmunoassay, can serve as a reference technique to evaluate the selectivity of other methods.

Because no one method appears to achieve the needed level of performance, the development of a hybrid technique or techniques combining the advantages of currently published techniques with additional improvement in sensitivity and selectivity may be useful. The objectives of a method development effort to obtain the optimum technique(s) would include

- A detection limit of 10 ppb or lower
- Automated and instrument-based consistent with performance requirements
- No significant interferences from the organic and inorganic materials found in wastewaters.

One possible approach is presented in Figure 2. Any of the three pathways through the analytical hierarchy shown in Figure 2 may achieve the above objectives. Each pathway differs in the number of sample-handling steps and the cost of analysis. The first approach involves the direct application of high performance liquid chromatography or pyrolysis/gas chromatography with minimal sample pretreatment. Detection is by fluorometric, photometric, or nephelometric means.

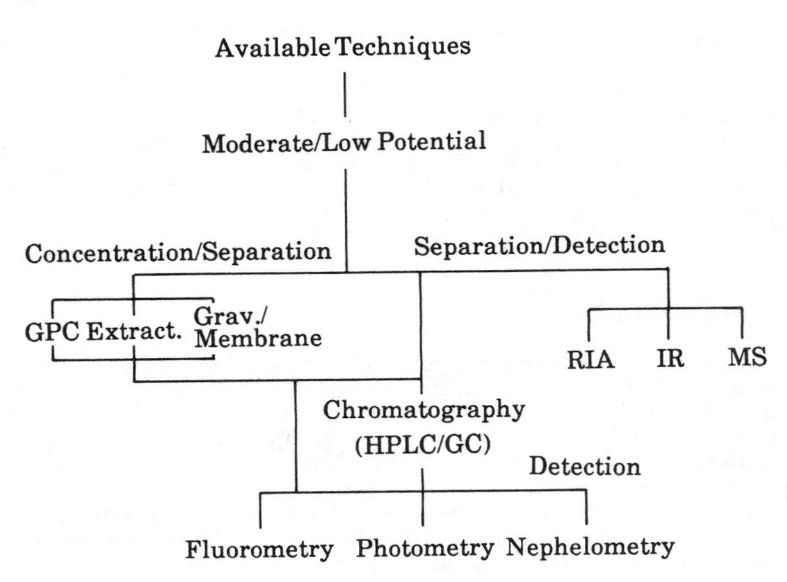

Figure 2. Analytical Methods Development Recommendations

The second approach utilizes preconcentration of the sample by gel permeation chromatography, membrane or gravimetric techniques followed by detection by fluorescence, photometry, or nephelometry. The final approach combines additional sample concentration/separation with chromatography to improve both selectivity and sensitivity, but at the expense of increased sample handling.

Finally, it is apparent that a qualitative confirmatory technique be used in conjunction with the early stages of development of the quantitative technique. None of the recommended quantitative techniques is capable of the absolute identification of the analyzed species. Therefore, a technique such as radioimmunoassay or other radiochemical methods (i.e., using radiolabeled polyelectrolytes and scintillation counting procedures) should be employed to confirm a positive detection of a polyelectrolyte in the quantitative technique under development.

REFERENCES

1. Alvarez-Roa, E. R., Prieto, N. E., and Martin, C. R., *Anal. Chem.*, 1984, *56*, 1939-1944.

2. Attia, Y. A. and Rubio, J., *Br. Polym. J.*, 1975, *7*, 135-138.

3. Bernhardt, H., Clasen, J., and Schell, H., *J. Am. Water Works Assoc.*, 1971, *63*, 355.

4. Berry, L. S., Lafayette, P. F., and Woodard, F. E., *J. Water Pollut. Control Fed.*, 1976, *48*, 2394.

5. Biesinger, K. E. and Stokes, G. N., *J. Water Pollut. Control Fed.*, 1986, *58* (3), 207-213.

6. Biesinger, K. E., Lemke, A. E., Smith, W. E., and Tyo, R. M., *J. Water Pollut. Control Fed.*, 1976, *48* (1), 183-187.

7. Cole, A. I. and Singer, P. C., *J. Environ. Eng. Div.*, Amer. Soc. Civil Eng., 1985, *111* (4), 501-510.

8. Crummett, W. B. and Hummel, R. A., *J. Amer. Water Works Assoc.*, 1963, *55*, 209-219.

9. Dey, A. N. and Palit, S. R., *Indian J. Chem.*, 1968, *6* (9), 538-539.

10. Drewes, P. A., Kamp, A. O., and Winkelman, J. W., *Experimentia.*, 1978, *34* (3), 316-318.

11. Furosawa, K., Onda, N., and Noriko, Y., *Chem. Ltrs. (Japan)*, 1978, 313-314.

12. Gramain, P. and Myard, P., *Polymer Bulletin (Germany)*, 1980, *3*, 627-631.

13. Habibian, M. R. and O'Melia, C. R., *J. Env. Eng. Div.*, Proc. Am. Soc. Civ. Eng., 1971, *101* (EE4), 567.

14. Hanasaki, T., Ohnishi, H., Nikaidoh, A., Tanada, S., and Kawasaki, K., *Bull. Environ. Contam. Toxicol.*, 1985, *35*, 476-481.

15. Hatanaka, T. and Yamaguchi, T., *Chiba Kogyo Daigaku Kenkyu Hokoku, Riko-hen (Japanese)*, 1976, *21*, 107-114.

16. Hendrickson, E. R. and Neuman, R. D., *Anal. Chem.*, 1984, *56* (3), 354-358.

17. Josephs, R. and Feitelson, J., *J. Polym. Sci.*, 1963, *1*, Part A, 3385-3394.

18. Jungreis, E., *Anal. Lett.*, 1981, *14* (A14), 1177-1183.

19. Kato, Y. and Hashimoto, T., *J. Chromat.*, 1982, *235*, 539-543.

20. Kavanaugh, M., Engster, J., Weber, A., and Boller, M., *J. Water Pollut. Control Fed.*, 1972, *49*, 2157.

21. Klyachko, Yu. A., Schneider, M. A., Kamenskaya, E. V., Topchiev, D. A., and Korshunova, M. L., *Izv. Wyssh. Uchebn. Zared., Pishch. Tekhnol. (Russian)*, 1984, *4*, 94-97.

22. Koene, R. S. and Mandel, M., *Macromolec.*, 1983, *16*, 220-227.

23. Koene, R. S., Nicolai, T., and Mandel, M., *Macromolec.*, 1983, *16*, 227-231.

24. Luttinger, L. B., in *Polyelectrolytes for Water and Wastewater Treatment*, W.K.K. Schwoyer, ed., 1981, Chapter 7, 211.

25. Michaels, A. S. and Morelos, O., *Ind. Eng. Chem.*, 1955, *47*, 1803.

26. Nagata, I. and Okamoto, Y., *Macromolecules*, 1983, *16*, 749-753.

27. Schempp, W. and Tran, H. T., *Wachenbl. Papierfabr (German)*, 1981, *109* (19), 726-732.

28. Schweigart, H.E.L.G., "Small Solute Concentration Differences Determined by Interferometry," CSIR Report CENG 424, Council for Scientific and Industrial Research, 1982, Pretoria, South Africa.

29. Schwoyer, W.K.K., in *Polyelectrolytes for Water and Wastewater Treatment*, W.K.K. Schwoyer, ed., 1981, Chapter 6, 159.

30. Scoggins, M. W. and Miller, J. W., *Soc. Pet. Eng. J.*, 1979, *19* (3), 151-154.

31. Shireman, H. C., *Water Works Wastes Eng.*, 1972, *9*, 34.

32. Tschobanoglous, G., *J. Water Pollut. Control Fed.*, 1970, *43*, 604.

33. Ueno, K. and Kina, K., *J. Chem. Ed.*, 1985, *62*, 627-629.

34. Wang, L. K. and Shuster, W. W., *Ind. Eng. Chem.*, Prod. Res. Deve., 1975, *14* (4), 312-314.

35. Wickramanayake, G. B., Vigon, B. W., and Clark, R., "Flocculation in Biotechnology and Separation Systems," *Process Technol. Proc.*, 1987, *4*, 125-147.

ADSORPTION BEHAVIOR OF A HYDROPHOBIC POLYMERIC FLOCCULANT AND ITS RELATION TO FLOCCULATION AND FILTRATION DEWATERING OF COAL SLURRIES

Y. A. Attia and S. Yu
The Ohio State University
Department of Materials Science and Engineering
143 Fontana Laboratories
116 West 19th Avenue
Columbus, Ohio 43210

ABSTRACT

This study investigated the adsorption behavior of a totally hydrophobic polymer, FR-7A, and its role in the flocculation and filtration of fine coal slurries. Adsorption of FR-7A on coal, pyrite, and shale minerals revealed that (a) FR-7A had a higher adsorption affinity to coal and pyrite than to shale; (b) FR-7A aided the flocculation of coal slurries and improved the moisture removal by filtration from 42.4 to 37 percent; and (c) an acidic slurry condition favored unselective adsorption of FR-7A on coal minerals, leading to improved total flocculation and filtration of fine coal slurries, whereas alkaline pH and the presence of sodium metaphosphate favored selective adsorption and flocculation of coal from associated minerals in the slurry.

INTRODUCTION

The use of polymeric flocculants as aids for solid/liquid separation has been studied and applied widely in many industries(5,7,8,10). The adsorption of the polymeric flocculant on the fine particles in suspension causes the aggregation of these particles, which significantly increases the rate of their separation from suspending liquids.

FR-7A is believed to be a totally hydrophobic polymeric flocculant. Under favorable adsorption conditions, excellent separation of coal from associated silicates can be achieved through selective flocculation using this polymer(4). The high affinity of this polymer to coal is believed to result from hydrophobic "bond" association. However, under different conditions, FR-7A adsorption becomes less selective and results in total flocculation of coal slurries. Since this polymer causes fine particles to agglomerate together into flocs, the filtration rate is expected to increase. In addition, since this polymer is also hydrophobic and is dispersed as an oil in water emulsion, the final moisture content of the filter cake is expected to be lower than that achieved with hydrophilic polymeric flocculants.

In this study, the objective was to investigate the adsorption behavior of FR-7A and predict conditions for the total flocculation and dewatering of fine coal slurries.

EXPERIMENTAL TECHNIQUES

MINERALS

The coal samples used for adsorption studies were from the Upper Freeport and Pittsburgh No. 8 seams. They were precleaned by gravity separation using a heavy liquid, 1,1,1-trichloroethane with a specific gravity of 1.33, to reject the majority of coarse-sized impurities (shale and pyrite). The float coal products were washed by acetone and then dried in an oven at about 105 C for 5 hours to remove any residual liquid. The coals were dry-ground in a stainless steel grinder. The cumulative weight of the size below 25 µ was 50 percent for the Upper Freeport coal and 75 percent for the Pittsburgh No. 8 coal.

Pure pyrite (FeS_2) crystals from Hunzala, Peru, and a sample of argillaceous shale minerals were supplied by Ward's Science Establishment, Inc., Rochester, New York. The samples of pyrite and shale were dry-ground to very fine size, about 80 percent by weight below 25 µ (-500 mesh).

Table 1 lists the analytical data of these samples. The surface areas of the samples were determined by an Accusorb Model 2100E produced by Micromeritics Corp., and the mean diameters of the particles were measured using a Microtrac II Particle Size Analyzer made by Leeds & Northrup Co. The surface electrical potentials of the minerals (see Figure 1) were measured using a Laser Zee™ Model 501 apparatus, manufactured by PenKem Inc.

The measured pzc (point of zero charge) of each mineral sample was very close to that reported in the literature (e.g., pzc of anthracite and bituminous coals were between pH 5 to 8(1), pzc of pyrite was at pH 6.2 to 6.9(6), and pzc of shale was at about pH 1.0 to 2.0(12)).

In the flocculation and filtration tests, only Pittsburgh No. 8 raw coal was used. The coal sample contained 10.8 percent ash, 2.65 percent total sulfur. The coal sample was crushed and then wet-ground down to very fine (>90 percent below 25 μ) with a ball mill.

Table 1. Analyses of the Samples Used in Polymer Adsorption Tests

Materials	Ash %	Total Sulfur %	Surface Area m^2/g	Mean Diameter um	pzc at pH
Upper Freeport Coal (pre-cleaned)	3.5	1.80	2.62	14.62	6.7
Pittsburgh No. 8 Coal (pre-cleaned)	4.2	1.67	3.50	9.46	7.1
Pyrite Crystal	---	----	3.58	4.12	7.1
Arigillaceous Shale	---	----	7.42	4.36	2.1

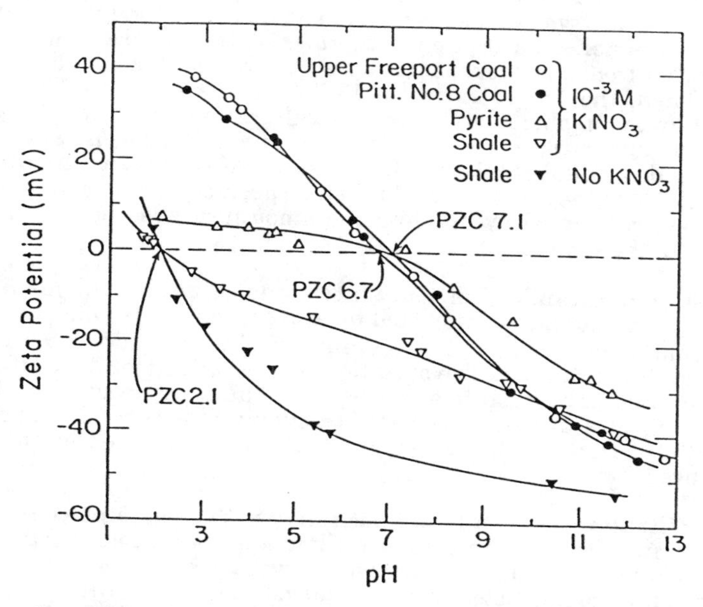

Figure 1. Zeta Potentials of the Upper Freeport and Pittsburgh No. 8 Coals, Shale, and Pyrite at Various pH

CHEMICALS

FR-7A, a totally hydrophobic polymer, was provided by Calgon Corp. (Pittsburgh, Pennsylvania). Since this polymer was specifically emulsified, it could disperse well in water. The molecular weight of FR-7A was believed to be slightly less than 1 million[4]. However, it was observed that the FR-7A polymeric colloid clouds tended to slowly migrate to anode in the electrophoresis cell. It seemed that FR-7A was weakly negatively charged.

Sodium metaphosphate (SMP) was used as a general dispersant, particularly in selective flocculation of coal from shale removal. This chemical was obtained from Fisher Scientific Co.

PROCEDURES

Nephelometry for Determination of Polymer Concentration. The nephelometric approach was originally developed by Attia *et al.*[2,9]. In their procedure, tannic acid and sodium chloride were added to the polymer solutions to produce turbid polymeric colloids. Since FR-7A is a naturally turbid emulsion, its turbidity can be measured directly without the addition of tannic acid and sodium chloride. This modified approach was employed in this study for the determination of the polymer concentration, using a nephelometer, Model 21 manufactured by Monitek Inc. (Hayward, California). The calibration curve, established using known concentrations of polymer in solutions, showed a linear relationship between turbidity and FR-7A concentration; this relationship was not influenced by pH of solution[3]. Using this calibration curve, the polymer concentration of an unknown solution could be determined from the measured turbidity.

Adsorption. The adsorption studies included the adsorption kinetics and isotherms of FR-7A on coal, pyrite, and shale, and the effects of pH and SMP on the adsorption of FR-7A. The aqueous suspensions at 2 percent solids by weight were mixed with the predetermined concentration of FR-7A for 5 minutes with a magnetic stirrer and then were allowed to stand for another period of time, which was varied from 0.5 to 24 hours for the kinetics study and for 12 hours to establish adsorption isotherms. After the adsorption, the solids were separated from the suspensions by a membrane (0.2 μ) filter under air pressure. Then, the concentration of FR-7A in the solid-free solution was determined by the nephelometry from the calibration curve of FR-7A concentration versus turbidity. From the difference in FR-7A concentration between the initial and residual solutions after adsorption of FR-7A, the quantity of adsorbed FR-7A on the minerals could then be derived. Finally, the adsorption densities were calculated by dividing the amount of adsorbed FR-7A by the surface area of the sample.

Flocculation. The coal slurry was diluted to about 5 percent weight solids content and was adjusted to desired conditions (various pH levels and SMP concentrations). The flocculant, FR-7A, was added to the slurry under mechanical agitation. The agitation was continued for 2 minutes at high speed and for another 1 minute at low speed. The slurry was then allowed to settle for 10 minutes, and the flocculated fraction was separated from the slurry by decantation. After drying, the weights of the floc and supernatant fractions were measured.

Filtration. A 50-mL coal floc slurry with about 5 percent solids was transferred into a membrane pressure filter apparatus connected to a compressed air cylinder. Ordinary filter papers were used instead of membranes. The air pressure was adjusted at the required value, and filtration was continued for a predetermined time. The filter cake was weighed before and after drying in an oven at 105 C to determine its moisture content.

RESULTS AND DISCUSSION

ADSORPTION OF FR-7A

Adsorption Kinetics. The purpose of studying the adsorption kinetics was to determine the time required for the adsorption to attain equilibrium. The adsorption equilibrium was determined when the adsorption reached a steady state. The results illustrated in Figure 2 indicated that the time for attaining adsorption equilibrium was about 7 hours for coal and pyrite and 5 hours for pyrite. However, the equilibration time for the following adsorption isotherm tests was 12 hours. The results also indicated that adsorption of FR-7A was much faster on coal than on pyrite or shale.

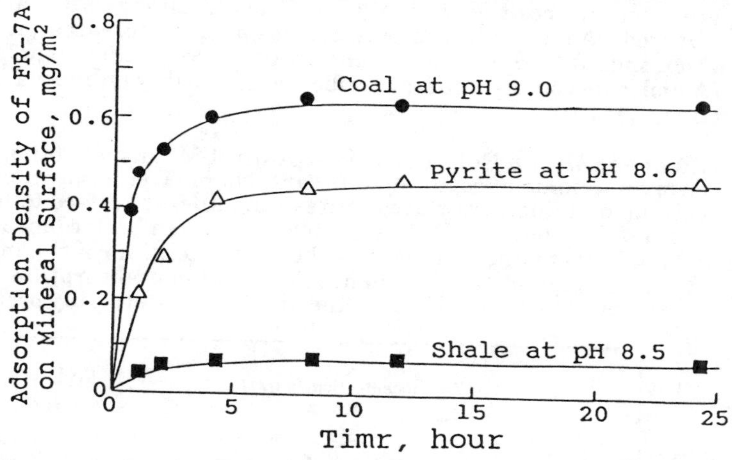

Figure 2. Adsorption Kinetics of FR-7A on Coal, Pyrite, and Shale at Initial FR-7A Concentrations of 100 mg/L and at pH 8.4 to 9.0

Adsorption Isotherms. Figure 3 illustrates the adsorption isotherms (i.e., the equilibrium adsorption densities) at pH range of 8.5 to 9.0. The figure shows that the equilibrium adsorption density of FR-7A on coal was about 7 times higher than that on shale, but the equilibrium adsorption density

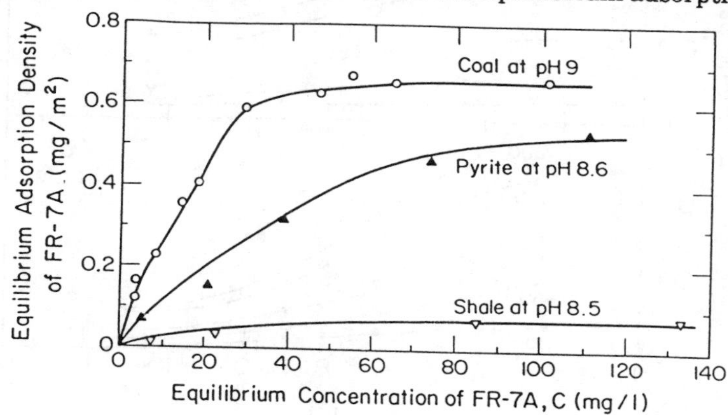

Figure 3. Adsorption Isotherms of FR-7A on Coal, Pyrite, and Shale at pH 8.5 to 9.0

of FR-7A on pyrite was also quite high. The high adsorption of FR-7A to coal and pyrite could be attributed to the hydrophobic bond between the hydrophobic polymer molecules and the naturally hydrophobic surfaces of coal and pyrite. For shale, the low adsorption resulted from both the naturally hydrophilic surface and the relatively high negative zeta potential of shale, which depressed the adsorption of FR-7A on shale surface.

Effect of pH on Adsorption. The effect of pH on the adsorption of FR-7A on the minerals was studied at various initial FR-7A concentrations. Figure 4 illustrates that, at increasing pH, the equilibrium adsorption density of FR-7A molecules on all the minerals generally decreased. This is understandable as both FR-7A and the minerals became negatively charged at alkaline pH, where electrostatic repulsion would be operative. Appropriately, these are the conditions where selective adsorption is favored. However, at acidic pH, coal and pyrite became positively charged and shale minerals became less negatively charged. As a result, the polymer adsorption increased on all the minerals, which should lead to general, unselective flocculation. The adsorption of FR-7A molecules on the mineral surface was probably enhanced by electrical attraction.

Effect of SMP on Polymer Adsorption. SMP dispersant dissolves in water to form an anionic phosphate group and Na+. The adsorption of the SMP anionic group on mineral surfaces causes the mineral surface to be more electrically negative, as shown in Figure 5. However, when SMP concentration was above 400 mg/L, no further increase in the zeta potentials of the minerals was observed. This implied that metaphosphate anionic absorption on the mineral surface stopped, probably because the active sites were fully occupied.

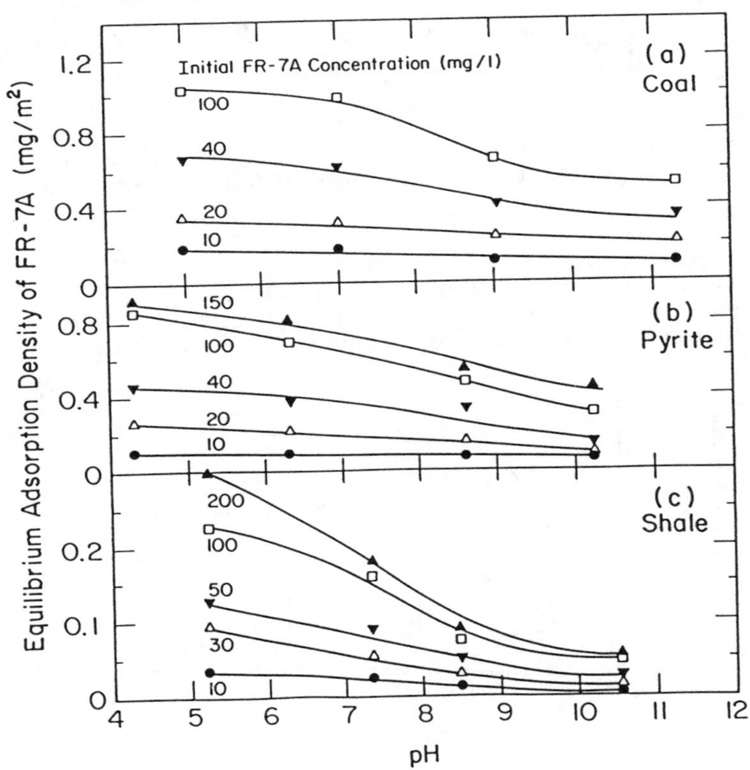

Figure 4. Effect of pH on the Adsorption Density of FR-7A on Coal, Pyrite, and Shale

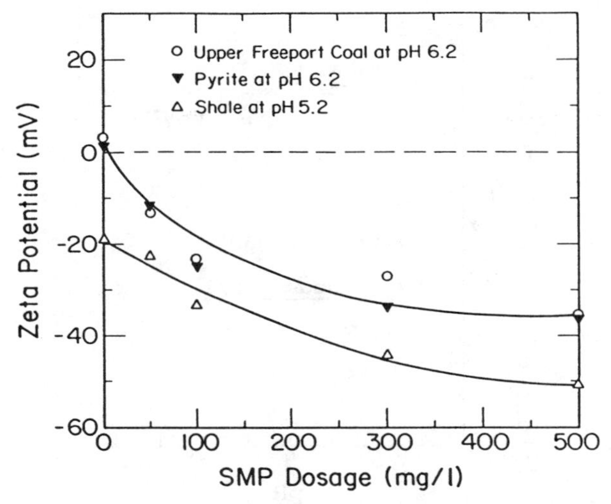

Figure 5. Effect of SMP Dosage on the Zeta Potentials of Coal, Pyrite, and Shale

Even though SMP could cause the mineral surfaces to be more electrically negative, the adsorption of FR-7A on the minerals was not significantly influenced, as shown in Figure 6. Evidently, the larger polymer molecules were able to displace the smaller metaphosphate anions from the mineral surface when enough time was allowed to achieve equilibrium adsorption. With shorter periods for adsorption, the smaller metaphosphate anions would be kinetically favored by their faster diffusion rate.

Flocculation of Coal Slurries Using FR-7A

Effect of pH and SMP on Flocculation. Figure 7 shows the effect of pH on flocculation using FR-7A. It indicated that, at pH 4.8, almost total

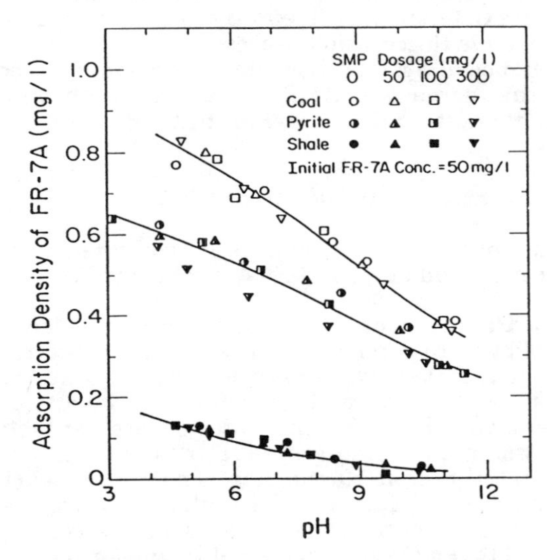

Figure 6. Effect of SMP on the Equilibrium Adsorption Density of FR-7A on Coal, Pyrite, and Shale at Various pH

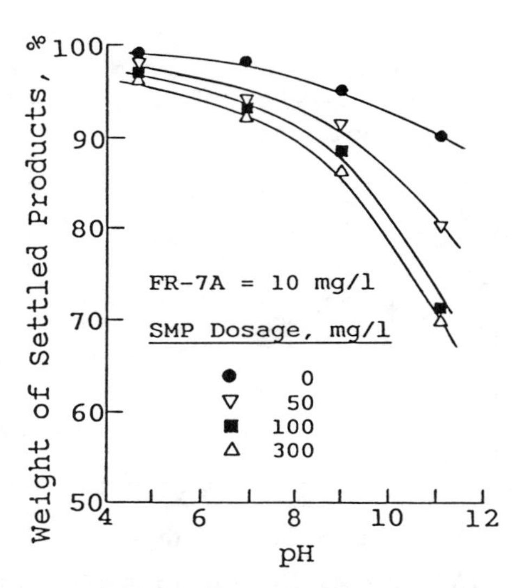

Figure 7. Effects of pH and SMP Dosage on Flocculation of Coal Slurry Using FR-7A

flocculation (over 98 percent weight) occurred and the solids content in the supernatant was reduced from 5 percent to less than 0.2 percent. This could be attributed to the relatively high and unselective adsorption of FR-7A molecules on the minerals. While at alkaline pH, due to the relatively highly negative surface zeta potential of the minerals, adsorption of FR-7A molecules and flocculation were reduced.

Figure 7 also shows the effect of SMP concentration on flocculation of coal slurries. Since the adsorption of metaphosphate anions rendered the minerals' (especially those in the shale group) surfaces more negative and thus more dispersible, increasing SMP concentration caused a decrease in the present weight flocculated. As mentioned earlier, SMP is able to compete with FR-7A at the shorter mixing time periods used for flocculation tests. It was noticed that the products dispersed by SMP consisted mainly of shale minerals. It confirmed that the existence of SMP enhanced the removal of shale from coal slurry through a selective flocculation process.

The influence of SMP on flocculation was more effective at alkaline condition and, for the increase in SMP concentration above about 100 mg/L, produced only a slight additional decrease in the flocculation.

Filtration of Coal Slurries with the Aid of FR-7A

In this set of tests, the variables investigated were pressure drop, time, pH, SMP concentration drop, and FR-7A concentration.

Effect of Pressure on Filtration Rate and Cake Moisture. The purpose of these tests was to determine the workable pressure and time for filtration tests. Two parameters were used for the evaluation: percent water removal and percent cake moisture. Figure 8 shows the effect of pressure drop on percent water removal. The flocculation tests were conducted with 10 mg/L of FR-7A, at pH 7 using the same procedure described earlier, and time for these tests was arbitrarily set at 10 minutes. The graph clearly indicated that the dewatering rate was much faster under higher pressure. However, because of the weakness of filter paper, the tests could not be conducted at pressure higher than 82.8 to 96.6 kPa (12 to 14 psig) before the formation of cake.

After formation of filter cake, the pressure could be adjusted higher up to about 276 kPa (40 psig). In testing the effect of pressure drop on cake

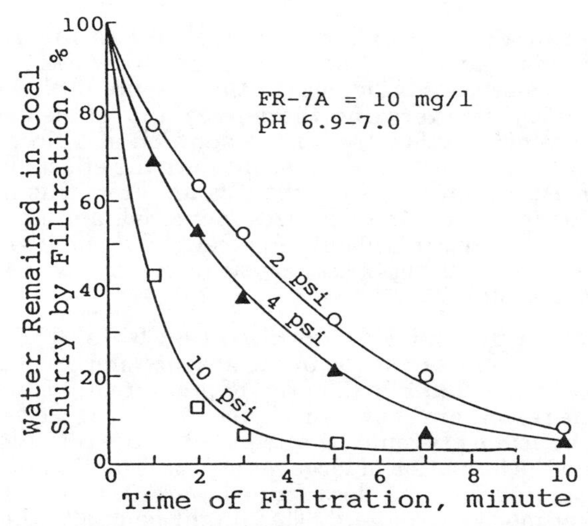

Figure 8. Water Removal from Coal Slurries by Press Filtration at pH 7, 10 mg/L FR-7A (1 psig = 6.9 kPa)

moisture, the conditions were set at pH 6.9 to 7.0, a dosage of 10 mg/L FR-7A, and 2-minute duration of filtration. From Figure 9, we can see that moisture content in the floc product decreased as the pressure drop was increased to about 138 kPa (20 psig). At higher pressure drops than 138 kPa, little additional moisture could be removed.

It seemed that the residual moisture content of the filter cake was still quite high, but it agreed reasonably well with published results in the literature for the feed slurries with sizes below -25 μ[11]. Since the fine particles have a very large surface area, the moisture content is expected to be higher than that of coarser particles.

From the results presented in Figures 8 and 9, the effective pressure drop after formation of cake was selected at about 138 kPa (20 psig), and the filtration time was set at 2 minutes.

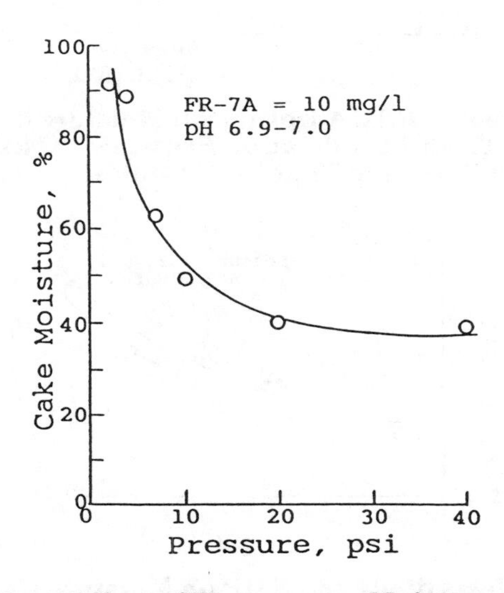

Figure 9. Effect of Pressure on Moisture Content of the Filter Cakes (1 psig = 6.9 kPa)

Effect of FR-7A Concentration on Filtration Rate. Filtration rate greatly depends upon the permeability of the filter cake, which is a function of the particle size distribution(8). In the absence of flocculant, since the particles in the slurry were very fine, the permeability of the filter cake and the filtration rate became very low. With the addition of a flocculant, filtration rate became significantly faster. Figure 10 shows the effect of FR-7A polymer dosage on the moisture content of the filter cake. From this figure, as FR-7A concentration was increased, the floc sizes increased, and therefore, the cake moisture content was reduced to about 37 percent. But, as FR-7A concentration was raised higher than 20 mg/L, the moisture content of cake was no more affected by the dosage of FR-7A.

Effects of pH and SMP on Filtration Rate. The tests examining the cake moisture content versus pH in the absence and presence of SMP were set at 10 mg/L of FR-7A, 138 kPa (20 psig) of pressure drop, and 2 minutes filtration time. The results are shown in Figure 11. The relatively low moisture content in cake at acidic pH could be attributed to the relatively high adsorption of FR-7A and the formation of large flocs. While at high pH, the adsorption of FR-7A was reduced and the flocs formed were very small in size, which resulted in higher moisture content. This might be due to the higher surface area of the smaller flocs at alkaline media compared with the larger flocs at acidic suspensions. Also, the permeability of the filter cake would be lower with the smaller flocs than with the larger flocs.

The addition of SMP to the slurry caused the particles to be more dispersible, thus affecting the formation of flocs and cake permeability. However, the presence of SMP did not appear to affect the final moisture content significantly.

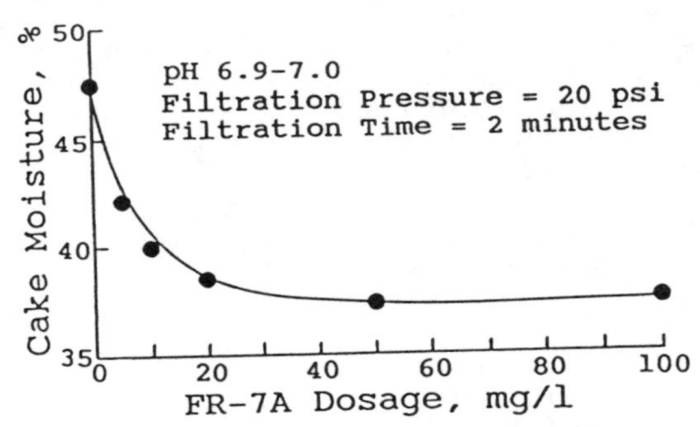

Figure 10. Effect of FR-7A Dosage on Moisture of Filter Cake at pH 6.9 to 7.0, Filtration Pressure 138 kPa (20 psig), and Filtration Time = 2 Minutes

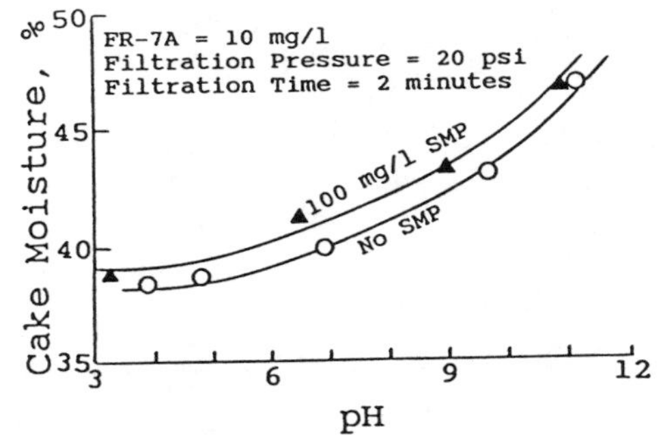

Figure 11. Effects of pH and SMP on Moisture Content of the Filter Cake at FR-7A = 10 mg/L, Filtration Pressure 138 kPa (20 psig), and Filtration Time = 2 Minutes

CONCLUSIONS

From the above results and discussion, the following conclusions can be drawn:

1. The adsorption of FR-7A on the minerals was very sensitive to suspension pH. In acidic conditions, the equilibrium adsorption densities of FR-7A on the minerals are relatively high, and correspondingly total flocculation can occur. Filtration is more efficient under these conditions because of increased cake permeability. In alkaline conditions (above pH 7), the adsorption of FR-7A on the minerals is greatly depressed, and the total flocculation and filtration of coal slurry become less effective.
2. The totally hydrophobic polymer, FR-7A, had a higher adsorption affinity to both coal and pyrite than to shale minerals. The equilibrium adsorption density of FR-7A on coal was about 7 times higher than that on shale at pH 8.5 to 9.0. These results implied that using FR-7A as a flocculant would be applicable for separation of shale from coal but not very effective for pyrite rejection from fine coal slurry.
3. The presence of SMP in the coal slurry causes the minerals' surfaces to be more electronegative, but it would not significantly affect the equilibrium adsorption of FR-7A on the minerals. However, at nonequilibrium shorter adsorption periods, SMP can affect the flocculation ability of FR-7A polymer, which could also decrease the efficiency of filtration of the coal slurries.

ACKNOWLEDGMENTS

The support by the Ohio Coal Development Office and The Ohio State University on the project RI1-87/91 is greatly acknowledged. The authors also acknowledge the fellowship to S. Yu from the Ohio Mining and Mineral Resources Research Institute.

REFERENCES

1. Aplan, F. F., "Coal Flotation," *Flotation--A. M. Gaudin Memorial Volume*, 2, M. C. Fuerstenau, ed., 1979.

2. Attia, Y. A. and Rubio, J., "Determination of Very Low Concentrations of Polyacrylamide and Polyethyleneoxide Flocculants by Nephelometry," *Brit. Polym. J.*, 1975, *7*, 135-138.

3. Attia, Y. A. and Yu, S., "Adsorption Behavior of a Totally Hydrophobic Polymeric Flocculant on Coal, Pyrite, and Shale," in *Proceedings*, 23rd Annual Meeting of Fine Particle Society, Boston, MA, August, 1989.

4. Attia, Y., Yu, S., and Vecci, S., "Selective Flocculation Cleaning of Upper Freeport Coal with a Totally Hydrophobic Polymeric Flocculant," in *Flocculation in Biotechnology and Separation Systems*, Y. A. Attia, ed., Elsevier, 1987, 547-564.

5. Chander, S., "A Review of the Use of Surfactants as Filter Aids," in *Filtration and Separation*, E. R. Frederick and L. F. Mafrica, eds., Proceedings of the American Filtration Society, Kingwood, TX, March, 1988, 345-352.

6. Kelly, E. G. and Spottiswood, D. J., *Introduction to Mineral Processing*, Wiley Interscience Publication, 1982, 101.

7. Krishnan, S. V. and Attia, Y. A., "Polymeric Flocculants," in *Reagents in Mineral Technology*, P. Somasundaran and B. M. Moudgil, eds., Marcel Dekker, 1988, 485-518.

8. Lewellyn, M. E. and Avotins, P. V., "Dewatering/Filtration Aids," in *Reagents in Mineral Technology*, P. Somasundaran and B. M. Moudgil, eds., Marcel Dekker, 1988, 559-578.

9. Pradip, Attia, Y. A., and Fuerstenau, D. W., "The Adsorption of Polyacrylamide Flocculants on Apatites," *Colloid and Polymer Sci.*, 1980, *258*, 1343-1353.

10. Richardson, P. F. and Connelly, L. J., "Industrial Coagulants and Flocculants," in *Reagents in Mineral Technology*, P. Somasundaran and B. M. Moudgil, eds., Marcel Dekker, 1988, 519-558.

11. Sandy, E. J. and Matoney, J. P., "Mechanical Dewatering," in *Coal Preparation*, J. W. Leonard, ed., 4th Edition, AIME Press, 1979, 12-15.

12. Wang, D. Z., *Mechanisms and Applications of Froth Flotation Reagents*, China Metallurgy Publisher, Beijing, 1982, 29.

ROLE OF POLYMER AND SURFACTANT ADSORPTION AND MICROSTRUCTURE OF ADSORBED LAYERS IN SOLID/LIQUID SEPARATION

P. Somasundaran,
K. F. Tjipangandjara, and C. Maltesh
Langmuir Center for Colloids and Interfaces
Henry Krumb School of Mines
Columbia University
911 S.W. Mudd Building
New York, New York 10027
(212) 280-2926

ABSTRACT

Removal of solids from suspension is achieved by making use of any difference in their natural or induced surface properties, which is normally achieved by controlled adsorption of polymers, surfactants, and inorganics. Success of the separation depends upon our ability to manipulate adsorption, which in turn is dependent upon a full understanding of adsorption mechanisms.

Principles governing adsorption mechanisms of surfactants and polymers on solids and the resultant interfacial processes such as flocculation and solid/liquid separation are discussed here. Emphasis will be on recent approaches using fluorescence, electron spin resonance (ESR), and size exclusion chromatographic techniques to understand the microstructure of adsorbed polymer and surfactant layers and competitive/cooperative interactions in them.

INTRODUCTION

Processing slow-settling mineral suspensions called slimes or sludges is a major problem in many industries[10]. Optimizing the settling rate of aggregates or flocs is of vital importance to improved efficiency of solid/liquid separation. It has been fairly well understood that the addition of a flocculant to the slurry improves the subsidence rate of the aggregates and enhances solid/liquid separation. However, different criteria are chosen for the study of sedimentation of suspensions, depending on the aim of the particular process[12]. For thickening, it is mostly the solid concentration of the pulp that is important, whereas for effluent treatment, the clarity of the supernatant becomes the prime criterion. In yet other processes, the yield strength of the solids or moisture content is of major concern. Understanding the role of several factors becomes essential to achieving the desired subsidence of stably dispersed suspensions. Surface properties play a very important role in determining the flocculation and dispersion of any suspension, and it is possible to manipulate the natural or induced interfacial properties by controlled adsorption of polymers, surfactants, and inorganics. The structure of the resultant adsorbed layers largely controls processes such as flocculation, and there is a need for accurate characterization of the micro- and nanostructure of such adsorbed layers. Until now, characterization of the adsorbed layers on a molecular scale was not possible, primarily because of a lack of a reliable technique to study the solid/liquid interface and the limitations in making use of the existing techniques due to difficulties in handling the solid/liquid samples under in-situ conditions.

We have initiated and developed the use of several spectroscopic techniques to probe in situ the microstructure of the adsorbed layers. Using these techniques along with the conventional methods, configuration of the adsorbed molecules has been successfully monitored to achieve maximum solid/liquid separation.

SURFACTANTS IN SOLID/LIQUID SEPARATION

Flotation is a process in solid/liquid separation technology where solids in suspension are recovered by means of their attachment to air bubbles. Most naturally occurring solid particles have surfaces with a strong affinity for water and invariably are not floatable. Even when certain minerals such as coal have natural hydrophobicity, it is often necessary to enhance their hydrophobicity because of marked changes in their surface properties upon weathering. This has to be accomplished selectively without making undesirable components, such as pyrite and clays, hydrophobic. Displacement of the water film from their surfaces is essential for the solid particles to be separated by flotation. This is usually accomplished by the adsorption of surface-active agents, which act at the solid/liquid interface and diminish the particle/water interactions. The particle surfaces are rendered, at least in part, hydrophobic. Collectors are usually long-chain hydrophobic molecules containing polar groups. The molecule adsorbs onto the solid surface via the charged group with the hydrocarbon chain presented to the aqueous phase.

The adsorption characteristics of ionic surfactants on charged solids, e.g., oxides, have been studied mostly by determining adsorption isotherms and zeta potentials. A model mineral/surfactant system studied widely is alumina-sodium dodecylsulfate. A typical adsorption isotherm for this system is shown in Figure 1. The isotherm is characterized by four distinct regions: region I is dominated by electrostatic adsorption, region II is marked by a sharp rise in the adsorption caused by surfactant aggregation, region III is characterized by a decreasing slope even though surfactant adsorption continues to increase, and region IV represents the maximum surface coverage and is marked by micelle formation in the bulk. The concept of surfactant aggregation at the solid/liquid interface has been employed by several authors, notably Fuerstenau and Somasundaran(11), to account for the sharp changes in interfacial properties, such as the amount adsorbed, hydrophobicity, and zeta potential observed above a critical surfactant concentration. For example, lateral aggregation among adsorbed dodecylsulfonate species on alumina results in drastic increase in the adsorption density of dodecylsulfonate and the settling rate, as well as the electrophoretic mobility of alumina-dodecylsulfonate system (Figure 2)(11). This process was termed hemimicellization by analogy to micellization. The term

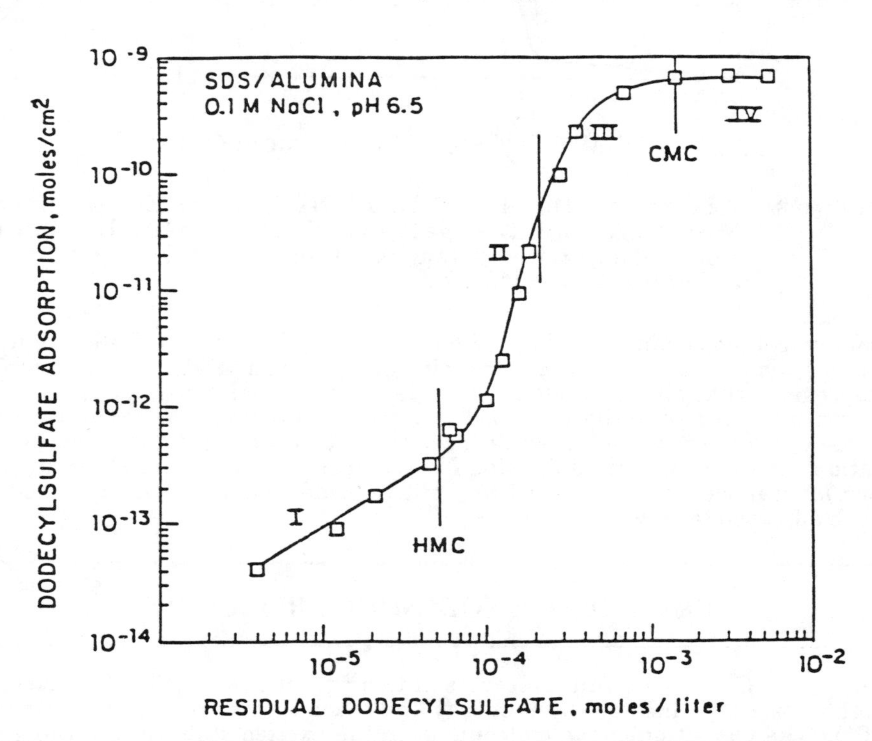

Figure 1. Adsorption Isotherm of Sodium Dodecylsulfate on Alumina at pH 6.5 in 0.1 kmol/m³ NaCl

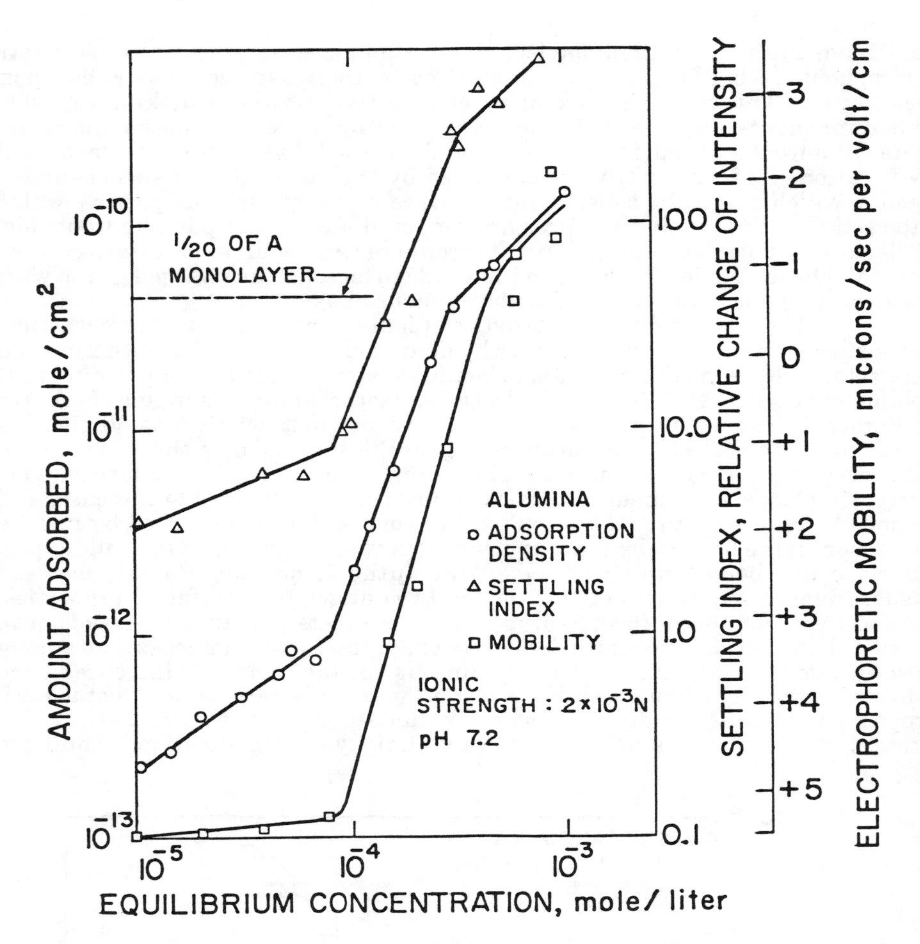

Figure 2. Adsorption Density of Dodecylsulfonate, Electrophoretic Mobility, and the Settling Rate of Alumina-Sodium Dodecylsulfonate System as a Function of the Concentration of Sodium Dodecylsulfonate

solloid (colloids on surface) is used for such surfactant hemimicelles as well as other polymer surfactant and inorganic aggregates on solids. The hemimicellar aggregates that form in region II are viewed as two-dimensional monolayered structures with the tails extended toward the solution side. The molecular structure of the adsorbed layer has been the subject of often controversial speculation. Spectroscopic probing using fluorescence and ESR methods conducted in our laboratories were designed to yield information on the microstructure of such adsorbed layers(13).

FLUORESCENCE SPECTROSCOPY(7)

Certain organic molecules in their ground state (P), when excited by light, absorb the incident light energy and reach an electronically excited state (P*). The transition of the molecule from the excited state (P*) to the ground state (P) is accompanied by the emission of light, which is referred to as *luminescence*. Complex organic molecules that are commonly used as luminescence probes typically absorb energy in the spectral range 250 to 650 nm, which corresponds to transitions with energy changes of 2 to 5 eV. Several features of luminescence, such as fluorescence, phosphorescence, excimer fluorescence, and delayed fluorescence, can be used as tools to obtain information at the molecular level on polymers and surfactants in bulk and at the solid/liquid interface. The luminescence experiment essentially involves measuring changes in the emission properties of a probe or its photochemical intermediates to determine the nature and rates of photophysical and photochemical processes in the system. In this respect, fluorescence is the most widely used technique. Fluorescence responses are sensitive to changes in the microenvironment around the probe so that a luminescence probe in different microenvironments will display experimentally distinct luminescence properties that will uniquely characterize each microenvironment.

The fluorescent probes we have utilized are pyrene and dinaphthyl propane. These are organic molecules whose fluorescence emission is highly sensitive to the polarity of the environment. A typical emission spectrum of pyrene is shown in Figure 3. The region from ≈ 370 to ≈ 400 nm of its spectrum is characterized by five well-resolved peaks (373, 379, 384, 389, and 393 nm). The intensities of the first (I_1) and third (I_3) peaks respond to the polarity of the medium[15]. This empirical finding, though lacking a precise quantitative understanding, is a useful tool in arriving at the micropolarities of inaccessible regions as in the interior of the micelles. The I_3/I_1 value of pyrene changes from 0.6 in water to a value of >1 in a nonpolar medium.

The excimer formation tendency of pyrene is widely exploited for analysis of aggregation and conformational studies[5]. An excimer is a dimer formed between an excited molecule and a molecule in the ground state. The photophysics of pyrene excimer can be represented as follows:

$$P + h\nu \dashrightarrow P^* \dashrightarrow P + h\nu' \text{ ...... MONOMER FLUORESCENCE}$$

$$P + P^* \dashrightarrow PP^*_{excimer} \dashrightarrow P + P + h\nu'' \text{ ..... EXCIMER FLUORESCENCE}$$

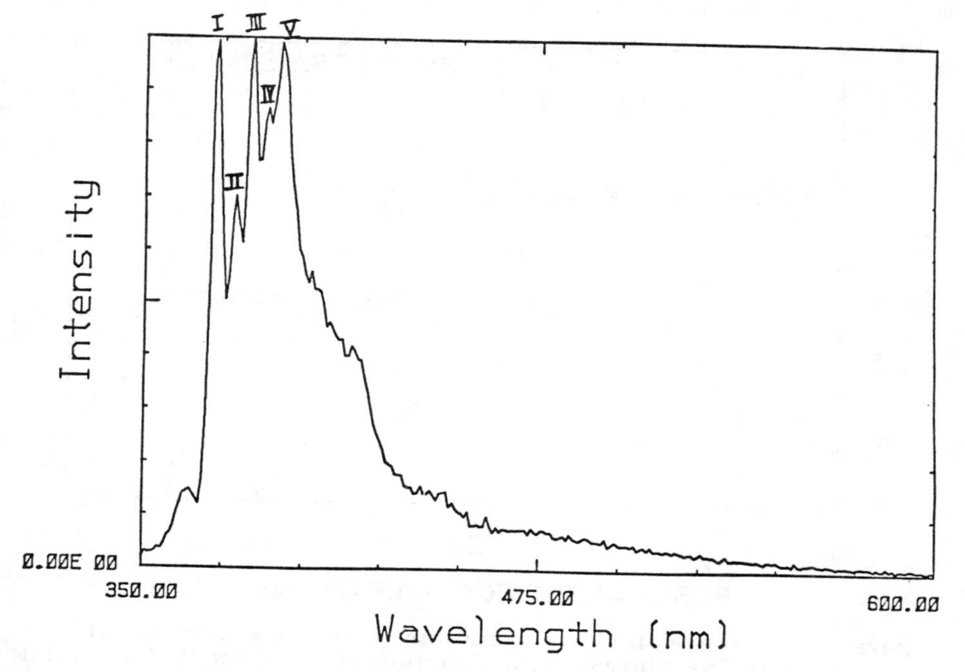

Figure 3. Typical Emission Spectrum of Pyrene

Excimer fluorescence in pyrene can be observed as a broad spectral band centered around 480 nm that is red-shifted with respect to the monomer emission. It is clear that the formation of the excimer would be a critical function of the proximity of the fluorescent moieties. Thus if two pyrene molecules are close to each other, the probability of formation of an excimer is higher than if they were far apart. The excimer complex is formed only when the aromatic rings approach each other within 0.4 to 0.5 nm[2].

PYRENE EMISSION IN SODIUM DODECYLSULFATE-ALUMINA ADSORBED LAYER[3]

The emission spectra, upon excitation of sodium dodecylsulfate-alumina slurries containing pyrene, showed the characteristic pyrene fine structure. In Figure 4 the polarity parameter (I_3/I_1) is plotted as a function of the adsorption density. The values of I_3/I_1 for pyrene in water and in sodium dodecylsulfate micelles are included for comparison. At low sodium dodecylsulfate values, the probe is mostly in the supernatant and the I_3/I_1 value is 0.6. As the sodium dodecylsulfate level increases, a sharp increase in I_3/I_1 to a limiting value of 1.0 is observed. This value, which indicates a micelle-like polarity, is measured at concentrations well below the critical micellization concentration. Importantly, this change in the polarity coincides with the transition from region I to II (see Figure 1) and is indicative of the existence of micelle-like aggregates at the solid/liquid interface. It is to be noted that the presence of pyrene had no effect on the adsorption density of sodium dodecylsulfate on alumina under the tested conditions.

PYRENE EMISSION DYNAMICS FOR AGGREGATION NUMBER

The time-resolved fluorescence emission of pyrene incorporated in the sodium dodecylsulfate hemimicelle studied under conditions of monomer

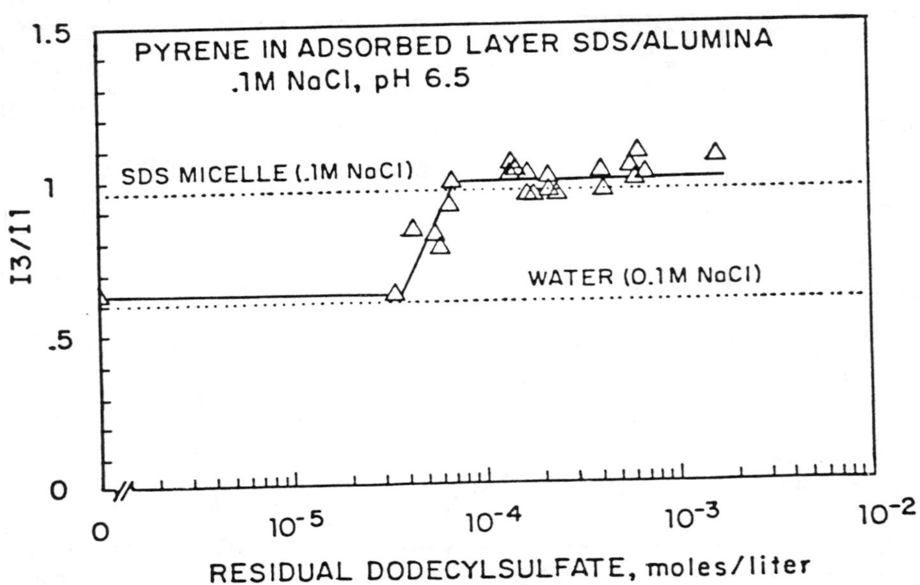

Figure 4. **Pyrene Polarity Parameter in Sodium Dodecylsulfate-Alumina Slurries as a Function of Residual Dodecylsulfate Concentration**

and excimer conditions can yield information about the aggregation number of the aggregates. A knowledge of the average occupancy number of the probes (n) leads to the aggregation number (N) from the following expression:

$$n = [P]/[Agg] = [P] \cdot N/([S] - [S_{eq}])$$

where [P] is the total pyrene concentration, [Agg] is the concentration of the aggregates, and ([S] - $[S_{eq}]$) is the concentration of the adsorbed surfactant. n is obtained from a kinetic scheme that assumes a Poisson distribution of the probe in fragmented micellar ensembles. The above methodology was applied to various regions along the sodium dodecylsulfate isotherm, and the calculated aggregation numbers are marked in Figure 5(4). A mechanistic model is suggested to represent the evolution and growth of the surfactant layers on alumina and is illustrated in Figure 6.

ELECTRON SPIN RESONANCE SPECTROSCOPY

Electron spin resonance (ESR) is well suited to the study of molecular structure and dynamics. It is a technique that records transitions between spin levels of molecular unpaired (paramagnetic) electrons in an external magnetic field in the form of absorption of microwave radiation. The intrinsic angular

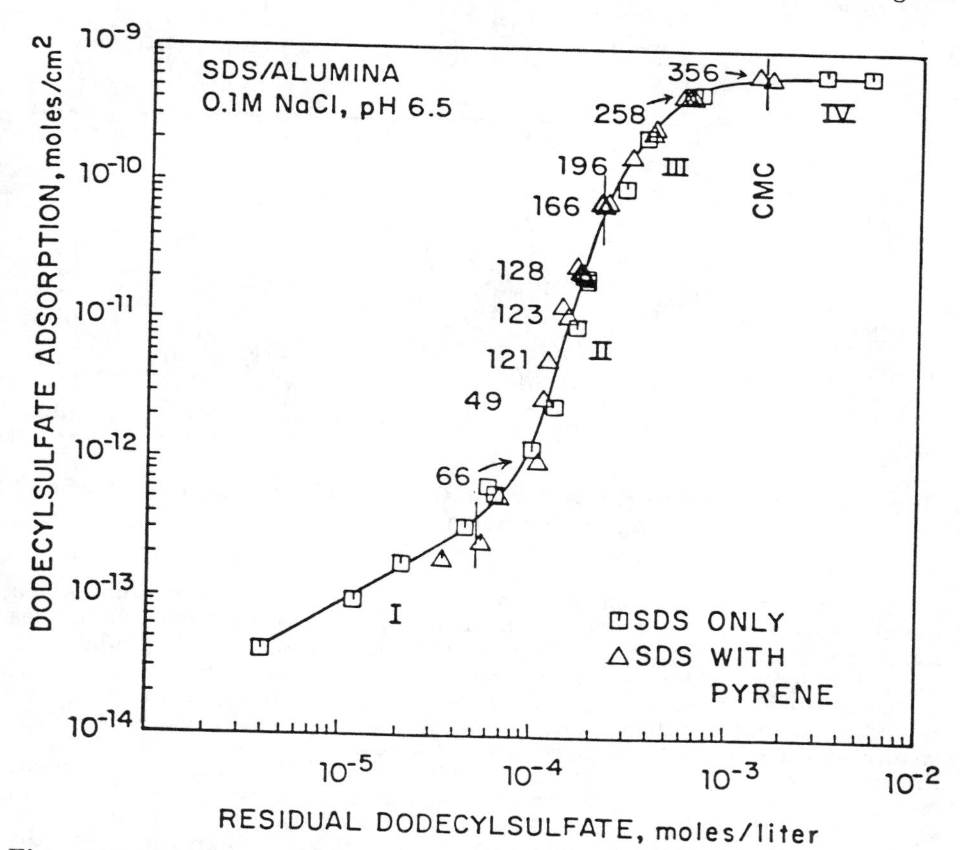

Figure 5. Surfactant Aggregation Numbers Determined at Various Adsorption Densities (Aggregation Numbers at Each Adsorption Density Shown Along Isotherm)

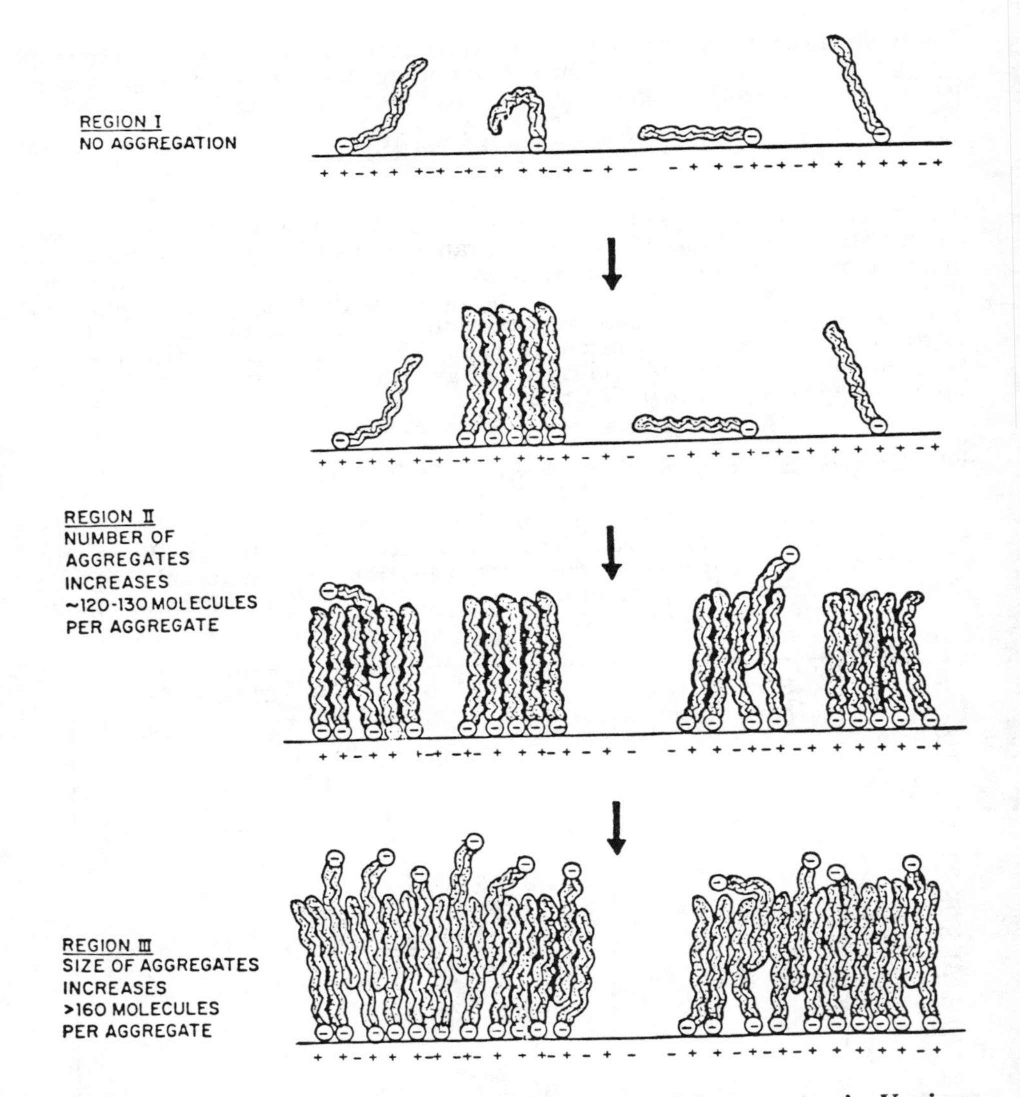

Figure 6. Mechanistic Model for the Growth of Aggregates in Various Regions Along the Isotherm

momentum of a free electron splits in an external magnetic field, which upon the influence of secondary magnetic moments of neighboring nuclei, undergoes hyperfine splitting. The range of applications can be extended by spin labeling methods whereby stably free radicals are incorporated into systems of interest to characterize their dynamic and physical properties(1). The nitroxide free radical is the most commonly used radical in spin labeling. In our study doxyl stearic acid (stearic acid labeled with a nitroxide-bearing moiety) was chosen as the probe because this structure resembles the dodecylsulfate surfactant molecules to some extent. Thus, the anionic functionality and long alkyl chain should enable the probes to coadsorb with the sodium dodecylsulfate molecules. Comparison of the spectra of the probe in the surfactant aggregates with the spectra of the probe in ethanol/glycerol mixtures of known viscosities indicated a microviscosity of 120 to 165 cP in the sodium dodecylsulfate hemimicelles(17). Figure 7 shows the spectra obtained with the 5-, 12- and 16-doxyl stearic acid

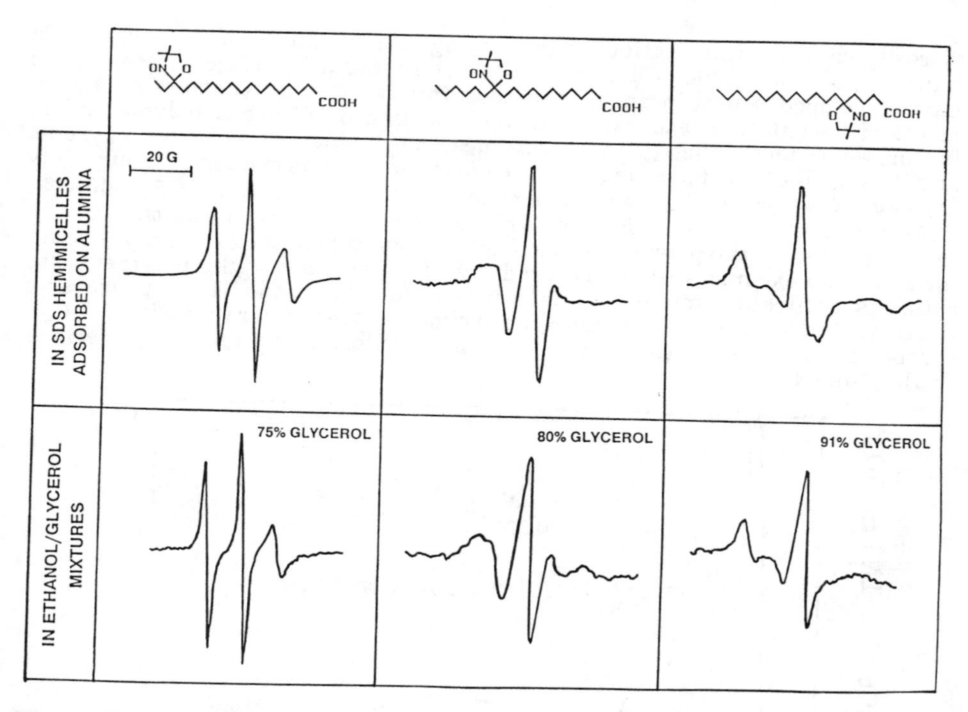

Figure 7. ESR Response of the Three Probes (5-, 12-, and 16-doxyl Stearic Acid) in Sodium Dodecylsulfate Hemimicelles Adsorbed on Alumina and in Ethanol-Glycerol Mixtures. From left to right, the probe describes an environment increasingly closer to the alumina surface.

probes in the hemimicelles and similar spectra obtained in ethanol/glycerol mixtures. The nitroxide in the 5th position is closer to the solid surface and is essentially immobile, whereas the nitroxide in the 16th is further away from the surface and exhibited greater mobility. This implies that the chain segments near the surface are tightly packed, whereas those near the end of the chain are considerably more loosely packed, thus more mobile(8).

The above results on the characterization of surfactant adsorbed layers provide us with an insight into the microstructure of the adsorbed surfactants on a molecular scale. Better manipulation of the adsorbed layers to achieve maximum efficiency in interfacial processes is possible if correlations between microstructures and efficiency of processes are also available.

POLYMERS IN SOLID/LIQUID SEPARATION

Dispersion or flocculation of suspended particles is influenced by macromolecular adsorption both in terms of the amount adsorbed and the configuration of the adsorbed species. Even though it has been recognized that configuration of polymers at the interfaces can lead to either flocculation or dispersion, what type of conformation (stretched versus coiled or flat versus dangled) is required or how one can manipulate a system to achieve the optimum configuration has never been established. This lack of knowledge has resulted from the nonexistence of reliable in-situ techniques with which one can determine conformation and orientation of species adsorbed on solids in liquids. Techniques such as gel permeation chromatography, light scattering, and viscosity can give information on the average conformation of the polymer in bulk, but are incapable of doing so at the solid/liquid interface. Recently we have developed a multipronged approach involving simultaneous measurements of flocculation responses and configuration of adsorbed polymer using luminescence techniques(6). In fluorescence, the polymer is labeled with pyrene, and the distribution of pyrene along the macromolecule is random. In our study we have used pyrene-labeled polyacrylic acid (molecular weight 88,000)(16). The conformation of the polymer is related to the excimer formation of the tagged pyrene. If the polymer were completely stretched, then the probes being far apart, no excimer will be observed. On the contrary, if the polymer were coiled, significant excimer formation will be observed. This process is illustrated in Figure 8. The ratio of the excimer and monomer intensities (I_e/I_m) can be taken as an indication of the polymer conformation and can be termed "coiling index."

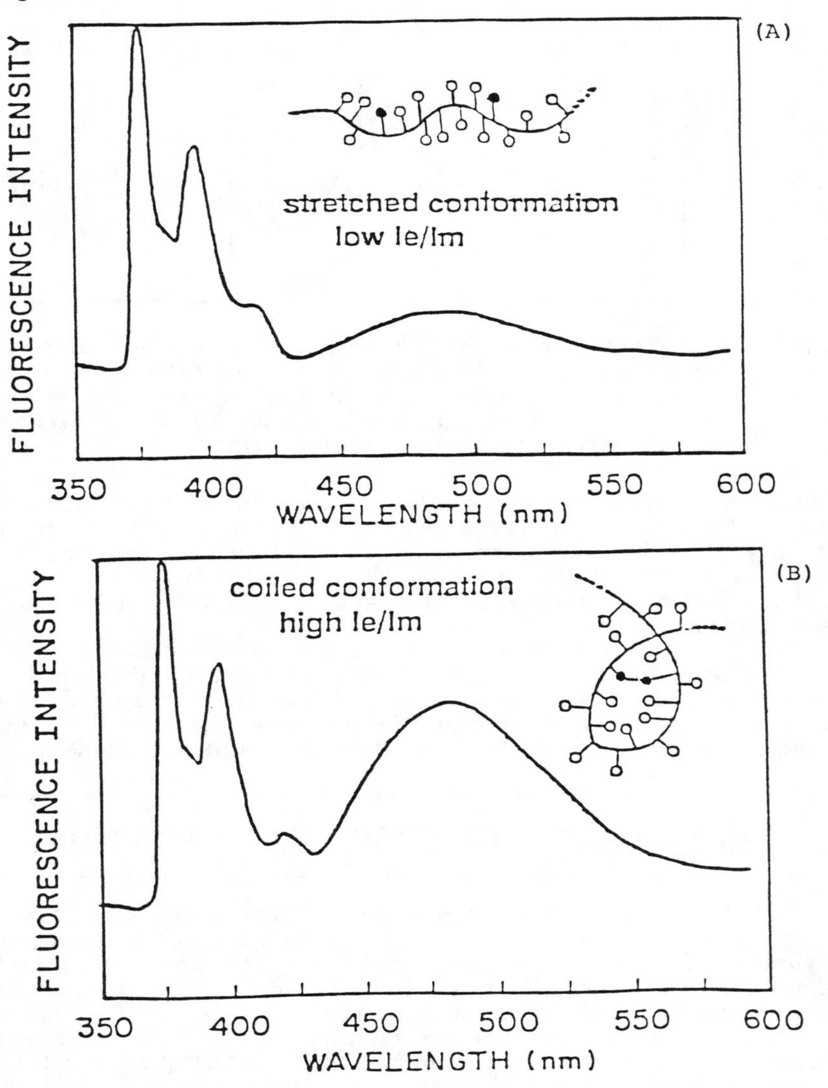

Figure 8. A Typical Fluorescence Spectrum of a Pyrene-Labeled Polymer in (a) Stretched and (b) Coiled Conformation

Flocculation of alumina suspensions by polyacrylic acid was studied, and the effect of polymer conformation on it was determined. The conformation of the polymer at the solid/liquid interface was monitored as a function of pH. At low pH (pH 4) the adsorbed polymer was found to be coiled, whereas at high pH (above pH 6), it was stretched. The conformational state of the polymer in solution was the same as that of adsorbed polymer on alumina. This suggested that either the polymer was loosely bound to the surface and the conformational changes reflected the changes in the degree of ionization, or that the adsorption of the polymer is rapid and the conformation is determined by the state of the polymer before adsorption. To differentiate between the two mechanisms, the pH was changed after the polymer adsorption. When the pH was lowered from 10 to 4 after adsorption, the conformational state of the polymer was not changed, as shown by the constant excimer-to-monomer ratio (Figure 9). Since polyacrylic acid is mostly ionized, able to attach itself with many segments onto the alumina via hydrogen bonding, and able to assume a stretched configuration, lowering the pH does not cause the adsorbed polyacrylic acid chain to coil. However, when the pH was raised from 4 to 10, the conformation of adsorbed polyacrylic acid changed from coiled to stretched. This implied that, at low pH, the coiled form of the polymer is less strongly attached to the alumina surface, and when the pH is raised, it adjusts its conformational state to allow greater contact with the alumina surface. A schematic representation of the conformational changes of the polymer is shown in Figure 10(a-f).

In our efforts to manipulate the polymer conformation at the solid/liquid interface for the best solid/liquid separation, it was discovered that if the polymer is adsorbed first at low pH, and then the pH is raised, excellent solid/liquid separation is obtained (Figure 11). Comparing this to the system under fixed pH conditions (Figure 12), it can be concluded that a change in the adsorbed polymer conformation from coiled to slightly extended will result in the best separation[6].

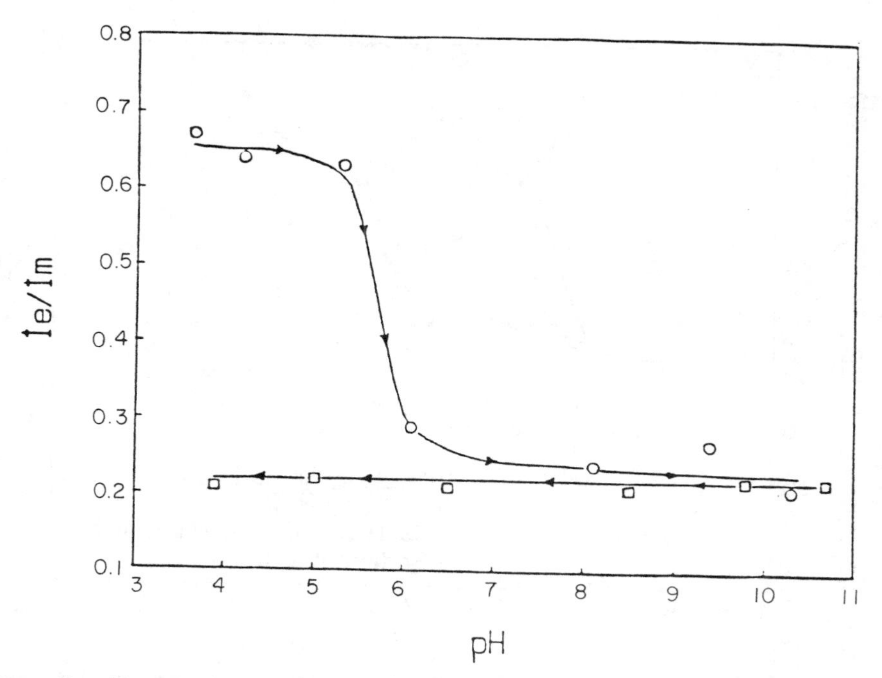

Figure 9. Excimer-to-Monomer Ratio (I_e/I_m, Coiling Index) of Polyacrylic Acid at the Alumina/Liquid Interface as a Function of Final pH Under Changing pH Conditions (I.S. = 0.03 M NaCl; S/L = 10 g/200 mL)

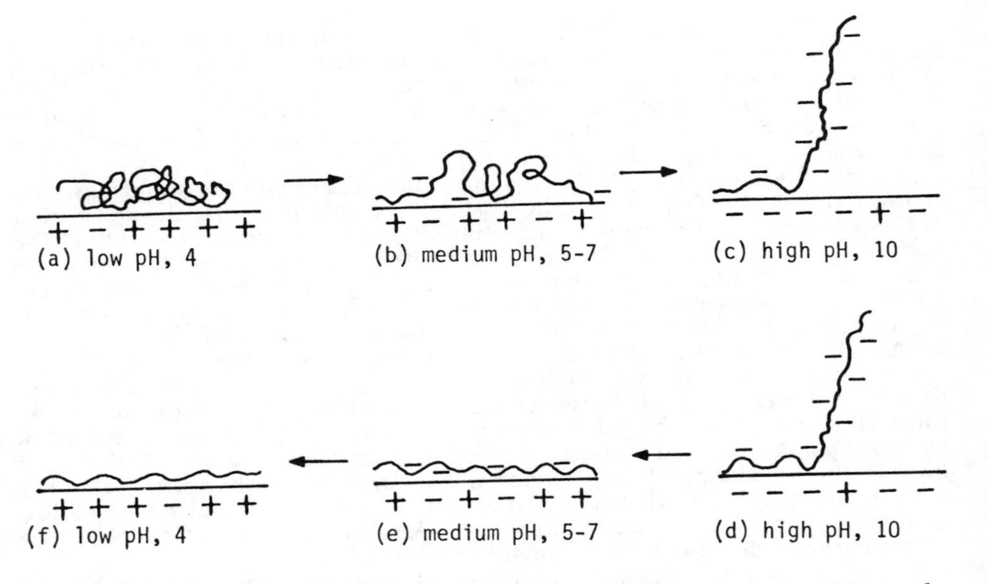

Figure 10 (a-f). Schematic Representation of Polyacrylic Acid on the Alumina/Liquid Interface Under Changing pH Conditions

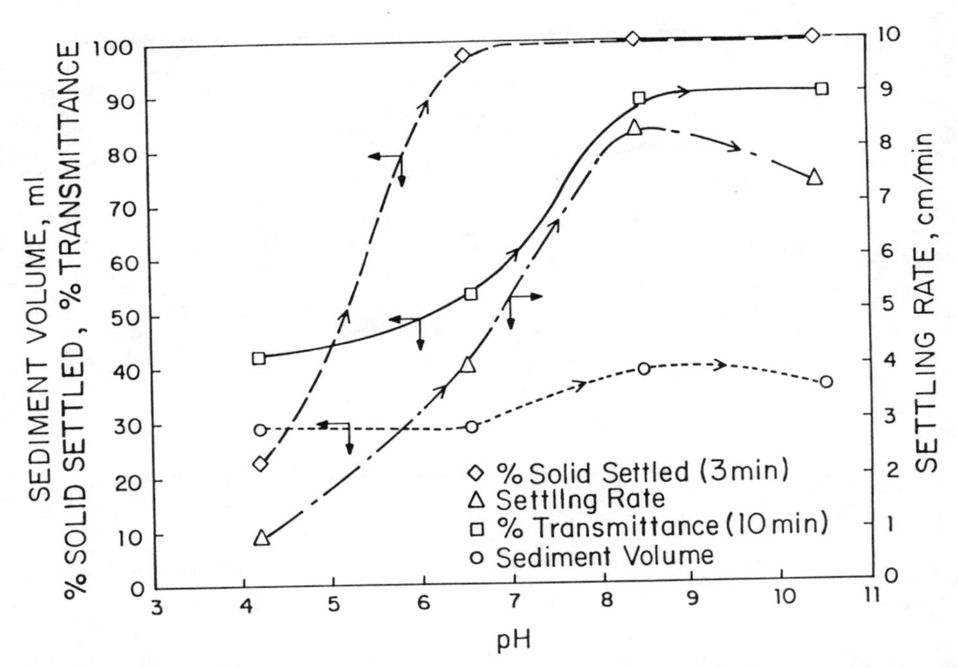

Figure 11. Flocculation Properties of Alumina with 20 ppm Pyrene-Labeled Polyacrylic Acid as a Function of Final pH Conditions Under Changing pH Conditions with an Initial pH of 4 (I.S. = 0.03 M NaCl; S/L = 10 g/200 mL)

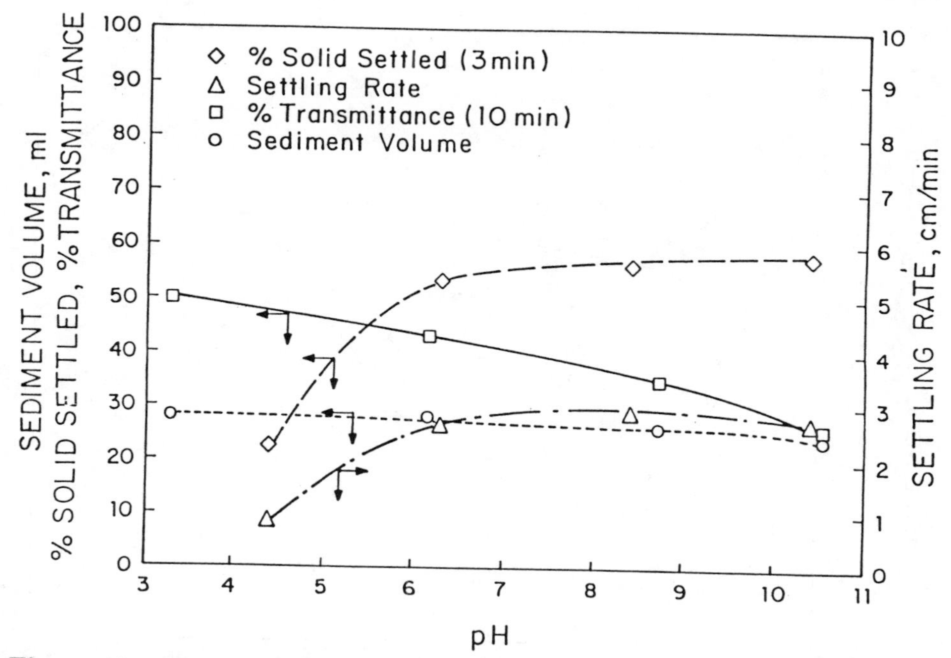

Figure 12. Flocculation Properties of Alumina with 20 ppm Pyrene-Labeled Polyacrylic Acid as a Function of pH Under Fixed pH Conditions (I.S. = 0.03 M NaCl; S/L = 10 g/200 mL)

EFFECT OF POLYDISPERSITY OF THE POLYMER

Interpretation of data for polymer adsorption and flocculation has often been made difficult by the inherent polydispersity of the polymers--the effect of the molecular weight differences often mask important phenomena that occur at the solid/liquid interface in response to changes in system parameters. We examined the adsorption of the anionic polyelectrolyte, sodium polystyrene sulfonate (MW 4,600 and 1,200,000), on hematite under competitive conditions using size exclusion chromatography(9).

The adsorption isotherms of monodispersed sodium polystyrene sulfonates (MW 4,600 and 1,200,000) are given as a function of residual polymer concentration in Figure 13. Evidently, in the absence of salt, slightly higher adsorption of the lower molecular weight sodium polystyrene sulfonate is seen. When the electrolyte concentration is increased to 0.1 kmol/m^3, significantly higher adsorption of 1,200,000 is seen (Figure 14). The higher adsorption at increased ionic strength is attributed to the coiled conformation of the polymer at high salt, both in bulk and at the solid/liquid interface. The kinetics of displacement at 0.1 kmol/m^3 NaCl is illustrated in Figure 15. It is clear that, in a very short time, there is significant adsorption of 4,600 sodium polystyrene sulfonate, and as the system is equilibrated, 4,600 sodium polystyrene sulfonate is displaced from the surface and substituted with 1,200,000 sodium polystyrene sulfonate.

The displacement and polymer/polymer interactions at the solid/liquid interface can significantly influence the flocculating characteristics of a suspension through conformational changes as a function of time. This would be particularly important in systems where one polymer is added before another to enhance flocculation by using the "piggyback" effect, in which the second polymer rides atop the first. This relatively recent approach is referred to as "double flocculation."

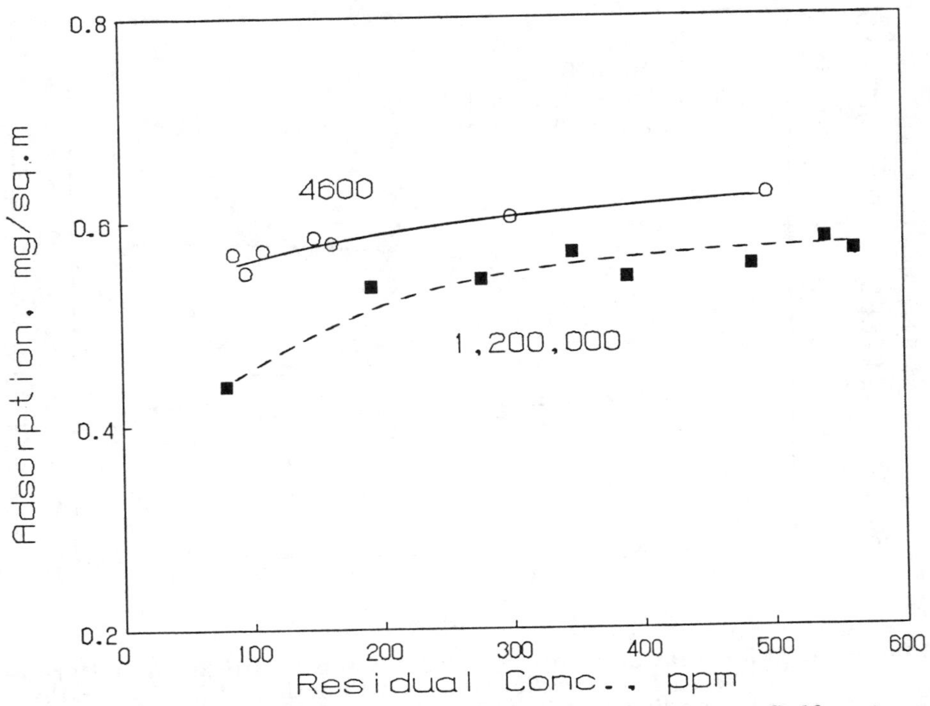

Figure 13. Adsorption Isotherm of Sodium Polystyrene Sulfonate on Hematite (pH 7.0; no salt)

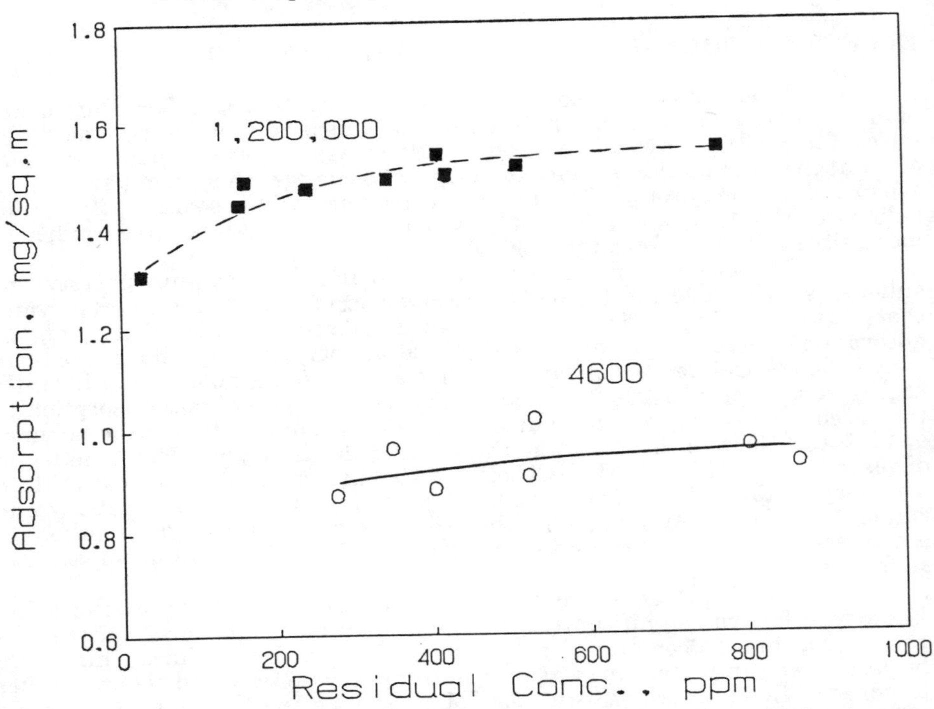

Figure 14. Individual Adsorption Isotherm of Sodium Polystyrene Sulfonate on Hematite (pH 7.0; 0.1 kmol/m^3 NaCl)

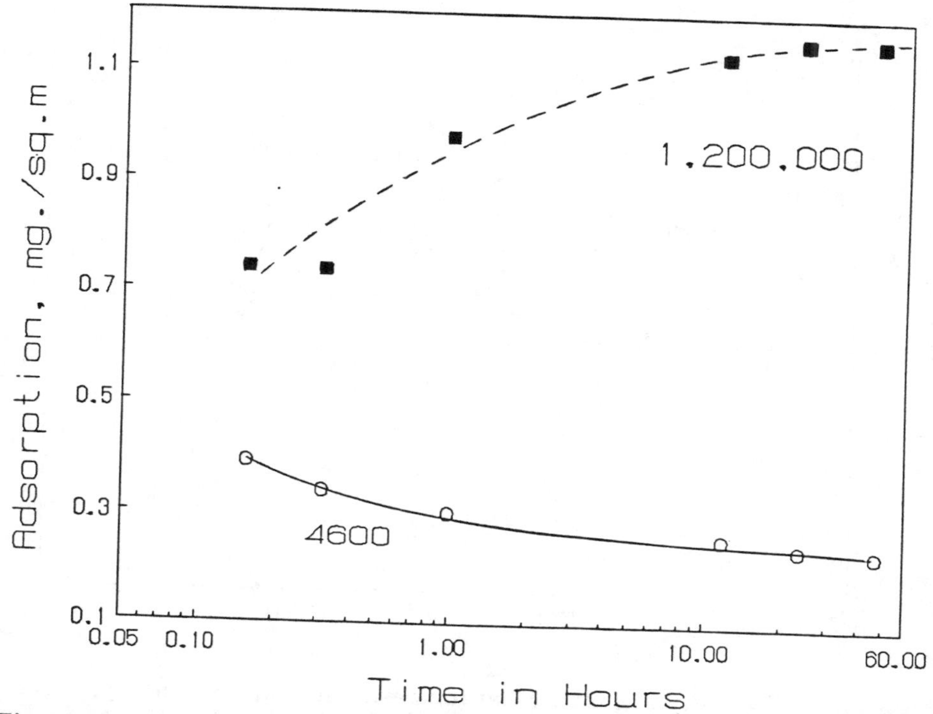

Figure 15. **Kinetics of the Displacement of 4,600 MW Polystyrene Sulfonate by 1,200,000 MW Polystyrene Sulfonate at 0.1 kmol/m^3 NaCl**

POLYMER/SURFACTANT INTERACTIONS

Polymers together with surfactants are being used in several fields including enhanced oil recovery, microelectronics, and mineral processing. Previously described results have clearly illustrated the dependence of solid/liquid separation processes on the conformation of the polymer at the solid/liquid interface. We have studied the effect of sodium dodecylsulfate on the conformation of polyethylene oxide both in bulk and at the solid/liquid interface[14]. The polymer was end-labeled with pyrene to be able to use fluorescence spectroscopy to study the interactions.

Figure 16 shows the change in the coiling of the pyrene end-labeled polyethylene oxide as a function of added sodium dodecylsulfate in bulk solution. The sharp rise in the coiling indicates the binding of the surfactant to the polymer, resulting in an increase in its hydrophobicity. The point at which the polymer is in its most coiled conformation coincides with the formation of free sodium dodecylsulfate micelles in bulk solution, as detected by surface tension measurements. After free micelles form in solution, the polymer stretches out completely.

This study was further extended to the solid/liquid interface. Polyethylene oxide does not adsorb on alumina, whereas sodium dodecylsulfate strongly does so. In bulk solution sodium dodecylsulfate micelles can stretch out a coiled polymer, so the effect of formation of sodium dodecylsulfate hemimicelles at the alumina/solution interface on the adsorption of polyethylene oxide onto alumina remained to be seen. Figure 17 shows the adsorption isotherm of sodium dodecylsulfate on alumina in the presence of 0.5 kmol/m^3

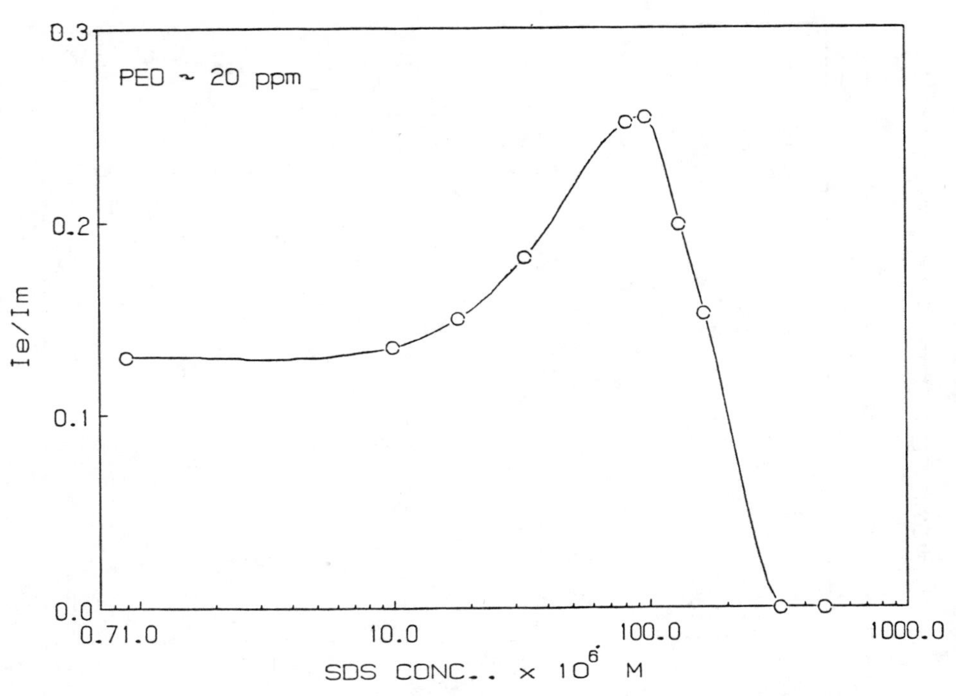

Figure 16. Polyethylene Oxide Conformation in Sodium Dodecylsulfate Solutions

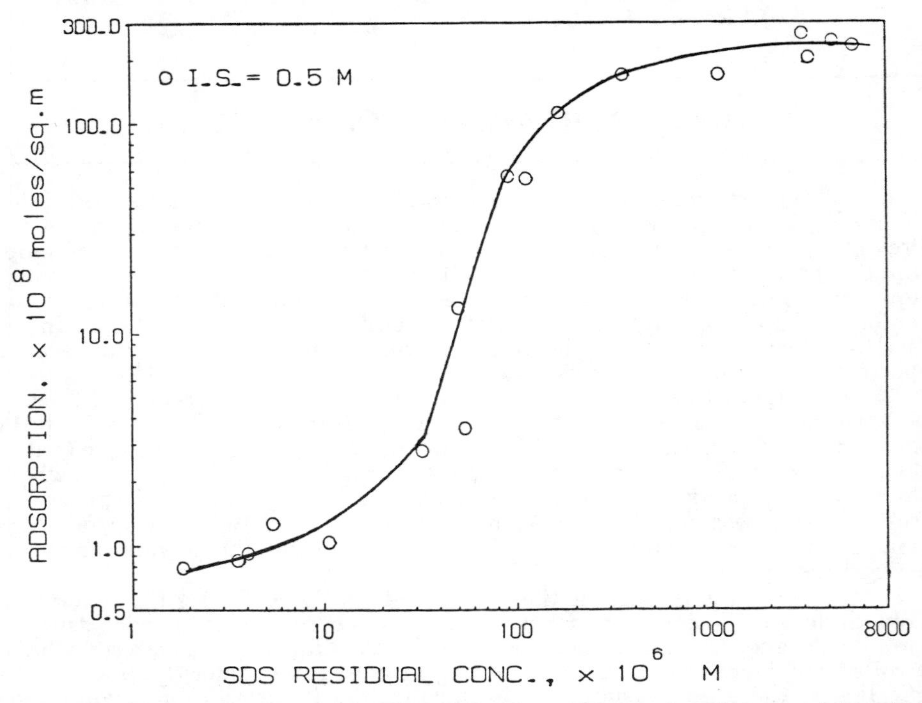

Figure 17. Adsorption Isotherm of Sodium Dodecylsulfate on Alumina (I.S. = 0.5 kmol/m^3 Na_2SO_4)

sodium sulfate. Alumina pretreated with sodium dodecylsulfate was further conditioned with polyethylene oxide. The results are shown in Figure 18. The presence of surfactant at the solid/liquid interface led to the adsorption of some polyethylene oxide, and the formation of surfactant hemimicelles caused the adsorbed polymer to change from a coiled to a stretched conformation. This finding could have important applications in the field of solid/liquid separation and selective flocculation of mixed mineral systems.

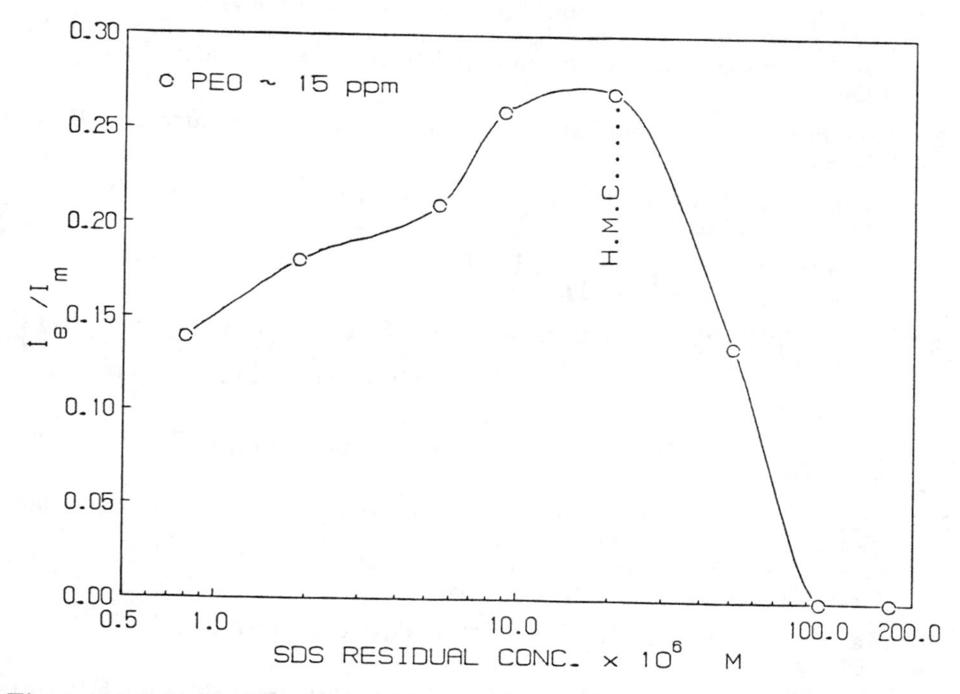

Figure 18. Coiling Index (I_e/I_m) of Polyethylene Oxide Adsorbed on Alumina in the Presence of Preadsorbed Sodium Dodecylsulfate

SUMMARY

The above results provide a means to investigate the microstructure of the adsorbed layers on a molecular scale under in-situ conditions. Structural characterization studies help in understanding the nature of the solid/liquid interfacial region and enable one to correlate the optimum performance of solid/liquid separation to structural features. Such knowledge provides enormous potential to develop reagent and process schemes for manipulating structures of adsorbed layers for desired system performance.

ACKNOWLEDGMENT

Financial support from the National Science Foundation (NSF-MSM-86-17183) and International Fine Particle Research Institute (IFPRI) is gratefully acknowledged.

REFERENCES

1. Berliner, L. J., *Spin Labeling I: Theory and Applications*, Academic Press, New York, 1979.

2. Chandar, P., *Fluorescence and ESR Spectroscopic Studies of the Structure of Dodecylsulfate Adsorbed Layer at the Oxide-Water Interface*, D.E.S. Thesis, School of Engineering and Applied Sciences, Columbia University, 1986.

3. Chandar, P., Somasundaran, P., and Turro, N. J., *J. Coll. Inter. Sci.*, 1987, *117*, 91.

4. Chandar, P., Somasundaran, P., Waterman, K. C., and Turro, N. J., *J. Phys. Chem.*, 1987, *91*, 150.

5. Chandar, P., Somasundaran, P., Turro, N. J., and Waterman, K. C., *Langmuir*, 1987, *3*, 298-300.

6. Huang, Y. B., Tjipangandjara, K. T., and Somasundaran, P., *Proceedings of the International Symposium on Production and Processing of Mineral Fines*, E. J. Plumpton, ed., Canadian Institute of Mining and Metallurgy, Montreal, 1988, 269.

7. Lakowicz, J. P., *Principles of Fluorescence Spectroscopy*, Plenum, New York, 1983.

8. Malbrel, C. A. and Somasundaran, P., to be published in *Proceedings of the 3rd International Conference on Fundamentals of Adsorption*, A. B. Mersmann, Y. M. Ma, and K. K. Unger, eds., AIChE Publication, Sonthofen, West Germany, May 1989.

9. Ramachandran, R. and Somasundaran, P., *J. Coll. Inter. Sci.*, 1987, *120*(1), 184-188.

10. Somasundaran, P., in *Mineral Processing*, 13th International Mineral Processing Congress, J. Laskowski, ed., Elsevier, Amsterdam, 1981, Vol. 2, Part A, 233-261.

11. Somasundaran, P. and Fuerstenau, D. W., *J. Phys. Chem.*, 1966, *70*, 90-96.

12. Somasundaran, P., Chia, Y. H., and Gorelik, R., in *Polymer Adsorption and Dispersion Stability*, E. D. Goddard and B. Vincent, eds., ACS Symposium Series, Vol. 240, American Chemical Society, New York, 1984, 393-410.

13. Somasundaran, P., Turro, N. J., and Chandar, P., *Colloids & Surfaces*, 1986, *20*, 45.

14. Somasundaran, P., Maltesh, C., and Ramachandran, R., presented at the 7th International Symposium on Surfactants in Solution, Ottawa, October 2-7, 1988.

15. Thomas, J. K., *The Chemistry of Excitation at Interfaces*, ACS Monograph, American Chemical Society, Washington, DC, 1984.

16. Turro, N. J. and Arora, K. S., *Polymer*, 1986, *27*, 783.

17. Waterman, K. C., Turro, N. J., Chandar, P., and Somasundaran, P., *J. Phys. Chem.*, 1986, *90*, 6828.

IN-SITU REMEDIATION

As a result of recent legislation and increased environmental awareness, thousands of sites have been identified for cleanup of both contaminated soils and groundwater. The U.S. Environmental Protection Agency has been given the task of identifying suitable technologies, performing demonstration projects, and implementing these technologies on superfund sites. For example, just recently, the U.S. Department of Energy announced that it will spend $19.5 billion to clean federal sites. Unfortunately, very few technologies exist to treat the soil and groundwater.

Although aboveground technologies are prevalent, excavation and treatment are prohibitively expensive, and performing them causes significant site disruption. It has been reported that some hazardous materials are found 1,000 feet below the land surface where excavation is almost impossible.

During this session, a number of interesting technologies will be discussed. Professor Shapiro, Massachusetts Institute of Technology, will discuss the use of electric fields to decontaminate soils underground. Although electroosmotic applications for dewatering are well known, electroreclamation of soils is rather recent. Dr. Hinchee of Battelle will discuss the electroacoustic decontamination of soils. This involves a combination of dc electric fields with acoustic fields to promote effective soil washing. Dr. Chawla of Howard University provides a critical analysis of the role of surfactants in in-situ treatment, and Dr. La Mori of Toxic Treatments provides insight on in-situ steam extraction of volatile organic compounds. Steam stripping was originally developed during tertiary oil recovery in the petroleum industry.

IN-SITU EXTRACTION OF CONTAMINANTS FROM HAZARDOUS WASTE SITES BY ELECTROOSMOSIS

Andrew P. Shapiro, Patricia C. Renaud,
and Ronald F. Probstein
Department of Mechanical Engineering
Massachusetts Institute of Technology
Cambridge, Massachusetts 02139
(617) 253-2240

ABSTRACT

A procedure is discussed for the in-situ extraction of contaminants in solution from hazardous waste sites. The contaminant is displaced by a harmless aqueous purge solution, which is moved through the soil, together with the contaminant, by electroosmosis. The dc electric field is set up through embedded porous electrodes. Laboratory experiments with compacted kaolin clay samples saturated with acetic acid and phenol solutions have shown very high degrees of removal. A theoretical model of the process, including ionic migration and diffusion, is in good agreement with the low energy costs shown by these experiments.

INTRODUCTION

Of the many different techniques proposed over the years for removing contaminants from hazardous waste sites, all have suffered from one or more technical or economic disadvantages. The excavation and subsequent treatment of contaminated soil, for example, by soil washing or incineration, is a costly technique and may expose workers to health risks. Moreover, in the case of soil washing, the procedure may not extract all of the contaminants attached to other clay or silt components of the soil, and in the case of incineration, a site pollution problem may be replaced by an air pollution problem.

In-situ collection and injection remediation techniques have been employed also. Collection techniques, such as the collection of a contaminant plume by pumping and/or drains, often suffer from dilution by surrounding groundwater during collection, thus increasing pumping and treatment costs. Further, effective control of the direction of flow is generally not possible because of soil heterogeneity and cracks. Injection techniques, such as injecting chemicals or biological agents in situ into the soil to detoxify the wastes, suffer from the difficulty of achieving a uniform distribution of the detoxifying materials throughout the soil. Moreover, both collection and injection techniques based on the use of pressure-driven liquid flows may be impossible to use in soils with low hydraulic permeability. In addition, for many in-situ remediation methods, including high-pressure soil flushing, vacuum or steam extraction, and radio frequency volatilization, many contaminant materials, and particularly heavy metals, cannot be removed because of the strong attachment forces that bind the metals to the soil particles.

In the present paper we discuss an electroosmotic procedure for the in-situ extraction of contaminants from hazardous waste sites. Electroosmosis describes the phenomenon whereby an ionic liquid under the action of an applied electric field moves past a charged surface or through a porous material, such as a soil or clay, whose internal surfaces are charged. The magnitude of this charge is usually measured by the zeta potential of the surface; this value is typically in the range of tens of millivolts. The flow is caused by the electric field acting on the charged double layer at the pore surface[8]. The double layer is a thin region near the pore wall in which the fluid possesses a net charge density that balances the surface charge on the wall. The magnitude of the double layer thickness is measured by the Debye length λ, which for an aqueous solution of a fully dissociated symmetrical electrolyte at 25 C, is given by

$$\lambda = \frac{9.61 \times 10^{-9}}{(z^2 c)^{1/2}}$$

where λ is in meters, c is the electrolyte concentration in mol/m^3, and z is the valence of the anions and cations[8]. For a univalent electrolyte at a concentration of 1 mol/m^3, the Debye length is thus about 10 nm.

The applied electric field produces a force on the charged fluid in the double layer that causes the fluid to move in a direction parallel to the electric field. The bulk of the fluid far from the wall is then set into motion by viscous drag interaction. The magnitude of the resulting electroosmotic velocity, when the double layer thickness is small compared to the pore radius and for a constant zeta potential, is given by the Helmholtz- Smoluchowski equation

$$U = \frac{\varepsilon \zeta E}{\mu}$$

Here, ε is the permittivity of the liquid, μ its viscosity, ζ the zeta potential, and E is the magnitude of the electric field assumed to be uniform.

In connection with the Helmholtz-Smoluchowski equation, we may observe that in a circular capillary of radius a, which can model a pore in soil, the volume flow rate is proportional to a^4 multiplied by the pressure gradient. The electroosmotic flow rate is equal to U multiplied by the cross-sectional area πa^2. Therefore, the ratio of electroosmotic to hydraulic flow rate will be proportional to $1/a^2$. Thus, for example, if we employ a capillary model for a porous soil, it is evident that as the average pore size decreases electroosmosis will become increasingly effective in driving a flow through the medium compared with pressure, so long as the double layer is thin. It is for this reason that electroosmosis is a process appropriate to low-permeability soils and clays and not very porous sandy soils.

Electroosmosis has been applied extensively and successfully in foundation engineering to dewater and consolidate soils for construction purposes(1,2), as well as for dewatering clay and mine tailings(5,6). In these procedures, an electric field is established in the soil or clay by means of embedded electrodes, causing the water to migrate toward and accumulate at or near one of the electrodes. The accumulated water is then removed by some appropriate means such as pumping.

Efforts to apply electrokinetics in general, and electroosmosis techniques in particular, to the removal of contaminants from a waste site are relatively recent(4). The movement of metal ions in fully dissociated salts within samples of saturated sands and soils was investigated by Hammet(3). For his particular conditions, the contaminant movement appeared to be governed by ionic migration, that is, charged particle movement in an electric field, rather than by electroosmosis. Renaud and Probstein(10) carried out limited laboratory studies on the use of electroosmosis for the removal of acetic acid from saturated clay samples. It was shown in their work that electroosmosis might be particularly useful in control and remediation at hazardous waste sites, especially when it is necessary to control the direction of flow and when the site has relatively low hydraulic permeability and hydraulic removal techniques have become ineffective.

PROPOSED METHOD

Because of the many possible advantages that electroosmosis techniques might bring to the problem of in-situ contaminant removal at hazardous waste sites, we have devised a new electroosmotic procedure(9), which preliminary laboratory experiments and theoretical modeling show to be very effective with great potential for field use.

The removal of chemical species from porous media by means of electroosmosis relies on the convection of the pore liquid containing the contaminant species toward one of the electrodes where the fluid is collected. Not all contaminant liquids (for example, hydrocarbons) will result in a sufficiently high zeta potential being developed on the internal surfaces of the porous soil to ensure significant electroosmotic transport. In addition, if the contaminant fluid is removed by electroosmosis from a portion of the soil, that portion will become dried, soil cracking may ensue, and in any case the electrokinetic effectiveness of electroosmosis to remove the contaminant liquid will be greatly decreased. To overcome these limitations, we employ the idea of simultaneously introducing a purging solution at the electrode opposite to the one where collection takes place.

It should be recognized that, in addition to electroosmotic convection, the species dissolved in the pore liquid will be transported by diffusion, and charged molecules will migrate in the electric field. These movements may not necessarily be in the same direction as the bulk convective flow. Effects that may modify these transport mechanisms include, for example, adsorption on the soil and electrode reactions that induce pH changes altering both the zeta potential and the charge on dissolved molecules.

Figure 1 is a schematic cross section of the extraction system proposed. Alternating porous anode and cathode structures are emplaced in the contaminated soil. In any actual arrangement, the electrodes will be emplaced in an appropriate manner over the site surface. In the example shown, it is assumed that the soil is negatively charged so that the bulk flow of the contaminant liquid is toward the porous cathode electrodes. Were the soil positively charged, the electroosmotic flow would be toward the porous anode electrode. We here term an electrode structure from which the flow emanates a "source electrode" and one to which the flow migrates a "sink electrode."

A noncontaminating purging liquid, such as water, is supplied to the source electrodes and, under the action of a dc electric field, moves by electroosmosis into the contaminated soil. The contaminated liquid in the soil is displaced by the purging liquid. Depending upon the nature of the contaminant, its motion may also be a consequence of electroosmotic action in addition to being driven by the purge liquid. The contaminant liquid which moves through the soil enters the porous sink electrode structure where it is then removed to the surface by a pumping or syphoning action. Contaminant materials of opposite charge to that predominating in the double layer may move toward the source electrodes where they accumulate and can be removed.

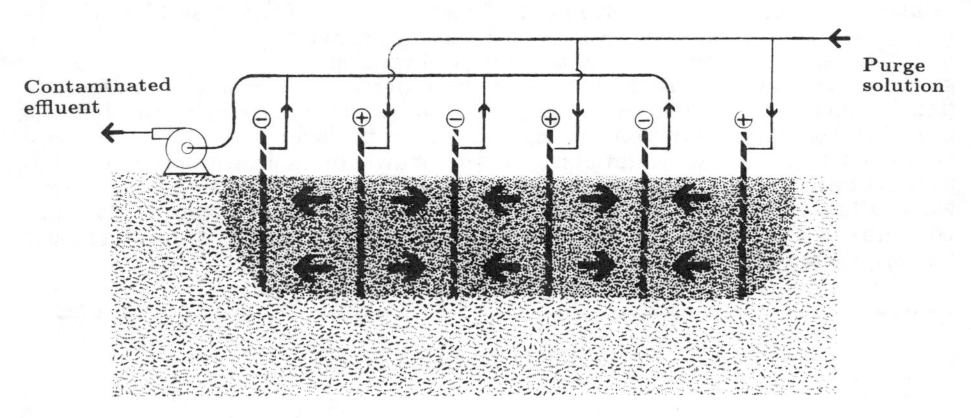

Figure 1. Schematic of Technique for Electroosmotic Extraction of Contaminants at Hazardous Waste Site(9)

EXPERIMENTAL EXTRACTION RESULTS

One-dimensional laboratory experiments have been carried out on kaolin clay saturated with acetic acid and phenol solutions to test the feasibility of the proposed electroosmotic purging technique to extract contaminant liquids from low-permeability soils. The test cell in which the experiments were made was a cylindrical acrylic tube of about 0.1 m diameter and 0.5 m length (Figure 2). Two sets of porous electrodes were used at each end. The voltage was applied to the active electrodes (typically carbon fiberboard), which were submerged in liquid and separated from the clay by perforated acrylic plates and filter paper. A pair of passive electrodes between the acrylic plates and the clay were used to measure the voltage at each end of the clay cylinder. Voltage probes were embedded along the length of the cell to enable measurement of the potential distribution. The current was measured by measuring the voltage drop across a resistor in series with the test cell. The effluent was collected in a cylinder having a pressure transducer at the bottom that enabled continuous volume measurements to be determined from the transducer output. The purging solution was introduced to the source electrode structure at atmospheric pressure. All data were automatically collected and converted with the aid of an analogue to digital converter. The chemical composition of the effluent was determined by means of batch HPLC separation and UV absorption or a differential refractometer detector. The degree of contaminant removal could then be determined from the organic acid concentration in the effluent and the accumulated volume as a function of time. For additional details, see Shapiro *et al.*(11).

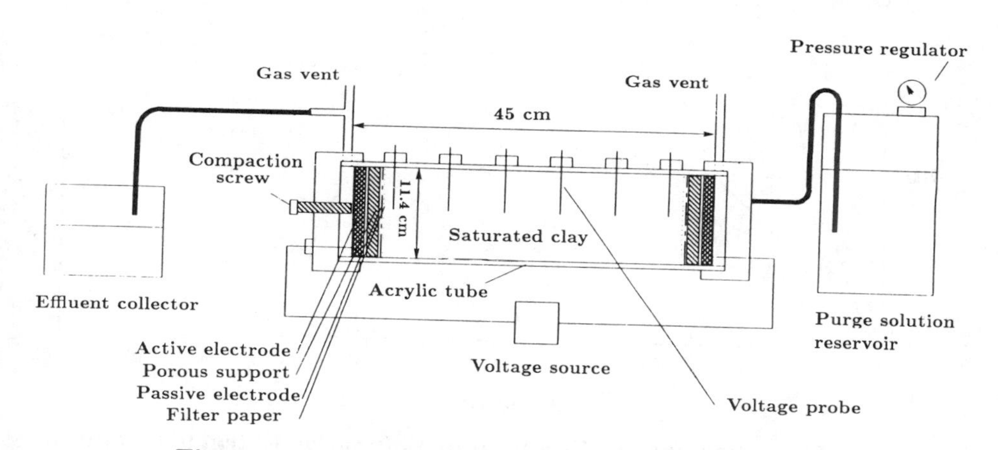

Figure 2. Laboratory Electroosmosis Apparatus

We report here two sets of experiments: one in which the saturating solution was 0.5 M acetic acid with a 0.1 NaCl purge solution, and the other a 450 ppm phenol solution with tap water as the purge solution. Figure 3 shows for both sets of experiments the measured cumulative volume of extracted solution versus time. Typical voltage gradients applied across the clay sample were on the order of 50 V/m, and hydraulic permeabilities were on the order of 10^{-16} to 10^{-17} m^2 (10^{-7} to 10^{-8} cm/s). For the ionic concentrations of the contaminant and purging solutions, the double layer thickness is small compared with the characteristic pore radius of the clay. Values measured for the electroosmotic permeabilities of between 10^{-9} and 10^{-8} m^2/V are in agreement with published results for kaolin clay with water(7). This permeability range corresponds to a time period of between 1 and 2 months to remove 1 pore volume for an electrode spacing of 0.5 m.

The most remarkable result found from these experiments is the high degree of contaminant removal. This may be seen in Figure 4, which is a plot of the fraction of contaminant extracted versus the pore volumes of effluent removed. For the 0.5 M acetic acid solution, 94 percent was extracted after 1.4 pore volumes of the 0.1 M NaCl purge solution had been moved through the sample. In the phenol experiment, 95 percent of the phenol was removed after 1.5 volumes of the purge solution of tap water had been moved through the solution.

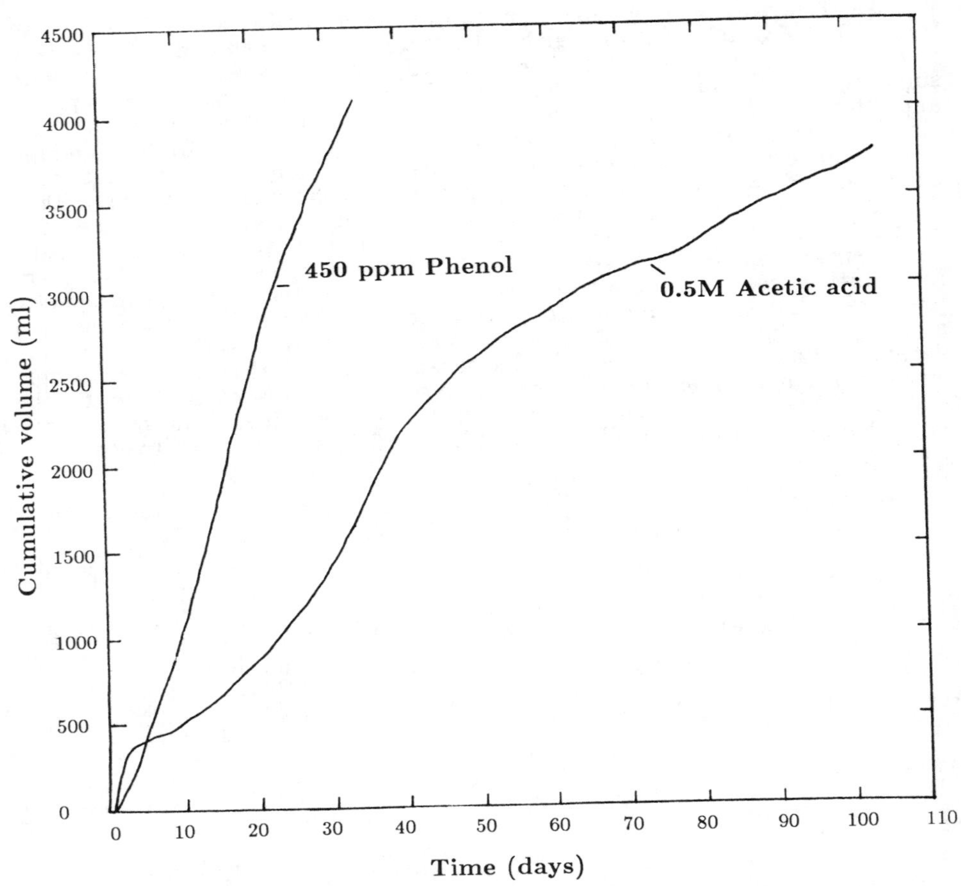

Figure 3. Measurements of Cumulative Volume of Extracted Solution Versus Time for Electroosmotic Removal of 0.5 M Acetic Acid and 450 ppm Phenol Solutions from Kaolin Clay Sample

It should be emphasized that the high degrees of removal shown in Figure 4 were not necessarily achieved in other tests where the experimental conditions were different. Among the reasons for this are that the degree of contaminant removal is a function not only of the clay type, which together with the contaminant solution determines the zeta potential, but also of the electrode reactions that can introduce significant pH changes. Changes in pH can lead to lower contaminant removal because they change the valence of the contaminant molecule and cause ion migration in the opposite direction of the bulk flow.

Another important result of these tests was the low energy cost obtained for the contaminant removal. For the experimental conditions described above, the energy costs were found to be \$5.3/tonne of acetic acid solution extracted and \$0.53/tonne of phenol solution extraction. The energy price assumed is \$0.10/kWh.

PROCESS MODEL

For a thin double layer, Shapiro *et al.*[11] have developed a one-dimensional, time-dependent model of the electroosmotic removal of chemical

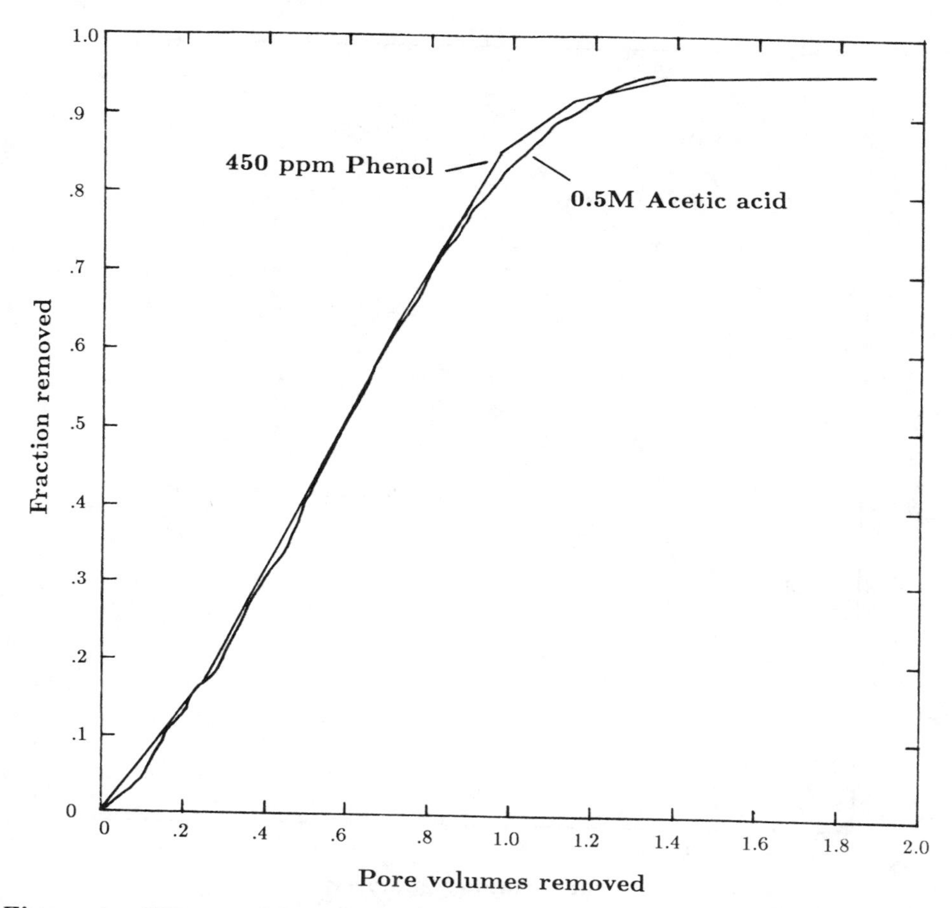

Figure 4. Measurements of Contaminant Fraction Extracted Versus Pore Volumes of Effluent Removed from Kaolin Clay Sample Initially Saturated with 0.5 M Acetic Acid and 450 ppm Phenol Solutions

species from porous media. The assumption of a thin double layer implies that the current and mass transport within the double layer are negligible in comparison with that in the bulk of a pore itself where electroneutrality holds. The porous medium is modeled using a capillary model wherein the medium is characterized by a series of uniformly distributed circular capillaries of equal radii. The model incorporates the electroosmotic bulk flow equation (2), ionic migration of the chemical species in the electric field, and molecular diffusion. Chemical equilibrium and equilibrium electrode reactions are accounted for by the introduction of finite production rate terms in the species conservation relations (convective-diffusion equations).

To compare the model with experiment, it is necessary to know the zeta potential. The zeta potential was determined from equation (2) using the measured average flow rate through the porous medium. This flow rate defines the superficial velocity through the medium, which is equal to the medium porosity multiplied by the electroosmotic velocity U.

In comparing the results of the model with the acetic acid experiments, six chemical species were assumed to be involved in the process: HAc, Ac^-, H^+, OH^-, Na^+, and Cl^-. To simplify the calculation, the diffusion coefficients of all species were assumed to be equal to 10^{-9} m^2/s. The initial NaCl concentration is selected to match the initial current density. Figure 5 is a

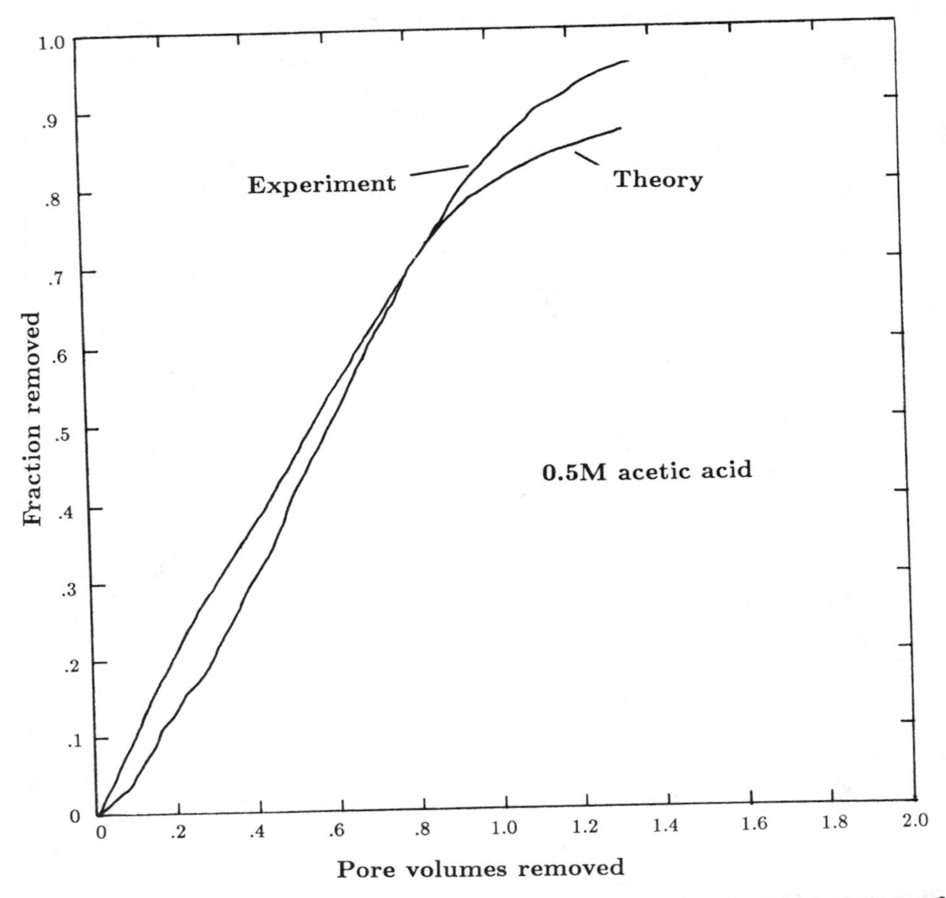

Figure 5. Comparison Between Theory and Experiment of Fraction of 0.5 M Acetic Acid Solution Extracted from Kaolin Clay Sample Versus Pore Volumes of Effluent Removed

comparison of the acetic acid extraction data from Figure 4 with the model results. The theory is seen to agree very well with the experimental data. Theoretical calculations of the energy cost are shown also to be in good agreement with that calculated from the measured power consumption.

CONCLUDING REMARKS

Preliminary one-dimensional laboratory experiments on the degree of contaminant extraction, time for removal, and energy cost by electroosmotic purging from saturated clay samples indicate high removal efficiency and low energy cost. A theoretical model of the extraction process agrees well with experiment. The results suggest that the proposed electroosmotic purging technique offers much promise for the in-situ extraction of contaminants in solution from hazardous waste sites in which soil permeability was not too high. However, additional laboratory work with different contaminants, purge solutions, contaminant levels, soil characteristics, electric field strengths, and electrode materials, as well as with three-dimensional geometries, is required before field test can be undertaken with confidence.

REFERENCES

1 Bjerrum, L., *et al.*, *Geotechnique*, 1967, *17*, 214-235.

2. Casagrande, L. J., *Boston Soc. Civil Engineers*, 1983, *69*(2), 255-302.

3. Hammet, R., "A Study of the Processes Involved in the Electroreclamation of Contaminated Soils," M.Sc. Thesis, University of Manchester, Manchester, England, 1980.

4. Herrmann, J. G., ed., "Proc. Workshop in Electro-Kinetic Treatment and its Application in Environmental-Geotechnical Engineering for Hazardous Waste Site Remediation" (unpublished), University of Washington, Seattle, WA, August 4-5, 1986, Hazardous Waste Engineering Research Laboratory, U.S. Environmental Protection Agency, Cincinnati, OH, 1986.

5. Lockhart, N. C., *Colloids and Surfaces*, 1983, *6*, 229-251.

6. Lockhart, N. C., *Int. J. Mineral Processing*, 1983, *10*, 131-140.

7. Mitchell, J. K., *Fundamentals of Soil Behavior*, Wiley, New York, NY, 1970, 355.

8. Probstein, R. F., *Physicochemical Hydrodynamics: An Introduction*, Butterworths, Boston and London, 1989.

9. Probstein, R. F., Renaud, P. C., and Shapiro, A. P., "Electroosmosis Techniques for Removing Materials from Soil," U.S. Patent Pending, 1989.

10. Renaud, P. C. and Probstein, R. F., *J. Physicochemical Hydro.*, 1987, *9*, 345-360.

11. Shapiro, A. P., Renaud, P. C., and Probstein, R. F., *J. Physicochemical Hydro.*, 1989, *11*(5), in press.

IN-SITU TREATMENT OF SOILS CONTAMINATED WITH HAZARDOUS ORGANIC WASTES USING SURFACTANTS: A CRITICAL ANALYSIS

R. C. Chawla, M. S. Diallo,
J. N. Cannon, and J. H. Johnson
Department of Chemical Engineering
Howard University
Washington, DC 20059
(202) 636-6625

and

C. Porzucek
Dow Chemical Corporation
Freeport, Texas

ABSTRACT

In-situ surfactant washing of soils contaminated with hazardous organic wastes has been the object of many laboratory and field studies in response to CERCLA regulations. Despite the success of this technique in the laboratory to remove a variety of contaminants from the contaminated soils, the field tests have shown only limited success.

This paper critically reviews the past work on surfactant-assisted in-situ washing of organic laden subsurface soils. This review indicates that the majority of past work was primarily an extension of tertiary oil (surfactant flooding) recovery to in-situ soil washing. As a result of this, only the removal of nonaqueous-phase liquid (NAPL) contaminants held to the pores of subsurface soils by capillary forces was addressed. Because forces other than capillarity (e.g., sorption) can play a significant role in pollutant emplacement in subsurface soils, it is critical to address the removal of pollutants bound by such forces. Therefore, a more fundamental understanding of contaminant removal mechanisms is critical to assess adequately the applicability and limitations of surfactant-assisted in-situ soil washing. Such assessment is essential in determining which types of contamination problems are suitable for this technology.

INTRODUCTION

Over the last 15 years, the contamination of soils and aquifers by hazardous chemicals has become a major concern in the United States. In 1980, Congress passed the Comprehensive Environmental Response, Compensation and Liability Act (CERCLA), otherwise known as Superfund, to abate this problem. Under this act, the U.S. EPA and various other federal agencies (DoD, DOE, etc.) have initiated the development of a series of countermeasures to mitigate the impact of hazardous materials released into the environment. A key component of these efforts is the Chemical Countermeasures Program, which was designed to evaluate in-situ methods for the remediation of contaminated subsurface soils[7,8]. These methods would be effective under the following conditions[8]:

- Contaminant is spread over a large volume, e.g., 100 to 100,000 m^3, at a depth of 1 to 10 m.
- Contaminant concentration is low (not over 10,000 ppm).
- Contaminant can be removed or immobilized by aqueous chemical solutions injected into the contaminated site.
- Hydraulic conductivity of the target contaminated area is greater than 10^{-4} cm/s.

Two in-situ techniques were considered under this program: in-situ remediation of soils contaminated with hazardous heavy metals using acid and chelating agents, and in-situ washing of soils contaminated with hazardous organic compounds using surfactants. Whereas the development of the former is still at the laboratory stage[25], the latter has been the object of many investigations including a field test[7,8,17,19,20,22,26-29].

Past work on surfactant-assisted in-situ washing included a number of laboratory and pilot studies[7,8,19,22,26] and a field test at a "site of opportunity" located at Volk Air National Guard Base in Wisconsin[17]. Although the laboratory and pilot experiments demonstrated the effectiveness of surfactant aqueous solutions to remove a variety of contaminants, the field test showed these solutions were ineffective in removing contaminants from the "site of opportunity."

This paper describes and critically evaluates past studies on surfactant-assisted in-situ soil washing. Its primary objectives are (a) to assess the applicability and limitations of the results of past investigations and (b) to show the importance of a good understanding of soil contaminant/surfactant interactions in determining which types of contamination problems are suitable for surfactant-assisted in-situ soil washing.

This paper is divided into five sections. The next section gives an overview of in-situ soil washing. The third section describes in detail the past work on surfactant-assisted in-situ soil washing. The fourth section critically evaluates past work and assesses the applicability and limitations of its results. Finally, this review is concluded by a summary of its main findings and a discussion of some of the research needed to turn surfactant-assisted in-situ soil washing into a viable remediation technology.

IN-SITU SOIL WASHING: PROBLEM OVERVIEW

Over the last 15 years, the contamination of groundwater sources by petroleum products, pesticides, and other toxic chemicals has become a major problem in the United States[1,6,10,14,21,30]. This contamination has been caused primarily by organic loading of groundwater formations following accidental spills and leakages from underground storage tanks of nonaqueous-phase liquid (NAPL) chemicals[1,6,10]. Following its infiltration in a subsurface soil, an NAPL contaminant migrates downward through the unsaturated soil zone under the influence of gravity. Initially, the extent of this migration is controlled by the contaminant's hydraulic potential and its physical properties--density, interfacial tension, and viscosity[1,10]. If the contaminant source (leakage or spill) is relatively small, the contaminant will exhaust its hydraulic potential before it reaches the subsurface soil-water table. However, if the contaminant source is sufficiently large and/or if the water table is not located very far from the surface, the contaminant will usually reach the water table.

Not all of a spilled or leaked NAPL contaminant will migrate to the water table. A significant part of it will remain trapped in the pores and fractures of subsurface formations by the action of capillary forces. The amount of NAPL contaminant that is trapped in a subsurface soil is referred to as residual saturation and primarily depends on the contaminant and soil properties[1,10]. In addition, some of the NAPL contaminant will also bind with the subsurface soil matrix according to a contaminant-soil interaction process commonly referred to as sorption. Figure 1 shows an example of contaminant distribution in a groundwater formation following an NAPL spill. Bound and trapped NAPL contaminants can be retained in subsurface formations for indefinite time periods and consequently serve as contaminant sources as they slowly dissolve into infiltrating rain water or a rising water table[1,10]. The only way to mitigate this problem is to remove trapped and bound contaminants from subsurface formations. Because excavation and on-site remediation of contaminated subsurface soils are rarely feasible, in-situ remediation has received considerable attention. Possible in-situ cleanup technologies that have been investigated include NAPL dissolution into extracted groundwater, vacuum extraction, steamflooding, and in-situ soil washing using organic solvents[8,10,16].

Dissolution into extracted groundwater has been suggested as a way of removing hydrophilic organic pollutants such as phenols and alcohols from contaminated subsurface soils[8]. The high water solubility of these compounds and their low adsorptivity in soil matrices make them suitable for this kind of process. Hunt *et al.*[11] have recently shown the possibility of using steamflooding to remove a number of volatile organic compounds (trichloroethylene, benzene, and toluene) from contaminated laboratory sand packs (2.5 percent residual saturation). Malmanis *et al.*[16] reported successful utilization of vacuum extraction to remove a variety of volatile organic compounds (VOCs), including trichloroethylene and tetrachloroethylene from a contaminated subsurface soil at a Superfund site in Michigan. Although the possibility of using

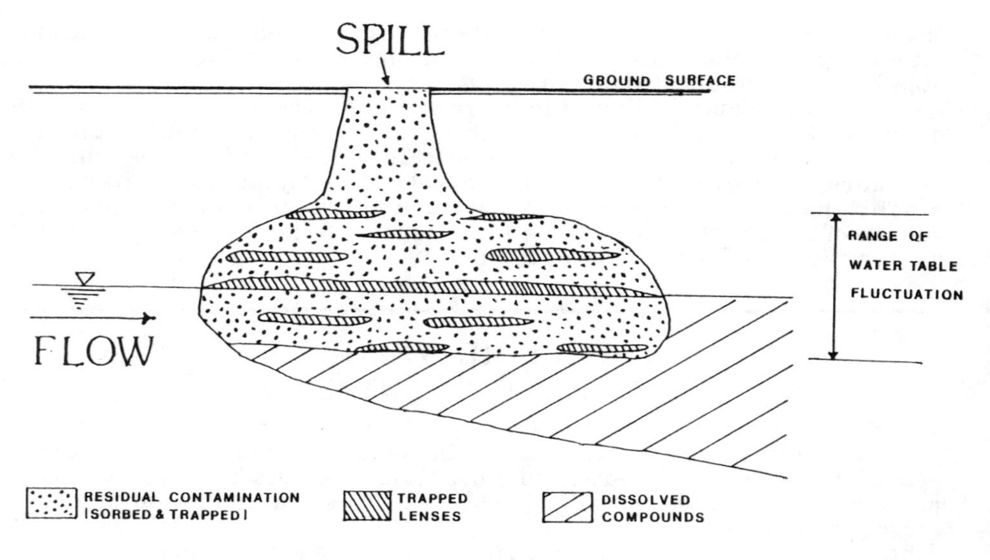

Figure 1. Soil and Aquifer Contamination by a Lighter-than-Water NAPL Contaminant[10]

steamflooding and vacuum extraction to remove VOCs has been demonstrated, their applicability to the cleanup of subsurface soils contaminated with other hydrophobic compounds (polychlorinated biphenyls, polynuclear aromatics, etc.) remains to be established.

Griffin and Chou[9] have examined the possibility of using organic solvents (acetone, dioxane, and methanol) to remove hydrophobic compounds such as polybrominated biphenyls (PBBs) from contaminated soil matrices. They found that PBBs, which usually are strongly sorbed onto soil matrices, are highly mobile in dioxane and to a lesser extent in acetone and methanol. Chou *et al.*[4] reported a slight decrease of hexachlorocyclopentadiene sorption onto soils treated with sodium hypochlorite. Although the possibility of using organic solvents and other chemicals to remove hydrophobic compounds from contaminated soils has been shown in the laboratory, their high cost and toxicity hinder their utilization in the cleanup of contaminated groundwater formations.

This situation has triggered a number of studies aimed at developing more efficient in-situ cleanup technologies to tackle the removal of hydrophobic organic compounds. Because the removal of organic pollutants from contaminated soils is similar to detergency (a process that relies upon the cleaning power of surfactants), the use of surfactant aqueous solutions to remove hydrophobic organic pollutants from groundwater formations has been the object of many investigations, described in detail in the following section.

SURFACTANT-ASSISTED IN-SITU SOIL WASHING: LITERATURE REVIEW

The initial work on surfactant-assisted in-situ soil washing was conducted by the Texas Research Institute (TRI)[26]. This research was an application of tertiary oil recovery technology to in-situ soil washing. It was mainly a laboratory evaluation of the effectiveness of using aqueous surfactant solutions to recover gasoline from a gasoline-contaminated subsurface soil by enhancing its flow to the subsurface soil water table, where it could be recovered using a primary oil recovery technique.

The first part of this study reviewed the flow of spilled gasoline in a groundwater formation. This review established that the forces retaining gasoline in the unsaturated zone were directly related to the interfacial tensions (IFT) between water and gasoline and between water and air. It concluded that a surfactant solution was needed to decrease these tensions so that the gasoline could be mobilized and displaced to the water table from which it could be recovered.

The second part consisted of three tasks. The first task was a survey of major manufacturers to obtain data and samples of commercially available surfactants. This included a preliminary selection of the most promising surfactants based on their solubility in water, toxicity, biodegradability, tolerance to hard water, tendency to strongly adsorb onto soils, and ability to markedly reduce interfacial tensions between water and gasoline. In the second task, surfactant-assisted displacement studies of gasoline from contaminated water wet-sand packs were carried out to determine the effectiveness of the selected surfactants. These studies permitted the identification of a mixture of 2 percent (by weight) each of an anionic surfactant (Richonate--YLA, Richardson) and a nonionic surfactant (Hyonic PE-90, Henkel), which was effective in recovering 80 percent of the sand-pack residual gasoline. The third task consisted of displacement studies in containers simulating a thin vertical section of the soil in the vicinity of a gasoline spill using the same surfactant solution[26]. These studies showed that the surfactant solution was able to drain residual gasoline from the capillary zone.

TRI[27] also conducted gasoline displacement studies in a large-scale aquifer model using three different application techniques for the surfactant solutions. The application techniques tested were: single application by percolation through the sand volume, multiple (daily) applications by direct injection, and multiple (daily) applications by percolation through the sand volume. Percentages of gasoline removed with the above techniques were 6, 76, and 83, respectively.

Shortly after the publication of TRI's results, Science Applications International Corporation (SAIC), under an EPA contract, initiated a series of studies to evaluate the possibility of using aqueous solutions of surfactants to remove hydrophobic and slightly hydrophilic compounds from contaminated subsurface soils[7,8]. This work was based on the results of the TRI study and included a literature search and review to evaluate the applicability of existing technology to in-situ soil washing, and laboratory experiments to evaluate the utilization of surfactant aqueous solutions for in-situ soil cleanup and treatment of the contaminated leachates.

A Freehold hapludult soil from Clarksburg (New Jersey) of low organic carbon content (0.12 percent) and cation exchange capacity (8.6 meq/100 g) was selected for the laboratory studies. The contaminants tested were: polychlorinated biphenyls (PCBs); a mixture of di-, tri-, and pentachlorophenol; and Murban crude oil (a mixture of intermediate and high molecular weight aliphatic and polynuclear aromatic hydrocarbons). These contaminants were selected based on their frequency of occurrence at Superfund sites, toxicity, and persistence in soils. The surfactant solutions were chosen based on their aqueous solubility and their ability to disperse Murban crude oil and minimize the dispersion of clay-sized particles. Minimization of clay particles dispersion was found to be an important criterion for surfactant selection because preliminary studies showed that a mixture of 2 percent Richonate YLA and 2 percent Hyonic NP-90, previously proven effective to remove gasoline from contaminated sand packs by TRI[26,27], was found to be ineffective by SAIC due to its "marked tendency to suspend the silt- and clay-size grains (less than 63 μm in diameter), which resettled in small pores, thereby inhibiting column flow"[7].

The laboratory tests consisted of shaker table experiments to determine the partitioning of the pollutants between soils and surfactant solutions, and column experiments to evaluate pollutant removal efficiencies under gravity flow. For both shaker table and column tests, contaminants were sprayed on the soil samples using an aerosol spray filled with contaminant mixtures dissolved in methylene chloride. To ensure uniform contaminant dispersion, soil samples were stirred after the methylene chloride had evaporated.

The laboratory experiments demonstrated the feasibility of removing PCBs, Murban crude oil, and phenols using aqueous mixtures of nonionic surfactants. For the PCBs, contaminant removal from the column experiments was 92 percent using an aqueous solution of 0.75 percent (by weight) each of Adsee 799 (Witco) and Hyonic NP-90 (Diamond Shamrock), whereas that for the Murban crude oil was 93 percent using a 2 percent (by weight) solution of the same surfactant. Treatability studies for separating the surfactants from the contaminated leachate were unsuccessful.

Although these studies showed the effectiveness of surfactant aqueous solutions to clean up contaminated laboratory columns, SAIC recognized that, because of the low organic carbon content (0.12 percent) and cation exchange capacity (8.6 meq/100 g) of the sandy soil tested, the contaminants were not probably strongly bound to the test soils. As a result, SAIC recommended testing soils with higher amounts of organic carbon and clay minerals to "expand the overall applicability of the program results to a broader variety of soil matrices"[8].

After their successful experiments, SAIC conducted another laboratory study to test their procedure on contaminated soil samples from a waste "site of opportunity" located at Volk Air National Guard Base in Wisconsin[17]. Soil samples were collected at various depths and locations and characterized. They were very sandy (98 percent α-quartz) and exhibited high permeabilities (5.2×10^{-4} cm/s to 1.7×10^{-2} cm/s), low cation exchange capacities (0.8 to 5.1 meq/100 g), and low to medium total organic carbon contents (365 to 14,900 mg/kg). Chemical analyses of contaminated samples showed that petroleum derived hydrocarbons (JP-4 jet fuel), chlorinated solvents (1,1,1-trichloroethane, dichloromethane, trichloroethylene, and chloroform), and a fire-fighting foam were the main contaminants.

Removal experiments were carried out in two different columns. In both cases, four passes (3.5 L each) of an aqueous solution of 0.75 percent (by weight) each of Adsee 799 and Hyonic NP-90 were applied to the top of each column, and the effluent from each pass was collected for leachate treatment studies. Total contaminant (aliphatic + aromatic and unresolved hydrocarbons) removal was 94 percent for one column and 89.5 percent for the other. Treatability studies for separating the surfactants from the contaminated leachate were also unsuccessful in these experiments.

After these successful laboratory tests, a field test was conducted at Volk Air National Guard Base by Mason & Hanger-Silas Mason Co., Inc.[19,20]. A fire training pit was selected as the test site. The subsurface soil of the test site was determined to be 85 to 95 percent sand and 5 to 15 percent fines, and the contaminants were found to be mainly JP-4 jet fuel, lubrication oils, and chlorinated solvents. The site water table was located 12 feet below the surface and extended continuously to 700 feet. Ten holes for application of surfactant aqueous solutions were dug at locations chosen to ensure that the washing test would be conducted in a uniformly contaminated area. Half of the holes were 2 foot × 2 foot × 1 foot and the five remaining holes were 1 foot × 1 foot × 1 foot. The hole depths were chosen to match the thickness of the soil layer, which would be excavated prior to full-scale washing at the "site of opportunity."

Three synthetic surfactants and three "natural" surfactants were tested in various combinations. The synthetic surfactants were a mixture of ethoxylated fatty acids (Witco), an ethoxylated alkylphenol (Henkel), and an anionic sulfonated alkylester (Henkel), whereas the "natural" surfactants were the by-products of indigenous biological transformations. Rates of addition of surfactant aqueous solutions were 1.87 gal/day/ft^2. These solutions were added four times a day over a period of 4 to 6 days. After three days of testing only seven test holes remained unplugged and operational. These holes were rinsed with well water (2 to 7 gallons) at the end of the test. After test completion, soil samples from 2 to 4 in. and 12 to 14 in. below the bottom of the holes were collected and analyzed for oil and grease using infrared spectrophotometry. Results showed that washing solutions were ineffective in removing contaminants.

After a critical evaluation of the test results, Mason & Hanger-Silas Mason Co.[19] suggested that the failure of surfactant aqueous solutions to remove contaminants from the test site soil was largely because the washing solutions bypassed the highly contaminated areas of the test site. Because these flow paths contained very small amounts of contaminants, overall contaminant removal was insignificant.

Recent studies on in-situ soil washing with surfactants were conducted by CH2M Hill and MTA Remedial Resources, Inc., (MTARRI)[28,29]. These studies consisted of laboratory and field experiments to evaluate the applicability of in-situ treatment techniques to the remediation of the Union Pacific Rail Road Tie Treating Plant site in Laramie, Wyoming. The site was mainly contaminated by a wood-preserving oil, which was a complex mixture of polynuclear aromatic hydrocarbons, tar acids, and tar bases. Site characterization studies in four test areas showed the basal sand and gravel layers contained the majority of the contamination. Some pockets of contamination were also found in the overlying material ,which consisted of discontinuous clay, silt, sand, and gravel lenses.

Because some areas of the site alluvium contained sufficient quantities of free wood-preserving oil to justify primary oil recovery, laboratory and field tests were conducted by CH2M Hill to determine the fraction of oil that could be recovered through pumping of oil and water from the alluvial aquifer[28]. The laboratory experiments consisted of three linear and three radial core flood tests. Average percentages of oil recovered were 51 and 27, respectively. Following the laboratory tests, a primary oil recovery pump test in the field was conducted using a recovery system consisting of two 15-foot drain lines spaced 20 in. apart, one directly above the other. Each drain line was provided with its own pumping and water treatment system. This test lasted 29 days and enabled the recovery of 10,363 gallons of oil. A total of 770,000 gallons of water was also pumped, treated, and recharged in the aquifer. Based on the success of these laboratory and field tests, CH2M Hill recommended primary oil recovery as an effective part of the site remediation plan.

Because primary oil recovery would remove only a fraction (less than 60 percent) of the total amount of contaminant, laboratory studies were conducted by SURTEK, an enhanced oil recovery testing firm that acted as a subcontractor to MTARRI, to evaluate the effectiveness of surfactant solutions to remove the residual oil that would remain in the site after primary recovery of the contaminant[19]. These studies slightly departed from methodologies used in previous investigations[7,8,17,19,20,26,27] and evaluated the effectiveness of alkali, polymer, and surfactant mixtures to remove residual wood-preserving oil from the site. This new approach to surfactant formulation was chosen after preliminary tests showed that lower IFT and better removal efficiencies could be achieved using mixtures of alkali and surfactants rather than aqueous solutions of surfactant alone. Because other tests showed that additional contaminant removal up to 50 percent could be obtained if a viscosifier* (xanthan gum biopolymer) was added to the displacing solutions, linear and radial coreflood tests were conducted by SURTEK to identify the most effective alkali/polymer/surfactant formulations.

Formulation A: 1.5 percent sodium carbonate, 1,600 mg/L xanthan gum biopolymer, and 1 percent Tergitol NP-9 (by weight).

Formulation B: 1 percent sodium carbonate, 160 mg/L xanthan gum biopolymer, and 1 percent Polystep A-7 (by weight).

---

* The utilization of a viscosifier in tertiary oil recovery (surfactant flooding) to provide a favorable mobility ratio is highly recommended in surfactant flooding. Its implications on surfactant formulation for in-situ soil washing will be discussed in the last section of this review.

These two formulations yielded best contaminant removal efficiencies, over 95 percent, in all cases. After these encouraging results, MTARRI recommended pilot studies to evaluate the effectiveness of these formulations under field conditions.

Recently, MTARRI conducted a pilot study for Atlantic Richfield Corporation to evaluate the applicability of surfactant-assisted in-situ soil washing to the remediation of an industrial plant site located in Florida and contaminated by a viscous oil (130 cp at 20 C)[22]. Site characterization studies showed that "solution cavity limestone comprises the major rock type at the site with cavities infilled by clay-and-organic rich sand"[22]. The contamination consists of a free oil layer (0.01 to 1.5 feet thick) floating on the water table over a 40,000-ft^2 area and trapped oil (20 percent residual saturation) retained above and below the free oil layer by capillary forces. Because the contaminated site was separated from a drinking water supply aquifer only by a semi-permeable silt layer, synthetic surfactants were not used in this study to avoid contamination of the drinking water aquifer. Instead, "chemicals used in water treatment and as food additives" and a "biodegradable polymer" were used as IFT-lowering agents and viscosifier, respectively. Blends of these chemicals were used as wash fluids and were introduced in the test zone through a slotted pipe placed in a lateral trench. Four wells dug on the perimeter of the test area were used to recover the wash fluids. After test completion, soil samples were collected at various depths and analyzed for oil. Results show that the washing solutions were able to remove up to 75 percent of the trapped oil.

SURFACTANT-ASSISTED IN-SITU SOIL WASHING: CRITIQUE AND ASSESSMENT

Past work on in-situ soil washing with surfactant was primarily an application of tertiary oil recovery technology (surfactant flooding) to in-situ soil washing. Tertiary oil recovery is used to mobilize and displace the residual oil ganglia trapped in the reservoir pores toward a recovery well after secondary recovery (waterflooding). There are several proven tertiary oil recovery techniques. Surfactant flooding, in its various forms (surfactant and/or polymer flooding, micellar solution flooding, etc.) has been among the most widely tested. In all surfactant-based tertiary oil recovery techniques, a surfactant slug is injected into the reservoir to promote the displacement of the residual oil ganglia and their coalescence into a continuous oil bank[15,18]. In addition to the slug, a polymer solution and water (brine) are injected into the reservoir to drive the slug and oil bank toward a recovery well. To avoid the instabilities at the interface of the surfactant slug and oil bank shown in Figure 2, a viscosifier--which usually consists of high molecular weight, water-soluble polymer--is usually added to the slug to provide a favorable mobility ratio[18]. The mobility ratio is the ratio of the mobility of the displacing fluid (surfactant solution) to the mobility of the fluid being displaced (oil bank)[18]. The mobility of a fluid in a porous medium is a measure of its resistance to flow through the medium. Mathematically the mobility (M_f) of a fluid f is expressed by the equation given below[18]:

$$M_f = \frac{K_{rf}}{\mu_f}$$

where K_{rf} is the relative permeability to the flow of fluid f, and μ_f is the viscosity of fluid f.

For a surfactant slug, the mobility ratio (M) is defined as follows[18]:

$$M = \frac{\beta_s}{\beta_0}$$

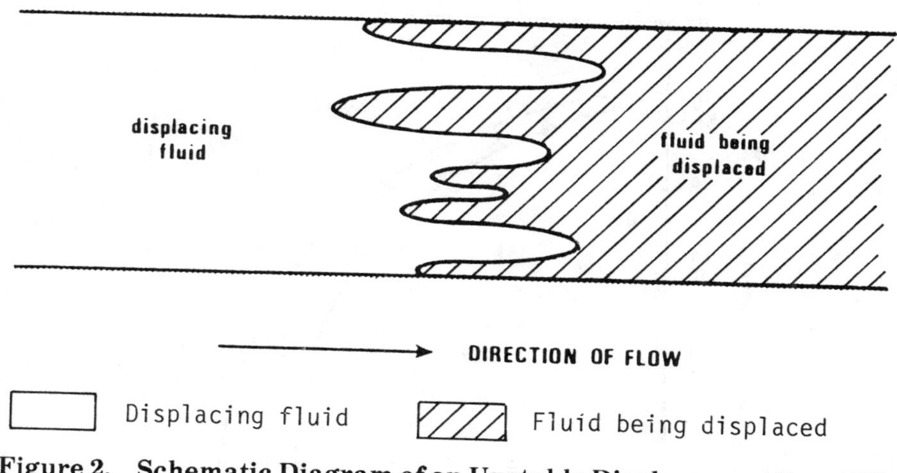

Figure 2. Schematic Diagram of an Unstable Displacement Front[18]

where β_s and β_o, respectively, are the mobilities of the slug behind and the oil ahead of the displacement front[18]. The most favorable mobility ratio is 1[18]. For this ratio, the resistance to flow through the porous medium for both the displacing fluid and the fluid being displaced are the same, thus decreasing the possibility of fingering associated with frontal instabilities. To avoid fingering that would have probably caused a poor soil washing sweep efficiency, MTARRI added a polymer to its alkali and surfactant mixtures. Although mobility considerations are important in surfactant flooding, it is the capillary number that is the critical parameter. The capillary number, N_{ca} (the ratio of the displacing fluid viscous forces to the capillary forces holding the residual oil ganglia to the pores of the formation), controls the extent of residual oil displacement by a surfactant flood and is expressed by the equation given below[15,18]:

$$N_{ca} = \mu V / \phi \; a$$

where μ is the viscosity of the displacing fluid, V is the velocity of the displacing fluid, a is the interfacial tension between the displacing fluid and the oil phase, and ϕ is the formation porosity.

Figure 3 shows the dependence of the residual oil on the capillary number. As indicated in this figure, the amount of residual oil recovered during surfactant flooding increases with the capillary number[15,18]. An increase in the capillary number can be obtained by increasing the displacing fluid velocity or viscosity or by decreasing the IFT between the residual oil and the displacing fluid. However, in practice, high capillary numbers corresponding to very high recovery efficiencies can only be achieved by lowering the IFT. Because of this, surfactants are primarily used in tertiary oil recovery for their IFT-lowering ability. TRI used the IFT-lowering ability as the main screening criterion for the choice of surfactants[26,27]. MTARRI used mixtures of alkali and surfactants because they yielded higher degrees of IFT lowering than surfactant aqueous solutions alone[28,29].

Because of the nature of the displacement mechanism involved, contaminant removal by IFT lowering between displacing fluid and contaminant should be effective to remove contaminant held to the soil pores by capillary forces. This situation is encountered when a contaminant-free phase is trapped in the external pores of a sandy soil[26-29]. Because the soil used in the SAIC study was sandy (95 to 98 percent α-quartz)[7,8,17], it was not surprising that surfactant aqueous solutions primarily formulated to disperse a free-phase contaminant (Murban crude oil) were able to remove virtually all contaminants from the tested soil. It was also not surprising that the wash fluids used in the pilot test conducted by MTARRI were able to remove up to 75 percent of the oil

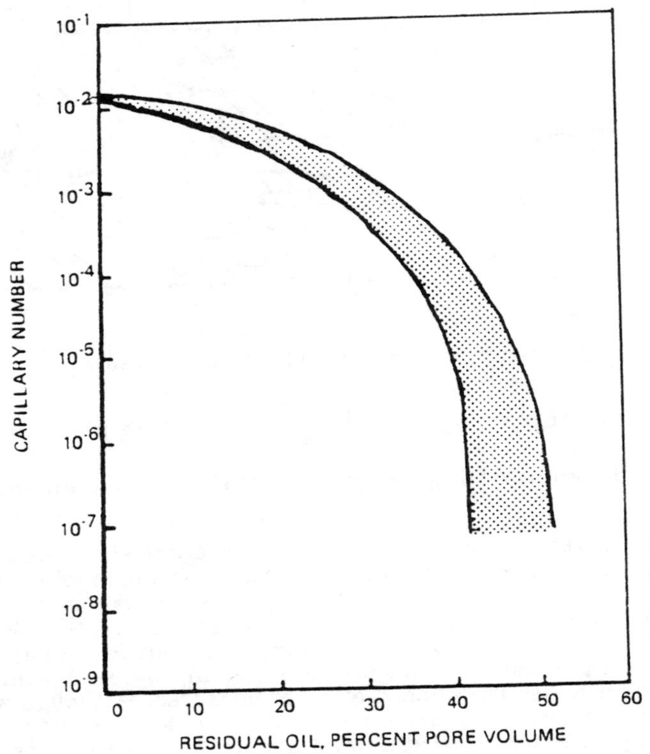

Figure 3. Dependence of Residual Oil Saturation on Capillarity[18]

held to the site by capillary forces[22]. However, for contaminants bound to soil matrices by forces other than capillarity, surfactant aqueous solutions primarily formulated for tertiary oil recovery may not be effective in removing them.

It is known from numerous studies of pesticide sorption in various soil substrates that organic compounds are bound to soil matrices by a variety of physical and chemical forces[13]. For instance, soil organic matter can bind with organic pollutants by hydrophobic bonding, hydrogen bonding, ion exchange, charge transfer, etc.[2]. Because it primarily consists of humic substances--which are polymeric amorphous compounds with varying external dimensions (coiling and stretching of the polymer chains), variable charges, and strong tendencies to form complexes with certain cations[2]--soil organic matter plays an important role in the sorption of a variety of cationic species, including many ionizable hydrophobic organic compounds (e.g., chlorinated phenols)[24]. Soil organic matter also plays an important role in the sorption of a variety of neutral hydrophobic compounds[12] and is the dominant sorbent in soils containing low amounts of clay minerals and amounts of organic carbon greater than 0.1 percent[13,24].

In the field test conducted by Mason & Hanger-Silas Mason, surfactant aqueous solutions failed to remove contaminants from a "real world" site[19,20]. Mason & Hanger-Silas Mason attributed this failure to the bypassing of the contaminated areas by the washing solutions[19]. An alternate explanation could be the ineffectiveness of the surfactant aqueous solutions tested to remove and displace the contaminants from the test site soil because they were bound to its organic fraction. Sampling studies conducted prior to the field test showed that soil samples from the "site of opportunity" contained low amounts of clay minerals and appreciable amounts of organic carbon, up to 1.5 percent (14,900 mg/kg) in some cases[19]. Thus, it is possible that there was enough organic matter in the subsurface soil to bind with the contaminants in such a way that surfactant aqueous solutions formulated for their IFT-lowering and oil dispersion ability would not have been able to remove and displace the contaminants from the subsurface soil.

The foregoing suggests that past work on surfactant-assisted in-situ soil washing did not address the removal of organic pollutants bound to subsurface soil matrices by forces other than capillarity. Only sandy soils with very low amounts of organic matter and clay minerals were tested in the past laboratory experiments. Because of this, pollutant emplacement in the tested soils was primarily due to capillary forces, and as a result, surfactant aqueous solutions for contaminant removal experiments were mainly chosen based on their abilities to lower the IFT between the displacing fluid (surfactant aqueous solutions) and the free-phase contaminants. Other properties of surfactant aqueous solutions such as solubilization and detergency, which are deemed necessary for the removal of unwanted substances from a solid surface[5,23], were not given much consideration.

Because soil surfaces can support numerous physical, chemical, and biological reactions[13], forces other than capillarity can play a significant role in pollutant emplacement in contaminated subsurface soils. It is even possible that these forces are the limiting forces that must be overcome to remove bound organic pollutants from contaminated subsurface soils. Past work jumped from laboratory tests to field tests without addressing the removal of organic pollutants bound to subsurface soil matrices by forces other than capillarity. This approach to process development was rather unusual and the failure of surfactant aqueous solutions to remove contaminants from the field--despite their excellent removal efficiencies (over 90 percent) during laboratory tests--can be partly attributed to this approach. Accordingly, it is essential to address the removal of organic pollutants bound to subsurface soil matrices by forces other than capillarity in order to assess adequately the applicability and limitations of surfactant-assisted in-situ soil washing. This assessment is particularly needed to determine which type of remediation problem is suitable for surfactant-assisted in-situ soil washing.

SUMMARY AND FUTURE WORK

A review of the current state-of-the-art of surfactant-assisted in-situ washing of soils contaminated with hazardous organic compounds was conducted. The review included a detailed description of past work on surfactant-assisted in-situ soil washing, along with an evaluation of the applicability and limitations of its results. This evaluation indicated that the past work on surfactant-assisted in-situ soil washing was primarily an extension of tertiary oil recovery technology. As a result of this, only the removal of NAPL contaminants held to the pores of subsurface soils by capillary forces was addressed. Because forces other than capillarity, e.g., sorption, can play a significant role in pollutant emplacement in subsurface soils, it is critical to address the removal of pollutants bound by such forces in order to assess adequately the applicability of surfactant-assisted in-situ soil washing. Research is currently under way (a) to assess the applicability and limitations of tertiary oil recovery based in-situ soil washing, (b) to determine the extent of surfactant-assisted desorption of sorbed organic pollutants, and (c) to test a new approach to surfactant formulations based on the detergency and solubilizing power of surfactant aqueous solutions[3]. The results of the overall research will help to determine which type of contamination problem is suitable for surfactant-assisted in-situ soil washing.

ACKNOWLEDGMENT

This work was partially supported by Los Alamos National Laboratory under Contract No. 9-X58-8080U-1.

REFERENCES

1. Abriola, L. M. and Pinder, G. F., *Water Res. Res.*, 1985, *21* (1), 11-18.

2. Bolt, G. H. and Bruggenwert, M.G.M., in *Soil Chemistry Basic Elements*, G. H. Bolt and M.G.M. Bruggenwert, eds., Developments in Soil Science 5A, Elsevier Scientific Publishing Company, New York, 1976, Chapter 1.

3. Chawla, R. C., Diallo, M. S., Cannon, J. N., and Johnson, J. H., "In-Situ Treatment of Soils Contaminated With Hazardous Organic Compounds Using Surfactants: Possibilities, Problems, and Research," Report to the Los Alamos National Laboratory for Contract No. 9-X58-8080U-1, December 1988.

4. Chou, S. J., Griffin, R. A., and Chou, M. M., in *Proceedings of the Sixth Annual Research Symposium*, Cincinnati, OH, 1980, EPA/600/9-80-010, 549.

5. Cutler, W. G. and Davis, R. C., in *Detergency: Theory and Practice*, W. G. Cutler and R. C. Davis, eds., Surfactant Science Series, Volume 5, Marcel Dekker, New York, 1987.

6. Dragun, J. and Kuffner, A. C., *Chem. Eng.*, 1984, *91* (24), 65-70.

7. Ellis, W. D., Payne, J. R., Tafuri, A. N., and Freestone, F. J., in *Proceedings of Hazardous Material Spills Conference*, Nashville, TN, 1984, 116-124.

8. Ellis, W. D., Payne, J. R., and McNabb, G. D., "Treatment of Contaminated Soils with Aqueous Surfactants," Final Report EPA/600/2-85/129, 1985, Hazardous Waste Engineering Research Laboratory, Office of Research and Development, U.S. EPA, Cincinnati, OH, NTIS, Springfield, VA, PB86-122561, 1986.

9. Griffin, R. A. and Chou, S. J., in *Proceedings of the Sixth Annual Research Symposium*, Cincinnati, OH, 1980, EPA/600/9-80-010, 291.

10. Hunt, J. R., Sitar, N., and Udell, K. S., *Water Res. Res.*, 1988, *24* (8), 1247-1258.

11. Hunt, J. R., Sitar, N., and Udell, K. S., *Water Res. Res.*, 1988, *24* (8), 1259-1269.

12. Karickhoff, S. W., *J. Hydraulic Eng.*, 1984, *110* (6), 707-733.

13. Khan, S. U., *Pesticides in the Soil Environment*, Elsevier Scientific Publishing Company, New York, 1980, Chapter 2.

14. Lindorff, D. E., *Ground Water*, 1979, *17* (1), 9-12.

15. Ling, T. F. and Shah, D. O., in *Industrial Applications of Surfactants*, Special Publications of The Royal Society of Chemistry, No. 29, The Royal Society of Chemistry, Burlington House, London, England, 1986.

16. Malmanis, E., Fuerst, D. W., and Pineiwski, R. J., in *Proceedings of the 6th National RCRA/Superfund Conference and Exhibition*, New Orleans, LA, 1989, 538-541.

17. McNabb, Jr., G. D., Payne, J. R., Ellis, W. D., Kirstein, B. E., Evans, J. S., Harkins, P., and Rotunda, N. P., "Chemical Countermeasures Application at Volk Field Site of Opportunity," Internal Report to EPA, 1985.

18. Miller, C. A. and Qutubuddin, S., in *Interfacial Phenomena in Apolar Media*, H. F. Eicke and D. Parfitt, eds., Surfactant Science Series, Volume 21, Marcel Dekker, New York, 1987, 117-185.
Figures 2 and 3 are reprinted from Ref. (18)
by courtesy of Marcel Dekker, Inc.

19. Nash, J. H., "Field Studies of In-Situ Soil Washing," Final Report to EPA on Contract No. 68-03-3203, 1986.

20. Nash, J. and Traver, R. P., in *Proceedings of Hazardous Material Spills Conference*, Nashville, TN, 1984.

21. *Pesticides in Groundwater: Background Document*, U.S. EPA, Office of Ground Water Protection, EPA 440/6-86-002, 1986.

22. Pouska, G. A., Trost, P. B., and Day, M., in *Proceedings of the 6th National RCRA/Superfund Conference and Exhibition*, New Orleans, LA, 1989, 423-430.

23. Rosen, M. J., *Surfactants and Interfacial Phenomena*, Wiley Interscience Publication, John Wiley and Sons, New York, 1978.

24. Schwarzenbach, R. P. and Westall, J., "Sorption of Hydrophobic Trace Organic Compounds in Groundwater Systems," *Water Sci. Technology*, 1985, *17* (9), 39-55.

25. *Technology Transfer: A Compendium of Technologies Used in the Treatment of Hazardous Wastes*, Center for Environmental Research Information, Cincinnati, OH, EPA/625/8-87/014, 1987.

26. Texas Research Institute, Inc., Final Report 7743 (1-3)-F, API Publication 4317, 1979.

27. Texas Research Institute, Inc., Final Report API Publication 4390, 1985.

28. Union Pacific Railroad: Milestone Report I, Volume I, DE/UPRR15/052, 1987.

29. Union Pacific Railroad: Milestone Report I, Volume 3/4, DE/UPRR16/018, 1987.

30. Vanloocke, R., DeBorger, R., Voets, J. P., and Venstraete, W., *Int. J. Environ. Stud.*, 1975, *8* (2), 99-111.

ELECTROACOUSTIC SOIL DECONTAMINATION PROCESS FOR IN-SITU TREATMENT OF CONTAMINATED SOILS

R. E. Hinchee, H. S. Muralidhara, F. B. Stulen,
G. B. Wickramanayake, and B. F. Jirjis
Battelle
505 King Avenue
Columbus, Ohio 43201-2693
(614) 424-4698

INTRODUCTION

Several sites are contaminated with nonaqueous phase liquids (NAPL) and heavy metals in the United States[23]. The U.S. Environmental Protection Agency (EPA) has estimated that 189,000 underground storage tanks are leaking at retail fuel outlets alone. The total number of leaking underground tanks is substantially higher than this, and NAPL contamination in the form of coal tars and petroleum sludges from the aboveground tanks is a significant problem. Following an NAPL spill or release, the liquid typically migrates to the water table, where it spreads out and floats since it is lighter than water. In a typical cleanup, the first phase is to recover the free-phase "floating" NAPL. The fraction of spill that is recoverable by conventional technology is very low, and residual contamination following drainage of this recoverable NAPL is very high, often in the percent range[12]. Heavy metals may enter the environment in many ways. At contaminated sites heavy metals may be found at depths ranging from surficial to hundreds of feet below the land surface.

This paper describes a novel separation process called electroacoustic soil decontamination (ESD) for in-situ removal of organics and heavy metals from contaminated soils. The following pages consist of technical background of the process mechanisms and the details of preliminary experimental study for the Superfund Innovative Technology Evaluation (SITE) program funded by the EPA.

BACKGROUND

The electroacoustic soil decontamination (ESD) process is based on application of a dc electric field and an acoustic field to contaminated soils to increase the transport of liquids through the soils. Figure 1 illustrates the operating principle of the process. Electrodes, one or more anodes and a

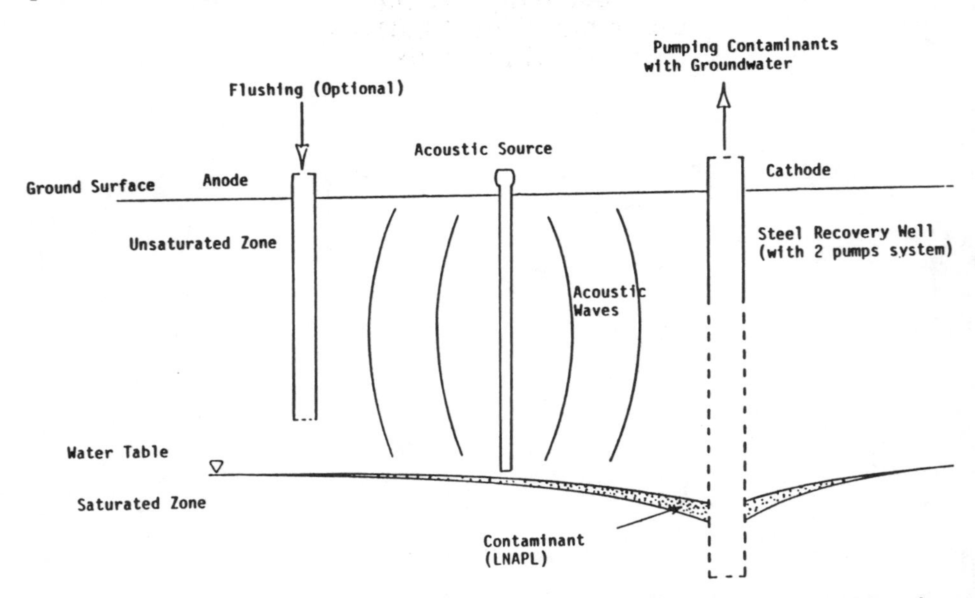

Figure 1. Conceptual Layout of Electroacoustic Soil Decontamination

cathode, and an acoustic source are placed in a contaminated soil to apply the electric and acoustic fields. The process is expected to be most effective for clay-type soils with small pores or capillaries in which hydraulic permeability is very low.

The dominant mechanism of the enhanced flow is electroosmosis. In-situ electroosmosis was first applied successfully to soils by Casagrande in the 1930s in Germany for dewatering and stabilizing soils(6,7). Recently, Muralidhara and coworkers at Battelle have discovered that the simultaneous application of an electric field and an acoustic field produces a synergistic effect and results in further enhancement of water transport(3,8,15-22). The Battelle process is called electroacoustic dewatering (EAD). Currently, Battelle is engaged actively in the development and commercialization of the EAD process for a variety of industrial and wastewater treatment applications.

Based on Battelle's research and development experience in the application of electric and acoustic fields to dewatering and proven soil dewatering technology by electroosmosis, Battelle is applying the principles of electroacoustic dewatering technology to in-situ soil decontamination. Background information on theories and operating principles is provided in the following sections.

ELECTRICALLY ENHANCED SEPARATION

Electroosmosis in porous media, such as clays, works by inducing an electrical double layer of negative and positive ions formed at the solid/liquid interface. For soil particles, the double layer consists of a fixed layer of negative ions that is held firmly to the solid phase and a diffuse layer of positive ions that is held more loosely. Application of an electric potential on the double layer results in the displacement of the two layers to respective electrodes, i.e., the positively charged layer to the cathode and the negatively charged layer to the anode.

Since the particles in the soils are immobile, the fixed layer of the negative ions is unable to move. However, the diffuse layer containing positive ions can move and drag water along with it to the cathode. This is the basic mechanism of electroosmotic transport of water through wet soils under the influence of an applied electric potential.

Some noteworthy examples of the prior work on soil leaching, consolidation, and dewatering by electroosmosis are summarized in Table 1. Numerous patents also have been issued on various applications of electric field for enhanced recovery of crude oil(1,2,5,9-11,14). The examples demonstrate the feasibility and practicality of electroosmosis in large-scale applications. The

Table 1. Applications of Electroosmosis in Soil Leaching, Consolidation, and Dewatering

	Application	Investigators	Scale of Operation	Voltage and Current	Results and Comments
1.	Leaching of Cr from soils	Banarjee	Laboratory	0.1 to 1.0 V/cm	• Obtained increased leaching rate with electric field
2.	Leaching of Cr from soils	Horng et al.	Laboratory and field	NA	• Obtained increased leaching rate with electric field; determined effect of anode materials
3.	Crude oil production	Anbah et al.	Laboratory	NA	• Obtained increased flow of oil-water mixture through porous media; determined beneficial effect of a small addition of electrolyte to kerosene to obtain increased electroosmotic flow

Table 1. Applications of Electroosmosis in Soil Leaching, Consolidation, and Dewatering (Continued)

	Application	Investigators	Scale of Operation	Voltage and Current	Results and Comments
4.	Soil dewatering (Salzgitter, Germany)	Casagrande	Field	180 V 9.5 A/well	• Electrodes placed 22.5 ft deep and 15-ft apart; flow rate increased by a factor of 150 from 10 gal/day/-well without electric field to 1500 gal/day/well with electric field; energy was 0.30 kwh/gal.
5.	Soil dewatering (Trondheim, Norway)	Casagrande	Field	40V 26 A/well	• Electrodes placed 60-ft deep and 15-ft apart; flow rate increased from 6-300 gal/day/well to 70-3040 gal/day/well; energy usage was 0.30 kwh/gal.

reported electrical energy consumption in the range of 0.3 to 0.4 kWh/gal is low and should be acceptable for soil decontamination applications ($0.015/gal to $0.020/gal power cost). The examples of metal leaching, oil recovery, and in particular Casagrande's work on soil dewatering clearly indicate that the application of an electric field has been demonstrated successfully at a large enough scale to suggest the real possibility of Battelle's ESD technology performing adequately at pilot-scale, and eventually full-scale, levels for soil decontamination.

ACOUSTICALLY ENHANCED SEPARATION

An acoustic field is one in which the pressure and particle velocity vary as a function of time and position. These pressure fluctuations form a wave that propagates from the source throughout the medium. Sinusoidal pressure fluctuations are characterized by their pressure amplitude and frequency. A particle velocity is imparted also to the medium by the action of the pressure wave, which also varies as a function of time, frequency, and position. Acoustic pressure and particle velocities are related through the acoustic impedance of the medium.

The pressure fluctuations are the result of the transmission of mechanical energy that can perform useful work to bring about desired effects. The type and magnitude of these effects depend on the medium. In acoustic leaching, the many forces that can contribute to the overall effectiveness include:

- Orthokinetic forces, which cause small particles to agglomerate
- Bernoulli's force, which causes larger particles to agglomerate
- Rectified diffusion, which causes gas bubbles to grow inside capillaries and thereby expel entrapped liquids
- "Rectified" Stokes' force, which causes an apparent viscosity to vary nonlinearly and forces the particle towards the source
- Decreased apparent viscosity, which may result from high strain rates in a thixotropic medium or localized heating, which in turn lowers both the viscosity and the driving force to move particles
- Radiation pressure, which is a second-order effect, static pressure that adds to the normal pressure differential.

To introduce high-energy acoustic signals into the ground, one must address the issues of elastic wave propagation in solids. The earth, for the purposes of in-situ leaching, can be treated as a semi-infinite half-space where the earth's surface is the boundary of the half-space. It is well known that a source acting normal to and on the surface not only produces acoustic waves (more properly referred to as compression waves in this case), but also two additional waves as well. These are shear waves, where particle velocity is perpendicular to the direction of propagation, and surface waves. Surface waves exist at the boundary; extend into the medium a given depth, which is inversely proportional to the wavelength; and produce elliptical particle motions.

Thus, the energy into the source is partitioned into these three types of waves with roughly 10 percent going into compression, 25 percent into shear, and 65 percent into surface waves. Likewise, as the signal propagates from the source, the intensity of the compression and shear waves decreases as the inverse of distance squared because they are propagating in the bulk of the material. Since the surface waves propagate beneath the surface material, their intensity only falls off as the inverse of the square root of distance. In addition, all three waves will be further reduced by soil attenuation. Soil attenuation generally increases by the square root of frequency. Therefore, lower frequency waves will propagate (i.e., penetrate) much farther.

COMBINED ELECTROACOUSTIC SEPARATION

Acoustics, when properly applied in conjunction with electroseparation and water flow, enhances dewatering or leaching. The phenomena that augment dewatering when using the combined technique are not understood fully. However, we have developed some hypotheses about possible mechanisms.

It is theorized that, in the presence of a continuum of liquid phase, the acoustic phenomena (e.g., inertial and cavitation forces) that separate the liquid from the solid into the continuum are facilitated by the electric field and a pressure differential to enhance dewatering by means of one or more of the electroseparation phenomena. There is also evidence of synergistic effects of the combined approach. In addition, as the cake densifies (by sequestration and electroosmosis), the liquid continuum normally would be lost, but it is believed that by channelling, on a macroscale, acoustic energy delays the loss of the continuum, making additional leaching possible. It is the carefully executed combination of techniques to augment the overall solid/liquid separation process that is the essence of Battelle's current EAD process. And because of this combined effect, EAD has been more effective than either electroseparation or acoustically enhanced separation. The same effectiveness is expected for ESD.

The combined effect of superimposing an acoustic field with an electric field for dewatering applications is discussed elsewhere[17].

Additional electroacoustic dewatering tests performed at Battelle on hazardous wastes are summarized in Table 2. The results indicate the effectiveness of applying electric and acoustic fields for removing heavy metals and oils from wet soils. Specifically, we note that ESD was effective for a sandy soil even though we fully expect that the process will be even more effective for clay-type soils. Each of these summary results (Table 2) has been validated by additional experiments and is fully consistent with Battelle's extensive experience with electroacoustic dewatering.

Besides electroosmosis, passage of dc current through a wet soil also produces other effects, such as ion exchange, development of pH gradients, electrolysis, gas generation, oxidation and reduction, and heat generation. It is conceivable that the heavy metals present in contaminated soils can be precipitated out of solution by electrolysis, oxidation and reduction reactions, or ionic migration. The contaminants in the soil may be cations, such as Cd^{++}, Cr^{+++}, and Pb^{++++}, or anions, such as $(CN)^-$, $(CrO_4)^{--}$, and $(Cr_2O_7)^{--}$.

The existence of these ions in their respective states depends upon the local pH and concentration gradients existing in the soil systems. Application of an electric field is expected to increase the leaching rate and precipitate the respective heavy metals out of solution by establishing appropriate pH and osmotic gradients. For example, $CdCl_2$ in solution can be precipitated out as $Cd(OH)_2$ at the cathode due to the generation of $(OH)^-$ ion at the cathode and swept out of the ground and separated by conventional techniques.

Table 2. Electroacoustic Dewatering and Decontamination of Hazardous Wastes

	Application	Results
1.	Dewatering and deoiling of petroleum sludge	Reduced oil content on the sludge from 25 percent to 4 percent.
2.	Leaching of heavy metals from electroplating sludge	Obtained up to 10-fold increase in Cu and Hg concentrations in the leachate.
3.	Removal of jet fuel from a sandy soil	Application of 100 V DC voltage across a ½-inch cake of as-received soil sample initially containing 10 to 11 percent moisture (water was added to give a moisture level of approximately 40 percent) resulted in substantial oil removal by application of electric and acoustic fields. The residual moisture content of the sandy soil was only about 2 percent.

EXPERIMENTAL PROGRAM

The experimental study described here was in part supported by a grant from the EPA, a Superfund Innovative Technologies program.

OBJECTIVE

The objective of the proposed program is to develop the in-situ electroacoustic leaching process, ESD, for decontaminating hazardous wastes. The goals of the proposed effort are to demonstrate the potential feasibility of the ESD process to

- Decontaminate soils containing hazardous organics in situ by the application of dc electrical and acoustic fields
- Decontaminate soils containing heavy metals by the application of dc and acoustic fields.

PARAMETRIC INVESTIGATION

A clay soil was contaminated with n-decane (organic) and zinc chloride (inorganic), which was used during all experimental trials. A typical laboratory test cell is shown in Figure 2. These compounds were chosen as surrogates for NAPL and heavy metals, respectively. Some of the experimental variables evaluated in this development program included applied voltage, treatment time, acoustic intensity, acoustic frequency, and contaminant type. A discussion of the significant results follows.

DECANE

Batch ESD tests were conducted on a decane-contaminated soil sample. The soil sample was spiked with 8 percent decane (dry weight). Tap

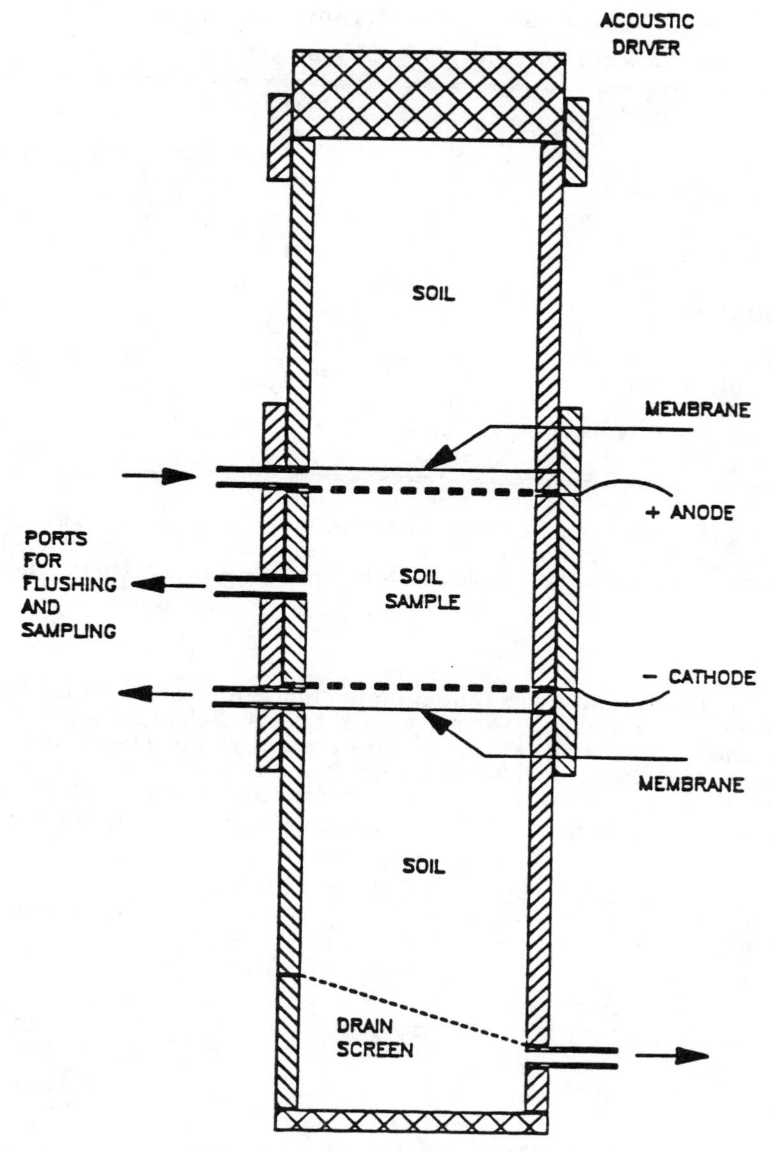

Figure 2. Laboratory Test Cell

water was added to moisten the soil. The final soil had the following approximate composition (by weight percent):

Soil solids	52.7
Decane	8.0
Water	39.3

The result was a soil saturated to just below its field capacity, i.e., it would not drain by gravity.

Table 3 summarizes the results of selected testing. It was found that the cell configuration utilized resulted in very little removal of free decane so the decision was made to divide the cake into layers for analysis. The total cake thickness was 2.5 in.; sample A was the upper (adjacent to the anode) 0.5 in., sample B was the center 0.5 in., and sample C was the lower (adjacent to the

Table 3. Results of Decane Testing

Concentrations are in %, dry weight basis

	Test			
Sample	26	27	28	30
Initial	8.0	8.0	8.0	8.0
A	5.6	6.0	6.1	6.0
B	6.1	6.8	6.2	5.1
C	6.3	7.0	5.6	5.8
Test Conditions				
Duration (hr)	2.0	2.0	2.0	2.0
Voltage (volts/inch)	37.5	45.0	25.0	37.5
Current (amp)	0.13	0.11	0.10	0.11
A continual power (watts)	0	0	0	0.697
Frequency (Hz)	--	--	--	400

cathode) 0.5 in. Some samples were analyzed by EPA in their laboratory and some by a commercial laboratory. Table 3 summarizes the results of the EPA's analysis of samples from those tests in which the cake was divided into layers.

ZINC

The soil sample was contaminated inorganically in the lab by spiking 2,000 mg/kg of zinc ($ZnCl_2$) on a dry weight basis. Before adding $ZnCl_2$, the soil was dried and ground. The moisture content of the prepared soil was increased to 42.2 percent. The prepared soil was sealed in an aluminum container and stored in a cooler. The results of the completed tests are shown in Table 4. Results confirm that ESD moved the zinc from the layer in contact

Table 4. Percent S.D. Zinc Data Analysis

Test[a]	Distance Between Electrodes, in.	Test Time, hr	Voltage, V/in.	Average Current, mAmps	Acoustic Power, W	Initial Zinc Concentration, mg/kg	Final Zinc Concentration, mg/kg	Anode[b] Flushing Solution, PN	Leachate[c], PN	Comment
3%	2.0	1.25	10	289	0	2,000	1,697	--	--	The whole cake was mixed and sent for analysis
4%	2.0	1.25	20	603	0	2,000	1,368	--	--	The whole cake was mixed and sent for analysis
5%1	8.0	191.3	2.5-12.5	7.5	0	2,000	2,110 2,168 (R)	--	--	Cake close to cathode (1/2 cake)
5%2	8.0	191.3	2.5-12.5	7.5	0	2,000	384 381 (R)	--	--	Cake close to anode (1/2 cake)
6%1	8.0	191.3	2.5-12.5	7.5	0	2,000	213 203 (R)	--	--	Cake close to anode (1/2 cake)
6%2	8.0	191.3	2.5-12.5	7.5	0	2,000	1,822 1,934 (R)	--	--	Cake close to cathode (1/2 cake)
7%A	4.5	50	1.4-4.3	50	0	2,000	180.4	3.56	11.65	Cake in contact with anode
7%D	4.5	50	1.4-4.3	50	0	2,000	687.1	3.56	11.65	Cake in contact with anode layer
7%C	4.5	50	1.4-4.3	50	0	2,000	1,846.8	3.56	11.65	Cake in contact with cathode layer
7%B	4.5	50	1.4-4.3	50	0	2,000	5,644.3	3.56	11.65	Cake in contact with cathode
8%A	4.5	25	1.3-5.3	50	0	2,000	818	3.82	11.00	Cake in contact with anode

Table 4. Percent S.D. Zinc Data Analysis (Continued)

Test[a]	Distance Between Elec- trodes, in.	Test Time, hr	Voltage, V/in.	Average Cur- rent, mAmps	Acous- tic Power, W	Initial Zinc Concen- tration, mg/kg	Final Zinc Concen- tration, mg/kg	Anode[b] Flushing Solu- tion, PN	Leach- ate[c], PN	Comment
8%D	4.5	25	1.3-5.3	50	0	2,000	1,542	3.62	11.00	Cake in contact with anode layer
8%C	4.5	25	1.3-5.3	50	0	2,000	2,066	3.62	11.00	Cake in contact with cathode layer
8%B	4.5	25	1.3-5.3	50	0	2,000	3,214	3.62	11.00	Cake in contact with cathode
9%A	4.5	100	0.3-20	50	0	2,000	118.8	3.37	11.25	Cake in contact with anode
9%D	4.5	100	0.3-20	50	0	2,000	174.7	3.37	11.25	Cake in contact with anode layer
9%C	4.5	100	0.3-20	50	0	2,000	204.6	3.37	11.25	Cake in contact with cathode layer
9%B	4.5	100	0.3-20	50	0	2,000	6,341	3.37	11.25	Cake in contact with cathode

(a) Test was performed in a graduated cylinder designed for flushing purposes.
(b) PN of flushing water solution at the end of test.
(c) PN of leachate at the end of test.

with the anode toward the layer in contact with the cathode. The initial tests 3Z and 4Z were performed using the same cell as that used for the decane tests. Samples of treated soil from the tests were sent for lab analysis. The soil samples consisted of cake in contact with anode as well as the cake in contact with cathode. The anode layer where the zinc was removed was mixed with the cathode layer where the zinc was accumulated, and hence, lab analysis did not show any zinc removal.

Zinc is soluble below pH 6. Above pH 6, zinc would exist as $Zn(OH)_2$ and precipitate in the soil. Zinc accumulated around the cathode in the form of $Zn(OH)_2$, since the soil around the cathode was basic (pH value of 9 to 11). Hence, it was decided to divide the ESD-treated soil into four 1-in. sections:

- Soil in contact with anode
- Soil in contact with anode layer
- Soil in contact with cathode
- Soil in contact with cathode layer.

Further, the test cell was modified so that the soil layer near the anode could be flushed with water.

A series of tests (7Z, 8Z, and 9Z) were conducted using the modified cell. In these tests, the treatment time was varied at constant applied electrical power. Acoustic power was not used during these tests. Results indicate about 90 to 95 percent zinc removal from more than 75 percent of the cake tested. The longer the ESD time, the higher zinc removal in the 75 percent of cake tested and more zinc accumulation at the cathode layer (1 in.). Results of the above tests are plotted in Figures 3 and 4. Figure 3 shows zinc concentration as a function of cake thickness where position 0 cake thickness is the cake in contact with anode and position 4 is cake in contact with cathode. Figure 4 shows percent zinc removed over the same thickness. Figures 3 and 4 show that a 25-hour test was not enough time to remove zinc from all layers. The first layer in contact with the anode had 818 mg/kg dry weight. This means 60 percent of zinc was removed from that layer. However, the second layer (layer D) shows only less than 25 percent removal. The 50-hour test shows 90 percent removal in the layer in contact with the anode (layer A); the next layer (layer B) shows 66 percent of the zinc removed. In the 100-hour test, layers A, D, and C show more than 90 percent zinc removal. In all of the three tests (7Z, 8Z, and 9Z), zinc accumulated at the cathode layer (layer B). The thickness of layer B is 1 in. Since there was zinc concentration reduction in layers A, D, and C, it is possible that layer B has a concentration reduction gradient, too. This means

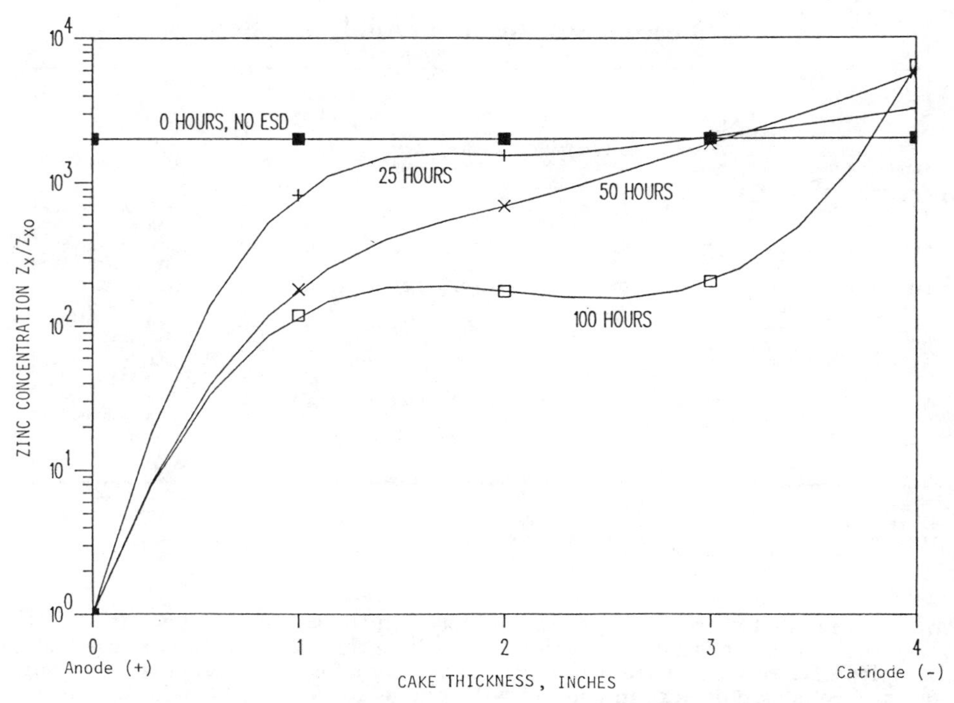

Figure 3. Zinc Concentration as a Function of Cake Thickness for Tests 7Z, 8Z, and 9Z

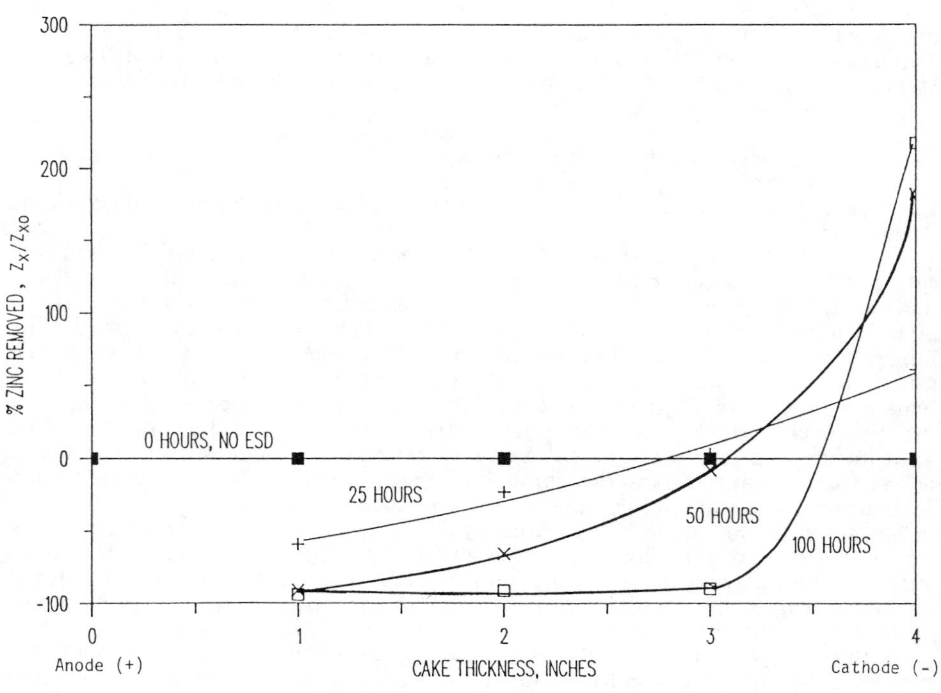

Figure 4. Zinc Removal as a Function of Cake Thickness for Tests 7Z, 8Z, and 9Z

that zinc could have accumulated in the last 0.25 in. in contact with the cathode. If this is true, then zinc was removed over 90 percent of the cake. However, more tests are needed to validate the above assumption. This batch testing analysis indicates that there is a very significant opportunity to achieve heavy metals removal by the ESD process.

TECHNICAL PERFORMANCE OF ESD WITH OTHER IN-SITU TECHNOLOGIES

Based upon the results of this limited study, it is not possible to make a direct quantitative comparison of the ESD technology to other technologies; however, a qualitative comparison is possible. Table 5 summarizes these comparisons.

ORGANICS TREATMENT

The most likely ESD application for treatment of organics is to enhance the recovery of such NAPLs as solvents and fuel oils. Another possible application is to enhance recovery of more soluble polar organics. This applica-

Table 5. Comparison of ESD with Other In-Situ Technologies

Technology	Status	Cost	Limitations
Organic Treatment			
ESD	Early Bench Scale	Low?	Unproven
Pump and Trent	Commercially Available	Low, initial cost but potentially high life cycle cost.	Never ending, limited to the saturated zone.
Soil Venting	Commercially Available	Low (without air treatment) Moderate (with air treatment)	Limited to volatiles in the vadose zone.
Heat Enhanced Soil Venting	Limited Commercial Availability	Moderate - High	Limited to semi-volatiles in the vadose zone.
Steam Injection	Limited Commercial Availability	Moderate - High?	Limited field experience.
RF Heating	Pilot Scale	Moderate - High	Limited field experience.
Direct Current Heating	Bench/Pilot Scale	Moderate - High	Limited field experience.
In-Situ Vitrification	Commercially Available	Highest	Very high temperatures and energy cost.
In-Situ Biodegradation	Limited Commercial Availability	Low - High	Not fully proven, limited to biodegradable compounds.
Inorganics Treatment			
ESD	Bench Scale	Low?	Unproven.
Direct Current	Pilot Scale	Low?	Unproven.
Pump and Trent	Commercially Available	Low initial cost but potentially high life cycle cost.	Never ending, limited to saturate zone.
In-Situ Vitrification	Commercially Available	High	Stabilizes metals inplace, rather than removing them.

tion would be more like the metals treatment. ESD has the potential to reduce NAPL concentrations from at or near saturation levels (approximately 5,000 to 50,000 mg/kg) to below saturation (approximately 100 to 1,000 mg/kg), but most probably not to low mg/kg or mg/kg levels. This discussion will focus on the potential for increased NAPL recovery.

Conventional technology for NAPL recovery consists of some form of groundwater and/or NAPL pumping followed by NAPL separation and/or water treatment. This technology typically can succeed in controlling groundwater and NAPL flow and decreasing the potential for off-site migration. However, success in substantially reducing residual contamination is limited. One limitation of pump-and-treat technology is that conventional NAPL recovery is dependent upon gravity drainage to bring the NAPL into a recovery well or trench for skimming. As water tables move up and down and vadose zone moisture levels change, the fraction of the NAPL in this free-floating phase changes. As a result, an NAPL recovery system may reduce or even remove the measurable NAPL phase, only to have it return under different hydrological conditions.

Under the new Resource Conservation and Recovery Act (RCRA) underground tank regulations (CFR 280.64), the minimum remediation requirements are "free product removal." Achievement of this level of remediation may be difficult using conventional pump-and-treat technology. ESD coupled to a conventional pump-and-treat technology has the potential to reduce (relatively rapidly) the residual NAPL concentrations to levels below those that would result in the free-phase NAPL or "free product" layer.

Soil venting, also known as soil vacuum extraction and in-situ volatilization, is a relatively simple and widely used technology for removing volatile organic compounds from the vadose zone. If off-gas treatment is unnecessary, costs are very low; if treatment is required, costs are moderate. Where off-gas treatment is required, ESD can be less expensive than soil venting and, in some cases, may prove to be a cost-effective pretreatment prior to soil venting. It is unlikely that ESD can achieve residual concentrations as low as those possible with soil venting for volatiles.

Some vendors of soil venting have begun to inject heated air to accelerate the process and extend treatment to less volatile or semivolatile organics. The cost of energy to heat the soils is moderately high depending, of course, upon the targeted temperature. Comparisons to ESD are similar to those discussed above for soil venting.

Injection of steam to treat volatiles, and some less volatile compounds, has been demonstrated on a limited number of sites. Sufficient data are not yet available to evaluate its feasibility fully; however, energy costs are high. Due to the increased heat capacity of wet soils, more heat, and therefore energy, is required than for other soil-heating technologies.

Radio frequency (RF) heating is an emerging technology for in-situ soil heating. Roy F. Weston, the licensed vendor, intends to couple it with soil venting to achieve accelerated remediation. The comparison to ESD would be very similar to those discussed above.

Direct current is being explored as a means of soil heating. As for all technologies that require increased soil temperature, more energy would be required than for ESD.

In-situ vitrification (ISV) is a commercially available technology in which a direct current is applied to the soils to achieve super heating. This results in soils melting to form a vitrified solid. This differs from direct current heating only in that much higher temperatures are achieved and correspondingly higher energy costs are incurred. ISV is applied typically to inorganics; however, limited data suggest that it is applicable to a wide range of organic compounds. The organics probably are either volatilized or oxidized. Due to the high cost, ISV most likely will be utilized only at very high hazard sites where very low cleanup levels are required. ESD alone most likely would not be applicable to these sites.

In-situ biodegradation is a technology that is receiving widespread attention. It has, to date, proven effective at a limited number of sites and for a limited number of compounds. The technology is applicable only to biodegradable organics. As the technology evolves, wider spread application may occur. At some sites, ESD may prove a cost-effective pretreatment prior to application of an in-situ biodegradation technology.

METALS TREATMENT

Using ESD for removal of metal ions is a distinctively different application of the technology than NAPL organics treatment. In this application, ESD may or may not be coupled with a more conventional pump-and-treat technology. ESD has the potential to reduce substantially residual metals concentrations to or below the low mg/kg or mg/kg level. Unlike organics treatment, the number of technologies for the in-situ treatment of metals is relatively limited.

Direct current has been applied to remove metals in situ. The Dutch Geokinetics process is a promising technology with a novel circulating fluid electrode to prevent metals deposition. The direct current technology is a part of the ESD technology; however, by combining electric and acoustic fields, ESD has the potential to improve treatment efficiency.

As discussed for organics treatment, the pump-and-treat technology is potentially successful at hydraulically controlling a plume of contaminated groundwater, but is frequently ineffective at substantially reducing residual soil contamination. ESD has the potential to improve this treatment substantially.

ISV was designed for and is applied typically to inorganic contaminants. Direct current is applied to heat soil to its melting point and vitrify the contaminated soil into an impermeable mass. This technology does not remove the metals, but rather immobilizes them in situ. The technology requires substantially more energy and cost than ESD.

CONCLUSIONS

Based upon this ESD study and Battelle's previous work with electroacoustics, the following conclusions are drawn:

1. The application of combined electric and acoustic fields can result in increased mobilization of oil (decane) and metals (zinc).
2. The limited decane testing makes definitive conclusions about the feasibility of ESD for NAPL treatment impossible; however, further work is suggested and the possibility of developing a feasible technology exists.
3. The zinc testing indicates that in-situ treatment may be feasible. Despite very limited optimization opportunities, zinc removals of up to 95 percent were achieved.

REFERENCES

1. Anbah, S. A., *et al.*, "Application of Electrokinetic Phenomena in Civil Engineering and Petroleum Engineering," *Annuals*, 1965, Volume 118, Art. 14.

2. Banerjee, S., "Electrodecontamination of Chrome-Contaminated Soils," *Land Disposal, Remedial Action, Incineration and Treatment of Hazardous Wastes*, Proc. Thirteenth Annual Research Symposium, July 1987, 193-201.

3. Beard, R. E. and Muralidhara, H. S., "Mechanistic Considerations of Acoustic Dewatering Techniques," Proc. IEEE, Acoustic Symposium, 1985, 1072-1074.

4. Bell, T. G., U.S. Patent No. 2,799,641, 1957.

5. Bell, C. W. and Titus, C. H., U.S. Patent No. 3,782,465, 1974.

6. Casagrande, L., "Review of Past and Current Work in Electroosmotic Stabilization of Soils," *Harvard Soil Mechanics Series*, 1957, No. 1, 45.

7. Casagrande, L., "Electroosmosis and Related Phenomena," *Harvard Soil Mechanics Series*, 1962, No. 66.

8. Chauhan, S. P., Muralidhara, H. S., and Kim, B. C., "Electroacoustic Dewatering of POTW Sludges," Proc. National Conf. on Municipal Treatment Plant Sludge Management, Orlando, FL, May 28-30, 1986.

9. Faris, S. R., U.S. Patent No. 3,417,823, 1968.

10. Gill, W. G., U.S. Patent No. 3,642,066, 1972.

11. Horng, J. J., Banerjee, S., and Hermann, J. G., "Evaluating Electrokinetics as a Remedial Action Technique," Second International Conf. on New Frontiers for Hazardous Waste Treatment, Pittsburgh, PA, September 27-30, 1987.

12. Houy, G. E. and Marley, M. C., "Gasoline Residual Saturation in Uniform Aquifer Materials," *J. Env. Eng.*, ASCE, 1986, *112* (3), 586-604.

13. Hunter, C. J., *Zeta Potential in Colloid Science Principles, and Applications*, Academic Press, 1981.

14. Kermabon, A. J., U.S. Patent No. 4,466,484, 1984.

15. Muralidhara, H. S., ed., *Recent Advances in Solid-Liquid Separation*, Battelle Press, Columbus, OH, November 1986.

16. Muralidhara, H. S., "Recent Developments in Solid-Liquid Separation," presented at the Trilateral Particuology Conference in Peking, China, September 1988.

17. Muralidhara, H. S. and Ensminger, D., "Acoustic Dewatering and Drying: State-of-the-Art Review," *Proceedings IV*, International Drying Technology Symposium, Kyoto, Japan, 1984.

18. Muralidhara, H. S. and Senapati, N., "A Novel Method of Dewatering Fine Particle Slurries," presented at International Fine Particle Society Conference, Orlando, FL, 1984.

19. Muralidhara, H. S., Senapati, N., and Parekh, B. K., "Solid-Liquid Separation Process for Fine Particle Suspensions by an Electric and Ultrasonic Field," U.S. Patent 4,561,953, December 1985c.

20. Muralidhara, H. S., *et al.*, "Battelle's Dewatering Process for Dewatering Lignite Slurries," Battelle Phase I Report to UND Energy Research Center/EPRI, 1985a.

21. Muralidhara, H. S., *et al.*, "A Novel Electro Acoustic Process for Separation of Fine Particle Suspensions," in *Advances in Solid-Liquid Separation*, H. S. Muralidhara, ed., 1986, Chapter 13, 374.

22. Senapati, N., Muralidhara, H. S., and Beard, R. E., "Ultrasonic Interactions in Electro-Acoustic Dewatering," presented at British Sugar Technical Conference, Norwich, UK, June 1988.

23. U.S. EPA, *1986 Underground Motor Fuel Storage Tanks: A National Survey*, Volume 1, U.S. EPA Technical Report 560/5-86-013, Washington, DC, 1986.

IN-SITU HOT AIR/STEAM EXTRACTION OF VOLATILE ORGANIC COMPOUNDS

P. N. La Mori
Senior Vice President and Technical Director
Toxic Treatments (USA) Inc.
151 Union Street, Suite 207
San Francisco, California 94111
(415) 391-2113

ABSTRACT

This paper presents the first detailed results on using a new technology for the in-situ removal of volatile organic compounds (VOCs) from soil. The technology injects hot air/steam into soil through two 5-ft-diameter counter rotating blades that are attached to drill stems. The soil is cleaned as the drill stems are advanced into and retracted from the ground. The evolved gases are trapped in a surface covering, and contaminants are captured aboveground via condensation and carbon absorption. The recovered contaminants can be reprocessed or destroyed.

The results of the initial use of the commercial prototype show 95 to 99 percent removals of chlorinated VOCs from clay soils. The technology was able to achieve its goal of less than 100 ppm over 80 percent of the time in a blind test. The mass of material recovered in the condensate was conservative (89 percent) of the mass determined removed from the soil by chemical analysis and physical measurements. The technology also removed significant quantities (80 percent) of semivolatile hydrocarbons (as quantified by EPA 8270). This was unexpected. Subsequent analysis has identified potential mechanisms for removal of semivolatile components. Further testing is planned to evaluate these mechanisms.

The process has not been found to cause undesirable environmental emissions as a result of its operation. Noise and air emissions during operation are below the limits set by regional environmental regulations in southern California. Soil hydrocarbon emissions during treatment are not increased from background before treatment. Soil adjacent to the treatment has not been found to have increased VOCs after the process is operated. Environmental emissions caused by operational problems can generally be cured without shutdown.

INTRODUCTION

This paper describes in detail the initial results on evaluating and proving a new technology to remove organic compounds from soil without excavation. The process removes volatile organic compounds (VOCs) from soil by actively mixing hot air and steam into the soil and capturing the evolved hydrocarbons. The technology is described in sufficient detail to understand its basic physical and operational principles. The remainder of the paper describes in detail all the field results to date (mid-1989) of a test program to evaluate the use of the technology to remove chlorinated VOCs plus some nonchlorinated hydrocarbons from a former tank storage facility in California. The unit is called "The Detoxifier" and is operated by Toxic Treatments (USA) Inc., (TTUSA).

The Detoxifier is a mobile treatment unit used in the in-situ remediation of contaminated soils and waste deposits. The soil is treated in place and is not excavated or removed to the surface. The Detoxifier is capable of a wide range of site remediation methods, including

- Steam-air stripping of volatile contaminants
- Solidification/stabilization and construction of containment structures by addition of chemicals or physical agents (e.g., pozzolanic materials)
- Neutralization or pH adjustment by addition of acids or bases
- Destruction or chemical modification of contaminants via use of oxidizing or reducing chemicals
- Addition of nutrients, microorganisms, and oxygen to promote in-situ biodegradation.

DESCRIPTION OF TECHNOLOGY

The Detoxifier consists of a process tower, a control unit, and a chemical process treatment train (Figure 1). These components are configured to meet site-specific requirements. The process tower is essentially a drilling and remediation agent dispensing system, capable of penetrating the soil medium to depths of 30 ft as currently designed. Remediation agents (in dry, liquid, vapor, or slurry form) are added to and mixed with the soil at various depths by the drill head assembly. A box-shaped shroud, under vacuum, covers the drill head assembly to isolate the treatment area and prevent any environmental release. The drill assembly is composed of two drill blades, each 5 ft in diameter, with injection dispensers (Figure 2).

Figure 1. The Detoxifier

OPERATING METHOD

The remediation of a large area is carried out by a block-by-block treatment pattern. The area to be remediated is divided into rows of blocks, with the process tower, control unit, and process treatment train being moved from one block to the next after the remediation of a block is completed. To assure complete coverage of the area to be remediated, the drill assembly is positioned with a 20 percent overlap of the previously treated block. The effective surface area of a treatment block is approximately 30 ft^2. The volume of each block is determined by the depth of remediation (Figure 3).

INSTRUMENTATION AND CONTROL

On-line analytical instruments continuously monitor and record the treatment conditions. The monitoring data are used to control the treatment process and determine the achievement of remediation objectives.

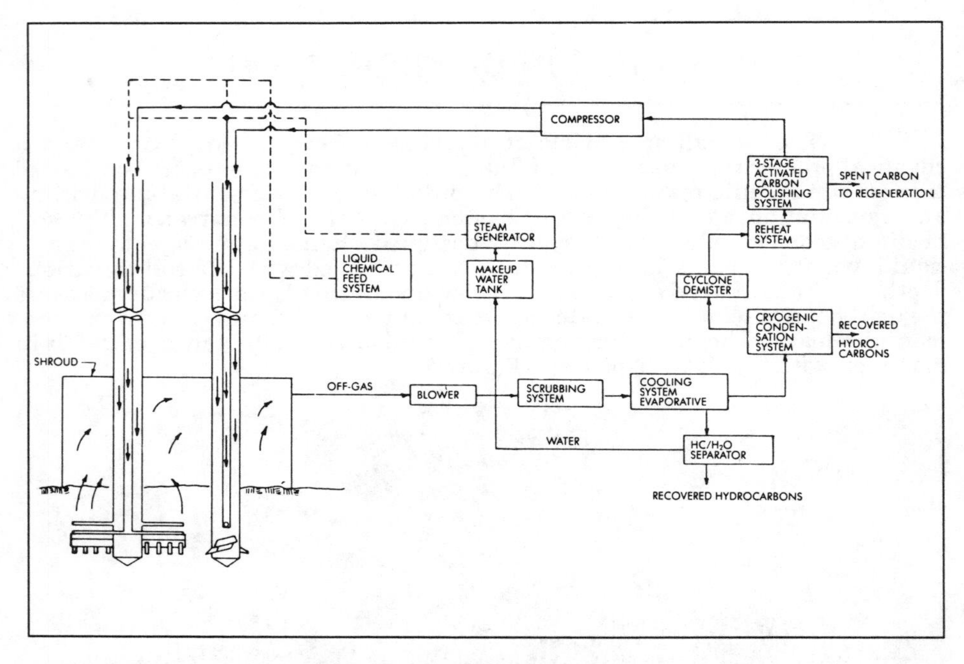

Figure 2. Process Diagram

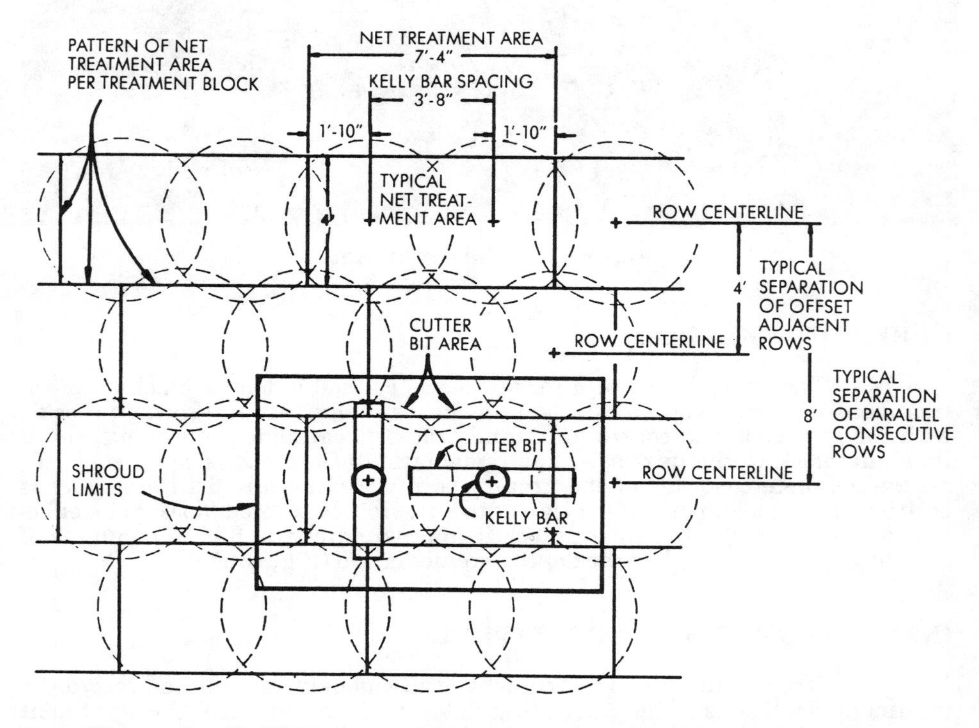

Figure 3. Drill Pattern of Detoxifier Treatment Block Pattern

In applications involving steam/air stripping of volatile contaminants, the off gas containing the contaminants is captured in the shroud and sent in a closed loop (to prevent any environmental release) to a trailer-mounted chemical process train for removal of water and chemical contaminants; the clean air is then recycled to the soil treatment zone. The level of contaminants in the off gas indicates the concentration of contaminants remaining in the ground (see "Baseline Calibration" below).

The control system consists of process monitoring and control instrumentation. Flame ionization detectors (FIDs) monitor the concentration of total hydrocarbons at selected process locations, including the process off gas from the shroud and the purified return air. A gas chromatograph (GC) provides a periodic check of the identification and concentration of specific compounds in the off gas stream. The output of the FID, temperature sensors, depth gauge, and other instrumentation is stored in a computerized data logging system, displayed on a terminal, and recorded on a strip chart recorder. This enables operator control of process parameters to achieve the most effective treatment in the least amount of time.

IN-SITU REMOVAL OF VOLATILE ORGANIC COMPOUNDS

The Detoxifier presently is remediating a site in San Pedro, California, contaminated mostly with chlorinated hydrocarbons. The activities at the site were undertaken at the request of a client and are under the jurisdiction of the California Department of Health Services (CDHS), Toxic Substances Control Division.

The remedial design requires that a baseline calibration and testing program be conducted to develop operational procedures and provide the data necessary to evaluate the effectiveness of the Detoxifier steam/air stripping process to remove VOCs from the soil at the San Pedro site.

The baseline calibration and testing program involved three phases and was conducted during the periods shown below:

Stage 1: Baseline Calibration
October 3 to December 11, 1987

Stage 2: Extended Baseline Calibration
July 6 to August 12, 1988

Stage 3: Baseline Testing
September 13-30, 1988.

STAGE 1: BASELINE CALIBRATION

Purposes:

- Establish site-specific operating parameters (correlate in-process off gas levels with VOCs in the soil as determined by sampling and chemical analysis)
- Conduct subsurface pressure monitoring tests
- Monitor level of emissions from the Detoxifier.

Treatment Protocol. Soil samples were collected prior to, during, and after the treatment process for each of 12 specified blocks. These samples were analyzed by EPA Method 8240 to measure the concentrations of VOCs present. The results of these analyses were compared with data generated by in-process FID measurements of process off gas during treatment of the 12 blocks to establish a correlation between the in-process data and the results of the soil analyses. The length of treatment cycles and degree of treatment for each block were varied to develop information regarding treatment time and the concentration of contaminants remaining in the soil.

(It is important to note that baseline calibration, as well as extended baseline calibration, operations were not designed to remediate the treated blocks to any particular level. As stated, the purpose was to establish operating parameters to correlate in-process FID data with soil chemical analyses. Because of this testing protocol, treatment times and efficiency during baseline calibration cannot be directly related to operating conditions.)

Tests were performed at the request of CDHS to evaluate whether the steam/air stripping process created pressures in the soil subsurface that were sufficient to cause lateral migration of contaminants from a block under treatment. These subsurface pressure monitoring tests were conducted at four locations under the guidance of the CDHS Alternative Technology Section.

Tests conducted to verify emissions generated by the internal combustion equipment in the chemical process train of the steam/air stripping equipment were performed at the request of the South Coast Air Quality Management District. These tests occurred while two treatment blocks were remediated under normal operating conditions.

Results. Baseline calibration (and extended baseline calibration) operations successfully established that: (a) the correlation of soil sample chemical analyses and process off-gas FID monitoring data establishes parameters that permit the determination of the amount of VOCs remaining in the soil by the level of VOCs in the off gas; (b) subsurface lateral migration of VOCs as a result of slight pressures generated by the Detoxifier process, if it occurs, should be minimal; and (c) emissions generated by the Detoxifier's internal combustion equipment are within the limits of the South Coast Air Quality Management District.

More detailed discussion of these results may be found in the following description of extended baseline calibration.

STAGE 2: EXTENDED BASELINE CALIBRATION

Purposes:

- To refine operational parameters
- To conduct additional subsurface pressure monitoring and soil chemical analyses to determine if lateral migration of contaminants occurs during treatment
- To simulate remedial operations.

Treatment Protocol. Soil samples were collected prior to and after the remediation process for selected treatment blocks. Pretreatment samples were collected from 22 treatment blocks, and posttreatment samples were collected from 16 of the treatment blocks. These samples were analyzed by EPA Method 8240 to measure the concentrations of volatile organics present. The results of these analyses were compared with data generated by in-process monitoring equipment to correlate an endpoint for remediation based on in-process data.

Of the 79 blocks treated, 40 were distributed among Treatment Areas A, B, D, and E. The remaining 39 blocks were located in Treatment Area C and were treated in sequence in order to simulate full-scale remedial operations.

Subsurface pressure monitoring and soil chemistry tests were conducted at five locations during extended baseline calibration. The tests were used to compile further information on the effect of the steam/air stripping process on subsurface lateral migration.

Results. Extended baseline calibration confirmed baseline calibration results in establishing a correlation between the on-board process instrumentation readings and the hydrocarbons remaining in the soil. This correlation is believed to be site specific and related to the chemical contaminants and soil types.

Specifically, it has been determined that after achieving the operating temperature of 160 to 170 F at an airflow of 5 to 600 scfm, FID readings of less than 1,000 ppm will achieve a soil chemistry of less than 100 ppm VOCs. The results in Areas A, B, and C averaged 12 and 30 ppm and ranged from 5 to 52 ppm when we reached 800 ppm on the FID. The results in Area D were different. The soils in Area D are wetter and contain more clay than the other areas, and the contaminant in Area D is mostly tetrachloroethylene. This compound has a very high boiling point (low vapor pressure). In this case, gas chromatography gives a more reliable correlation between the soil concentration of tetrachloroethylene than the FID. This last result was deduced after baseline testing.

STAGE 3: BASELINE TESTING

Purposes:

- To evaluate the effectiveness of the Detoxifier in removing hydrocarbons from the soil at the San Pedro site
- To measure soil vapor emissions before and during remediation.

Treatment Protocol. Sampling and Chemical Analysis. Soil samples were collected prior to and after the treatment of 10 highly contaminated preselected blocks. All drilling, sampling, and chemical analysis were conducted according to legally defensible QA/QC procedures approved by CDHS. (Sampling and chemical analysis protocols are available upon request.)

Pretreatment Sampling and Analysis. Soil samples were collected from 24 borings located in the 10 treatment blocks and in 4 uncontaminated locations west of the treatment areas (for background data). A total of 91 soil samples were taken, including

- 68 samples collected from the 20 borings distributed among 10 treatment blocks located in Treatment Areas A, B, and D
- 12 samples collected from the 4 borings located to the west of the current treatment areas (as a control)
- 11 duplicate samples were selected from the 20 borings.

The 68 samples collected from treatment blocks in Areas A, B, and D, and the 12 samples collected from the 4 borings located to the west of the current treatment areas were analyzed for VOCs by EPA Method 8240. Eight of the duplicate samples were also analyzed for VOCs by EPA Method 8240; these duplicates were labeled in a manner such that the laboratory could not distinguish them from the previously analyzed samples. At the request of the CDHS, the remaining three duplicate samples were analyzed for semivolatile hydrocarbons (SVHs) by EPA Method 8270 and for priority pollutant metals by EPA Method 6020.

Following the above-described analyses, additional chemical analyses were performed on soil samples that had been kept in refrigerated storage. At the request of the CDHS, 65 soil samples were analyzed for SVHs by EPA Method 8270. These 65 samples were the remainder from the 68 samples collected from the 10 treatment blocks located in Treatment Areas A, B, and D.

The composite average pretreatment concentrations of VOCs and SVHs for each block are presented in Table 1. (Posttreatment concentrations, also shown in Table 1, are discussed below.) The distribution of VOCs and SVHs varied among the treatment areas. Treatment blocks in Area A had the greatest number of species or groups of VOCs and SVHs detected, and treatment blocks in Area D had the least. The average concentration of VOCs and SVHs was also dissimilar. Area D exhibited the highest concentration of VOCs and Area A the lowest; Area B had the highest concentration of SVHs and Area D the lowest.

Analyses for physical properties were also performed on soil samples collected from the 10 treated blocks sampled in Treatment Areas A, B, and D. A total of 68 soil samples were analyzed to evaluate the density and moisture content of the sample.

Table 1. Summary of Chemical Analyses

Treatment Block	Concentration (ppm) Pre-Treatment	Post-Treatment	% Reduction
Volatile Organic Compounds:			
A-8-g	1,149	18	98
A-9-g	824	7	99
A-10-g	1,368	11	99
B-50-n	1,123	23	98
B-51-n	1,500	13	99
B-51-m	1,872	55	97
B-52-m	917	29	97
D-92-b	2,305	53	98
D-93-b	3,720	163	96
D-94-b	5,383	203	97
Semi-Volatile Hydrocarbons:			
A-8-g	1,794	637	64
A-9-g	2,510	653	74
A-10-g	7,020	592	92
B-50-n	22,829	1,670	93
B-51-n	14,924	2,304	85
B-51-m	10,040	2,495	75
B-52-m	669	594	11
D-92-b	707	55	92
D-93-b	1,465	90	94
D-94-b	869	111	87

Treatment. The remediation protocol was designed to achieve the reduction of VOCs to less than 100 ppm. Some of the key data recorded during treatment included

- Duration of treatment
- Steam injected (total pounds)
- Hot air injected (cubic feet per minute)
- Air temperature
- Concentration of volatile organics in the off gas from the shroud to the process system
- Depth of treatment.

TTUSA did not have knowledge of the pretreatment soil analyses of the 10 blocks until completion of baseline testing. The decision to terminate treatment of each test block was made when the concentration of VOCs measured by the FID decreased to a value previously determined as equating with less than 100 ppm VOCs in the soil. This concentration level had been determined during baseline calibration and extended baseline calibration. TTUSA did not have the benefit of any other information to determine when to cease treatment.

Posttreatment Sampling and Analysis. The same procedures were used for posttreatment sampling and analysis. A total of 55 soil sample tubes collected from 14 borings distributed among the 10 treatment blocks located in Treatment Areas A, B, and D were analyzed for VOCs and SVHs by EPA Methods 8240 and 8270, respectively.

Results. Based on a comparison of the pre- and posttreatment chemical analyses, the major effects of treatment were

- A very substantial reduction in the concentration of VOCs
- A significant and unexpected reduction in the concentration of SVHs.

Reduction of VOCs. A very substantial reduction in the concentration of VOCs was calculated from chemical analysis in all test blocks (see Table 1 for summary data). This conclusion is reinforced by the fact that the mass of VOCs collected in the chemical process train was roughly equivalent to the calculated mass removed from the soil (see "Mass Balance" below). The mechanism that caused the reduction appears to be volatilization. The treatment techniques used were designed to reduce the concentration of VOCs to less than 100 ppm. This level was achieved in 8 of the 10 treatment blocks remediated (see Table 1 for summary data). As indicated in Table 1, all treatment blocks in Areas A and B were remediated to substantially below this level, but only 1 of 3 treatment blocks in Area D was similarly remediated. The average pre- and posttreatment concentrations (ppm) for VOCs are

Treatment Area	Pretreatment	Posttreatment
A	1,114	12
B	1,353	30
D	3,954	139

The reduced effectiveness of the process in achieving a level of 100 ppm in Area D is believed to be caused by the high initial concentration of tetrachloroethylene and the presence of more clayey soil than encountered in Areas A and B. (Following baseline testing, the hot air/steam-stripping equipment was modified to improve its effectiveness. Preliminary testing indicates that the equipment in its modified configuration has the potential to remediate the soil more effectively to less than 100 ppm VOCs, even in Area D.) We will soon conduct a controlled test to demonstrate this.

Reduction of SVHs. An unexpected and major reduction in the concentration of SVHs was indicated in all treatment blocks remediated (see Table 1). The average pre- and posttreatment concentrations (ppm) for SVHs are

Treatment Area	Pretreatment	Posttreatment
A	3,775	627
B	12,116	1,766
D	1,014	85

Mechanisms that may account for this reduction include volatilization, steam distillation, hydrolysis, and oxidation. The reasons for the reductions, however, have not been confirmed by the chemical analyses. A small quantity of SVHs (2.1 lb) was collected in the process train; however, the mass of SVHs collected does not account reasonably for the mass reduction (1,018 lb) indicated by posttreatment soil sample analysis (see "Mass Comparison" below). A testing program to identify the exact mechanism(s) for the SVH reduction has been developed based on potential chemical reactions of the semivolatile component (SVC) present in a steam/oxygen/clay environment. The results of this program will be made available when it is completed.

Soil Gas Emissions During and After Treatment. One concern expressed by the CDHS was the potential for increased emissions of VOCs into the ambient air during and after treatment by the Detoxifier. This would occur, for example, if the heat used to volatilize hydrocarbons caused increased soil emissions adjacent to the remediation or emissions occurred from just treated soil after removal of the shroud. An evaluation of this was made during baseline testing. The test used was to be qualitative in nature as a quantitative test would be expensive and time consuming, per agreement with the CDHS. A background value for soil vapor emissions was established for each area prior to treatment, and emissions were measured around the perimeter of the shroud during baseline testing. A comparison of these data and the data obtained during baseline testing indicated that the overall effect of the hot air/steam-stripping process was to increase slightly the rate of soil gas emissions within and directly adjacent to the treated area. Emissions from the just treated uncovered blocks were always higher than background for several hours afterward.

A more recent test (under the approval of the CDHS) using improved process vacuum and soil covering of remediated blocks showed no increased emissions to the ambient air. In fact, the levels remained below background during and after treatment.

We believe this means that soil gas emissions will not be caused by the treatment process.

Mass Balance. The mass of VOC and SVH liquids collected by the hot air/steam-stripping process was measured directly. The mass of VOCs and SVHs removed from the soil also was calculated using measured values based on the chemical and physical analyses. The mass of VOCs collected is essentially equivalent to the calculated mass of VOCs removed from the soil. The following results present the treatment blocks remediated, the mass of VOCs collected by the process, the mass of VOCs removed from the soil, and the percent VOCs recovered. (The six identified blocks were analyzed separately because the other four blocks had relatively low concentrations of SVHs prior to remediation.)

Treatment Blocks	VOCs Collected by the Process, lb	Calculated VOCs Removed from the Soil, lb	% VOCs Recovered
6 (A-8-g, A-9-g, A-10-g, B-50-n, B-51-n, B-51-m)	124.5	142.5	87.4
10 (All Blocks)	264.8	296.6	89.3

This comparison indicates that the Detoxifier treatment removes VOCs from the soil and captures them in the process. The closure of the VOC mass balance is within what is possible for the constraints of the experimental measurements. The reasons for the difference in the mass of VOCs collected and the mass removed from the soil can be attributed to the following:

- The size of the test block group compared with the size of the process system (i.e., mass possibly lost through process variables)
- Uncertainties in sampling
- Other unidentified causes.

The mass of SVHs in the collected hydrocarbon liquid was not equivalent to the mass of SVHs removed from the soil. After remediation of six treatment blocks (as shown below), chemical analyses indicated that only 2.1 lb of SVHs had been collected, whereas it was calculated that 1,018 lb of SVHs had been removed from the treatment blocks.

Treatment Blocks	SVHs in Process, lb	SVHs Removed from the Soil, lb	% SVHs Recovered
6 (A-8-g, A-9-g, A-10-g, B-50-n, B-51-n, B-51-m)	2.1	1,018	0.21

The reason for this difference in mass has not been formally demonstrated. However, examination of the compounds present, their potential chemical fate in the presence of hot air/steam, and chemical analysis suggest that hydrolysis catalyzed by the clay in the soil plays an important role in their disappearance. It is also believed that the hydrolysis reactions produce compounds that are bound to the clay and cannot be identified by the analytical techniques employed.

SUMMARY

This paper has presented the first detailed information on a new technology for the in-situ removal of VOCs from soil. The data indicate that VOCs are removed substantially from soil and that they are captured aboveground for disposal without adverse effects to the ambient environment. No information is given regarding the economics of this process, although it is believed to be competitive with alternate remediation schemes. This technology may be an attractive alternative to landfill disposal as landfills become unavailable and passive vapor extraction schemes are slow or uncertain in their efficiency. The technology has demonstrated also that SVCs are removed from soil by the action of steam and air in the soil environment. The mechanisms for this are not well understood. The removal of SVC may add to the utility of this technology in the future.

BENEFICIAL RE-USE OF SLUDGES

Nitrogen, phosphorus, and potassium are the major crop nutrients. Optimal use of commercially available fertilizers containing these nutrients has resulted in an abundance of food production for human and domesticated animal consumption. All raw materials from which the major nutrients are derived are nonrenewable, and economically recoverable reserves are expected to last only until the year 2020. Almost 60 percent of these raw materials are limited to certain geographical regions of the world that are considered politically volatile. In addition, processing these raw materials to finished fertilizer products is highly energy intensive, so an energy crisis could also result in a serious worldwide food crisis.

Alternative renewable sources containing the nutrients required for crops must be found and strategies must be formulated to avoid a potentially serious food crisis. Animal manures, food processing wastes, and municipal wastes are potential renewable alternatives to commercially available fertilizers. However, their high water content makes them bulky, thereby making the application of these fertilizers to crops technically difficult and economically nonviable.

In this session, we highlight the fertilizer and food scenario and the potential for food processing, animal, and municipal wastes to provide a viable renewable source of crop plant nutrients. The first article focuses on this area and addresses the need for innovative separations technologies for making economical use of the wastes. Subsequent articles focus on the sludge wastes mainly because of the current use of sludge wastes in agricultural practices and the potential these wastes present for increased use in the future. The articles highlight the importance of sludge wastes in crop productivity and present modeling approaches to address the quantities required for economic crop yields.

This session includes several distinguished speakers such as Professors Chang of the University of California, O'Connor of New Mexico State University, and Logan of The Ohio State University. Dr. Velagaleti, Dr. Gavaskar, and their colleagues provide some useful insight into the real application and markets of such products.

SLUDGE METAL BIOAVAILABILITY

Terry J. Logan, Ph.D.
Professor of Soil Chemistry
Department of Agronomy
The Ohio State University
2021 Coffey Road
Columbus, Ohio 43210
(614) 292-9043

ABSTRACT

Over the last decade, research on sludge metal chemistry and plant uptake of sludge-applied metals has suggested that trace metals are bound tightly within the sludge matrix. Trace metals, including Cd, Cu, Cr, Co, Hg, Mn, Ni, Pb, and Zn, usually occur in sludges at concentrations between 1 and 2,000 mg/kg. At these levels, the trace metals are probably coprecipitated with Al, Fe, Ca, Mg, and Si, which can occur at concentrations as high as 10,000 mg/kg or more. Some metals, such as Cu, are also strongly associated with the organic sludge fraction as strong humic complexes. Long-term field studies with sewage sludge suggest that sludge metal bioavailability is generally low and does not change markedly after the first year of application. A model of sludge metal bioavailability is discussed and a new extractant to predict sludge Cd bioavailability is proposed.

INTRODUCTION

Application of municipal sewage sludge to agricultural land has increased markedly in the last 15 years. Along with this increase has been concern for widespread contamination of the food chain by trace elements, including the so called "heavy metals." Sludges contain a wide range of concentrations of most elements, and several of these have been shown to pose health problems to livestock (*e.g.*, Cu, Se, and F) and humans (primarily Cd and Pb). Previous experience with nonsludge sources of some metals, chiefly Cu, Ni, and Zn, also suggests that trace elements at high soil concentrations can be phytotoxic to crops[15].

It is important to note that prior to the mid-late 1970s, almost all inferences to sludge metal bioavailability (plant and animal uptake) were based on studies of inorganic metal sources (pot studies with metal salts, deposition from metal ore smelters, pesticides such as lead arsenate), or with sludge spiked with inorganic salts. Early plant uptake studies were also conducted in pots in the greenhouse, and Logan and Chaney[15] have pointed out the potential errors of studies with small pots. With the advent of long-term field studies with sludge on a variety of crops, more valid data on sludge metal bioavailability began to emerge. In this paper, I will discuss several questions that have most often been raised regarding sludge metal bioavailability, and the extent to which current information has provided answers to these questions:

1. What are the dominant forms of metals in sludges and how are these related to sludge metal solubility and hence bioavailability?
2. What fraction of sludge metal is associated with the organic fraction versus mineral forms, and what happens to sludge metal bioavailability when the sludge organic matter decomposes?
3. What determines sludge metal bioavailability in soil: the chemical characteristics of the soil, the sludge, or both?
4. How can sludge metal bioavailability best be quantified?

SLUDGE METAL CHEMISTRY

Digested sewage sludge is a complex, heterogeneous material comprised of humified organic matter in intimate chemical association with mineral forms. Each sludge is chemically unique, and its chemical makeup reflects not only the diverse sources of chemicals discharged into the sewer system, but also the chemical transformations that occur during sludge treatment. Only a few studies have attempted to characterize sewage sludge, and these offer only a limited view of this complex material. Lake *et al.*[14] reviewed the literature on sludge metal characterization. They found that the most common approach was to use sequential chemical extraction to partition metals into broad chemical classes, including soluble, exchangeable, adsorbed, organically bound, "available," carbonate, sulfide, occluded, and residual. None of the studies employing sequential chemical extraction were able to provide direct chemical evidence for the various forms reported. Other methods used to characterize sludge metals include physical separation by elutriation and filtration[10,11]. Essington[8] used density and magnetic separation to concentrate a metal-rich fraction of sewage sludge and examined single particles with scanning transmission electron microscopy. Trace metals were associated (presumably as coprecipitates) with minerals of Si and Fe. Corey *et al.*[4], Logan and Feltz[18], and Corey *et al.*[5] have hypothesized that much of the trace metals in the mineral fraction of sewage sludge may exist as coprecipitates of Fe, Al, Si, Ca, and other cations as sulfides, carbonates, oxides, phosphates, and silicates. Many of these mineral solids may be thermodynamically stable and thus not subject to extensive transformation when applied to soil. Based on some of the direct mineralogical evidence of Essington[8], I hypothesize that many of the mineral solids in sewage sludge may be precipitated at the point of discharge into the sewer when acidic metal wastes mix with domestic sewage. This is supported by the work of Elenbogen *et al.*[6], who found that metals in raw sewage entering a treatment plant are bound to the solids in solid:solution ratio similar to that of the activated sludge.

Of particular importance in the solubility of sludge metals is the role of sulfide. Fractionation studies suggest that anaerobically digested sludges contain significant quantities of sulfide[23] and sulfide-bound trace metals. The trace metals are likely coprecipitated with Fe^{2+} rather than existing as discrete sulfides, although the latter are thermodynamically possible for metals like Zn and Pb, which can occur at fairly high levels in some sludges[15]. Metal sulfides are very insoluble and are not acid labile. Lake *et al.*[14], in reviewing the literature on sludge metal extraction, found that water or dilute electrolyte extractability was reduced over time when sludges were allowed to become anaerobic. Feltz and Logan[9] found that aeration greatly increased the acid extractability of Cd in a Cd-contaminated anaerobically digested sludge. The oxidation of sulfide when anaerobic sludges are land applied suggests that sludge metal solubility should increase after application. Whether this increased solubility results in increased bioavailability in the soil will be determined by the subsequent interactions of the metal with soil and sludge constituents.

The sludge fractionation studies conducted by Stover *et al.*[22] and others[14] suggest that a large fraction of the trace metals in municipal sewage sludges may be inorganic, not organic. This brings into question the assumption often made in the past that as sludge organic matter decomposed, the organically complexed metal would be released into the soil and metal bioavailability would increase. Long-term field studies have not shown this to be the case (as discussed below), giving credence to the dominant role of mineral rather than organic sludge metal forms.

REACTIONS OF SLUDGE METALS IN SOIL

As indicated previously, many of the initial inferences made about sludge metal behavior in soil were based on studies with metal sources other than sludge (smelter emissions, pesticides), pot studies with sludge, or short-term field studies. As experience with long-term field studies accumulated, it became increasingly apparent that these inferences were often incorrect and that only long-term field data with sludge itself should be relied on to determine the ultimate fate of sludge metals in soil. This was discussed in detail by Logan and Chaney[15] and again by Chaney *et al.*[2], Corey *et al.*[5], and Chang *et al.*[3]. It is unfortunate that the EPA, in developing new proposed regulations for sludge disposal[7], ignored the cautions raised by Logan and Chaney[15] and Page *et al.*[19] against using data on plant metal uptake other than from the long-term field sludge studies[20].

At the heart of research on sludge metal bioavailability in soil are the following questions:

1. Does sludge metal bioavailability change with time after sludge application?
2. What determines sludge metal bioavailability: metal retention by the sludge, by the soil, or equilibrium between the sludge and soil?

Chang *et al.*[3] examined the results of long-term (more than 5 to 8 years) studies on plant uptake of sludge-applied metals. In particular, they summarized results of a cooperative study conducted in 15 U.S. states with a single high-metal sludge over 10 years. They concluded from these data that sludge metal uptake by crops tends to decline in the first 1 to 2 years after sludge application. There was little change in uptake of most of the metals studied after this period, with small declines observed in most cases. The more rapid early changes probably result from sludge organic matter decomposition, which affects both solid and solution sludge metal speciation, and changes in soil properties with sludge addition. The slower, long-term reduction in metal uptake could result from continuing adsorption of sludge metals by soil, immobilization in stable soil organic matter, or dispersion of sludge metals by tillage, runoff-erosion, earthworm mixing, and other processes. These conclusions regarding long-term sludge metal uptake were based on the assumption that soil pH changes little with time after sludge application. It has been well documented[15,21] that solubility and plant uptake of all of the cationic trace metals increase when soil pH falls much below 5. Although large or repeated applications of many sludges tend to buffer soil pH at values around 6 to 7, Chaney[1] has argued that some sludges low in base cations and applied to soils with low pH buffer capacity can result in very low soil pHs.

The second question--the importance of sludge versus soil properties in metal retention--has been much more difficult to answer for many reasons. Because the trace metals by definition occur at relatively low concentrations in sludge (and even lower levels in the soil-sludge mixture), the study of this system is greatly limited analytically. Both soil and sludge are heterogeneous, multicomponent, multiphase systems, and interactions between the two are not well known. As indicated previously, our knowledge of sludge metal chemistry is indirect. The same is true for the behavior of trace metals in soil, although it appears that the mechanisms are similar. Trace metals are most likely bound in sludge to the mineral fraction as coprecipitates or adsorbed to reactive surfaces such as the oxides, carbonates, and phosphates. Trace metals may also be complexed to the humic and fulvic acid fractions of sludge organic matter, although the latter represents a small percentage of the total metals[14]. When sludge is added to soil, most of the organic matter decomposes over the first 2 years, releasing organically bound metal. Sulfides in anaerobic sludge are oxidized, probably within the first few months of application, although there is no literature on sludge-sulfide oxidation in soil. This will also result in release of trace metals. Metal carbonates are unstable in acid soil and will decompose. The oxides and phosphates are thermodynamically quite stable in soil and should persist for years. The net effect, then, should be some release of sludge metals in the first year or so after sludge application; this is supported by the long-term sludge studies which show greatest metal uptake by crops in the first 2 years after sludge application[3]. The degree to which the sludge metal is released into an available form will be determined to a large extent by the thermodynamic stability of the mineral trace metal forms in the sludge. The ultimate bioavailability of sludge metal in soil will be determined by the thermodynamic stability of either the sludge form or soil reaction product, whichever is greater.

In recent years, a growing body of field data on metal uptake by crops from sewage sludge suggest that uptake does not increase linearly with sludge metal application rate, but rather tends to plateau at higher rates[3-5,15,16]. This effect is shown graphically in Figure 1. This plateauing does not represent an uptake maximum by the crop because the observed uptake rarely approaches maximum values obtained with metal salts. Rather, the binding capacity of the sludge for the metal increasingly dominates the overall metal solubility as the sludge application rate increases. In other words, the sludge is both a *source* and a *sink* for metals. Figure 1 suggests that soil properties such as pH will affect metal solubility at low sludge rates, but at high rates (as a single application or over repeated applications) metal solubility and, hence, bioavailability will be determined by the chemical characteristics of the sludge. It would be invaluable to use the plateau concept in a risk assessment of food-chain contamination by sludge metals by comparing the plateau value for crop uptake with maximum food-chain exposure[16]. In this way, a sludge with a plateau value for crop uptake less than that required to protect the food chain could be applied to cropland regardless of application rate and perhaps with minimal effect on soil properties. Extending this approach further, it would be invaluable to identify the sludge characteristics that control the plateau uptake value, or to develop a sludge test to predict this value.

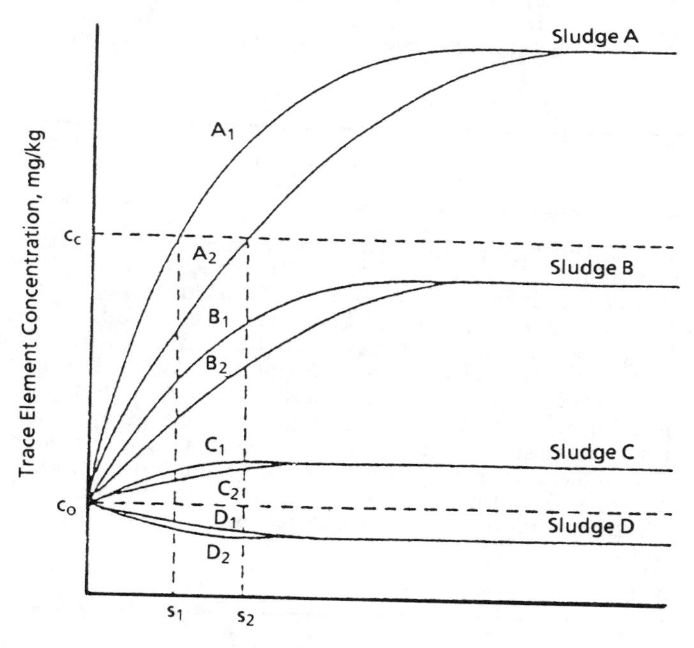

Figure 1. Conceptual Relationship Between Soil and Sludge Properties and Trace Element Uptake. Four sludges (A, B, C, D) and two soils (s_1, s_2) with differing trace element retention capacities are represented. C_o is the trace element concentration in the plant at zero sludge application for the two soils, and C_c is a theoretical threshold for food-chain contamination[5].

For several years, researchers[15,19] have argued that the major factor affecting the bioavailability of a trace metal in sludge is its total concentration. While other factors may affect metal retention[5], it is theorized that the lability (the fraction that can be solubilized or easily displaced) of a metal is proportional to its content. This would be expected whether the mechanism of metal retention were coprecipitation, adsorption, or complexation. This theory is supported by the work of King and Dunlop[13], Chaney [cited in Corey *et al.*[5]], and Logan and Feltz[17], who found that Cd uptake by plants increased with Cd content of the sludge. In a recent pot study, Jing and Logan[12] measured the uptake of Cd by Sudax (a sorghum-sudangrass hybrid) from 16 different anaerobically digested sludges applied at the same Cd rates (2.5 and 5 mg Cd/kg soil) to an acid sandy soil. Cd uptake was correlated with total Cd content of the sludges ($r^2 = 0.49$) and with resin-extractable sludge Cd ($r^2 = 0.87$). Resin-extractable Cd was shown to be a good indicator of labile sludge metals and may be used to screen sludges for impact on crop uptake.

The findings that sludge metal bioavailability may be related to the metal content of sludges lend credence to the emerging concept of a "clean" sludge[19,20]. A clean sludge would be one in which the concentrations of metals (this concept can also be extended to organics) would be below the threshold necessary to cause food chain contamination even at very high application rates. The specific values for each contaminant would have to be established for the various sludge disposal options on the basis of an appropriate risk assessment. Once established, the clean sludge would be subject to minimal regulation with resulting disposal cost savings to the POTW. The incentives for aggressive industrial pretreatment with this approach are obvious, but it also emphasizes the EPA's stated commitment to beneficial sludge use and should reduce the regulatory burden for a significant percentage of our nation's municipal sewage sludge.

CONCLUSIONS

Recent findings suggest that trace metals in sewage sludge are retained tightly by the sludge solids and that this retention greatly reduces the bioavailability, particularly with respect to plant uptake, of sludge metals compared with other metal sources. Long-term field studies indicate little change in sludge metal bioavailability with time after initial sludge application, which suggests that sludge metal forms may be as thermodynamically stable as those in soil. Field and greenhouse studies also suggest that the increase in metal uptake by crops is not linear with sludge application rate but rather reaches a plateau, the magnitude of which is a function of sludge properties. The most important sludge property related to sludge metal bioavailability appears to be total metal concentration. However, labile metal, as estimated by resin extraction, may prove to be a more precise indicator of sludge metal bioavailability.

REFERENCES

1. Chaney, R. L., Personal Communication, USDA-ARS, Beltsville, MD, 1989.

2. Chaney, R. L., Bruins, R.J.F., Baker, D. E., Korcak, R. F., Smith, J. E., and Cole, D., in *Land Application of Sludge--Food Chain Implications*, A. L. Page, T. J. Logan, and R. A. Ryan, eds., Lewis Pubs., Chelsea, MI, 1987, 67-99.

3. Chang, A. C., Hinesly, T. D., Bates, T. E., Doner, H. E., Dowdy, R. H., and Ryan, J. A., in *Land Application of Sludge--Food Chain Implications*, A. L. Page, T. J. Logan, and R. A. Ryan, eds., Lewis Pubs., Chelsea, MI, 1987, 53-66.

4. Corey, R. B., Fujii, R., and Hendrickson, L. L., *Proc. Fourth Annual Madison Conf. Appl. Res. Pract. Munic. Ind. Waste*, University of Wisconsin Extension, Madison, 1981, 449-465.

5. Corey, R. B., King, L. D., Lue-Hing, C., Fanning, D. S., Street, J. J., and Walker, J. M., in *Land Application of Sludge--Food Chain Implications*, A. L. Page, T. J. Logan, and R. A. Ryan, eds., Lewis Pubs., Chelsea, MI, 1987, 25-51. Figure 1 is reprinted with permission.

6. Elenbogen, G., Sawyer, B., Rao, K. C., Zenz, D. R., and Lue-Hing, C., "Studies of the Uptake of Heavy Metals by Activated Sludge," Res. and Dev. Dep. Rep. 84-7, The Metropolitan Sanitary District of Greater Chicago, Chicago, IL, 1984.

7. Environmental Protection Agency. 40 CFR Parts 257 and 503. *Standards for the Disposal of Sewage Sludge; Proposed Rule*, Federal Register, February 6, 1989, 5746-5902.

8. Essington, M. E., Ph.D. Thesis, University of California at Riverside, March, 1985.

9. Feltz, R. E. and Logan, T. J., *J. Environ. Qual.*, 1985, *14*, 483-488.

10. Gould, M. S. and Genetelli, E. J., *Proc. 30th Ind. Waste Conf. Purdue Univ.*, Ann Arbor Sci. Pubs., Ann Arbor, MI, 1975, 689-699.

11. Hayes, T. D. and Theis, T. L., *J. Water Poll. Cont. Fed.*, 1978, *50*, 61-72.

12. Jing, J. and Logan, T. J., *Agronomy Abstracts*, 1989.

13. King, L. D. and Dunlop, W. R., *J. Environ. Qual.*, 1982, *11*, 608-616.

14. Lake, D. L., Kirk, P.W.W., and Lester, J. N., *J. Environ. Qual.*, 1984, *13*, 175-183.

15. Logan, T. J. and Chaney, R. L., in *Utilization of Municipal Wastewater and Sludge on Land*, A. L. Page, T. L. Gleason, J. E. Smith, I. K. Iskander, and L. E. Sommers, eds., University of California at Riverside, 1983, 235-326.

16. Logan, T. J., Chaney, R. L., in *Heavy Metals in The Environment, Vol. 2*, S. E. Lindberg and T. C. Hutchinson, eds., CEP Consultants, Ltd., Edinburgh, UK, 1987, 387-389.

17. Logan, T. J. and Feltz, R. E., *J. Environ. Qual.*, 1985, *14*, 495-500.

18. Logan, T. J., Feltz, R. E., *J. Water Poll. Cont. Fed.*, 1985, *57*, 406-412.

19. Page, A. L., Logan, T. J., and Ryan, R. A., eds., *Land Application of Sludge--Food Chain Implications*, Lewis Pubs., Chelsea, MI, 1987.

20. Peer Review Committee. *Standards for the Disposal of Sewage Sludge.* U.S. EPA Proposed Rule 40 CFR Parts 257 and 503, W-170 Regional Res. Comm. USDA-CSRS, Washington, DC, 1989.

21. Sommers, L. E., Volk, V. V., Giordano, P. M., Sopper, W. E., and Bastian, R., in *Land Application of Sludge--Food Chain Implications*, A. L. Page, T. J. Logan, and R. A. Ryan, eds., Lewis Pubs., Chelsea, MI, 1987, 5-24.

22. Stover, R. C., Sommers, L. E., and Silviera, D. J., *J. Water Poll. Cont. Fed.*, 1976, *48*, 2165-2175.

SLUDGE ORGANICS BIOAVAILABILITY

G. A. O'Connor
Department of Agronomy and Horticulture
New Mexico State University
P.O. Box 30003
Las Cruces, New Mexico 88003-0003
(505) 646-2219

and

G. A. Eiceman, C. A. Bellin, and J. A. Ryan
U.S. EPA
RREL
Cincinnati, Ohio

ABSTRACT

Concern over the bioavailability of toxic organics that can occur in municipal sludges threatens routine land application of sludge. Available data, however, show that concentrations of priority organics in normal sludges are low. Sludges applied at agronomic rates yield chemical concentrations in soil-sludge mixtures 50 to 100 fold lower. Plant uptake at these pollutant concentrations (and at much higher concentrations) is minimal. Chemicals are either (1) accumulated at extremely low levels (PCBs), (2) possibly accumulated, but then rapidly metabolized within plants to extremely low levels (DEHP), or (3) likely degraded so rapidly in soil that only minor contamination occurs (PCP and 2,4-DNP).

SLUDGE ORGANICS BIOAVAILABILITY

Land application is an efficient, cost-effective method of sludge disposal that also recycles essential nutrients to the soil. Sludge additions can also improve soil physical and chemical properties, albeit at high initial or cumulative application rates. Thus, land application of sludge in the agricultural sector is responsive to the U.S. EPA's policy of promoting beneficial use of sludge. However, lack of knowledge about the fate of toxic organics that can occur in sludges causes concern and threatens routine use of the practice.

The number of potentially hazardous organics that may be present in wastewaters and sludges is large[10]. Of particular concern in the United States, however, are some of the organic compounds referred to as priority pollutants. Fortunately, data available in the literature[6,10,21] and from sludge analysis conducted in our laboratory suggest that the concentrations of these toxic organics is low. Municipal sludges normally contain toxic organics in the 1.0 to 100 mg/kg range, with ≥ 90 percent of the toxic organics present at concentrations ≤ 10 mg/kg[10]. When these sludges are applied at normal (agronomic) rates (20 to 45 mg/ha), the resulting soil-sludge mixtures have organic concentrations 50 to 100 fold lower. Thus, toxic organic concentrations in sludge-amended soils are similar to (or lower than) routine pesticide concentrations of 0.02 to 1 mg/kg[17].

Scientists[5,10,17,18] considered these numbers along with the toxicity rankings of priority pollutants and tentatively concluded that median sludges applied to soils at normal rates would minimize toxic organic risks to crops and the food chain receptor (human or animal). However, until recently, insufficient crop uptake data existed to adequately assess food chain risks. Thus, the U.S. EPA sponsored our study of the plant availability of toxic organics typically reported, or expected, in sludge. Simultaneously, methodologies were developed based on simple chemical properties (half-life, Kow and Henry's constant) to screen organic chemicals for potential plant uptake[22].

According to the Ryan *et al.*[22] model, compounds with log Kow values of 1 to 2 are most likely to have significant transport to aboveground tissue (although plant metabolism of chemicals in the root may negate translocation[16]). Compounds with a log Kow of >5 would be strongly bound in soil, or on the root, and not translocated. Compounds with half-lives of <10d would likely be lost from the system before they could be taken up by plants. Persistent chemicals (half-life longer than the growing season of crop) exist long enough to affect plants. Volatile chemicals (HC $> 10^{-4}$ unitless) are lost from soils, but may contaminate plants by vapor phase uptake. The Ryan *et al.*[22] model is intended as a screening, rather than as a predictive, tool. Plant uptake of soil-borne organic pollutants is a complex process and is not easily predicted from simple models. The model is also limited by uncertainty in several of the chemical parameters utilized, especially HC and half-life values. Nevertheless,

Ryan *et al.*[22] were able to eliminate about half of the EPA organic priority pollutants from further consideration based on the simple model. The data reported herein were intended partially to validate the model.

Compounds studied to date, along with pertinent chemical properties, are identified in Table 1. Diethylhexylphthalate (DEHP) is typically the most highly concentrated and most frequently detected toxic organic in U.S. municipal sludges. It is unique in that its median concentration is ~100 mg/kg dry weight[17]. It is moderately long-lived and nonvolatile.

Table 1. Selected Chemical Properties of Toxic Organics Studied

Compound	Log Kow	T1/2 d	Hc	pKa	Conc.* mg/kg
Diethylhexylphthalate (DEHP)	4 - 5	10 - 50	10^{-5}	--	100
Polychlorinated biphenyls (PCBs)	5 - 6	>50	10^{-1}	--	1
Pentachlorophenol (PCP)	5	10 - 50	10^{-4}	4.7	5
2,4-dinitrophenol (DNP)	1.5	>50	10^{-8}	4.1	--

* Median sludge concentrations.

Polychlorinated biphenyls (PCB) are a family of compounds of considerable public notoriety. PCB is the only organic compound addressed in current EPA sludge guidelines (40 CFR 257). Materials containing >50 mg/kg total PCBs are classified as hazardous wastes and may not be applied to land. Sludges containing >10 mg/kg must be incorporated in soil to reduce potential PCB transfer to grazing animals. The median PCB concentration in municipal sludge is about 1 mg/kg, and the composition of PCBs in most digested sludges is similar to Aroclor 1254 or 1260[6]. Some PCBs are reportedly volatile, but strong adsorption in the presence of sludge greatly reduces volatilization[8]. Higher chlorinated PCB congeners are very persistent in soil; lower chlorinated (mono- and dichloro species) degrade much more quickly[23].

Pentachlorophenol (PCP) is a wood preservative, but can occur in municipal sludges. In one study of 223 industrial sludges, PCP was detected at a median concentration of 5 mg/kg[10]. PCP is a weak acid (pKa = 4.7), so pH is expected to significantly affect PCP soil-plant behavior.

2,4-dinitrophenol (DNP) is another weak acid (pKa = 4.1) we have studied. It is much more polar than the other compounds and has chemical characteristics that suggest high plant availability[22]. Thus, despite the lack of reported DNP occurrence in sludges, it was included for study because of its presumed plant availability and its reported[28] toxicity to plants and animals.

Toluene was orginally intended to be included in plant studies as an example of a volatile (HC = 0.28) chemical. However, soil incubation studies[31] showed the chemical was lost from shallow soils so rapidly that plant studies would have been fruitless. Toluene was completely volatilized from sludge-amended or unamended, coarse and fine textured soils at field capacity or saturation in less than 10d.

MATERIALS AND METHODS

Greenhouse studies have been conducted with all of the chemicals identified in Table 1, involving a variety of treatments, e.g., soil type, sludge rate, chemical loading rate, etc. For the sake of uniformity and brevity, data summarized here represent treatments common to all compounds.

The Glendale clay loam (pH 8.0, 6.5 g organic carbon/kg, and 132 g $CaCO_3$/kg) is a fine-silty, mixed thermic typic torrifluvent. It is typical of calcareous, low-organic matter and in productive irrigated land in river valleys of the southwestern United States.

Crops grown included tall fescue grass (*Festuca arundinacea*), lettuce (*Lactuca sativa*), carrots (*Daucus carota*), and chile peppers (*Capsicum annum*). The vegetables were chosen to allow evaluation of chemical uptake in different plant parts (leaves, roots, and fruiting bodies), and to represent different times (to harvest) of chemical residence in soil (fescue ~30d, lettuce ~50d, carrots ~75d, and chile ~115d).

The chemicals were supplied either indigenously in the sludge (PCBs), as ^{14}C-labeled chemical equilibrated with sludge (DEHP), or as ^{14}C-labeled reagent grade chemical spiked to soil-sludge mixtures (PCP and DNP). In the PCB and DEHP studies, chemicals were supplied both indigenously (in the sludge) and as spikes to compare plant availabilities.

All ^{14}C-labeled soils were combusted for total ^{14}C determination and extracted with appropriate reagents for parent compound identification. Plant samples were also dried, combusted (as appropriate), and extracted for GC/MS or HPLC identification of the parent compound. Compound confirmation tremendously increased experimental complexity and completion time, but was critical to obtaining meaningful results(1,11,12,27).

Chemical loadings resulted in final soil-sludge rates from 0 (controls) to as high as 10 mg/kg. The higher rates exceeded (typically, far exceeded) normally expected pollutant concentration associated with median sludges applied annually at agronomic rates.

Plant contamination with a chemical may be expressed variously. We utilized a popular expression known as the bioconcentration factor (BCF). BCFs are the ratios of compound concentration (or ^{14}C content) in biomass to the concentration of chemical in the soil initially. The choice of contamination expression is not without consequence as demonstrated for DEHP in lettuce (Table 2). If compound concentrations in plants are expressed on a fresh weight basis, BCFs are typically 5 to 10 fold less than BCFs expressed on a dry weight basis, depending upon the harvested tissue moisture content. If final (harvest) soil DEHP concentrations are used in the denominator rather than initial soil values, BCFs increase in proportion to the degree of compound disappearance (degradation, volatilization, etc.). The effect is small for DEHP in lettuce because little DEHP soil degradation typically occurs over the lettuce growth period. However, with more rapidly degraded compounds, the effect may be large. Unfortunately, literature BCFs include values calculated on various bases and must be carefully evaluated.

Table 2. Net Lettuce Bioconcentration Factors for DEHP Based on Dry or Fresh Weight and Initial or Final Soil Concentration(1)

	Plant Concentration Basis			
	Dry wt. Basis		Fresh wt. Basis	
	Soil Concentration Basis			
Treatment	Initial	Final	Initial	Final
1	0.53	0.64	0.03	0.04
2	0.39	0.47	0.02	0.03
3	0.00	0.00	0.00	0.00
4	0.38	0.45	0.03	0.03
5	0.39	0.47	0.03	0.03
Mean+	0.42 $\pm$ 0.13++	0.51 $\pm$ 0.24	0.03 $\pm$ 0.01	0.03 $\pm$ 0.01

\+ Overall means across treatments (three replicates in each).

++ Standard deviation based on variance of the ratio dpm/g plant to dpm/g soil.

BCFs calculated on the basis of ^{14}C detected in plants and soils (combustion data) should be considered maximum values and interpreted very cautiously. ^{14}C may exist in plants in forms other than as parent compounds, representing root uptake of ^{14}C-labeled metabolites or foliar absorption of $^{14}CO_2$ from degraded parent compound. Confirmation of parent compound presence in plant extracts is absolutely necessary to accurately characterize plant uptake. Evidence is given below.

RESULTS AND DISCUSSION

BCF for the four compounds in various plant and plant parts are represented in Table 3. BCFs presented for DEHP, PCP, and DNP are calculated on the basis of ^{14}C in dry plant material divided by initial soil chemical concentrations (dpm ^{14}C/g).

Table 3. Crop Bioconcentration Factors Based on Plant Dry Weights and Initial Soil Concentrations (Sludge-Amended)[1,3,15,19]

Chemical				Carrots			Chile	
Name	Loading	Fescue	Lettuce	Tops	Roots	Peels	Plant	Fruit
	mg/kg							
DEHP	2.57	0.30	0.53	0.30	0.15	--	0.14	0.07
	7.15	0.15	0.39	0.16	0.12	--	0.10	0.06
PCBs	1.00	<0.02*	<0.02*	<0.02*	<0.02*	<0.02*	--	--
	2.50	<0.01*	<0.01*	<0.01*	<0.01*	0.09	--	--
PCP	1.10	0.90	0.16	0.49	0.80	0.99	0.21	0.61
	5.10	0.91	0.13	0.22	0.52	0.27	0.14	0.15
DNP	1.00	0.84	0.16	0.14	0.00	0.01	0.06	0.02
	10.00	1.05	0.38	0.06	0.00	0.05	0.07	0.01

* Maximum value, concentration in plant <LOD = 20 µg/kg.

The ratios for DEHP are not large (<1, indicating passive uptake), but are not negligible. Values <0.01 imply minimal contamination and little risk to animals consuming the plant material. The DEHP values are ⩾0.01, but need further consideration. With short season crops such as these, little soil degradation of DEHP occurs[7], and ^{14}C detected in the soil can be safely assumed to represent intact ^{14}C-DEHP[1]. ^{14}C in plants, however, may not exist as DEHP. Indeed, several workers[11,12,27] report plant metabolism of absorbed DEHP. Limited GC/MS analysis of methanol extracts of these plant tissues detected no ^{14}C-DEHP[1]. Thus, even if DEHP is taken up by these plants, it is apparently rapidly metabolized. The metabolic products have been tentatively identified as natural plant constituents from assimilated $^{14}CO_2$ and polar D-glucosyl conjugates of ^{14}C-DEHP[12]. The accumulated evidence thus strongly suggest that little or no intact DEHP persists in mature plants grown in DEHP-contaminated soils. Actual BCFs for DEHP are essentially zero for these plants[1].

Experiments with PCBs did not depend on ^{14}C analyses. The sludge used in this study[19] was highly contaminated (52 mg/kg) with a PCB mixture similar to Aroclor 1248. Such a sludge would not be acceptable for land application at any rate because it exceeds the 50 mg/kg limit for PCBs of EPA 40 CFR 257. We used application rates as high as 112 mg/ha, yielding a soil concentration of 2.5 mg/kg. This rate of such a highly contaminated sludge constituted a severe test of perceived PCB hazards to food chain crops.

Detailed GC/MS analysis allowed identification of individual congener classes (tri-, tetra-, and pentachloro species) as well as total PCBs. The individual class and total PCB limit of detection in plants was 20 µg/kg.

Even with such analytical sophistication, no PCB contamination of any crop or plant part was detected except in carrot peels. BCFs (Table 3) for PCBs are thus recorded as less than some number, based on the limit of detection. True values would be lower (essentially zero). Peels contained all of the PCB detected in carrots, and likely represented lipophilic tissue into which PCBs partitioned from the soil-sludge matrix. It is also possible that peel contamination with PCBs simply represented adhering soil (despite careful washing). However, an analysis of the relative abundance of the separate PCB congeners in peels versus those in soil-sludge mixtures suggest otherwise. Total PCBs in the peels are enriched with the trichloro species and slightly depleted in the tetra- and pentachloro species relative to the distributions in the soil[19]. This preferential contamination of plant material with the more soluble (less chlorinated) biphenyls is widely reported[9,24,25,29]. The difference in class distribution in peels versus soil supports the partitioning theory of plant contamination and argues against simple soil contamination.

Carrot peel contamination data from several studies[19] are summarized in Figure 1 as plant response to soil PCB loading. The data suggest a plateauing of plant contamination at high soil PCB concentrations (>2 µg/g), and a more or less linear relationship at low (0 to 1 µg/g) PCB soil loadings.

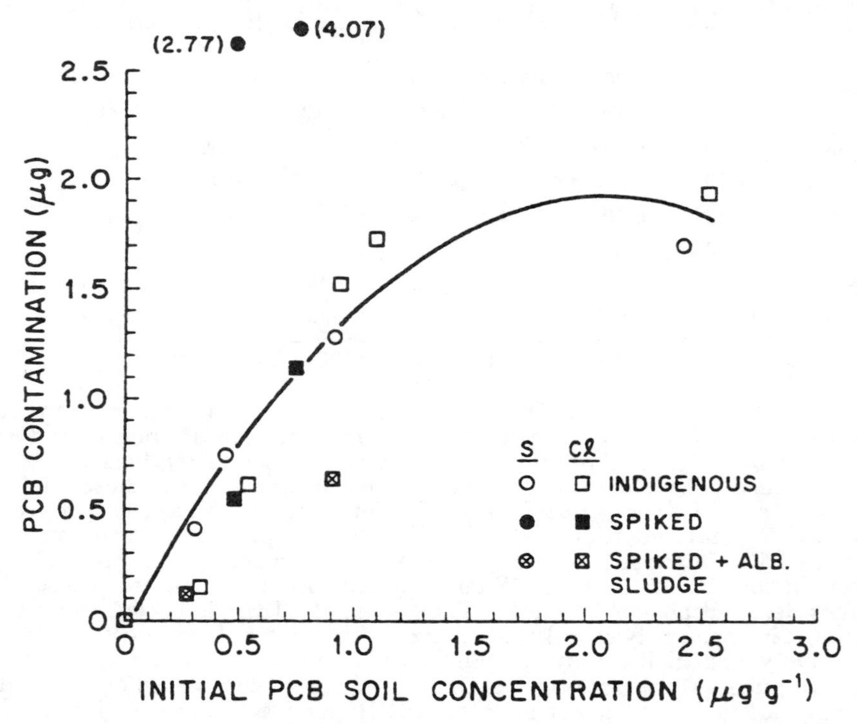

Figure 1. PCB Contamination of Carrot Peels as a Function of Soil PCB Loading and Various Treatments

More data at high PCB loadings would admittedly strengthen the plateau argument, but such high concentrations are not expected under normal (and even abnormal) circumstances. A median sludge (1 mg PCB/kg) applied at agronomic rates would result in PCB soil loadings of 0.01 to 0.1 mg PCB/kg soil. Even a hazardous waste sludge as defined by EPA 40 CFR 257 (>50 mg/kg) applied at agronomic rates would yield soil concentrations of about 0.1 to 1.0 mg PCB/kg. Thus, guidelines for land application of sludges containing PCBs should be based on loading rate rather than on sludge PCB content.

Further, allowing soil PCB concentrations of ~1 mg/kg (EPA's 40 CFR 761 reference concentration for cleanup of PCB spills on garden sites) seems reasonable. Minimal carrot contamination occurs at this soil concentration, and no detectable PCB contamination occurred in any other crop tested. Maximum PCB "availability" was observed in a sandy soil extremely low (1.2 organic carbon g/kg) in organic matter. Adding sludge to this soil reduced PCB "availability" to the same level measured in more normal soils. Apparently, as long as there is some reasonable amount of organic matter present in a soil, PCB bioavailability is extremely low and independent of soil.

Pentachlorophenol is a weak acid (pKa = 4.7) that exists primarily as the phenolate anion in neutral and high-pH soils. Anion adsorption by these soils is typically low, so PCP is expected to be readily soluble and available for plant uptake. The reported log Kow (≐ 5) applies to undissociated PCP at low pH. In high-pH soils, the effective log Kow is about 3[3]. Thus, Ryan's model[22] would suggest greater bioavailability in high-pH soils. Sludge additions increase soil retention of PCP[3], but the reaction is reversible and does not affect bioavailability[2].

The PCP BCFs for various crops (Table 3) are based on ^{14}C contents. The values suggest much greater plant uptake than with DEHP or PCBs. Extracts of plants, however, suggest little or no intact PCP in the plants[2]. Figure 2 shows various measures of plant contamination (as BCFs) as a function of soil-sludge organic carbon contents. BCFs based on total (combustion) ^{14}C decrease with organic carbon contents. This is expected if PCP adsorption

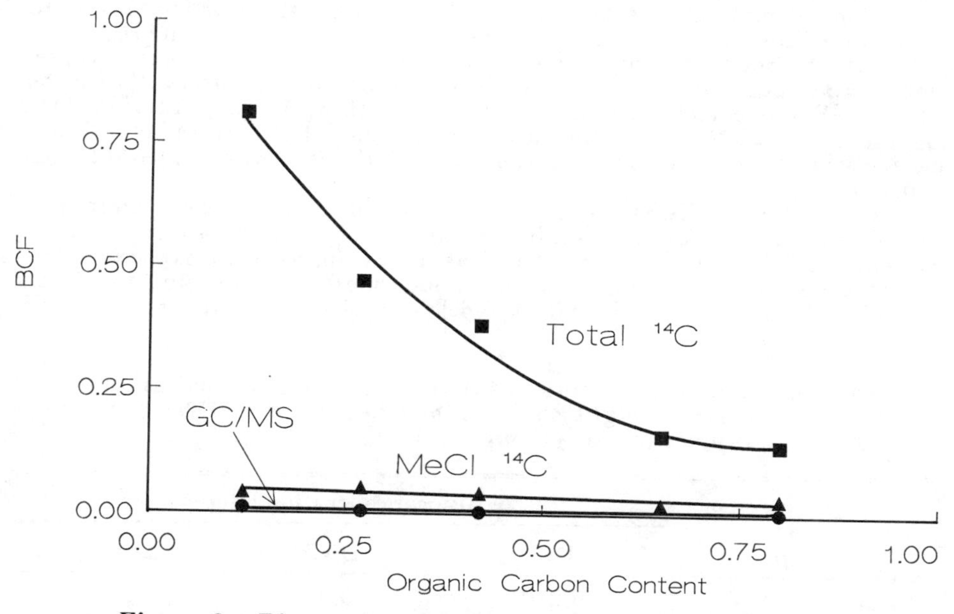

Figure 2. Bioconcentration Factors for PCP in Lettuce as a Function of Soil Organic Carbon Content

(reduced availability) increases with increasing organic carbon content. More important is the change in BCF values when they are calculated from ^{14}C extracted in methylene chloride (MeCl). Methylene chloride extracts PCP and its immediate degradation products, e.g., lower chlorinated phenols and anisols[4]. Apparently, considerable amounts of ^{14}C exist in the plants as compounds that are not PCP or its immediate degradation products. Selected GC/MS analysis of the MeCl extracts confirmed that extremely small amounts of intact PCP existed in the plants. BCFs calculated on the basis of actual PCP were <0.01 (dry weight)[2].

Extracts of soil (at harvest) also contained very low concentrations of intact PCP. Subsequent degradation studies[3] confirmed rapid PCP degradation in this soil. Apparently, PCP is degraded very rapidly in high-pH soils, leaving little chemical available for plant uptake. Recent literature[26,30] also suggests significant plant metabolism of PCP. Thus, even if some PCP were taken up, it may be metabolized, similar to the situation with DEHP.

Various reports[4,13] support the long half-life reported in Table 1. These reports, however, appear to be for acid soils where PCP is more persistent. Degradation is favored at high pH[3], and Lagas[14] reported short half-lives (10 to 15d) in high-pH systems. Casterline *et al.*[4] reported significant PCP uptake (BCF ~1) for spinach and soybean foliage grown in an acid soil (pH not reported). We have greenhouse studies underway with an acid soil (unamended and limed) to determine the effect of pH on PCP plant availability.

Work with 2,4-DNP has many similarities to the work with PCP. DNP is a weak acid (pKa = 4.1), but has a low log Kow (1.5). It is only slightly adsorbed in high-pH soils, and adsorption is only mildly influenced by sludge[20]. DNP is regarded as slowly degradable (Table 1), but degradation is favored at neutral or higher pHs[28]. Our degradation studies[20] suggest very rapid dissipation; almost complete degradation in 8d in the Glendale soil. Thus, the BCF values based on ^{14}C reported in Table 3 are again misleading. Extracts of plants were analyzed for intact DNP, but none was detected. Unfortunately, the limits of detection in most cases resulted in BCFs based on actual DNP contents that exceeded BCFs based on ^{14}C because limited plant material was available for extraction. Data for the highest initial DNP concentration (10 mg/kg), however, are useful. These data and BCFs based on ^{14}C are given in Table 4. BCFs are calculated on a fresh weight basis and represent soils both amended and unamended with sludge[15]. The last column represents our best guess of the true BCFs. Even this estimate is conservative, as no intact DNP was detected in any analysis. Further, our degradation data suggest minimal DNP existed in the soil long enough for plant uptake. Plant metabolism of even the small amounts of DNP accumulated could also occur[28].

All of the BCFs are low, suggesting minimal plant contamination with DNP. Contamination was minor regardless of sludge treatment, at DNP concentrations were at least an order of magnitude higher than expected under normal conditions. Municipal sludges do not normally contain detectable levels of DNP. The high water solubility and low log Kow of DNP promote DNP

Table 4. Various Estimates of DNP Bioconcentration Factors (Fresh Weight Basis) at the Highest Initial DNP Soil Concentration (10 mg/kg)[15]

Crop		Bioconcentration Factor		
		^{14}C	DNP*	Best Guess
Fescue		0.210	<0.040	<0.040
Lettuce		0.034	<0.010	<0.010
Carrot	Tops	0.010	<0.009	<0.009
	Peels	0.003	<0.050	<0.003
	Roots	0.000	<0.030	<0.000
Chile	Foliage	0.012	<0.008	<0.008
	Fruit	0.001	<0.019	<0.001

* These values assume that the concentration in plant tissue = detection limit.

residues in sewage effluent rather than in sewage sludge. Thus, despite having some chemical characteristics that seem to favor plant uptake, DNP is not likely to contaminate soils via sludge additions. Even if DNP existed in sludge-amended soils, its bioaccumulation by plants would be negligible (BCF ≤ 0.01).

SUMMARY AND CONCLUSIONS

We have shown that the concentrations of priority organic pollutants in normal municipal sludges are typically low (≤10 mg/kg). Sludges applied at agronomic rates result in chemical concentrations in the soil-sludge mixture 50 to 100 fold lower. Plant uptake at these pollutant concentrations (and at much higher concentrations) is minimal.

Chemicals were either (1) accumulated at extremely low levels (PCBs), (2) possibly accumulated, but then rapidly metabolized within the plants to extremely low levels (DEHP), or (3) likely degraded so rapidly in the soil that only minor contamination (BCF ≤0.01) of the plant occurred (PCP and 2,4-DNP).

Data both collected in our study and available in the literature suggest minimal plant contamination with sludge-borne organics. We conclude that the priority organic content of normal sludges should not limit normal land application practices.

REFERENCES

1. Aranda, J. M., O'Connor, G. A., and Eiceman, G. A., *J. Environ. Qual.*, 1989, *18*, 45-50.
   Table 2 is reprinted from Ref. (1) by courtesy of the American Society of Agronomy, Inc.

2. Bellin, C. A. and O'Connor, G. A., submitted for publication in *J. Environ. Qual.*

3. Bellin, C. A., O'Connor, G. A., and Yan, Jin, submitted for publication in *J. Environ. Qual.*

4. Casterline, J. L., Barnett, N. M., and Ku, Y., *Environ. Res.*, 1985, *37*, 101-118.

5. Connor, M. S., *Biocycle*, 1984, January/February, 47-51.

6. Davis, R. D., Howell, K., Oake, R. J., and Wilcox, R., in *Proc. Internat. Conf. Environ. Contamin.*, CEP Consultants, Edinburgh, UK, 1984, 73-79.

7. Fairbanks, B. C., O'Connor, G. A., and Smith, S. E., *J. Environ. Qual.*, 1985, *14*, 479-483.

8. Fairbanks, B. C., O'Connor, G. A., and Smith, S. E., *J. Environ. Qual.*, 1987, *16*, 18-25.

9. Iwata, Y. and Gunther, F. A., *Arch. Environ. Contamin. Toxicol.*, 1976, *4*, 44-49.

10. Jacobs, L. W., O'Connor, G. A., Overcash, M. R., Zabek, M. J., and Rygwiecz, P., in *Land Application of Sludge*, A. L. Page, T. J. Logan, and J. A. Ryan, eds., Lewis, Chelsea, WI, 1987, 101-143.

11. Kato, K., Nakaoka, T., and Ikeda, H., *Chem. Abstr.*, 1981, *95*, 60034k.

12. Kloskowski, R., Scheunert, I., Klein, W., and Korte, W., *Chemosphere*, 1981, *10*, 1089-1100.

13. Kuwatsuka, S. and Igarashi, M., *Soil Sci. Plt. Nutr.*, 1975, *21*, 405-414.

14. Lagas, P., in *Environmental Contamination*, A. A. Orio, ed., CEP Consultants, Edinburgh, UK, 1988, 264-266.

15. Lujan, J. R. and O'Connor, G. A., submitted for publication in *J. Environ. Qual.*

16. McFarlane, C., Nolt, C., Wickliff, C., Pfleeger, T., Shimabuku, R., and McDowell, M., *Environ. Toxic. Chem.*, 1985, *6*, 847-856.

17. Naylor, L. M. and Loehr, R. C., *Biocycle*, 1982, July/August, 18-22.

18. Naylor, L. M. and Loehr, R. C., *Biocycle*, 1983, November/December, 37-42.

19. O'Connor, G. A., Kiehl, D., and Eiceman, G. A., *J. Environ. Qual.*, in press.

20. O'Connor, G. A., Lujan, J. R., and Yan, Jin, submitted for publication in *J. Environ. Qual.*

21. Roger, H. R., Technical Report No. PRD 1593-M, WRC Environment, Medmenham, England, 1987.

22. Ryan, J. A., Bell, R. M., Davidson, J. M., and O'Connor, G. A., *Chemosphere*, 1988, *17*, 1199-2323.

23. Sawhney, B. L., in *PCBs and the Environment*, J. S. Waid, ed., CRC, Boca Raton, FL, 1988, Volume 1, Chapter 2.

24. Sawhney, B. L. and Hankin, L., *J. Food Protect.*, 1984, *47*, 232-236.

25. Sawhney, B. L. and Hankin, L., *J. Food Protect.*, 1985, *48*, 442-448.

26. Schafer, W. and Sandemann, Jr., H., *J. Agric. Food Chem.*, 1988, *36*, 370-377.

27. Schmitzer, J. L., Scheunert, I., and Korte, W., *J. Agric. Food Chem.*, 1988, *36*, 210-215.

28. Shea, P. J., Weber, J. B., and Overcash, M. R., Residue Rev., 1983, *87*, 2-41.

29. Suzuki, M., Alzawa, N., Okano, G., and Takahashi, T., *Arch. Environ. Contamin. Toxicol.*, 1977, *5*, 343-352.

30. Weiss, U. M., Moza, P., Scheunert, I., Haque, A., and Korte, F., *J. Agric. Food Chem.*, 1982, *30*, 1186-1190.

31. Yan, Jin and O'Connor, G. A., submitted for publication in *J. Environ. Qual.*

USING WASTE MANAGEMENT PRODUCTS AS AN ALTERNATIVE TO DEPLETING NONRENEWABLE FERTILIZER SOURCES

R. Velagaleti and T. McClure
Battelle Columbus Division
505 King Avenue
Columbus, Ohio 43201-2693

ABSTRACT

Use of nitrogen (N), phosphorus (P), and potassium (K) fertilizers are essential for sustaining the agricultural productivity required for feeding the global population. However, the raw materials for the manufacture of these three essential nutrients for crop plants are nonrenewable and are concentrated in few geographical regions. In addition, manufacture of these fertilizers is a highly energy intensive process. This paper identifies the fertilizer raw material sources, their geographical distribution, and the potential for a world food crisis in the event of an energy crisis or political turmoil in the regions where the raw materials are concentrated. Waste management products are suggested as renewable alternatives. Potential of animal, food processing, and municipal wastes to meet the N, P, and K demands of the crop plants are discussed. High water content of these wastes is the major constraint limiting their use in productive agriculture, and development of innovative separations technologies will be required to remove this constraint.

INTRODUCTION

Agriculture is the major industry in the United States (Table 1) and many other countries of the world. Use of fertilizers is an important component in sustaining the agricultural industry and feeding the global population. Figure 1 shows a striking correlation between the increased use of fertilizers and crop yields for the period 1950 to 1985[2,20,FAO]. An estimated 133 million metric tons of fertilizer was consumed worldwide during the year 1987[20] (Table 2), compared with 14 and 63 million metric tons in 1950 and 1970, respectively[2].

Use of nitrogen (N), potassium (K), and phosphorus (P) fertilizers is essential for achieving optimal crop productivity. World fertilizer resources are nonrenewable, and the reserves have restricted geographical distribution.

Currently, a number of countries have serious food deficiencies and, consequently, experience malnutrition because of low food productivity resulting from inadequate fertilizer use. Such countries are either not endowed with these raw materials required for manufacturing fertilizers, or do not have a sufficient financial resource base to import fertilizers or develop their own limited, but potential, fertilizer raw material base. Although the world outlook for nonrenewable resources required for fertilizer production may not sound alarming in this short-term scenario, the outlook for some countries, especially in Africa, is discouraging because of their lack of financial resources to purchase fertilizers required for food production

In the long term, the worldwide reserve of raw materials needed for nitrogen and phosphorus fertilizers is not comfortable, although potassium reserves may last longer than nitrogen and phosphorus reserves. In addition, as described below, the manufacturing of finished fertilizer products from their raw materials is highly energy intensive. Therefore, in addition to the availability of the raw materials, the energy costs involved in manufacturing and transporting fertilizers will play a significant role in their uninterrupted availability and affordability throughout the world.

With predicted population increases in many of the poorer countries of the world and the consequent demands on food production, there is an urgent need to look at alternative sources that could substitute for the fertilizer needs of crops. Manures, municipal sewage, industrial effluents, and food processing wastes are some renewable alternatives that contain some N, P, or K, and therefore have a potential for use as fertilizers in vegetable and food crop plants. The solid/liquid separations technologies could improve the efficiency of utilizing these waste uses in agriculture.

Table 1. A Statistical Profile of Agriculture and Food in the United States

- Agriculture and Food (Production, Processing, and Distribution) Industry is the Largest Industry in the United States.
- Population = 244.6 Million on January 1, 1988.
- Farm Assets = $709 Billion on December 31, 1987.
- Food and Beverage Consumption = $552 Billion in 1988; 17% of Total Personal Consumption Expenditures.
- Employment = 21 Million People Employed, Largest U.S. Employer
- Number of Farms = 2.2 Million Farms Employing 2.8 Million Workers in 1987.
- Other Agriculture Dependant Industries = 18.3 Million Workers .
- Land in Farms = 999 Million Acres in 1988.
- Average Size of Farm = 459 Acres.
- Total Fertilizer (N,P,K) use = ~19 Million Metric Tons in 1987.

  N = ~10 Million Metric Tons
  P = ~4 Million Metric Tons
  K = ~5 Million Metric Tons

Sources: CAST, 1988; USDA, 1988 and 1989; World Almanac, 1989.

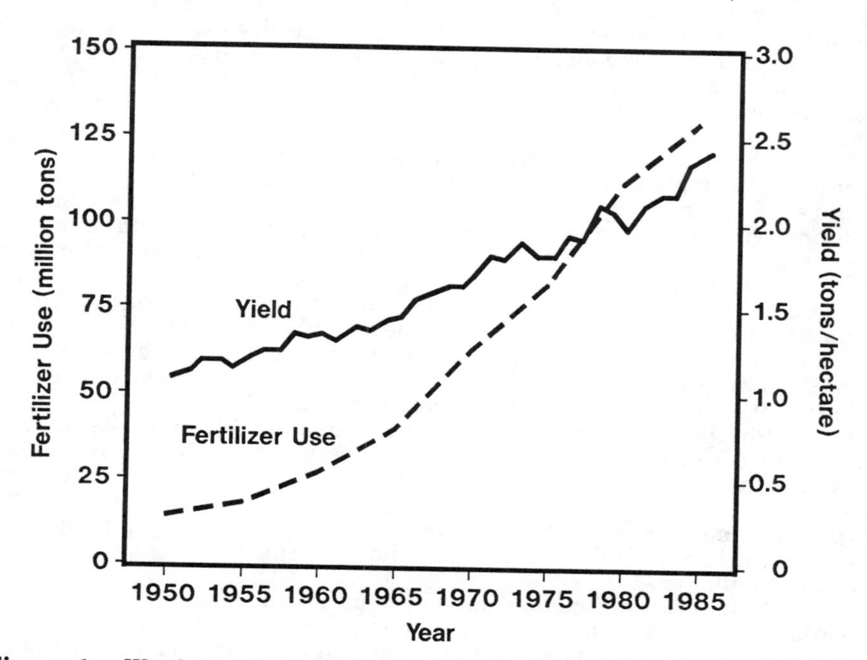

Figure 1. World Fertilizer Use and Grain Yield per Hectare (1950-1985)

Table 2. World Fertilizer Consumption

Nutrient	Million Metric Tons		Percent Change
	1986	1987	(1986 to 1987)
N	70.0	72.4	+3
P	33.2	34.7	+5
K	25.6	26.1	+2
TOTAL (N+P+K)	128.8	133.2	+3

Source: USDA, 1989.

In this paper, we present the importance of fertilizers in food production, their availability, and enumerate the various alternative organic waste sources and their potential for supplying the N, P, and K required for crop growth.

THE IMPORTANCE OF N, P, AND K IN FOOD PRODUCTION

Both the quality and quantity of the food produced by agricultural crops is dependent on the adequate use of fertilizers, especially N, P, and K. Food production would be significantly and drastically reduced without the application of these three major elements. Since a number of animal products (poultry, meat, and dairy products) are derived from animals fed with grain or other feed derived from the crop plants, food crop production is critical for human sustenance.

The U.S. Department of Agriculture stated "the use of nitrogen fertilizer alone is credited with providing one-third of the productive capacity of the crops"[18]. For example, if the use of nitrogen fertilizer is to be limited to 50 pounds per acre, an additional 18 million acres of cropland would be required to obtain productivity equal to that obtained without this restriction on nitrogen use. Thus, nitrogen ranks high among the plant nutrients that most often limit crop production. Supplemental nitrogen fertilization is essential for efficient production of nonlegumes. Among the crop plants, only the legume crops are endowed with an ability to fix atmospheric nitrogen in symbiosis with the soil bacterium, *Rhizobium*, and because of this property, the need for supplemental nitrogen fertilization in these crops is minimal.

Phosphorus is an important component of the life processes in plants and animals because it is an important constituent of the molecules involved in the genetic and energy transfer processes[12]. Phosphorus deficiency would therefore limit the efficiency and the productivity of the organisms. Phosphorus deficiency, in addition to limiting the productivity of crop plants, also may affect the human and animal health whose phosphorus nutrition is derived from the crop plants[14].

Potassium deficiency may disrupt photosynthesis, respiration, translocation, and a number of enzyme system functions, resulting in reduced crop growth and crop quality[21]. An adequate supply of potassium is necessary for maximizing crop productivity.

RESOURCE POSITION OF N, P, AND K FERTILIZERS

The limited global and regional availability of the N, P, and K fertilizers and the inability of the countries to generate resources to manufacture or import fertilizers will drive the search for alternative sources of plant nutrients. Although global shortages are not imminent in the next decade, the fact that the raw materials for N, P, and K fertilizers are concentrated in a handful of countries (Table 3) creates a potential for unanticipated shortages in the event of a political crisis in these countries[3,8,15].

NITROGEN FERTILIZER

Nearly all the nitrogen present in the harvested crops in the United States until the early 1900s was derived from residue decomposition and the natural processes in the soil, as well as conversion of atmospheric nitrogen to

Table 3. Geographical Distribution of Raw Materials for N, P, and K Fertilizers (60 Percent or Above of World Total Reserves)

Fertilizer	Geographical Distribution and Major Producer Countries
N	Middle East Saudi Arabia Kuwait Iran Iraq United Arab Emirates Libya
P	Northwest Africa Morocco Western Sahara
K	North America Canada Asia and Europe Soviet Union

Sources: N (Flavin, 1986)
K (Sheldrick, 1985)
P (Cathcart, 1980)

ammonia and other plant-available forms of nitrogen by biological nitrogen fixation or lightening. Until the late 1940s rotation farming involving a legume was practiced to take advantage of the biologically fixed nitrogen. Use of synthetic nitrogen fertilizers derived from fossil fuels came into being only after World War II. Profound increases in feed and food grain and oilseed crop (corn, wheat, soybean) acreage and production, and a concomitant decrease in legume green manure crops, have been noticed since then[6]. World consumption of nitrogen fertilizers rose from 6 million tons in 1955 to 28.5 million tons in 1970[9] to 74 million tons in 1988[20], an average increase of almost 8 percent per year over the past 33 years.

Almost all the nitrogen fertilizer is derived from the nonrenewable, fossil fuels. According to one estimate, the U.S. domestic oil and natural gas supplies are likely to be depleted by the year 2020. World natural gas reserves are expected to last only until the year 2040, and oil reserves are expected to last until approximately 2020[7], at the 1979 use rate. Nitrogen fertilizer shortages are expected to follow fossil fuel shortages closely. In addition, the United States and a number of other countries in the world are heavily dependent on the fossil fuel reserves located in often politically volatile areas. Thus, alternative sources of plant nutrient nitrogen should be in place to maintain crop productivity.

PHOSPHORUS FERTILIZER

World phosphate fertilizer consumption has increased from 3.3 million metric tons of P in 1955 to 8.15 in 1970[9] to 35.8 million metric tons in 1988[20]. According to one source, at the 1974 consumption rate, phosphate rock reserves would last for 128 years[11]. However, using the U.S. Bureau of Mines and Geological Survey's projected 3.6 percent annual increase in phosphate use, the known reserves are adequate for only 61 years and known-plus-anticipated reserves are adequate for only 88 years[7]. In addition, large phosphate reserves are located in northwest Africa, a region of political instability[3,7]. Thus, the phosphate reserve position also appears precarious, warranting immediate research into alternative P resources.

POTASSIUM FERTILIZER

World potassium consumption has increased from 5.33 million metric tons of K in 1955 to 13.12 in 1970[9] to 26.5 million metric tons in 1988[20]. The world supplies of potash reserves are expected to last 3,638 years according to the 1974 consumption rate. However, almost 90 percent of the world potash reserves are located in Canada and the Soviet Union[15], presenting difficulties of availability in the free markets in the event of political exigencies[7].

ENERGY NEEDS OF FERTILIZER MANUFACTURE

Fertilizer manufacture, from raw materials to finished products, and their subsequent transportation and application are all highly energy intensive processes. Nearly 646 million barrels of oil equivalent are used worldwide for fertilizer manufacture (second only to tractor fuel with worldwide consumption of 739 million barrels of oil equivalent), constituting almost a third of the total energy use in world agriculture[1]. In the United States, fertilizer manufacture far exceeds other farming operations in terms of energy consumed[5].

Fertilizer manufacturing is, therefore, a highly energy intensive process. Socioeconomic impacts during an energy crisis, such as those that occurred in the 1970s, could seriously affect the manufacturing process itself and result in spiralling costs. Disruptions and economic upheavals during an energy crisis could lead subsequently to a serious food crisis.

WASTE MANAGEMENT PRODUCTS AS ALTERNATIVE SOURCES OF N, P, AND K

By far the most important of the organic wastes from an agricultural perspective is the crop residue, which accounts for almost 390 million metric tons in the United States and contains almost 7 million metric tons of nitrogen. Other sources of organic wastes include animal manure, sewage sludge, food processing wastes, industrial organic wastes, municipal refuse, and logging and wood manufacturing wastes. Annual production of these wastes and patterns of current use on land are presented in Table 4. A brief discussion on the potential of animal manures (Table 5) and food processing wastes (Table 6) is presented here. Other presentations in this session deal in detail with the crop productivity potentials and limitations of municipal wastes. Excellent reviews are published on the availability of N by Smith and Peterson[16], P by Sommers and Sutton[17], and K by Xie and Hasegawa[23] in various organic wastes.

Table 4. Estimated Annual Production of Organic Wastes in the United States

Type of Waste	Total Production (thousands of dry tons)	Percent of Total	Current Use on Land (%)
Animal Manure	158,730	21.8	90
Crop Residues	391,009	53.7	68
Sewage Sludge and Septage	3,963	0.5	23
Food Processing	2,902	0.4	(13)
Industrial Organic	7,452	1.0	3
Logging and Wood Manufact.	32,394	4.5	(5)
Municipal Refuse	131,519	18.1	(1)
TOTAL	727,969	100	

Source: Parr et.al.
() Values in parenthesis are Estimates Based on Insufficient Data.

ANIMAL MANURES

According to a 1976 estimate, domestic animals in the United States produced about 2 billion tons of wet manure or 300 million tons of dry manure annually, of which about 50 percent is collectible[10]. The collectible manure was enough to fertilize 70 million acres or 18 percent of the cropland in the United States in 1975[5]. Collectible manures have estimated available nitrogen, phosphorus, and potassium equivalent to 19, 38, and 61 percent,

Table 5. Average Moisture (Percent) and N, P, and K (Pounds of Nutrient per Ton of Manure) Content of Animal Manures

Manure From	Moisture in * Fresh Manure	Nitrogen	Phosphorus	Potassium
Dairy Cattle	79	11	2	10
Beef Cattle	80	14	4	9
Swine	75	10	3	8
Horses	60	14	2	12
Sheep	65	28	4	20
Poultry	80	31	8	7

Sources: *Smith and Peterson (1982)
Doane's Agricultural Report (1989)

Table 6. Nitrogen Content of Wastes from Food Processing Industry

Wastes From	Total Nitrogen Content(mg/liter)
Vegetable Processing	
Snapbeans	31.2
Sweet Corn	61.9
Brussels-sprouts	5.7
Beets	66.4
Peas	44.7
Tomatoes	6.8
Cabbage	31.3
Corn	27.3
Potatoes	55.0
Fruit Processing	
Apple	2.2
Pear	2.6
Grapes	49.9
Citrus	7.8
Meat Packing Wastes	
Catch Basin Efflueut	124
Extended Aeration Influeut	92
Nontreated Slaughter Waste	69
Dairy Manufacturing Waste	
Whey	685
Fresh Milk Packaging	24
Curds	18

Source: Smith and Peterson (1982)

respectively, of the amounts of these materials in manufactured commercial fertilizers[4]. In addition, animal manures are beneficial to the structure and water-holding capacity of the soils. Nitrogen, phosphorus, and potassium contents of various liquid manures and other data are provided in Table 2.

Animal wastes contain nearly 75 to 90 percent water. The energy and the labor costs involved in transportation and application of these wastes are the major constraints. Solid/liquid separations technologies should be applied to the separation of the nutrients from these animal wastes.

FOOD PROCESSING WASTES

Both liquid and solid wastes are produced from food processing. The liquid wastes include effluents from the processing of potatoes, sugarbeets, vegetables, fruits, meats, or dairy products; or they may be peel wastes, pulp, etc. Whey or meat processing wastes have the highest nitrogen content among food processing wastes (Table 6). Almost all wastes from food processing could be used for fertilizing agricultural land. In one instance, 160 to 490 cm of wastewater applied by the potato processors supplied 1,080 to 2,200 kg N/ha[16]. Wastewater from sugarbeet processing liquid waste supplied nearly 1,330 kg N/ha from 142 cm of water. Solid wastes from the fruit processing industry could supply from 1.2 to 1.8 kg/N metric ton of waste. Meat and dairy processing waste have high nitrogen content; up to 124 mg N/L and 685 mg N/L, respectively, have been reported[16]. Sommers and Sutton[17] suggested that up to 55 cm/ha of industrial waste effluent will be required to supply 50 kg P/ha containing 10 mg of P/L. These wastes are high in water content, and the separations technologies may help reduce the water content and increase the economy of application.

MUNICIPAL WASTES

Sewage and sludge effluents are used in significant amounts on agricultural lands (Table 4). The removal of water and concentration of the nutrients by appropriate technologies will facilitate rapid expansion of the use of this renewable waste resource for agricultural use. This area will be extensively discussed by other conference participants.

SUMMARY AND CONCLUSIONS

Nitrogen, phosphorus, and potassium are essential crop nutrients. Optimal use of commercially available fertilizers containing these nutrients, along with improved plant varieties, has resulted in an abundance of food production required for human and animal consumption. Barring some exceptions, these increases in worldwide food production have helped revitalize the economies of many nations, prevented malnutrition, and created food surpluses to buffer the occasional food shortages from bad weather and plant pests. However, current agricultural production is overdependent on nonrenewable resources. The reserves are limited and geographically located in politically sensitive zones. Manufacture of fertilizers is a highly energy intensive process and is heavily sensitive to an energy crisis. Even at current rates, many developing countries cannot afford commercial fertilizers. Renewable organic wastes are a source of N, P, and K, and are constantly generated by human civilization. However, most of these wastes are high in water content, limiting their transportation and economical use. Solid/liquid separations technologies should evolve to address this need.

REFERENCES

1. Brown, L. R., "Sustaining World Agriculture," in *State of the World*, World Watch Institute, Washington, DC, 1987, 122-138.

2. Brown, L. R., "Reexamining the World Food Prospect," in *State of the World*, World Watch Institute, Washington, DC, 1989, 41-58.

3. Cathcart, J. B., "World Phosphate Reserves and Resources," in *The Role of Phosphorus in Agriculture*, American Society of Agronomy, Madison, WI, 1980, 1-18.

4. Council for Agricultural Science and Technology (CAST), *Utilization of Animal Manure and Sewage Sludge in Food and Fiber Production*, Ames, IA, 1975.

5. Council for Agricultural Science and Technology (CAST), *Energy Use in Agriculture: Now and for the Future*, Ames, IA, 1977.

6. Council for Agricultural Science and Technology (CAST), *Organic and Conventional Farming Compared*, Ames, IA, 1980.

7. Council for Agricultural Science and Technology (CAST), *Long-Term Viability of U.S. Agriculture*, Ames, IA, 1988.

8. Flavin, C., "Moving Beyond Oil," in *State of the World*, World Watch Institute, Washington, DC, 1986, 79-97.

9. Harre, E. A., Garman, W. H., and White, W. C., "The World Fertilizer Market," in *Fertilizer Technology and Use*, R. A. Olson *et al.*, eds., Soil Society of America, Madison, WI, 1971, 27-55.

10. Heichel, G. H., "Agricultural Production and Energy Resources," *Amer. Scientist*, 1976, *64*, 64-72.

11. Landsberg, H., Tilton, J., and Haas, R., "Nonfuel Minerals," in *Current Issues in Natural Resource Policy*, P. Portney, ed., Resources for the Future, Washington, DC, 1982.

12. Ozanne, P. G., "Phosphate Nutrition of Plants--A General Treatise," in *The Role of Phosphorus in Agriculture*, American Society of Agronomy, Madison, WI, 1980, 559-590.

13. Parr, J. F., Miller, R. H., and Colacico, D., "Utilization of Organic Materials for Crop Production in Developed and Developing Countries," in *Organic Farming, Current Technology and Its Role in a Sustainable Agriculture*, American Society of Agronomy, Madison, WI, 1984, 83-96.
   Table 4 is reprinted from Ref. (13) by courtesy of the American Society of Agronomy, Inc.

14. Reid, R. L., "Relationships Between Phosphorus Nutrition of Plants and the Phosphorus Nutrition of Animals and Man," in *The Role of Phosphorus in Agriculture*, American Society of Agronomy, Madison, WI, 1980, 847-886.

15. Sheldrick, W. F., "World Potassium Reserves," in *Potassium in Agriculture*, R. D. Munson, ed., American Society of Agronomy, Madison, WI, 1985, 1-28.

16. Smith, J. H. and Peterson, J. R., "Recycling of Nitrogen Through Land Application of Agricultural, Food Processing, and Municipal Wastes," in *Nitrogen in Agricultural Soils*, F. J. Stevenson, ed., American Society of Agronomy, Madison, WI, 1982, 791-831.
Table 6 is reprinted from Ref. (16) by courtesy of the American Society of Agronomy, Inc.

17. Sommers, L. E. and Sutton, A. L., "Use of Waste Materials as a Source of Phosphorus," in *The Role of Phosphorus in Agriculture*, American Society of Agronomy, Madison, WI, 1980, 515-543.

18. United States Department of Agriculture (USDA), "Energy to Keep Agriculture Going," Energy Letter, December, 1973.

19. United States Department of Agriculture (USDA), "Agricultural Statistics," U.S. Government Printing Office, Washington, DC, 1988.

20. United States Department of Agriculture (USDA), "Agricultural Resources, Situation and Outlook Report," Economic Research Service, USDA, Washington, DC, 1989.

21. Usherwood, N. R., "The Role of Potassium in Crop Quality," in *Potassium in Agriculture*, R. D. Munson, ed., American Society of Agronomy, Madison, WI, 1985, 490-514.

22. World Almanac, *The World Almanac and Book of Facts*, An Imprint of Pharos Books, Scripps Howard Company, New York, NY, 1989.

23. Xie, J-C and Hasegawa, M., "Organic and Inorganic Sources of Potassium in Intensive Cropping Systems: Experiences in the People's Republic of China and Japan," in *Potassium in Agriculture*, R. D. Munson, ed., American Society of Agronomy, Madison, WI, 1985, 1177-1199.

SEWAGE SLUDGE APPLICATION TO AGRICULTURAL LAND

Arun R. Gavaskar, Mick F. Arthur,
Barney W. Cornaby, G. B. Wickramanayake,
and Thomas C. Zwick
Battelle Memorial Institute
505 King Avenue
Columbus, Ohio 43201-2693

ABSTRACT

Application of sewage sludge to agricultural land has the advantage of providing an economical disposal alternative, in addition to supplying plant needs of nitrogen and phosphorus. Application rates are generally based on nitrogen requirements of crops. Phosphorus accumulation in soil and runoff to surface waters may limit the amount of sludge applied and the type of application. A model developed to predict annual application rates in some of the major crop-producing states, based on nitrogen requirement and phosphorus runoff potential, is also presented.

Sewage sludge is a nutrient-rich resource that has the potential to enhance the quality of cropland and orchard soils for increased fertility and improved soil structure. Sludge application to agricultural lands can also provide an alternative disposal method for an otherwise underused, valuable resource. However, as with commercial fertilizers, sludge must be applied to agricultural soils using proper management practices.

Sludge application rates are primarily based on the nitrogen requirement of a crop and the rate at which nitrogen in the sludge becomes available to the crop. However, phosphorus build-up in soil or surface runoff to streams may limit the amounts of sludge applied and the type of application in a watershed. These and other land management factors must be taken into account for economical and environmentally safe sludge application.

SEWAGE SLUDGE CHARACTERISTICS

One of the important processes in raw sludge treatment is sludge digestion, which can be accomplished in several ways: anaerobic or aerobic digestion, sludge lagoons, and Imhoff tanks[3]. These stabilization processes reduce the volume of raw sludge by 25 to 40 percent. The treated (digested) sludge is dark in color, has a homogenous texture, and is less odorous than primary and secondary sludges. Different treatment methods result in different sludge characteristics. Treated sludges contain nutrients (e.g., nitrogen, phosphorus, potassium) that are required for plant growth and may be used as soil conditioners and fertilizers.

POTENTIAL BENEFITS OF SLUDGE APPLICATION TO LAND

The benefits of applying sludge to croplands are two-fold. First, when properly used, sludge is a valuable resource as a soil conditioner and fertilizer. Based on 1983 fertilizer prices in the south central United States, a metric ton of dry sludge contains approximately $9.08 worth of nitrogen, $28.33 worth of phosphorus, and $0.66 worth of potassium[8]. The nutrients in sludge may be utilized by virtually all crops and can be as effective as commercial fertilizers. Studies conducted with 16 different types of barley indicated that sewage sludge was as valuable a nutrient source as a commercial fertilizer[1]. In addition to nitrogen, phosphorus, potassium, and sulfur, sewage sludge contains varying concentrations of micronutrients such as calcium, magnesium, sodium, copper, zinc, boron, molybdenum, iron, manganese, and cobalt. Land-applied sludge also acts as a soil conditioner to facilitate nutrient uptake, increase water retention, permit easier root penetration, and improve soil texture.

Second, land application is an effective method of managing the ever increasing quantities of municipal sewage sludge. In the United States, municipalities annually generate approximately 6.5 million dry tons of wastewater sludge[8]. This is equivalent to 56 dry pounds of sludge per person per year. By the year 2000, it is anticipated that the amount of sludge generated will double to approximately 13 million dry tons per year. Farmland to which sludge is applied represents a low-cost disposal alternative as opposed to the higher costs of incineration, landfilling, and dedicated land disposal.

POTENTIAL PROBLEMS WITH SLUDGE APPLICATION

In addition to plant nutrients, sewage sludge contains materials that are either nonessential to plant growth or harmful to the environment if available in excessive amounts. Potential contaminants can be transported from sludge to the environment. With relatively high rates of sludge application, plant uptake and leaf adsorption can either result in phytotoxicity or affect animals or humans who consume the plant. Other problems possible from excessive sludge application are surface water and groundwater pollution.

Chemical pollutants in sludge include nutrients, heavy metals, and toxic organic compounds. Nutrients such as nitrogen and phosphorus, when present in excess, can cause groundwater and surface water pollution. Increased nutrient levels in receiving waters may lead to elevated growth of algae and other aquatic plants. Although the accumulation of toxic metals is not expected to be a problem with most sludges, the potential for metal availability can be minimized further by maintenance of soil pH at 6.5 or greater. For surface-applied sludge on pastures and forages, it is recommended that sufficient time be allowed for incorporation of the sludge into the soil prior to harvesting or grazing. The time necessary will vary but should be approximately 1 month unless sufficient rainfall occurs to wash the sludge from the plant leaves.

Pathogens in sludges are significantly reduced during sludge digestion, but are not totally eliminated. However, the concentration of harmful microorganisms in sludges can be reduced to extremely low levels by disinfection or other treatment techniques prior to land application. Survival of pathogens contained in sludge after its application to land is an important consideration. The survival times for common bacterial pathogens and viruses on crops are reported to be less than 2 months[7]. Protozoan cysts and helminth eggs deposited on plant surfaces are expected to die off rapidly from exposure to sunlight and desiccation. According to an EPA report[7], no serious disease problems have resulted from the application of stabilized sludges on agricultural cropland.

Runoff to streams is also a potential problem. It is recommended that sludge be incorporated into the soil whenever possible instead of being applied to the surface. Since heavy rainfall can cause significant surface erosion, sludge applied with the intention of incorporation into soils should be incorporated within 24 hours of spreading. In addition, other established criteria relevant to sludge runoff should be followed. Whenever sludge is applied, normal farming practices to control erosion should be followed; for example, contour farming, terracing, sod waterway construction, and vegetative fence rows.

Despite these potential problems, sewage sludge has been used successfully in many land application systems. The key to successful use in agriculture has been and will continue to be proper management, coupled with application rates that are based on the fertilizer needs and tolerance of the crop, instead of rates to maximize sludge disposal.

SLUDGE APPLICATION RATES

The annual sludge application rate can be based on the nitrogen requirements for a specific crop because nitrogen is the major plant nutrient available in sludge. The actual nitrogen availability from sludge varies considerably depending on the type of sludge, the soil type, the method of application, and the rate at which organic nitrogen in sludge is converted to plant-available nitrogen.

Nitrogen in sludge occurs as ammonia, nitrate, or organic nitrogen. However, the rate at which nitrogen from sludge becomes available varies considerably. Organic nitrogen in sludge becomes available to crops at a fairly predictable rate[4]. However, some ammonia nitrogen will volatilize depending on the type of soil and the type of application, whereas nitrate-nitrogen is considered almost totally available. The plant-available nitrogen in sludge can be estimated as follows:

1. Ammonia-N availability:

	Soil		
	Coarse	Medium	Fine
Surface Application	50%	50%	50%
Incorporated Application	50%	80%	85%

2. Organic-N availability: 20 percent of the organic-N applied in the first year is available immediately; in the second, third, fourth, and fifth year after application, 10, 5, 2.5, and 1.25 percent, respectively, of the residue from the first year is assumed to be available.

3. Nitrate-N availability: 100 percent of the nitrate-N is assumed to be available.

Thus, the amount of sludge required in each following year decreases as organic-N from previous years becomes available. After the fifth year, additional organic nitrogen is not available to the crops from sludge that was applied 5 or more years earlier. An equilibrium situation is thus reached in the fifth year. Hence, the fifth year rate can be taken as a conservative rate of sludge application for all the 5 years and for all following years. The remaining nitrogen requirement in the first 4 years can be made up with fertilizers. This ensures maximum long-term utilization of both sludge as well as land. If metals must be taken into account, this conservative approach may also extend the life of the site by reducing heavy metal accumulation in the soil.

The addition of nitrogen to soils in excess of crop needs results in the potential for NO_3^- contamination of groundwater. In this approach, because the amount of plant-available nitrogen applied to soils in sewage sludge is consistent with state fertilizer recommendations for the crop, the potential for nitrate $(NO_3)^-$ contamination should be no greater than that caused by conventional nitrogen fertilizers[7].

Many sludges may contain appreciable phosphorus concentrations, and applying sludges at a rate to provide all the nitrogen needed by a crop could add more phosphorus than is required by the crop. Therefore, it is important to monitor soil phosphorus by annual soil fertility tests. Generally, phosphorus accumulates in the soil and is available in future years. If soil phosphorus reaches maximum levels allowed by state environmental regulations, then annual sludge applications will have to be reduced or terminated for a period of time.

An additional consideration concerns phosphorus runoff potential. Sludge that runs off to surface water contains phosphorus, which in excessive amounts may cause eutrophication. The phosphate-phosphorus limit for streams is 100 µg/L to prevent growth of nuisance aquatic organisms (eutrophication)[5]. Because relatively uncontaminated surface waters contain from 10 to 30 μg/L phosphate-phosphorus, no more than 70 μg/L of additional phosphate-phosphorus may be added to surface waters due to sludge application and fertilization.

The potential for phosphorus runoff from sludge for a typical watershed can be estimated. First, the sludge runoff from an acre of land of the watershed can be calculated using the Universal Soil Loss Equation (USLE):

$$Y = R \times K \times LS \times C \times P \times S_d$$

where

Y = sediment loading from surface erosion, tons/acre
R = the average rainfall factor; it is a summation of individual storm rainfall kinetic energy, in hundreds of foot-tons per acre, times the maximum 30-min rainfall intensity, in inches per hour, for all significant storms, on an average annual basis
K = the soil erodibility factor, commonly expressed in tons per acre per R unit
L = the slope-length factor (dimensionless ratio)
S = the slope-steepness factor (dimensionless ratio)
C = the cover factor (dimensionless ratio)
P = the erosion control practice factor (dimensionless ratio)
S_d = the sediment delivery ratio (dimensionless).

The rainfall erosivity factor (R) has been compiled for the USLE[6]. The R factor is a measure of the erosive force of specific rainfall events. The soil erodibility factor (K) is in dry tons/acre/unit R and varies from 0.12 to 0.7 tons/acre/R depending on soil texture[5]. The LS factor accounts for the slope length and steepness effect[6].

The C factor accounts for the relative protection of ground covers against erosion. For small grains such as rye, wheat, barley, and oats it ranges from 0.07 to 0.5[6]. For large-seeded legumes like soybeans the range is 0.1 to 0.65; while for row crops like cotton, tobacco, corn, and sorghum, the range is 0.1 to 0.7. The practice factor, P, is the ratio of soil loss with contouring, strip-cropping, or terracing to the soil loss from straight-row farming up and down slopes. The P values may vary from 0.75 for cross-slope farming without strips to 0.37 for contour farming or cross-slope farming with strips[6].

The sediment-delivery ratio, S_d, accounts for the fraction of the gross erosion in a watershed that is delivered to a stream. For example, for a second-order stream having a drainage area of 3,000 acres (4.7 sq miles) and an average length of 2 miles[2], the S_d is 0.45. This means that approximately 45 percent of the total eroded material in runoff may potentially reach the stream.

SLUDGE APPLICATION RATE MODEL

A sludge application model was developed for some of the major crop-producing states using the approach outlined above[9]. The following average sewage sludge composition[7] was assumed to remain constant in the calculations:

	Anaerobic	Aerobic
Total-N, percent	5.0	4.9
Ammonia-N, mg/kg	9,400	950
Nitrate-N, mg/kg	520	300

The remainder of the total-N is organic-N.

Two conditions--0 to 3 percent and 3 to 6 percent--for slope of the land were selected for the model. Two types of application--surface and incorporated--have been considered for model inclusion. Incorporated application is assumed to include all forms of incorporation, including injection, plowing, discing, and rototilling. Soil types used in the model are coarse, medium and fine-textured, which have cation exchange capacities (CEC) of roughly 0 to 5, 5 to 15, and >15 meq/100 g and correspond roughly to a sandy-loam, loam, and silty-clay loam.

Tables 1 and 2 are examples of the sludge application rates recommended for two crops, corn and apples, in the state of Iowa. Corn has a nitrogen requirement of 1.3 lb/bushel, whereas apples have a nitrogen requirement of 1 lb/tree. Note that in all cases, the amount of sludge required by incorporated application is less than that required by surface application. For surface application, soil texture did not seem to make a difference. However, sludge amounts varied depending on soil texture for incorporated application.

Table 1. Annual Sludge Application Rate, dry tons/acre

			Slope 0 to 3%			Slope 3 to 6%		
Sludge Type	Type of Application	Expected Yield Trees/Acre	Coarse Soil	Medium Soil	Fine Soil	Coarse Soil	Medium Soil	Fine Soil
Aerobic	Surface	100	3.1	3.1	3.1	3.2	3.2	3.2
	Surface	130	3.9	3.9	3.9	4.1	4.1	4.1
	Incorp.	100	2.9	2.8	2.8	2.9	2.8	2.8
	Incorp.	130	3.7	3.7	3.7	3.7	3.7	3.7
Anaerobic	Surface	100	2.8	2.8	2.8	3.0	3.0	3.0
	Surface	130	3.6	3.6	3.6	3.8	3.8	3.8
	Incorp.	100	2.6	2.3	2.2	2.6	2.3	2.2
	Incorp.	130	3.4	3.0	2.9	3.4	3.0	2.9

* For wet tons/acre, divide the dry rate by the total solids fraction in the sludge.

Table 2. Annual Sludge Application Rate, dry tons/acre

			Slope 0 to 3%			Slope 3 to 6%		
Sludge Type	Type of Application	Expected Yield Bushels/Acre	Coarse Soil	Medium Soil	Fine Soil	Coarse Soil	Medium Soil	Fine Soil
Aerobic	Surface	125	5.1	5.1	5.1	5.3	5.3	5.3
	Surface	179	7.0	7.0	7.0	7.3	7.3	7.3
	Incorp.	125	4.7	4.6	4.6	4.7	4.6	4.6
	Incorp.	179	6.7	6.6	6.5	6.7	6.6	6.5
Anaerobic	Surface	125	4.6	4.6	4.6	4.9	4.9	4.9
	Surface	179	6.4	6.4	6.4	6.7	6.7	6.7
	Incorp.	125	4.2	3.7	3.6	4.2	3.7	3.6
	Incorp.	179	6.1	5.3	5.2	6.1	5.3	5.2

* For wet tons/acre, divide the dry rate by the total solids fraction in the sludge.

Whereas most of the land may receive incorporated sludge, only a fraction of the acreage in a watershed may receive surface-applied sludge because of phosphorus runoff potential. Based on runoff calculated by the USLE, the percentage of land in a second-order stream watershed that can receive surface-applied sludge is presented in Table 3 for various states. In the case of Arkansas, for example, which receives substantial rainfall, only 1 percent of the land with a slope of 3 to 6 percent in a watershed may be amended with surface-applied sludge. Similarly, only 2 percent of land with a slope of 0 to 3 percent in a watershed may be amended with surface-applied sludge. However, the remaining land in the watershed may receive incorporated sludge. Similarly, the model was run for 36 crop-producing states, considering the major crops within each state. The manual resulting from application of the model is available from the National Technical Information Service[9].

Table 3. Percentage of a Watershed that May Receive Surface-Applied Sludge

State	Crop: Apples		Barley		Citrus		Corn	
	Slope: 0-3%	3-6%	0-3%	3-6%	0-3%	3-6%	0-3%	3-6%
Arizona	36	20	28	16	36	20	19	11
Arkansas	4	2	--	--	--	--	2	1
California	51	29	40	23	51	29	28	16
Colorado	23	13	18	10	--	--	13	7
Connecticut	11	6	--	--	--	--	--	--
Delaware	8	4	6	3	--	--	4	2
Florida	--	--	--	--	4	2	2	<1
Georgia	4	2	--	--	--	--	2	1
Idaho	91	52	71	41	--	--	51	29
Illinois	8	4	--	--	--	--	4	2
Indiana	9	5	--	--	--	--	5	2
Iowa	9	5	--	--	--	--	5	2
Kansas (east)	8	4	6	3	--	--	4	2
Kansas (west)	14	8	11	6	--	--	7	1
Louisiana	--	--	--	--	--	--	2	<1
Maryland	8	4	6	3	--	--	4	2
Massachusetts	11	6	--	--	--	--	--	--
Michigan (south)	13	7	10	5	--	--	7	4
Minnesota	13	7	10	5	--	--	7	4
Mississippi	--	--	--	--	--	--	2	<1
Missouri	7	3	--	--	--	--	3	1
Montana	--	--	40	23	--	--	28	16
Nebraska	--	--	13	7	--	--	9	5
New Jersey	11	6	8	4	--	--	6	3
New York	17	9	--	--	--	--	9	5
North Carolina	6	3	5	2	--	--	3	1
North Dakota	--	--	18	10	--	--	13	1
Ohio	11	6	--	--	--	--	6	3
Oklahoma (east)	--	--	4	2	--	--	2	1
Oklahoma (central)	--	--	6	3	--	--	4	2
Oklahoma (western)	--	--	8	4	--	--	6	3
Oregon	60	34	47	27	--	--	33	19
Pennsylvania	13	7	10	5	--	--	7	4
South Dakota	--	--	17	9	--	--	12	6
Texas (east)	--	--	3	1	4	2	2	<1
Texas (central)	--	--	5	2	7	3	3	1
Texas (west)	--	--	13	7	17	9	9	5
Utah	91	52	71	41	--	--	51	29
Virginia	8	4	6	3	--	--	4	2
Washington	60	34	47	27	--	--	33	19
Wisconsin	13	7	10	5	--	--	7	4

Table 3. Percentage of a Watershed that May Receive Surface-Applied Sludge (Continued)

Crop:	Cotton		Forages		Grapes, Vineyards		Horticulture		Oats	
State / Slope:	0-3%	3-6%	0-3%	3-6%	0-3%	3-6%	0-3%	3-6%	0-3%	3-6%
Arizona	19	11	30	17	36	20	36	20	--	--
Arkansas	2	1	4	2	4	2	4	2	3	1
California	28	16	44	25	51	29	51	29	40	23
Colorado	--	--	20	11	--	--	23	13	18	10
Connecticut	--	--	9	5	--	--	11	6	--	--
Delaware	--	--	7	3	--	--	8	4	--	--
Florida	2	<1	3	1	--	--	4	2	--	--
Georgia	2	1	4	2	--	--	4	2	3	1
Idaho	--	--	77	44	--	--	91	52	71	41
Illinois	--	--	7	3	--	--	8	4	6	3
Indiana	--	--	8	4	--	--	9	5	7	4
Iowa	--	--	8	4	--	--	9	5	7	4
Kansas (east)	4	2	7	3	--	--	8	4	6	3
Kansas (west)	7	4	12	6	--	--	14	8	11	6
Louisiana	2	<1	3	1	--	--	4	2	--	--
Maryland	--	--	7	3	--	--	8	4	6	3
Massachusetts	--	--	9	5	--	--	11	6	--	--
Michigan (south)	--	--	11	6	13	7	13	7	10	5
Minnesota	--	--	11	6	--	--	13	7	10	5
Mississippi	2	<1	3	1	--	--	4	2	--	--
Missouri	3	1	6	3	--	--	1	3	5	2
Montana	--	--	44	25	--	--	51	29	40	23
Nebraska	--	--	14	8	--	--	17	9	13	7
New Jersey	--	--	9	5	--	--	11	6	8	4
New York	--	--	14	8	17	9	17	9	13	7
North Carolina	3	1	5	2	--	--	6	3	5	2
North Dakota	--	--	20	11	--	--	23	13	18	10
Ohio	--	--	9	5	11	6	11	6	--	--
Oklahoma (east)	2	1	4	2	--	--	5	2	4	2
Oklahoma (central)	4	2	7	3	--	--	8	4	6	3
Oklahoma (western)	6	3	9	5	--	--	11	6	8	4
Oregon	--	--	51	29	--	--	60	34	47	27
Pennsylvania	--	--	11	6	13	7	3	7	10	5
South Dakota	--	--	18	10	--	--	22	12	17	9
Texas (east)	2	<1	3	1	--	--	4	2	3	1
Texas (central)	3	1	6	3	--	--	7	3	5	2
Texas (west)	9	5	14	8	--	--	17	9	--	--
Utah	--	--	77	44	--	--	91	52	71	41
Virginia	4	2	7	3	--	--	8	4	6	3
Washington	--	--	51	29	--	--	60	34	47	27
Wisconsin	--	--	11	6	--	--	13	7	10	5

REFERENCES

1. Day, A. D., Thompson, R. K., and Tucker, T. C., "Effects of Dried Sewage Sludge on Barley Genotypes," *J. Environ. Qual.*, 1983, *12*, 213-215.

2. Keup, L. E., Flowing Water Resources, *Water Resources Bulletin*, 1985, *21*, 291-296.

3. Metcalf and Eddy, *Wastewater Engineering: Collection Treatment and Disposal*, 1978, McGraw Hill, New York.

4. Schwing, J. E. and Puntenney, J. L., "Denver Plan: Recycle Sludge," *Water and Wastes Engineering*, September, 1974.

5. U.S. Environmental Protection Agency, *Quality Criteria for Water*, U.S. Environmental Protection Agency, Washington, DC, 1976, 186-190.

6. U.S. Environmental Protection Agency, Environmental Protection Technology Series, "Loading Functions for Assessment of Water Pollution from Nonpoint Sources," EPA-600/2-7-151, Washington, DC, 1976.

7. U.S. Environmental Protection Agency, Process Design Manual, "Land Application of Municipal Sludge," Office of Research and Development, EPA-625/1-83-016, Washington, DC, 1983.

8. U.S. Environmental Protection Agency, Environmental Regulations and Technology, "Use and Disposal of Municipal Wastewater Sludge," Intra-agency Sludge Task Force, Washington, DC, 1984.

9. U.S. Environmental Protection Agency, *Manual for Sewage Sludge Application to Croplands and Orchards*, Criteria and Standards Division, Office of Water Regulations and Standards, NTIS Publication PB89 110662/AS, National Technical Information Service, Springfield, VA, 1988.

LONG-TERM ENVIRONMENTAL EFFECTS ASSOCIATED WITH LAND APPLICATION OF MUNICIPAL SLUDGES: A REVIEW

A. C. Chang and A. L. Page
Department of Soil and Environmental Sciences
University of California
Riverside, California 92521

ABSTRACT

This presentation examines the long-term environmental effects associated with the land application of municipal sludges by evaluating how the sludge constituents may be assimilated in the soil through biochemical cycling of C, N, P, and trace elements and by reviewing related experimental data. Our analysis indicates that through the biochemical cycles the majority of the sludge constituents are readily assimilated in the soils during the course of a land application. As the ability of the biochemical cycles to accommodate trace metals is limited, the trace metals invariably accumulate in the soil layer where the sludges are mixed and will be present in the soil essentially indefinitely. Because of their long degradation half-life and their strong adsorption by the soils, many organic chemicals such as DDT and PCBs are assimilated into the C cycle at a very slow rate. It may take as long as 40 to 75 years for >99 percent of the added chemicals to be degraded. With time, parasite ova are expected to be inactivated. However, because these ova are able to survive the adverse environmental conditions, the sludge-treated soils may remain infectious for several years after the termination of sludge application.

INTRODUCTION

Community-wide wastewater collection and treatment have been practiced in the western hemisphere for over 100 years[14]. This practice, once it became widespread, was instrumental in improving the environmental sanitation of our community and helped to arrest the water-borne disease epidemics that plagued western Europe and North America for a good part of the 19th century. Although the composition of the municipal wastewater and the technology of treating it have changed considerably over time[31,35,38,45], the fundamental need to remove undesirable substances from the spent water prior to the discharge has remained unchanged. Today, publicly owned treatment works (POTW) form our society's frontline defense against water pollution. In addition to pathogens (bacteria, virus, and parasites), the wastewater treatment also must remove suspended solids, biochemical oxygen demands (BOD), nutrients, toxic substances, etc.

The wastewater collection network in urban communities receives discharges from numerous sources, each having its own characteristics. It is difficult, if not impossible, to anticipate the types and amounts of potential pollutants that may appear in the wastewater stream. All municipal wastewaters and the by-products of their treatment, however, contain biodegradable organic matter (BOM), disease-causing microorganisms, essential plant nutrients, toxic chemical constituents, and dissolved minerals, which are impurities added into the water through use. The types of chemicals present in the wastewater and their concentrations, to a large extent, reflect the nature of the community served. Through wastewater treatment, impurities are concentrated into the sludge fraction. The processing and disposal of sludges are by far the most difficult and costly operations for any POTW.

Municipal sludges have been land applied since the advent of modern wastewater treatment; today it remains a technically viable, cost-effective, and publicly accepted alternative for the sludge's ultimate disposal[15,33]. More than one-third of the sludge produced in the United States is channelled through this route for disposal[46]. Through this practice, large amounts of potential pollutants may be released to the environment. It is essential that these materials are properly assimilated by the environment. In the past two decades, land application of sludges has attracted a great deal of attention, and research results relating to land application appeared frequently in the technical literature[54,55]. Although technology is adequate to design engineered land application systems, the long-term impacts of adding potential pollutants to the soil is still not fully understood[42]. Any of the impurity categories outlined in the previous paragraph may limit the ability of soils to receive wastes. The purpose of this presentation is to evaluate the long-term environmental effects associated with the land application of municipal sludges through a review of the biochemical and chemical cycles of the soil and an examination of the research data.

CYCLES OF THE SOIL

The soil is a chemically and biologically complex porous medium consisting of weathered mineral fragments, organic matter, soluble inorganic and organic constituents, microorganisms, water, and air. Its constituents may be divided into four physical states: gaseous, aqueous, biotic, and solid. A chemical constituent introduced into the soil will be partitioned into these four compartments according to the physical state it is in and the chemical and biological reactions to which it is subject. These physical, chemical, and biological processes that act upon the soil constituents are fundamental to the fate and the behavior of potential pollutants added into the soils. Mechanisms responsible for partitioning constituents in the soil--such as the cation and anion exchange, surface sorption, precipitation, volatilization, degradation (chemical and biochemical), and bioabsorption--are essentially the same as those acting in other segments of the environment[8,39].

In land application, it is essential that the sludge constituents added are immobilized effectively or attenuated. Through these processes, anthropogenic constituents in the soil may be transformed into those in the stationary phase (solids and biota) and those in the mobile phase (soil solution and gases). The reaction kinetics, the hydraulic loading, and the diffusion rate will then determine the rates of transport of the mobile constituents via various pathways[34]. Ideally, the mass and hydraulic loadings at the application site should be matched with rates of pollutant attenuation so that no undesirable constituent escapes the waste-receiving medium. Although evaluating the behavior of pollutants in the soil allows one to design a land application system[46,52], it does not address the ultimate fate of sludge constituents that are deposited in the soil. To delineate the long-term effects of applying sludge on land, attention must be focused on the biochemical cycles of the soils. These cycles depict the flow of biomass-associated elements in the soils and are microbially driven. Since organic debris are decomposed by microorganisms in soils, the C, N, P, S, and trace elements all undergo the corresponding transformations. Discussion of biochemical cycles of the soil is beyond the scope of this presentation; readers are referred to the voluminous printed literature on this subject[5,17,72]. The cycling of C, N, P, S, and trace elements, however, will determine which sludge constituents may be assimilated by the soil and which sludge constituents will accumulate in the soil. C, N, P, S, and trace element cycling should, therefore, be the basis of evaluating long-term effects of land application.

BIODEGRADABLE ORGANIC MATTER AND CARBON CYCLING

Municipal sludges consist primarily of organic solids that are removed from the wastewater. Measured in terms of the 5-day BOD, the organic matter (OM) content of sludge probably varies from 10,000 to 30,000 mg/kg dry weight (DW). In land application, the BOD and the total suspended solid (TSS) loadings seldom limit the system's capacity. The soil clogging (due to application of liquid sludge) and anaerobic soil environment are the most common problems associated with OM overloading.

Generally, soils contain between 1 and 5 percent OM. It is believed that the OM content of the soil may be doubled without creating a reducing soil environment. A soil containing 1 percent OM will be able to receive 2×10^4 kg/ha OM (double the OM content of the surface 15 cm of soil). If 50 percent of the applied sludge is OM, 4×10^4 kg/ha sludge (40 mT/ha) may be applied. Drawing from the experience of applying food processing wastes (high BOD and high TSS wastes) on land, annual BOD loadings of 9.5 to 270×10^3 kg/ha have been reported(47). Using land application of food processing wastes as the guide and assuming BOD of the sludge is 30,000 mg/kg, between 320 and 9,000 mT/ha of sludge may be applied. As typical sludge application on agricultural land is <40 mT/ha, the organic matter loading will hardly become a limiting factor.

The OM added to soils through sludge application undergoes microbial decomposition as described by the C cycle. Different organic constituents decompose at different rates. Simple sugars, amino acids, proteins, and polysaccharides degrade rapidly and completely. Complexed organic molecules must be enzymatically hydrolyzed into simpler molecules prior to their mineralization. Miller(44) reported that, at a 90 mT/ha sludge application rate, over 20 percent of the sludge C in soil evolved as CO_2 following 6 months of incubation. OM decomposition in the soil is expected to be affected by the climate(76). More rapid OM decay is expected to occur in warm and humid regions. Through the cycling of C in soils, the sludge-borne organic C can be readily assimilated(73,74). The organic solids added into soils will significantly change their agronomic properties(11) (Table 1).

Table 1. Effects of Organic Sludge Solids on Agronomic Properties of a Composted Sludge-Treated Greenfield Sandy Loam Soil

Year	Sludge treatment: Annual	Sludge treatment: Cumulative	Bulk density	Water-holding capacity	Hydraulic conductivity	Modulus of rupture
	$mT\ ha^{-1}$	$mT\ ha^{-1}$	$g\ cm^{-3}$	%	$cm\ hr^{-1}$	$kg\ cm^{-2}$
1975	0	0	1.53	12.9	0.4	1.2
	22.5	0	1.55	12.9	0.3	1.1
	45.0	0	1.56	13.8	0.2	1.2
	90.0	0	1.56	13.4	0.2	1.3
1977	0	0	1.67	14.5	<0.1	4.2
	22.5	45	1.56	15.2	0.1	2.0
	45.0	90	1.47	17.2	0.5	1.1
	90.0	180	1.33	17.1	2.0	0.6
1979	0	0	1.57	14.8	0.8	1.8
	22.5	90	1.42	16.0	3.2	0.9
	45.0	180	1.32	18.2	4.3	0.6
	90.0	360	1.19	21.3	3.0	0.4

Data from Chang et al. [24].

DISEASE-CAUSING MICROORGANISMS IN SOILS

The pathogenic microorganisms related to domestic wastewater always are associated with fecal discharges and are transmitted by the fecal/oral route. Specific types and numbers of pathogens present in a wastewater may fluctuate with the community composition and the season, but pathogenic bacteria, enteric virus, protozoa, and parasites can all be isolated from any municipal wastewater and sludges in the United States[23,24,29,36,61]. Pathogens in the municipal sludges may be eliminated if they are subject to composting, heat treatment, or irradiation, but land-applied sludges seldom receive this type of treatment.

Potential public health risks associated with land-spreading of sewage wastes always have been a concern[62-65]. Diseases resulting from direct contact with soil and/or vegetation at the waste disposal sites were reported[62].

In nature, microorganisms always are present as a mixed population. A variety of microbes are known to exist in the soils[43]. Under normal conditions, bacteria and actinomycetes are by far the most abundant microorganisms of any soil. In addition, there are also fungi, algae, protozoa, and nematodes. Compared to the number of indigenous soil microorganisms, the number of pathogens added through land application of sludges is rather insignificant (probably 5 to 6 orders of magnitude lower). Pathogens, however, are well adapted to the growth environment of the intestinal tract of warm-blooded animals. In the soil, both population dynamics and environmental conditions are unfavorable for pathogen survival. With microbial antagonism and a hostile environment, pathogen die-off in the soil is expected to be rapid; it follows first-order reaction kinetics[60]. Nevertheless, a small number of them will survive for an extended period of time and may travel a substantial distance in soils[7,25,26,60]. Generally, cold temperature and moist soil conditions enhance pathogen survival, whereas sunlight, heat, and desiccation are most effective in deactivating the disease-causing microorganisms[4,21,30]. Bacteria of sanitary significance (fecal coliform and fecal streptococci), bacterial pathogens, and protozoa cysts are known to survive from as briefly as 1 day to as long as 200 days[23]. The survival time of viruses and parasitic ova could be 1 to 2 orders of magnitude longer than those of bacteria and protozoa cysts[23,60] (Table 2). Despite numerous laboratory demonstrations of pathogen survival in soils, no disease outbreaks associated with land application of municipal sludges have been reported[17,22,40,79].

Although the long survival time of parasite ova in soils may make the sludge-treated soils infectious for a rather long period of time, the pathogens in sludge-treated soils will gradually die off as soon as the sludge application is terminated. The short-term public health implications may be serious, but no long-lasting adverse effects of pathogens are expected to affect soils receiving municipal sludges.

N AND P CYCLES AND NUTRIENTS OF MUNICIPAL SLUDGES

Besides OM, municipal sludges also contain plant nutrients such as N and P. During the course of OM decomposition, N and P also undergo transformations. The cycling of N and P in the soil coincides with the cycling of C[69]. In fact, the ratios of organic C and N (C/N) and organic C and P (C/P) in the soil are the deciding factors of their mineralization. Through this mechanism, the N and P of the sludge may be assimilated by the soils, like other sources of organic N and P. In a soil system that receives normal inputs, the biochemical cycling of N and P often is in a steady state. The inputs of N and P through sludge application, however, could be significantly greater than the soils normally receive. Overloading the soil with N and P will no doubt accelerate the rate of the N and P cycling in soils and produce larger amounts of reaction products.

For P, the mineralization end product is the orthophosphate that reacts strongly with oxides of Al and Fe (in acid soils) or calcium carbonate (in calcareous soils) to form sparingly soluble inorganic P minerals; these are the sink for the global cycling of P[57] (Table 3). Considering the amounts of Al, Fe,

Table 2. Pathogen Survival Times in Soils

Organism	Survival time (days)
Coliform	38
Fecal streptococci	26-77
Salmonellae spp.	1-280
Enteroviruses	8-175
E. histolytica cysts	6-8
Ascaris ova	up to 7 years

Modified from Frankenberger [29].

Table 3. Phosphorus Accumulation in Sludge-Treated Greenfield Sandy Loam Soil

Sludge treatment	Total P input	Total soil P
	kg ha^{-1}	mg kg^{-1}
	Composted sludge application (1976-1981)	
0 (mT ha^{-1} yr^{-1})	0	515
22.5 (mT ha^{-1} yr^{-1})	1,650	1,092
45 (mT ha^{-1} yr^{-1})	3,304	1,657
90 (mT ha^{-1} yr^{-1})	6,582	2,617
	Liquid sludge I application (1976-1981)	
0 (cm yr^{-1})	0	489
3.75 (cm yr^{-1})	1,767	874
7.5 (cm yr^{-1})	3,482	1,184
15 (cm yr^{-1})	6,880	1,689
	Liquid Sludge II application (1976-1978)	
0 (cm yr^{-1})	0	514
3.75 (cm yr^{-1})	811	670
7.5 (cm yr^{-1})	1,592	939
15 (cm yr^{-1})	3,046	1,306

From Table 2 in Chang et al. [48].

and Ca that mineral soils contain, a large amount of P may be adsorbed before there is any significant change in the P's chemical equilibrium in soil. Hinedi *et al.*(28) reported that organic P in sludge-amended soils may be mineralized in less than 120 days. Chang *et al.*(10) fractionated the P in sludge-amended soils using a modified Chang and Jackson(13) extraction procedure and found that the majority of the P is accumulated as inorganic P. Based on the P solubility measurement of a sludge-amended calcareous soil, O'Connor *et al.*(49) reported that the P solid phase in the sludge-treated soil is a tricalcium phosphate-like mineral. Hinedi and Chang(27) examined the P solid phase of two sludge-amended California soils using the ^{31}P magic angle spinning nuclear magnetic resonance spectroscopy and found that carbonate substituted for hydroxy apatite in pyrophosphate phases.

The end product of the N mineralization (nitrate) is readily soluble and available to plants. If the sludge application introduces greater quantities of organic N than the soil normally receives, there will be more N mineralization resulting in more nitrate (Figure 1). Despite the greater potential for denitrification, nitrate leaching will take place if the amount of nitrate produced is greater than the plant is able to absorb. The transformations of N in sludge-amended soils have been extensively studied(12,66,75). The results invariably

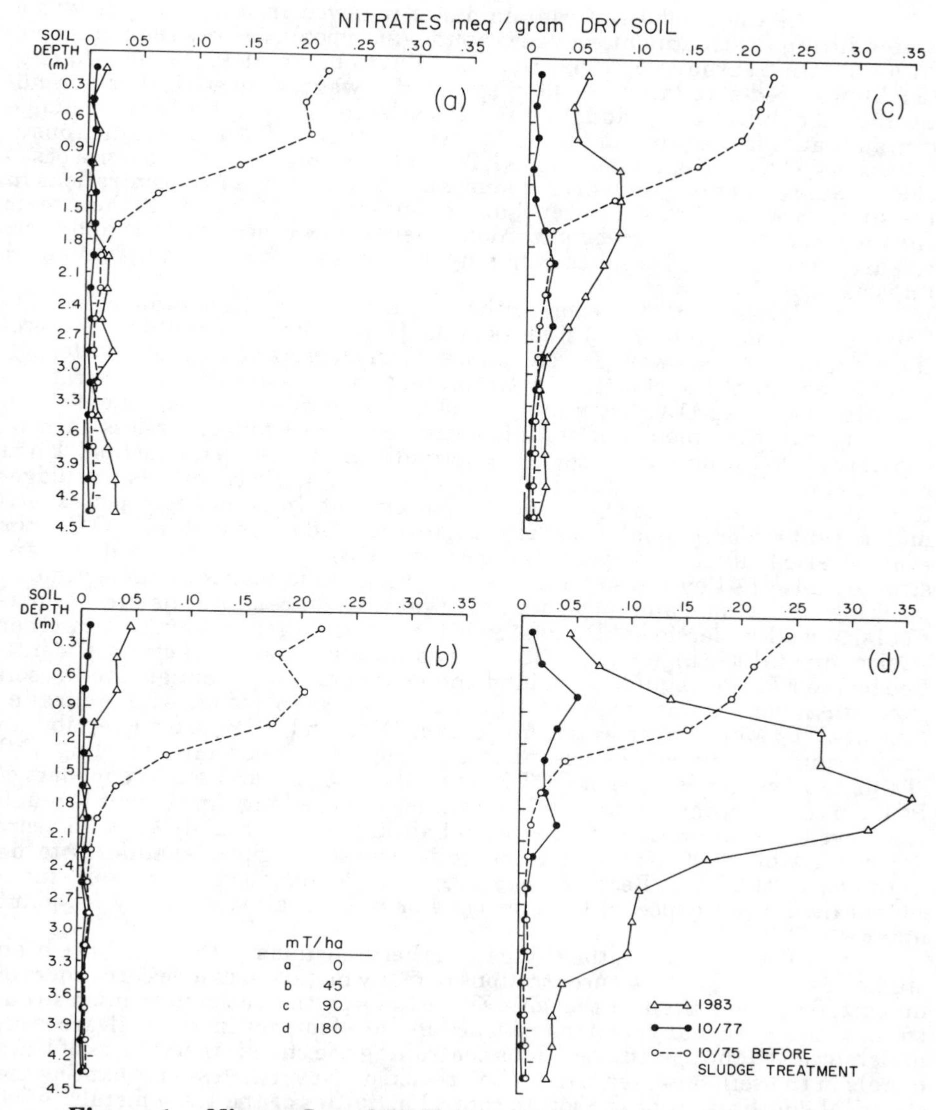

Figure 1. Nitrate Leached Underneath a Composted Sludge-Treated Greenfield Sandy Loam(12)

point out the importance of managing the input of N properly. If the sludge application is adjusted according to the cycling of N in soils, organic N in the sludge can be assimilated by the soil and the leaching of nitrate can be prevented[59,68]. The increased N content of the sludge-amended soils may provide the N requirement of the growing plants for an extended period of time[56] (Table 4).

TOXIC CHEMICALS OF MUNICIPAL SLUDGES AND THEIR FATE IN THE SOIL

As the catchall of contaminants removed from municipal wastewater during the treatment, sludge is abundant in potential environmental pollutants. On February 6, 1989, the U.S. Environmental Protection Agency (EPA) proposed standards for the disposal of sewage sludge[71]. For agricultural land application of sludges, the EPA selected 22 potentially toxic sludge contaminants for regulation (Table 5). Among them, 10 are inorganic constituents (As, Cd, Cr, Cu, Pb, Mo, Hg, Ni, Se, and Zn), and the remainder are pesticide residues and toxic industrial chemicals. Although their concentrations in the sludge are significantly lower than its contents of OM, N, and P, these toxic chemical constituents may be hazardous even at low concentrations and their release through land application may result in adverse environmental impacts[32,41,51,53].

Jacobs *et al.*[32] reported that no fewer than 149 organic chemicals have been found in municipal sludges in the United States. About two-thirds of the organic chemicals were detected in less than 50 percent of the sludge tested, and 90 percent of the chemicals were present in the sludge at concentrations less than 10 mg/kg DW. Assuming the sludge is applied at an agronomic rate (~10 mT/ha), the amount of organic chemicals added to soils is less than or comparable to the amount of pesticide normally used for crop production. Until recently, few have investigated the fate of trace organic chemicals in sludge-treated soils. Bellin and O'Connor[3] reported that, in greenhouse pot experiments, pentachlorophenol volatilized rapidly in sludge-treated soil. O'Connor *et al.*[50] also found that polychlorobiphenyls (PCB) in sludge-amended soils are strongly adsorbed by the soil and are not available to plants or susceptible to leaching. Reactions and movement of organic chemicals in soils are not particularly well understood[67]. Based on the fate of organic chemicals in other environmental settings (e.g., pesticide application), organic chemicals identified by the EPA for regulation in land application may be attenuated in the soil through mechanisms such as volatilization, adsorption, and degradation[1,58,78]. More importantly, all (except DDT and PCB) are susceptible to biodegradation and will be assimilated through the biochemical cycling of C (Table 6). The persistence of DDT (and its derivatives) and some congeners of PCBs in the environment are well known; they take a long time to flow through the C cycle. Based on their degradation half-life, approximately 40 to 75 years are needed for >99 percent of the soil-adsorbed organic chemicals to be degraded in the soil. Because they are strongly adsorbed by the soil, these chemicals are not expected to be leached or to be available readily for plant uptake[2,48].

Trace metals in the sludge and their reactions in the soils have been studied extensively[41]. Concentrations of many metals in sludges are orders of magnitude higher than in the soils[53]. Although the basic mechanisms controlling the chemical reactions of trace metal elements in the soil are well understood[39], the specific reactions controlling the chemical activities of trace metals in the soil, however, cannot be predicted. Nevertheless, one expects the chemical equilibrium of the soil to control activities of the trace metals[18-20]. Many trace metal elements are micronutrients. At low concentrations, they are essential to all biological growth, but they become toxic at only slightly

Table 4. Organic N Accumulation in Sludge-Treated Greenfield Sandy Loam Soil (0 to 15 cm Soil Depth)

Treatment	1975	1976	1977	1978	1979	1980	1981	1982	1983
	$Mg \cdot kg^{-1}$								
	Composted sludge application (1976-1981)								
0 (control)	460	513	514	489	530	488	496	539	482
22.4 ($mT\ ha^{-1}\ yr^{-1}$)	393	521	620	650	734	737	886	1008	903
45 ($mT\ ha^{-1}\ yr^{-1}$)	362	663	750	854	977	1125	1235	1536	1438
90 ($mT\ ha^{-1}\ yr^{-1}$)	453	734	988	1218	1371	1794	1828	2828	2877
	Liquid sludge I application (1976-1981)								
0 (control)	230	287	423	408	454	455	451	511	486
3.75 ($cm\ yr^{-1}$)	348	296	546	571	698	754	820	935	906
7.5 ($cm\ yr^{-1}$)	322	448	608	725	936	1076	1157	1197	1204
15 ($cm\ yr^{-1}$)	247	508	868	988	1230	1569	1592	1572	1589
	Liquid Sludge II application (1976-1978)								
0 (control)	435	510	520	504	558	508	517	560	542
3.75 ($cm\ yr^{-1}$)	438	400	588	603	730	640	702	728	782
7.5 ($cm\ yr^{-1}$)	390	488	667	684	748	764	833	819	842
15 ($cm\ yr^{-1}$)	416	554	864	1005	990	1094	1094	1036	1050

Table 3 in Chang et al. [54].

Table 5. Sludge Contaminants Selected by the EPA for Regulating Land Application of Municipal Sludges

Inorganic constituent	Organic substance
As	Aldrin/dieldrin
Cd	Benzo(a)pyrene
Cr	Chlordane
Cu	DDT/DDE/DDD
Pb	Heptachlor
Hg	Hexachlorobenzene
Mo	Hexachlorobutadiene
Ni	Lindane
Se	PCBs
Zn	Toxaphene
	Trichloroethylene

Derived from U.S. Environmental Protection Agency [58].

Table 6. Degradation Half-Life of Selected Organic Chemicals in Soils

Compound	Degradation half-life in soils* (days)
Aldrin/dieldrin	365/868
Benzo(a)pyrene	100-170
Chlordane	430-3500
DDT/DDE/DDD	3800
Heptachlor	2000
Hexachlorobenzene	>385
Hexachlorobudiene	--
Lindane	56-600
PCBs	1440 (for Arochlor 1242)
Toxaphene	100-3650
Trichloroethylene	360

*Data derived from Laskowski et al. [79], Jury et al. [14], and U.S. Environmental Protection Agency, Office of Water, Regulations and Standards reports on Environmental Profiles and Hazard Indices for Constituents of Municipal Sludges, June 1985.

higher concentrations. For this reason, only small amounts of trace metals may be accommodated by the biochemical cycling(72). Excessive amounts of trace elements in biological tissue cannot be metabolized readily, and they frequently exhibit bioaccumulation as they move up the food chain.

In sludge-treated soils inputs of trace metals always are greater than the amounts that can be assimilated (that is, annual trace metal inputs through sludge application are less than annual plant uptake), and the unassimilated trace metals usually remain in the surface soil where the sludges and the soils are mixed. The trace metal concentration in the sludge-treated soils, therefore, increases with each application of sludge(9,16). Although phytotoxicity caused by trace metals in the sludge-treated soils has not been documented, some metals added to soils as insecticides (As, Pb, Zn, etc.) were found to cause plant injuries years after cessation of the practice(6,77). It has been demonstrated that crops harvested from sludge-treated land contain higher amounts of trace metals than those from untreated land. Unlike the other sludge contaminants we have reviewed so far, the trace metals cannot be fully assimilated through the biochemical cycles. Once they are present, trace metals are expected to remain in the soil essentially indefinitely(70). Several recent literature reviews on sludge land application suggested that upper limits be set on allowable trace metals in the soils(41,55,71). There is, however, no agreement on what these threshold levels should be.

DISSOLVED INORGANIC MINERALS

Through each cycle of domestic use, total dissolved solids (TDS) of the water will increase by 150 to 300 mg/L. This degradation in water quality usually is characterized by increases in hardness (Ca^{++} and Mg^{++}), Na^{+} and Cl^{-} in water. Because the municipal wastewater treatment process has no significant effect on these constituents, their concentrations in the liquid sludge are not expected to differ much from those in the wastewater. The dissolved mineral content of sludge determines the salinity level of the material. Assuming the dissolved minerals are 5 percent of the sludge solids, the amounts added through a normal sludge application (10 to 20 mT/ha) will be 500 to 1,000 kg/ha. If the annual water input to the soil is 1.2 m (the typical annual precipitation of a humid climatic region or the typical annual water input through irrigation in the arid region), the amount of dissolved minerals added will amount to an increase of the water's TDS by 40 to 85 mg/L. This amount of salinity increase will not result in any serious environmental degradation. In the semiarid West, the annual salinity input via irrigation will be 4 to 5 times greater than the amount from an annual sludge application.

The background chemical matrix of the soil solution is derived from the composition of dissolved inorganic constituents and the weathering of minerals in the soil. Because the chemical matrix of dissolved minerals in sludge may be different from that of the receiving soil, the sludge application may affect temporarily or permanently the dissolved mineral composition of the soil. For a sludge-treated Greenfield sandy loam soil in southern California, the dominating ions of the soil saturation extracts gradually changed from Na^{+} and Cl^{-} to Ca^{++} and $SO_4^{=}$ (Table 7). The ionic strength of the soil solution (measured as the electrical conductance of the soil saturation extracts) increased with the amount of sludge added. Obviously, a change in the background matrix of the chemical system will bring about a new chemical equilibrium in the soil.

Table 7. Ionic Concentrations in Soil Saturation Extracts of Sludge-Treated Greenfield Sandy Loam Soil (mol/kg Dry Soil, Except EC-dS/m at 25 C)

Application	EC	Ca^{+2}	Mg^{+2}	K^{+}	Na^{+}	Cl^{-}	NO_3^{-}	HCO_3^{-}	SO_4^{-2}
Composted sludge application (1976-1981)									
0 (t ha^{-1} yr^{-1})	2.1	0.244	0.078	0.027	0.508	0.469	0.024	0.047	0.154
22.5 (t ha^{-1} yr^{-1})	2.5	0.634	0.14	0.027	0.367	0.351	0.033	0.040	0.714
45 (t ha^{-1} yr^{-1})	3.0	1.686	0.334	0.030	0.450	0.386	0.109	0.042	1.906
90 (t ha^{-1} yr^{-1})	4.0	2.71	0.508	0.038	0.522	0.459	0.230	0.036	2.888
Liquid sludge I application (1976-1981)									
0 (cm/y)	1.9	0.222	0.072	0.026	0.412	0.406	0.028	0.044	0.217
3.75 (cm/y)	1.8	0.45	0.116	0.023	0.382	0.352	0.086	0.023	0.486
7.5 (cm/y)	2.0	0.586	0.17	0.023	0.368	0.331	0.130	0.020	0.564
15.0 (cm/y)	2.2	0.928	0.224	0.021	0.340	0.231	0.133	0.021	1.094
Liquid Sludge II application (1976-1978)									
0 (cm/y)	1.9	0.226	0.079	0.025	0.423	0.426	0.021	0.040	0.226
3.75 (cm/y)	1.8	0.26	0.09	0.026	0.394	0.415	0.036	0.036	0.216
7.5 (cm/y)	1.8	0.30	0.094	0.021	0.405	0.425	0.037	0.030	0.262
15.0 (cm/y)	1.8	0.382	0.114	0.024	0.406	0.433	0.064	0.31	0.30

CONCLUSIONS

The majority of sludge constituents are naturally occurring substances and, in the course of land application, may be assimilated through the biochemical cycling of C, N, and P, and/or attenuated by the chemical and physical processes in the soil. Because rates of the biochemical cycling are controlled by the environmental conditions, and because the capacity of each soil compartment (the gaseous, the aquatic, the biotic, and the solid phases of the soil) to accommodate end products of the cycling is not unlimited, the key to operating a successful land application system is to maintain a balance of the mass flow through the biochemical cycles. Overloading the soil with sludge constituents may cause temporary or permanent buildup of the sludge constituents (or the end products of the cycling) at a soil compartment. For example, nitrate leaching will take place if the N input of the sludge application becomes excessive. A reducing soil environment undoubtedly is the result of organic matter overloading during the land application.

Based on our knowledge of the biochemical cycling of C, N, P, S, and trace elements and the experimental data collected from sludge land application sites, the long-term impacts associated with land application of municipal sludges may occur in the following areas:

1. Trace metals are present in all municipal sludges. Because their concentrations in the sludge are considerably higher than in the soils, and because the amount of trace elements flowing through the biochemical cycles is very small, long-term repeated sludge application invariably increases the concentrations of trace metals in the soil. These constituents will remain in the soil for many years to come, even long after sludge application is terminated. The future phytotoxic potential and bioavailability of the accumulated trace metals in sludge-treated soils cannot be predicted and will be determined by the chemical equilibrium of these soils.
2. DDT and PCBs are no longer in use, but they are ubiquitous in the environment and frequently will be found in municipal sludges. Because of their long degradation half-life, the rates at which DDT (and its degradation products), some PCBs, and other organic chemicals flow through the C cycle are significantly lower than the naturally occurring organic matter. Based on their degradation half-life, we estimated that it takes approximately 40 to 75 years for these soil-adsorbed organic chemicals to be reduced by >99 percent.
3. Ova of parasites are resistant to adverse environmental conditions. Field observations showed that viable ova may still be found in the soils as long as 7 years after land application. Although they are eliminated ultimately, they will remain infectious in the soil for extended periods of time even after termination of sludge application.

REFERENCES

1. Alexander, M., *Science*, 1981, *211*, 132-138.
2. Aranda, J., O'Connor, G. A., and Eiceman, G. A., *J. Environ. Qual.*, 1989, *18*, 45-50.
3. Bellin, C. A., O'Connor, G. A., *Agron. Abstr.*, 1988, *36*.
4. Bitton, G., Farrah, S. R., Pancorbo, O. C., and Davidson, J. M., in *Viruses and Wastewater Treatment*, M. Goddard and M. Butler, eds., Pergamon Press, NY, 1981, 133-136.
5. Bolin, B. and Cook, R. B., eds., *The Major Biogeochemical Cycles and Their Interactions*, John Wiley and Sons, New York, NY, 1983.

6. Brown, L. L. and Rasmussen, P. E., *Agron. J.*, 1971, *36*, 874-876.

7. Burge, W. D. and Marsh, P. B., *J. Environ. Qual.*, 1978, 7, 1-9.

8. Callahan, M. A., Slimak, M. W., Gabel, N. W., May, I. P., Fowler, C. F., Freed, J. R., Jenning, P., Dufee, R. L., Whitemore, F. C., Maestri, B., Mabey, W. R., Holt, B. R., and Gould, C., *Water-Related Environmental Fate of 129 Priority Pollutants (Volumes I and II)*, EPA-440/4-79-029A,B, U.S. Environmental Protection Agency, Washington, DC, 1979.

9. Chang, A. C., Warneke, J. E., Page, A. L., and Lund, L. J., *J. Environ. Qual.*, 1983, *12*, 391-397.

10. Chang, A. C., Page, A. L., Sutherland, F. H., and Grgurevic, E., *J. Environ. Qual.*, 1983, *12* (2), 286-290.

11. Chang, A. C., Page, A. L., and Warneke, J. E., *J. Environ. Engr. Div.*, Am. Soc. Civil Engr., 1983, *109* (3), 574-583.

12. Chang, A. C., Page, A. L., Pratt, P. F., and Warneke, J. E., in *Proceedings 1988 National Conference on Irrigation and Drainage Engineering*, D. R. Hay, ed., Lincoln, NE, July 19-20, 1988, 455-467.

13. Chang, S. C. and Jackson, M. L., *Soil Sci.*, 1987, *84*, 133-144.

14. Chase, E. S., *Water Works Waste Engr.*, 1964, *34* (6), 56-59, 86.

15. Chase, E. S., *Water Works Waste Engr.*, 1964, *34* (7), 48-49, 79.

16. Dowdy, R. H. and Volk, V. V., in *Chemical Mobility and Reactivity in Soil Systems*, D. W. Nelson, K. K. Tanji, and D. E. Elrich, eds., Am. Soc. Agron., Special Publication 11, Madison, WI, 1982, 229-240.

17. Elliott, L. F. and Stevenson, F. J., eds., *Soils for Management of Organic Wastes and Wastewater*, Special Publication, Am. Soc. of Agron., Madison, WI, 1977.

18. Emmerich, W. E., Lund, L. J., Page, A. L., and Chang, A. C., *J. Environ. Qual.*, 1982, *11* (2), 174-178.

19. Emmerich, W. E., Lund, L. J., Page, A. L., and Chang, A. C., *J. Environ. Qual.*, 1982, *11* (2), 178-181.

20. Emmerich, W. E., Lund, L. J., Page, A. L., and Chang, A. C., *J. Environ. Qual.*, 1982, *11* (2), 182-186.

21. Farrah, S. R., Bitton, G., Hoffman, E. M., Lanni, O., Pancorbo, O. C., Lutrick, M. C., and Bertrand, J. E., *Appl. Environ. Microbiol.*, 1981, *41*, 459-465.

22. Fitzgerald, P. R., in *Proc. 8th Natl. Conf. Municipal Sludge Management*, 214, Information Transfer, Inc., Silver Spring, MD, 1979.

23. Frankenberger, W. T., Jr., in *Irrigation with Reclaimed Municipal Wastewater, A Guidance Manual*, G. S. Pettygrove and T. Asano, eds., California State Water Resources Control Board Report No. 84-1wr, Lewis Publishers, Inc., Chelsea, MI, 1985.

24. Gerba, C. P., in *Proceedings of the Workshop on Utilization of Municipal Wastewater and Sludges on Land*, A. L. Page, ed., University of California, Riverside, CA, 1983, 187.

25. Gerba, C. P., Wallis, C., and Melnick, J. L., "Fate of Wastewater Bacteria and Viruses in Soil," *Amer. Soc. Civil Engr. Irr. Drainage Div. J.*, 1975, *IR3*, 157-174.

26. Hagedorn, C. E., McCoy, L., and Rahne, T. M., *J. Environ. Qual.*, 1981, *10*, 1-8.

27. Hinedi, Z. R. and Chang, A. C., *Soil Sci. Soc. Amer. J.*, 1989, in press.

28. Hinedi, Z. R., Chang, A. C., and Lee, R.W.K., *Soil Sci. Soc. Am. J.*, 1988, *52*, 1593-1596.

29. Hoadley, A. W. and Goyal, S. M., in *Land Treatment and Disposal of Municipal and Industrial Wastewater*, R. L. Sanks and T. Asano, eds., Ann Arbor Science, Ann Arbor, MI, 1976, 101-132.

30. Hurst, C. J. and Gerba, C. P., eds., in *Appl. Environ. Microbiol.*, 1979, *37*, 626-632.

31. Iskandar, I. K., in *Proceedings International Conf. on Heavy Metals in the Environment*, University of Toronto, Toronto, Canada, 1975, 417-432.

32. Jacobs, L. W., O'Connor, G. A., Overcash, M. R., Zabik, M. J., and Rygiewicz, P., in *Land Application of Sludge: Food Chain Implications*, A. L. Page, T. J. Logan, and J. A. Ryan, eds., Lewis Publishers, Inc., Chelsea, MI, 1987, 168 pp.

33. Jewell, W. J. and Seabrook, B. L., in *A History of Land Application as a Treatment Alternative*, Technical Report, EPA 430/9-79-012, U.S. Environmental Protection Agency, 1979, 83 pp.

34. Jury, W. A., Spencer, W. F., and Farmer, W. J., in *Hazard Assessment of Chemicals: Current Development*, Academic Press, NY, 1983, Volume 2.

35. Klein, L. A., Land, M., Nash, N., and Kirscher, S. L., *J. Water Pollution Control Fed.*, 1974, *46*, 1563-1662.

36. Kowal, N. E., in *Health Effects of Land Treatment: Microbiological*, EPA-600/1-82-007, U.S. Environmental Protection Agency, Cincinnati, OH, 1982.

37. Laskowski, D. A., Goring, C.A.I., McCall, P. J., and Swann, R. L., in *Environmental Risk Analysis for Chemicals*, R. A. Conway, ed., Van Nostrand Reinhold Co., New York, 1982, 558 pp.

38. Levins, P., Adams, J., Brenner, P., Coons, S., Harris, G., Jones, C., Thrun, K., and Wechsler, A., in *Sources of Toxic Pollutants Found in Influents to Sewage Treatment Plants: VI. Integrated Interpresentation*, EPA 68-01-3857, U.S. Environmental Protection Agency, Cincinnati, OH, 1979.

39. Lindsay, W. L., in *Recycling Municipal Sludges and Effluents on Land*, National Association of State Universities and Land Grant Colleges, Washington, DC, 1973, 91-97.

40. Liu, D., *Water Res.*, 1982, *16*, 957-961.

41. Logan, T. J. and Chaney, R. L., in *Proceedings of the Workshop on Utilization of Municipal Wastewater and Sludge on Land*, A. L. Page *et al.*, eds., University of California, Riverside, CA, 1983, 235-326.

42. Logan, T. J. and Miller, R. H., in *Environmental and Solid Wastes: Char-*

acterization, Treatment, and Disposal, C. W. Francis and S. I. Auerbach, eds., Ann Arbor Science, Ann Arbor, MI, 1983, Chapter 21.

43. Martin, J. P. and Focht, D. D., in *Soil for Management of Organic Waste and Wastewaters*, L. F. Elliott and F. J. Stevenson, eds., Amer. Soc. of Agron., Madison, WI, 1977, 114-169.

44. Miller, R. H., *J. Environ. Qual.*, 1974, *3*, 376-380.

45. Minear, R. A., Ball, R. O., and Church, R., in *Data Bases for Influent Heavy Metals in Publicly Owned Treatment Works*, EPA-660/281-220, U.S. Environmental Protection Agency, Cincinnati, OH, 1981.

46. Municipal Environmental Research Laboratory, *Process Design Manual, Land Application of Municipal Sludge*, EPA-625/1-83-016, U.S. Environmental Protection Agency, Cincinnati, OH, 1983.

47. Nemerow, N. L., *Industrial Water Pollution, Origins, Characteristics and Treatment*, Addison-Wesley Pub., Reading, MA, 1978, 735 pp.

48. O'Connor, G. A., in *Proc. 3rd Intl. Conf. Environmental Pollution*, Venice, September 26-28, 1988, 180-183.

49. O'Connor, G. A., Knudtsen, K. L., and Connell, G. A., *J. Environ. Qual.*, 1986, *15*, 308-312.

50. O'Connor, G. A., Kiehl, D., and Eicemann, G. A., in *Agron. Abstr.*, 1988, *44*.

51. Overcash, M. R., in *Proceedings of the Workshop on Utilization of Municipal Wastewater and Sludge on Land*, A. L. Page *et al.*, eds., University of California, Riverside, CA, 1983, 199-231.

52. Overcash, M. R. and Pal, D., *Design of Land Treatment Systems for Industrial Wastes: Theory and Practice*, Ann Arbor Science, Ann Arbor, MI, 1979, 684 pp.

53. Page, A. L., *Fate and Effects of Trace Elements in Sewage Sludge when Applied to Agricultural Land: A Literature Review Study*, EPA-670/2-74-005, National Environmental Research Center, U.S. Environmental Protection Agency, Cincinnati, OH, 1974.

54. Page, A. L., Gleason III, G. L., Smith Jr., J. E., Iskandar, I. K., and Sommers, L. E., eds., *Proceedings of the Workshop on Utilization of Municipal Wastewater and Sludge on Land*, University of California, Riverside, CA, 1983.

55. Page, A. L., Logan, T. J., and Ryan, J. A., eds., *Land Application of Sludge: Food Chain Implications*, Lewis Publishers, Inc., Chelsea, MI, 1987, 168 pp.

56. Parker, C. F. and Sommers, L. E., *J. Environ. Qual.*, 1983, *12*, 150-156.

57. Pastene, A. J. and Corey, R. B., in *3rd Annual Madison Conf. of Applied Research and Practice on Municipal and Industrial Waste*, Madison, WI, 1980, 63-74.

58. Pignatello, J. J., in *Reactions and Movements of Organic Chemicals in Soils*, B. L. Sawhney and K. Brown, eds., Soil Sci. Soc. of Agron., Special Publication 22, Madison, WI, 1989.

59. Pratt, P. F., Martin, J. P., and Broadbent, F. E., *Calif. Agriculture*, 1973,

27, 10-13.

60. Reddy, K. R., Khaleel, R., and Overcash, M. R., *J. Environ. Qual.*, 1981, *10*, 255-266.

61. Reimers, R. S., Little, M. D., Englande, A. J., Leftwich, D. B., Bowman, D. D., and Wilkson, R., *Parasites in Southern Sludges and Disinfection by Standard Sludge Treatment*, EPA-600/52-81-166, U.S. Environmental Protection Agency, Cincinnati, OH, 1981.

62. Rudolfs, W., Frank, L. L., and Ragotzkie, R. A., *Sew. Indust. Wastes*, 1950, *22*, 1261-1281.

63. Rudolfs, W., *et al.*, *Sewage Ind. Wastes*, 1951, *23*, 253-268.

64. Rudolfs, W., *et al.*, *Sewage Ind. Wastes*, 1951, *23*, 478-485.

65. Rudolfs, W., *et al.*, *Sewage Ind. Wastes*, 1951, *23*, 656-660.

66. Sabey, B. R., in *Proc. of the 1977 Cornell Agricultural Waste Management Conf. on Food, Fertilizer, and Agricultural Residues*, R. C. Loehr, ed., Ann Arbor Science, Ann Arbor, MI, 1977.

67. Sawhney, B. L. and Brown, K., eds., *Reactions and Movement of Organic Chemicals in Soils*, Soil Sci. Soc. of Amer., Special Publication 22, Madison, WI, 1989.

68. Sommers, L. E. and Nelson, D. W., in *Proc. 4th Ann. Madison Conf.*, 1981, 425-448.

69. Sommers, L. E., Nelson, E. W., and Silviera, D. J., *J. Environ. Qual.*, 1979, *8*, 287-294.

70. Sposito, G. and Page, A. L., in *Metal Ions in Biological Systems*, H. Sigel, ed., Marcel Dekker, Inc., 1984, Volume 18, 287-332.

71. "Standards for the Disposal of Sewage Sludge," U.S. Environmental Protection Agency, *Federal Register*, 1989, *54*, 5746-5902.

72. Stevenson, F. J., *Cycles of Soil*, John Wiley and Sons, New York, NY, 1986, 380 pp.

73. Terry, R. E., Nelson, D. W., and Sommers, L. E., *J. Environ. Qual.*, 1979, *8*, 342-347.

74. Terry, R. E., Nelson, D. W., and Sommers, L. E., *Soil Sci. Soc. Amer. J.*, 1979, *43*, 494-499.

75. Terry, R. E., Nelson, D. W., and Sommers, L. E., *Soil Sci. Soc. Am. J.*, 1981, *45*, 506-513.

76. Thomas, G. W., in *Soils for Management of Organic Waste and Wastewaters*, L. F. Elliott and F. J. Stevenson, eds., Amer. Soc. of Agron., Madison, WI, 1977, 492-507.

77. Vandecaveye, S. C., Horne, G. M., and Keaton, C. M., *Soil Sci.*, 1936, *42*, 203-216.

78. Wu, S. and Gshwend, P. M., *Environ. Sci. Technol.*, 1986, *20*, 717-725.

79. Zenz, D. R., Peterson, J. R., Brooman, D. L., and Lue-Hing, C., *J. Water Pollut. Contr. Fed.*, 1976, *48*, 2332-2342.

ELECTROTECHNOLOGIES

At the present time, solid/liquid separation processes are becoming increasingly important to the treatment of waste streams. Due to stringent environmental restrictions, new solid/liquid processes are becoming developed.

Generally, most solid/liquid separation processes are based on a single property or driving force. During normal practice, mechanical forces such as gravity, vacuum, or pressure are employed. Separation involving other forces such as ultrasonics or acoustics and nonmechanical forces such as electric or magnetic fields are also employed; however, commercial practice is somewhat limited.

At the present time, emphasis is increasing on combined field separation techniques, which exploit two or more properties in a single operation. The general interest is reflected in the active technical literature and in the upsurgence of a number of commercial processes. During this session, a number of electrically enhanced separation processes will be discussed. The electrotechnology processes provide a higher degree of solids separation or a higher rate of dewatering.

Mr. Ryan of Electropure Systems describes a novel electrocoagulation concept. Dr. Fleet of the University of Toronto and Dr. Borzone of EXPORTech discuss two concepts applied to detoxification of industrial effluent and fine coal process. Dr. Senapati of Battelle discusses the application of ultrasonics to sludge dewatering. Dr. Chauhan of Battelle updates the scale-up developments of Battelle's electroacoustic dewatering process. Dr. Muralidhara of Battelle, organizer of this conference and editor of the *Proceedings*, offers an overall picture of recent developments and an outlook on the future of electrotechnology in solid/liquid separation processes.

RECENT DEVELOPMENTS IN SOLID/ LIQUID SEPARATION INCLUDING ELECTROTECHNOLOGIES

H. S. Muralidhara
Battelle Columbus Division
505 King Avenue
Columbus, Ohio 43201-2693
(614) 424-5018

INTRODUCTION

Solid/liquid separation is an important operation practiced widely throughout the chemical, mineral, pulp and paper, biotechnology, food, and other industries. It involves the removal of either a liquid or a solid from solid/liquid systems. Solid/liquid separations span the range from high-value biotech products to low-value waste dewatering. This paper provides an overview of recent developments in solid/liquid separations including electrotechnologies. In addition, it describes some applications and presents some results of dewatering on waste-using electroacoustics.

At the present time, this operation is receiving wide attention in the treatment of waste streams and hazardous wastes, and in solid fuel processing. Because of stringent environmental restrictions, mounting cost of fuels, and lower grades of raw materials, new solid/liquid separation schemes are being developed. In industry, the focus in recent times has changed because materials are being processed with finer and finer particles. In addition, interest has increased in recovering supernatant, which often is a valuable product.

Most solid/liquid separation processes are usually based on a single property or approach. The driving forces generally act on local or surface phenomena. During normal practice mechanical means such as gravity, vacuum, or pressure are employed. Separations involving ultrasonic, acoustic, and non-mechanical forces such as electric or magnetic are also used; however, commercial practice is somewhat limited.

At the present time, combined separation techniques are being emphasized because of the potential benefits in a broad area of applications. This technique exploits two or more properties in a single operation. The field forces generally act on the system as a whole, as opposed to the generally local or surface phenomenon characteristic of a single property approach. The general interest in combined fields separation is reflected in active technical literature and the upsurgence of several commercial processes[9]. Figure 1

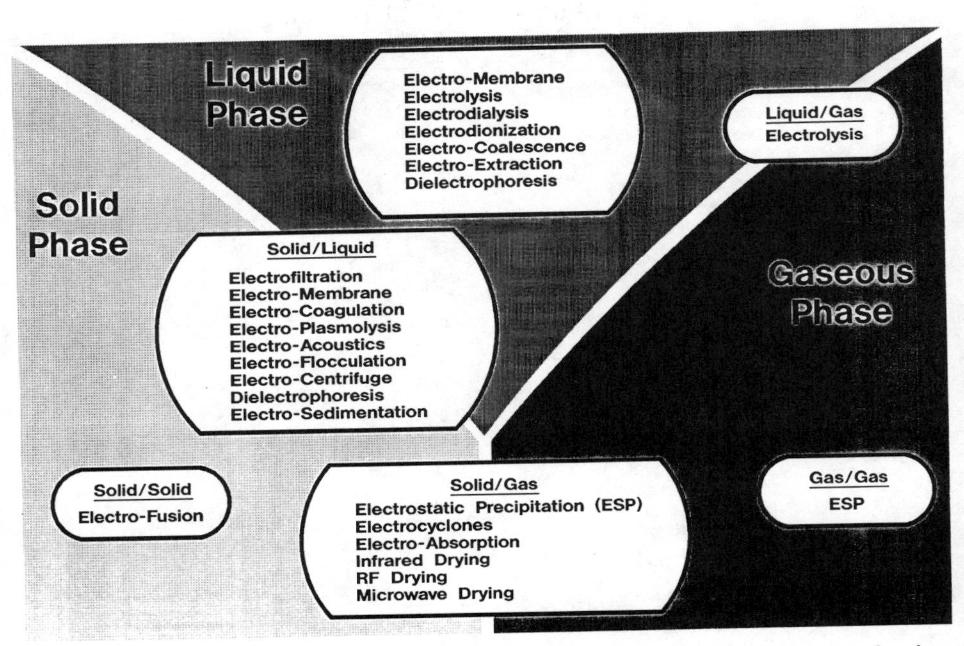

Figure 1. Summary of Electrically Enhanced Separation Technologies

shows a general schematic of a variety of combined fields separation techniques (electrotechnology) being developed at the present time[10].

Although there are a number of different solid/liquid separation systems such as filtration, thickening, centrifugation, hydrocycloning, etc., this paper will address only advanced dewatering concepts which include the presence of body forces such as electric, ultrasonic, magnetic, or combined fields.

ELECTRIC FIELDS DEWATERING

Electroosmosis and electrofiltration are two processes that use a dc electric field to enhance dewatering.

ELECTROOSMOTIC DEWATERING[1,6,15]

The electroosmosis phenomenon was first discovered by Reuss in 1809. In this process, the movement of water occurs through a porous membrane with the application of a dc electric potential. This phenomenon results from the presence of a double layer at the walls of the capillary. Many colloidal materials have a significant amount of water in open-ended pores and trap a film of water by counterions on the capillary walls. If the electric field is applied on the ends of the capillary containing water, the mobile diffuse layer containing molecules of water moves toward the cathode. The flow of water stops in the absence of an electrical field. Many workers have shown that electroosmotic flow is independent of surface area. Electroosmosis is a surface diffusion process.

A number of electroosmotic dewatering processes have been reported in literature by

- Bureau of Mines, USA
- Commonwealth Scientific and Industrial Research Organization (CSIRO), Australia
- Central Electricity Generating Board (CEGB), UK
- Battelle Memorial Institute, Switzerland, USA, West Germany
- Fuji Electric Co., Ltd., Japan (belt press)
- Monsanto Enviro-Chem Systems, Inc., USA
- Sinko Faudrer, Japan (plate and frame).

For example, Neville Lockhart, of CSIRO, is doing a significant amount of work in this area; some of his research is summarized in Table 1.

Recently, a belt press manufactured by Fuji Electric Co. has been introduced on the market. This press exploits the combined effect of mechanical pressure and electroosmosis. A schematic is shown in Figure 2. The equipment has a revolving anode and a moving belt that slides above the cathode.

Table 1. Electroosmotic Dewatering Test Results[6]

Sample	Solids Concentration, weight percent		Water pH	Electrical Efficiency
	As Received	After Electro Dewatering		
Coal Tailings (Bellambi)	13	67-74	7.5	Low
Sand Tailings (Bendigo)	30	70-79	6-6.5	High
Model Suspensions Kaolinite Na	15-19	65-70	6	Medium
Printing Works Sludge	6.2-8.5	25-35	5.5	Medium-Low
Red Mud	47	58	--	Low

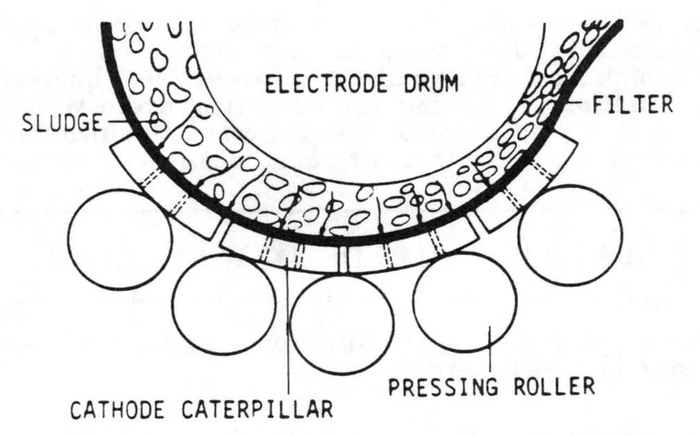

Figure 2. Schematic of Fuji's Electroosmotic Belt Press Unit

Typical operating pressure is about 3.5 kg/cm^2 and applied voltage is 90V. Typical energy consumption of this type of machine is about 0.15 to 0.2 kWh/lb of water removed.

ELECTROFILTRATION(5,15,17)

The electrofiltration process involves predominantly the electrophoretic phenomena, i.e., movement of charged particles toward electrodes in the presence of a dc electric field. The electrophoretic process occurs in the separation chamber. Figure 3 shows the coupled forces in an electrofilter type of separator and the phenomena occurring in various regions.

The application of electrofiltration to commercial scale, industrial processes is emerging. Different types of processes have been tested on a pre-

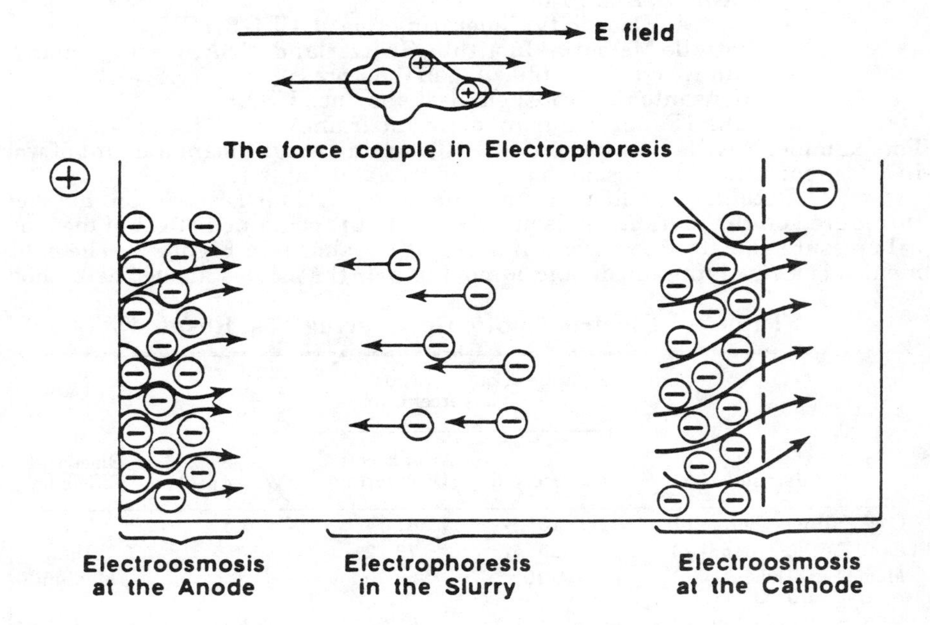

Figure 3. The Force Couple in Electrokinetics and the Phenomena in Various Regions

liminary basis, and a few have been developed to commercial reality. Typical materials that have been reported as candidates for electrofiltration are bentonites, coal fines, latex, and TiO_2.

A typical application of the electrofiltration process in the polyvinyl chloride (PVC) industry is shown in Figure 4. About 15 percent of PVC in the United States is manufactured by an emulsion polymerization process. The final resin is sold as a dry product. The polymer is formed in an aqueous suspension containing fine colloidal particles. Some manufacturers use the ultrafiltration to preconcentrate the particles to 50 percent before spray drying the suspension. This thermal drying is an energy intensive process.

Two electrofilters are commercially available: Elephant (Karl Handle and Sohne, West Germany) and EVAF^R (Dorr-Oliver, USA). The designs are totally different and each has unique applications. A schematic of the Dorr-Oliver unit is shown in Figure 5. Details of Elephant process can be obtained elsewhere[15].

In the EVAF^R process (Dorr-Oliver), the resin is dewatered to about 80 percent solids and can still be maintained in suspension using suitable additives. In this mode, the capacity of the spray dryer can be reduced and significant energy savings can be achieved.

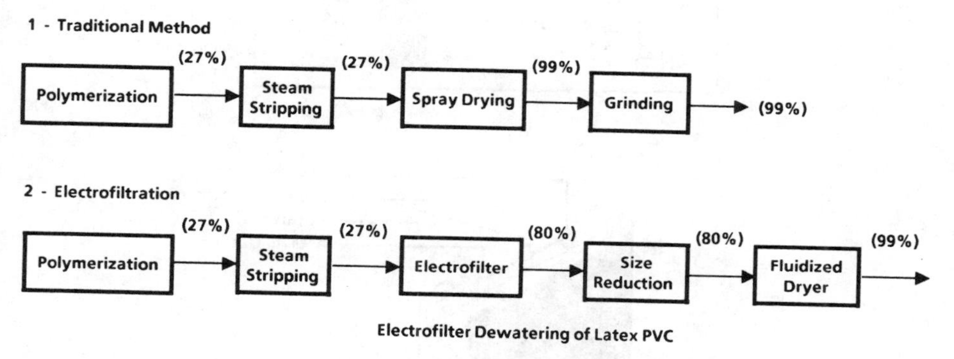

Figure 4. Flow Sheet for the PVC Processing Application

MAGNETIC FIELDS DEWATERING[7,8,13,14,16,17]

Magnetic fields (generally employing permanent magnets) also have been used in solid/liquid separation. This technology has been employed to separate highly magnetic materials such as scrap iron during beneficiation of iron ores. Recent developments in the use of high magnetic fields have stimulated new applications for this type of separation.

The magnetic force is dependent on the magnetic susceptibility of the material. The magnetic separation force is directly proportional to the difference between fluid susceptibility and particle susceptibility. Sometimes certain paramagnetic salts ($MnCl_2$, $FeCl_3$, etc.) are added to enhance the magnetic separation. Some of the typical high-magnetic-field separation techniques are HGMS (high gradient magnetic separation), MGM (magnetic gravimetric separation), and MHS (magnetohydrostatic separation).

Some of the important parameters which affect an HGMS process are:

- Magnetic field strength
- Magnetic susceptibility

THE ELECTROFILTRATION PROCESS

ELECTROPHORESIS: Particles migrating toward the anode element.
ELECTROOSMOSIS: Filtrate pumped out at the cathode element.
Product cake densified at the anode element

CATHODE
$2H_2O \rightarrow 2OH + H_2\uparrow$
ANODE
$H_2O \rightarrow 2H + 1/2\ O_2\uparrow$
Anode Membrane
Filtrate
Filter Cloth
Anode Cake
Anolyte Out
Dispersed Solids
Dissolved Ions
Anolyte In

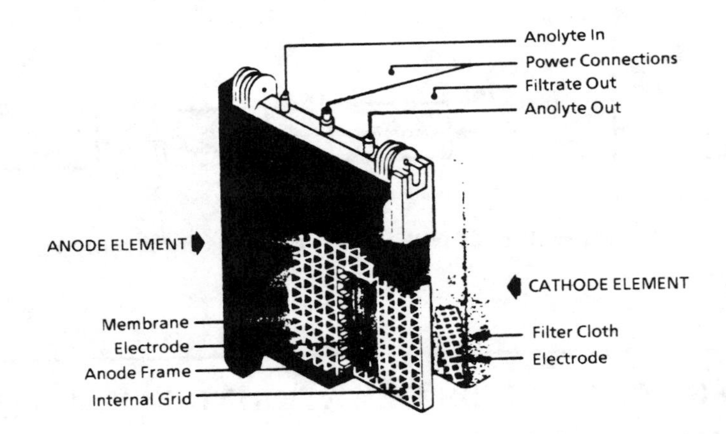

Figure 5. Schematic of Dorr-Oliver Unit

- Particle size
- Flow velocity
- Field gradient exerted
- Length of separation device.

The HGMS technique has many applications. However, initial applications were commercialized in the kaolin industry. Kaolin consists of two main impurities: hematite and TiO_2 with particle size ranging from 1 to 10 µm. Table 2 shows that at 20 kOe the brightness of the clay increases from 83.3 to 89.1 after processing with HGMS. Table 3 shows other applications in steel mill processing.

The use of high magnetic separation technology is being expanded slowly. This technique should find increased importance in the area of blood processing, submicron particulate separation, and recovery of spent catalysts, for example.

Table 2. Impurity Removal from Kaolin (7,8,14,17)

Impurity: Fe_2O_3, TiO_2
Particle Size 1 to 10 μm
Susceptibility 10^{-5} to 5×10^{-4} [SI]

	Recovery, weight percent	Brightness, percent	Fe_2O_3, weight percent
Feed	100	83.3	0.81
Product	88	89.1	0.55
Mags	12		2.72

Pacific Electric Motor 84" (210 cm) (HGMS)

- Feed — Solid Content 12 to 15 weight percent; Flow Velocity 10 to 60 cm/min; (Dry Basis, 5 to 10 tons/hour)
- HGMS — Field 20 kOe; Matrix 6 volume percent

Table 3. Various Types of Wastewater from Steel Mill Processes (13,16)

	Suspension				HGMS*	
Water Type	Formula	Size, μm	Solids, ppm	Solid Treated, ppm	Magnetic Field, kOe	Flow Velocity, m/min
Basic Oxygen Furnace Scrubber Water	Fe_3O_4, Fe, FeO	5-20	150-200	<15	3	3
Blast Furnace Scrubber Water	α-Fe_2O_3, Fe_3O_4, C	1-40	200	10	3	8
Hot Rolling Mill Water	Fe_3O_4, α-Fe_2O_3, Oil	20-1,000	100-150	15-20	3-5	8
Continuous Casting Process Water	Fe_3O_4, α-Fe_2O_3	40-150	150-200	5	1-2	15
Vacuum Degassing Scrubber Water	Fe_3O_4, Mn_3O_4	5-20	80-100	20-25	3-5	2.5-4

* Matrix: 0.1- to 1-mm diameter stainless steel wire or 1-mm expanded metal mesh. Packing density is 5 to 20 volume percent.

ULTRASONIC DEWATERING (3,11)

Using less energy through the application of ultrasonics for dewatering or drying materials was suggested many years ago. Ultrasonics was shown to be especially effective if the material to be dewatered or dried was of a fibrous nature. Although the ultrasonic method was effective, it was expensive and its usefulness was limited to more exotic applications. However, with the advances in piezoelectrics and improved energy efficiency, ultrasonics should now show considerable promise as a potentially economical means of removing water from certain types of products, especially when it is used in combination with other techniques. Table 4 summarizes the results obtained by a number of researchers using ultrasonics to dewater various materials.

The synergistic effect of ultrasonics when used in combination with other techniques such as electroosmosis offers new and interesting possibilities. Ultrasonic or acoustic dewatering in the past has been associated with air-coupled systems, which are very inefficient and, hence, expensive. Ultrasonics in combination with other techniques employs the mechanism in new ways and is potentially economical.

Many mechanisms are associated with ultrasonic radiation. The mechanisms that are beneficial to dewatering in an acoustic field vary with the type of material being dewatered. It is important to understand these mechanisms to apply the ultrasonic energy effectively.

A critical factor in the effectiveness of dewatering mechanisms is the intensity of the sound wave in the regions from which water is to be released. Attenuation occurs because of interparticle friction or scattering, or by viscous absorption. Attenuation of ultrasonic energy in the medium is a function of frequency, geometry, and acoustic properties of the material. Also, the coupling of energy into the material depends on the acoustic impedance of the material.

ELECTROACOUSTIC DEWATERING[2,4,12]

The electroacoustic dewatering (EAD) approach combines dc electrical and acoustic fields in the presence of vacuum or pressure to promote synergistic effects. This effect is shown in Figure 6, where the combination of ultrasonics plus a dc electric field and a vacuum provides both accelerated rates of dewatering and higher percentages of solids than in the presence of either vacuum plus electric or ultrasonics plus vacuum alone.

The relative effects of the three fields depend on the electrokinetic, rheological, surface chemical, and physical properties of the suspensions. For example, the performance of the electrokinetic field depends upon the water-particle relationship (open capillary or closed capillary), zeta potential behavior, etc.

Over the past 6 years, Battelle has tested about 50 different types of suspensions and established wide application of this process. Some typical applications are shown in Table 5, and the representative energy requirements for waste-activated sludge are shown in Table 6.

The EAD process is currently near the commercialization stage of development. It is adaptable to state-of-the-art solid/liquid separation equipment. At the present, the process is adapted to small commercial presses, belt presses, and a screw press nominally designed to handle approximately 20 tons/day feed. This EAD process scale-up has demonstrated that the following benefits can be realized:

- Higher degree of dewatering
- Enhanced recovery of product.

SUMMARY AND CONCLUSIONS

As a result of stringent environmental restrictions and interest in recovery of supernatant liquid for valuable products, the development of processes that depend upon application of body forces has greatly increased. This review demonstrates that these processes should no longer be viewed as laboratory curiosities, but are indeed approaching successful commercial reality. It is

Table 4. Summary of Past Research in Acoustic Dewatering and Drying*

Authors	Year	Type of Material Tested	Frequency, kHz	Intensity, dB	Particle Size, μm	Time of Drying, min	Initial Moisture, percent	Amount of Material Tested, kg	Moisture Reduction, percent	Type of Sonic Device	Remarks
Brun and Boucher	1957	Metal hydroxide	10	--	Colloidal	240	--	0.75	65	Multiwhistle	Control sample gave only 5% reduction.
		Carboxy-methyl cellulose	34	--	Powder	240	--	0.75	40	Multiwhistle	Control sample gave 5% reduction.
Boucher	1959	Fermentation sediment	8	138	Fine	14	118	0.02	100	Multiwhistle 0.95×10^5 -4.83×10^5 Newtons/m^2	Air was drawn through the material kept in a thin layer, without resonator, gave better results than 33 kHz.
		Fermentation sediment	33	138	Fine	14	--	0.02	55		
		Hormone	8	152	NA	20	--	0.03	17	Monowhistle	With vacuum drying, only 5.7% moisture was removed.
		Silica gel	6-8	145	NA	15	--	0.04	100	Monowhistle	Vacuum drying gave only 25% reduction.
		Fiberglas®	8	144	250 (thick)	2-5	--	--	100	Monowhistle	The material was placed 254 mm from source. Best results obtained at 8 kHz.
		Asbestos paper	8	144	6.850 (thick)	5	--	--	100	Monowhistle	Drying by mechanical draft took 15 min.
Boucher	1961	Gelatine	12	143-145	--	120	80	22.2 kg/hr	52.5	Monowhistle	Control gave only 19% reduction.
		Yeast cake	12	148	NA	20	70	22.2 kg/hr	18	Monowhistle	Control gave 10% reduction wet at 37 C.
		Granulated sugar	12	152	NA	20	1.59	22.2 kg/hr	98.6	Monowhistle	--
Borisov, Gynkina	1962	Filter paper	1.08	150-163	Sheet-10×10×0.4 mm thick	2	41	--	40	--	Varied from 4.93×10^{-3} at 150 db to 47.2×10^{-3} at 163 db.
Greguss	1963	Silica gel	8	152	--	2.5	--	--	20	Ultrasonic horn	Direct radiation gave better results than through the membrane.
White	1964	Paper sheet	10.5-12.2	148	Sheet-11.1 kg filter paper	8	--	--	--	Stem-jet whistle	--
Huxsoll Hall	1970	Wheat	11.5	165	NA	40	30	0.1	27	Stem-jet whistle	Experiments performed in Drum Dryer Control gave 13% reduction.
Wilson *et al.*	1971	Fine coal	10	250 watts power	73	10	9	0.45	89	Lead zirconate transducer	Lower frequencies produced larger increase in drying rates.
Fairbanks	1967	Fine coal	20	150-170	147	5	30	0.5	72	Ultrasonic horn	Improper mode of contact, rotary dryer was used for experiment.
		Sand	20	150-170	833	55	--	--	5	Ultrasonic horn	--
Kowalska *et al.*	1978	Mineral sludge	20	800 watts power	Fine	2	97	--	26	Piezo electric	The technique appears to be dewatering rather than drying, but it is not mentioned anywhere in the paper.
		Organic sludge	20	800 watts power	Fine	4	83	--	5	--	--
Swamy *et al.*	1983	Calcium carbonate	9.8	139	300-150	--	22	0.75	85.5	Stem-jet whistle	Dewatering in combination with centrifuge.
		Sawdust	--	--	420-1,680	--	--	0.9	94.5	--	--
Muralidhara, Ensminger	1986	Green rice	12	132	Coarse	100	20	0.4	40	Whistle	Lower frequency gave better results.
			19	140	Coarse	180	20	0.4	40	Whistle	

* Adapted from Muralidhara *et al.*(11). Reprinted with permission by Marcel Dekker, Inc.

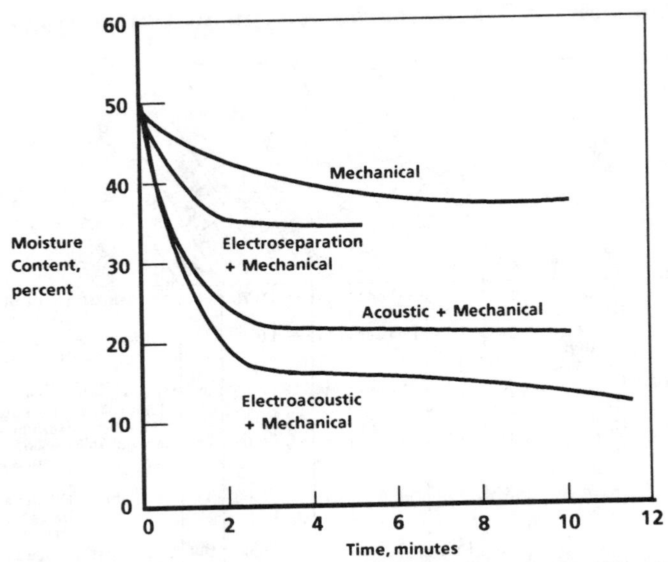

Figure 6. Electroacoustic Dewatering of a Typical Fine Particle Slurry

Table 5. Typical Results for Electroacoustic Dewatering of Wastes

Material	Initial Solids, weight percent	Product Solids After Dewatering, weight percent		Extra Water Removed, lb/lb dry solids
		Conventional	EAD	
Sewage sludge				
– Primary plus secondary	2	15 to 22	43	4.3 to 2.2
– Waste-activated, centrifuged	5	15 to 22	38[a]	4.0 to 1.9
– Primary plus secondary, flocculated and centrifuged	16	NA[b]	38	NA
– Anaerobically digested	2	15 to 22	45	4.4 to 2.3
Wastewater sludge from pharmaceutical plant	4	22	40	2.0
Phosphate slimes	5	NA	40	17.5
Mineral sludge	20	NA	48	2.9
Sludge from pulp/paper mill	3	18	50	3.6
Lagoon bottom sludge	39	NA	70	1.1

(a) Continuous, bench-scale.
(b) NA: not applicable.

Table 6. EAD Conditions for Minimum Energy Use on Waste-Activated Sludge[12]

	Final Solids, percent	
	28	35
Initial Solids, percent	19	19
Residence Time, min	1.1	1.9
Solids Loading, kg DS/m^2	1.0	1.0
Belt Speed, m/min	3.5	2.0
Throughput, kg/hr/m	215	124
Energy Use, kWh/lb filtrate	0.051	0.070

the author's belief that, with the advancement in new types of materials and improvements in manufacturing methods, these processes will have a significant role in the future of solid/liquid separations.

ACKNOWLEDGMENTS

Support from the U.S. Department of Energy and Ashbrook-Simon-Hartley on sludge dewatering is gratefully acknowledged. Special thanks to Ms. Laurie Peters for typing the manuscript.

REFERENCES

1. Casagrande, L., "Review of Past and Current Work," in *Electro-Osmotic Stabilization of Soils*, Harvard Soil Mechanics Series, 1937, *45*, 1962, *66*.

2. Chauhan, S. P., *et al.*, "Electroacoustic Dewatering (EAD) - A Novel Process," presented at 1987 Summer National AIChE, Minneapolis, MN, August 1987.

3. Ensminger, D. E., "Acoustic Dewatering," in *Advances in Solid-Liquid Separation*, H. S. Muralidhara, ed., Battelle Press, Chapter 13, 1986, 321-334.

4. Kim, B. C., *et al.*, "Electro-Osmotic and Electroacoustic Dewatering," presented at Advanced Separation Conference, Battelle, April 11-12, 1989.

5. Krishnaswamy, P. R. and Klinkowski, P., "Electrokinetics and Electro Filtration," in *Advances in Solid-Liquid Separation*, Battelle Press, Chapter 12, 1986, November, 291-320.

6. Lockhart, N. C., "Electro-dewatering of Fine Suspensions," in *Advances in Solid-Liquid Separation*, H. S. Muralidhara, ed., Battelle Press, Chapter 10, 1986, November, 241-274.

7. Lofthouse, C. H., "The Beneficiation of Kaolin Using Commercial HGMS Intensity Magnetic Separation," *IEEE Transactions on Magnetics*, 1982, *6* (17), 3302-3304.

8. Lubrosky, F. E., "High Gradient Magnetic Separation: A Review," *AIP Conf. Proc.*, 1975, *29*, 633-638.

9. Muralidhara, H. S., "The Combined Fields Approach to Separations," *Chem. Tech.*, 1988, April, 229-235.

10. Muralidhara, H. S. and Petty, S. E., "Emerging Electrotechnologies for Separations," presented at Advanced Separations Conference, Battelle, April 11-12, 1989.

11. Muralidhara, H. S., Ensminger, D., and Putnam, A., "Acoustic Dewatering and Drying (High and Low Frequency): State of the Art Review," *Drying Technology*, 1985, *3* (4), 529-566.

12. Muralidhara, H. S., Senapati, N., and Beard, R. B., "A Novel Electroacoustic Separation Process for Fine Particle Suspensions," *Advances in*

Solid-Liquid Separation, Battelle Press, Chapter 14, 1986, November, 335-374.

13. Nagashima, T., "Recent Status of Magnetic Separators," *Kagaku-Kogaka*, 1981, *45* (4), 226-234.

14. Oder, R. R., "High Gradient Magnetic Separation Theory and Applications," *IEEE Transaction on Magnetics*, 1976, *12*, 552-555.

15. Porta, A., "Scale Up Aspects in Electro Osmotic Dewatering," in *Advances in Solid-Liquid Separation*, Battelle Press, Chapter 11, 1986, November, 275-290.

16. Shinoda, T. and Takino, K., "Treatment of Steel Mill Waste Water With HGMS," *Kagaku-Kogaka*, 1981, *45* (4), 235-239.

17. Takayasu, M. and Kelland, D. R., "Use of Magnetic Fields in Solid-Liquid Separation," in *Advances in Solid-Liquid Separation (Supplement)*, H. S. Muralidhara, ed., Battelle Press, 1987.

Figures 3, 4, and 5 are reprinted by courtesy of New Logic Research, Inc.

SOLID/LIQUID SEPARATION USING ALTERNATING CURRENT ELECTROCOAGULATION

P. E. Ryan
Electro-Pure Systems, Inc.
10 Hazelwood Drive, Suite 106
Amherst, New York 14150
(716) 691-2600

and

T. F. Stanczyk
Recra Environmental, Inc.

and

B. K. Parekh, Ph.D.
Kentucky Energy Cabinet Laboratory

ABSTRACT

The efficient removal of suspended solids, organics, metals, and soluble oils from aqueous solutions has recently been demonstrated using alternating current. The principles of electrostriction (charge neutralization) and electroflocculation (catalytic precipitation with aluminum) facilitate coagulation, rapid settling, oil separation, and improved dewatering without the use of expensive polyelectrolytes, chemical aids, or sophisticated mechanical filtration systems.

The electrocoagulator has no moving parts, is small, and can usually be integrated with existing processes to facilitate source reduction and/or purification, while reducing the volume, as well as hazard potential, of the waste generated. It can also be used as a stand-alone unit process for end-of-pipe treatment or as a mobile on-site treatment step for remediation projects. Treatment of process wash waters is an excellent application. The coagulator has been operated in both continuous and batch flow modes in sizes ranging from 1 to 750 gal/min. Because of residual effectiveness and short contact times, often less than 50 percent of the solution requires treatment, thus minimizing pumping and space requirements.

This paper provides an overview of the technology and discusses applications and benefits in the areas of in-plant processing, industrial wastewater treatment, site remediation, and water purification.

INTRODUCTION

In our industrial society there is an ever-increasing awareness of the adverse impacts of inorganic and organic chemicals on the quality of water. Human health and environmental concerns dictate process improvements, substitutes, and control systems effective in preventing and minimizing related risks. Wastewaters are of prime concern and warrant reassessment in terms of both volume reduction and pollutant removal. Research must continue to search for effective and efficient solutions to contamination resulting from specific industrial sources as well as contamination related to groundwater and surface runoff.

Advancements in analytical chemistry permit scientists to detect and quantify hazardous chemicals at extremely low concentrations in aqueous media. This capability is expected to support new and stricter regulatory standards that take into account the cumulative impacts of chemical exposure, as well as the potential for chemical transport and transformation.

Wastewater treatment technologies are required to provide optimum removal of suspended and soluble pollutants. Technical strategies emphasizing the reduction of pollution before wastewater generation have been adopted in an attempt to achieve these higher performance levels. These strategies include the separation of fine and ultrafine solid products that were previously wasted in waterwash operations. These solid products may be suspended, emulsified, and/or partially solubilized in aqueous media to an extent that could be deemed significant in terms of potential hazard and toxic impact. Most of these washwaters require chemical addition to enhance solid agglomeration and settling before conventional mechanical dewatering systems can be employed. Unfortunately, the addition of these chemicals add to the volume of waste generated.

As an alternative to chemical conditioning and flocculation, recent developments indicate that liquid/liquid and solid/liquid phase separation can be achieved using alternating current electrocoagulation (AC/EC). The AC electrocoagulator has been used to flocculate and settle fine solids without the use of chemical aids[1,13,15,16]. Recent pilot-scale data[3] demonstrated phase separation of wastewater containing suspended and emulsified oils, thus minimizing potentially toxic pollutants. AC/EC may be easily integrated with conventional process and control systems to enhance solid product recovery and water purification. Waste reduction goals may be accomplished by integrating this technology with a variety of operations that generate contaminated water.

This paper discusses the theory of electrocoagulation with alternating current. Operating variables are reviewed and potential advantages and benefits are highlighted. The current stage of development and plans for future research are discussed in light of applications dealing with water purification, wastewater treatment, and site remediation.

STATEMENT OF PROBLEM

Contaminants influence the physical, chemical and electrical properties of water. These properties, in turn, are used to identify environmental concerns requiring control to ensure regulatory compliance. Water is used as a universal solvent and its properties vary as a function of use. Solid products subjected to waterwash operations create suspensions of finely divided colloidal matter in aqueous media. These wastewaters are generally difficult to phase separate, and the suspended solids contribute to the loadings of inorganic and organic pollutants soluble in the aqueous matrix.

Regardless of origin, wastewater may also contain the various forms of colloidal matter summarized below:

- Solid particles in the form of colloidals with a maldistribution of electrons in magnetic suspension in the water media
- Solution components present as water-soluble fractions and suspended by magnetic forces
- Chemically stable and soluble "salts" displaying an intermediate stable existence in the form of colloidal suspensions of unstable matter
- Suspended inert matter that is colloidal as well as susceptible to precipitation.

Molecular hydrogen bonding is also a consideration. It has a major impact on the bridging between the water and solid molecules of aqueous sludge. The mechanisms influencing the water associated with solid particles are interior adsorption, surface adsorption, capillary absorption, interparticle absorption, and adhesion water.

Interior and surface adsorption is referred to as "free" water, which is usually removed by mechanical techniques. The other three mechanisms require energy-intensive techniques such as thermal drying for phase separation.

The presence of an electrical charge on the surface of particles is often a prerequisite to their existence as stable colloids. This surface charge also depends on the properties of the aqueous phase because adsorption or binding of solutes to the surface of the colloids may increase, decrease, or reverse the effective charge on the particle. The adsorption may occur as a result of a variety of binding mechanisms: electrostatic attraction or repulsion, covalent bonding, hydrogen-bond formation, van der Waals interaction, or hydrophobic interaction.

Flocculation and filtration destabilizes suspended colloids by enhancing aggregation or the attachment tendency of these colloids.

BACKGROUND AND RELATED THEORY

Several studies(7,17-19) suggest that most solid particles suspended in aqueous media carry electrical charges on their surface. When the particles are larger than atomic or molecular dimensions, they will tend to separate from the aqueous media under gravitational force unless they are stabilized by electrical repulsion or other forces. Such forces can prevent aggregation into larger particle masses or flocs, which are more prone to settling. These surface charges may exist as an ionic double layer or a neutralized electric dipole, as conceptually depicted in Figure 1.

Generally, the gravitational force on small particles is weaker than other forces. Collisions between particles due to Brownian motion often result in aggregates held together by van der Waals forces, and coagulation may occur in the following ways:

- The particle crystal lattice may contain a net charge resulting from lattice imperfections or substitutions. The net charge is balanced by compensating ions at the surface such as zeolites, monmorillonites, and other clay minerals.
- The particle solids may contain ionizable groups.
- Specific soluble ions may be absorbed by surface complexes or compounds formed on the particle surface.

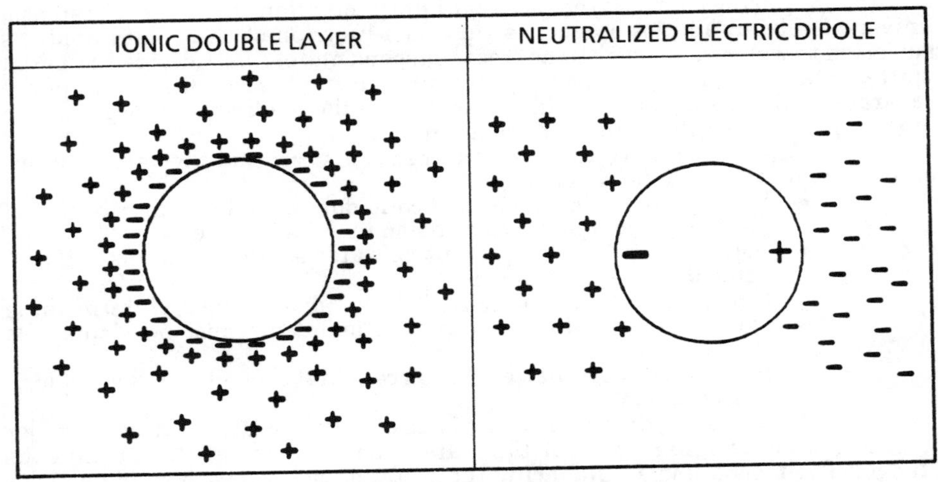

Figure 1. Surface Charge Distributions

PRINCIPLES OF AC ELECTROCOAGULATION

The electrocoagulator process, invented by Moeglich *et al.*(10-12), is based on colloidal chemistry principles using AC power and electrophoretic metal hydroxide coagulation. The process employs two main principles:

- Electrostriction, whereby the suspended particles are stripped of their charges by subjection to alternating current electrical field conditions in a turbulent stream
- Electroflocculation, whereby minute quantities of metal hydroxides are emitted from the electrodes to assist in flocculation of the suspended particles.

The theory of electroflocculation, or metal ion flocculation, is well established. Iron and aluminum ions have been widely used to clarify water. Recently, Parekh *et al.*(14,15) developed a coagulation system involving the use of metal hydroxide and fine particles. They reported that the optimum coagulation of a metal ion/particle system takes place at the isoelectric point of the metal hydroxide precipitate. Jensen(9) suggests that optimum coagulation may not necessarily occur at the exact point of zero charge, since other mechanisms such as bridging are also important.

A better understanding of the mechanisms that underlie the operation of AC/EC is expected to result from research initiated by the State University of New York at Buffalo in June, 1989. The current hypothesis for AC/EC operation is summarized as follows:

- Polar molecules adsorbed on the surface of small particles are neutralized by an equivalently charged diffuse layer of ions around the particle. A zero net change results.
- Nonspherical particles have nonuniformly distributed charges (dipoles) and elongated neutralizing charge clouds surrounding them.
- These dipoles come into play when the charge clouds are distorted by external forces or close proximity of other charged particles.
- External forces such as electric fields can (a) cause dipolar particles to form chains and (b) unbalance electrostatic forces resulting in dramatic phase changes (coagulation).
- AC electric fields do not cause electrophoretic transport of charged particles, but do induce dipolar chain-linking and may also tend to disrupt the stability of balanced dipolar structures.

PROCESS DESCRIPTION

INTRODUCTION

AC/EC system designs will vary depending on the characteristics and quantity of waste or process steams being treated, treatment objectives, and location. Characteristics, such as particle size, conductivity, pH, and chemical constituent concentrations, dictate the operating parameters of the coagulator. The quantity and flow rate of the raw solution will effect total system sizing, coagulator retention time, and mode of operation (recycle, batch, or continuous). Treatment objectives will establish the type of gravity separation system to use, establish recovery criteria, identify the utility of side stream treatment, define effluent standards to be met, and determine the advantages of recycle or multiple staging. Treatment objectives may include product recovery or simply preconditioning prior to using an existing process or as a polishing step after treatment. Location will affect design by imposing physical size constraints and pumping requirements. In-plant industrial applications, for example, may be configured differently than a mobile on-site system used for remediation or treatment of ponded water.

BASIC PROCESS

A basic process flow diagram for AC/EC is presented in Figure 2. Coagulation and flocculation occur simultaneously within the coagulator and in the product separation step. The redistribution of charges and onset of coagulation occur within the coagulator as a result of exposure to the electric field and catalytic precipitation of aluminum from the plate electrodes. This reaction is usually completed within 30 seconds for most aqueous suspensions. The solution may be transferred by gravity flow to the product separation step.

Product separation may be accomplished in conventional gravity separation and decant vessels. Coagulation and flocculation continue in this step until the desired degree of phase separation is achieved. Generally, the rate of separation is faster than in methods that employ chemical flocculants or polyelectrolytes, and for some applications the solid phase is denser than the solids resulting from chemical treatment. A recent feasability study[5] demonstrated 95 to 99.5 percent recovery of submicron fines from a 0.6 percent stable suspension after 1.5 hours settling time. Alternative treatment achieved only 80 percent removal after 1.5 hours.

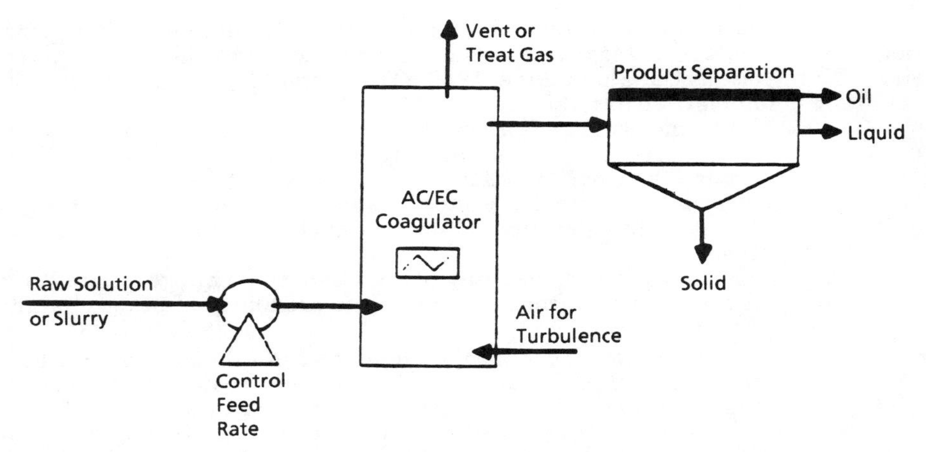

Figure 2. AC/EC Process Flow Diagram

In many applications, the electrocoagulator retention time may be reduced and performance improved by agitating the solution as it passes through the electric field. This turbulence can be induced by using a static aerator concept or simply diffusing small bubbles of air or nitrogen through the solution in the space between the plates. Air has been used in full-scale applications treating pond waters and removing fines from coal washwaters. Bottled nitrogen and bottled air have been used in the laboratory to conduct treatability tests. Since the gas used to create turbulence may also strip volatile organics, it is necessary to analyze the vent gas stream, especially when treating hazardous wastes. When appropriate, the vent gases may be collected and treated using available conventional technologies and thus control air emissions within acceptable limits.

After the product separation step, each phase (oil, water, solid) is removed for reuse, recycle, further treatment, or disposal. A typical hazardous waste decontamination application, for example, would result in a water phase that could be discharged directly to a stream or to a local wastewater treatment plant for further treatment. The solid phase, after dewatering, would be shipped off-site for disposal, the dewatering filtrate being recycled. Any floatable material would be reclaimed, re-refined, or otherwise recycled or disposed.

OPERATING REQUIREMENTS

The AC/EC operates on low voltage, generally below 110 VAC. It is designed to work at atmospheric pressure and is vented to alleviate any problems with gas accumulation. As previously mentioned, air abatement apparatus may be added, if necessary.

The internal geometry allows for free passage of particles smaller than 0.25 in. Although normal operation is relatively maintenance free, some problems can be encountered if process upsets allow heavy particulates to enter the lines inadvertantly. In this case, material buildup could restrict passage and thus retard flow. No permanent damage has been experienced in these cases, and the problem has always been alleviated by reverse flushing or minor disassembly and cleanout.

Although there has been some question about electrode deterioration, in practice none of consequence has been noted. Minor etching occurs on the electrode skins. As nearly as can be theorized, the alternating current cyclic energization retards the normal mechanisms of electrode attack that are experienced in DC systems, and reasonable electrode life has been proven. Electrodes were replaced after 4 months of continuous operation (20 hours per day) in a 250-gal/min commercial unit.

Electrical energy costs vary based on the solution being treated and the specific application. Commercial units have treated coal washwaters for \$0.40 per 1,000 gallons at power costs of \$0.05 per kWh. This cost is more than offset by savings in the chemical costs associated with alternative methods that require the use of polyelectrolytes and chemicals to adjust pH.

RESIDUAL EFFECTIVENESS

Bench-scale tests[5] and full-scale field applications[1] have demonstrated a phenomenon referred to as residual effectiveness. Once the solution has passed through the coagulator and settling is complete in the product separation stage, the separated products can be remixed and subsequent phase separation will occur without further treatment through the electrocoagulator. It appears as if the charge redistribution and coagulating forces remain effective for extended periods of time. This phenomenon is important in that mixing and pumping can be accommodated after coagulation, if so dictated by other system design conditions, without losing the phase separation effectiveness. This also indicates that in some applications only a portion of the total contaminated solution would have to be treated. For example, in removing constituents from a pond, a portion of the solution may be treated and returned to the pond until the desired phase separation results. Phase separations have been accomplished by passing as little as 25 percent of the total volume through the electrocoagulator.

The data presented in Table 1 were measured by treating a mine drainage stream sample obtained from an eastern Kentucky mine and blending it with untreated sample. Results show a 90 percent reduction in turbidity of this solid suspension after 4 hours settling when only half the slurry was treated with AC/EC.

Table 1. Propagation Test Data from Mine Drainage Stream

Percent of Slurry		Turbidity, NTU Settling Time, hr[a]				
Treated	Untreated[b]	0	1	2	4	24
100	0	700	70	--	50	40
75	25	700	250	80	50	40
50	50	700	500	250	75	40
25	75	700	600	300	80	40
0	100	700	700	700	700	700

Notes:

(a) NTU = Net turbidity units.
(b) Untreated Sample Characteristics:
Turbidity = approximately 700 NTU
Electrophoretic mobility of solids was about -2.3 μm/sec/volt/cm
Conductivity = 2,000 μmhos/cm.

EFFECTS AND APPLICATIONS

Studies suggest that alternating current coagulation causes the following effects on the resulting by-products:

- The magnetic forces associated with liquid suspensions are destroyed.

- Sludges tend to dewater and densify, suggesting a disruption and/or destruction of the hydrogen bonding of water molecules.
- Electronic or ion exchange creates an electrochemical environment, causing various reactions dependent on contaminating constituents.
- Oils, soap, detergents, and cellulosic material can be phase-separated from water.
- Inert clay colloidals can be removed from aqueous media.
- The generation of OH^-, as well as the potential for O_2, H_2, O_3, and H_2O_2, may influence soluble pollutants by chemical oxidation.
- Soluble oils phase-separated from aqueous media can extract other toxic constituents that are preferentially soluble in the oil fraction.
- Water characteristics created by electrocoagulation and subsequent clarification result in a long-lasting demagnetized effluent that resists contamination.

The following advantages were identified as a result of using AC/EC in the coal industry:

- Improved fine coal recovery
- Improved dewatering rates
- Reduced filtration time
- Reduced recirculation of coal and clay fines in closed loop water
- Reduced buildup of fines and clays on dewatering screens
- Neutralization of plant water pH
- Removal of heavy metal and organic carbon from water
- Reduced plant maintenance
- Increased plant availability
- Increased coal yields without sacrificing quality
- Increased quality at the same or increased yield
- Reduced freezing of treated coal.

Test data such as presented in Tables 2 and 3 support the cited advantages, as well as many of the pertinent principles of the technology.

Table 2. Applications of Electrocoagulation

Application	Results	Reference
1. Particulate removal		
a. Water from contaminated soil wash	>washwater clean enough to recycle to the ground	4
b. Clay colloids in ponded water	>99% removal of suspended solids	16
c. Removal of coal fines	Improved performance without chemicals	1-7, 9-19
d. Micron size particulates	99 to 99.5% recovery	5
2. Removal of soluble organics		
a. Ponded water	>99% removal of TOC*	16
b. Creosote suspension	>98% removal of TOC	3
c. Emulsion of rolling coolant and waste oils	>90% removal of TOC	3
d. Lake water (DC electrocoagulation)	>95% removal of TOC	19
3. Metals		
a. Ponded water	>99% removal of Fe, Mn, Al	16
b. Acid mine wastes	>98% removal of Fe, Cu, Al	8
4. Enhanced dewatering		
a. Coal fines	>dewatering rate increased by 30 to 50%	1, 16

* TOC = total organic carbon.

Table 3. Field Test Results of Electrocoagulation of Ponded Water

Parameter	Pond A, ppm except for pH: Raw Water	Pond A, ppm except for pH: After Treatment	Pond B, ppm except for pH: Raw Water	Pond B, ppm except for pH: After Treatment
pH	6.4	7.7	7.3	8.3
Suspended Solids	197	1	195,000	15
Dissolved Solids	--	--	7,212	3,344
Soluble Iron	88	0.13	--	--
Total Iron	285	263 (in sludge)	3,500	0.18
Manganese	3	1.9	104	0.02
Aluminum	--	--	304	0.08
Alkalinity	--	--	48,500	400
TOC	--	--	11,000	30

Adapted from Ref. (16) by courtesy of Westinghouse.

SUMMARY

The use of alternating current electrocoagulation to break emulsions and phase-separate aqueous solutions has been successfully demonstrated without using chemical aids. Based on pilot-scale results and an assessment of potential physicochemical reactions, the applicability of this technology to various industrial and hazardous waste management applications has been identified. Research activities are under way and/or planned to investigate this technology further and to define better its applications and benefits. Ongoing field demonstrations and treatability studies of emulsions, slurries, and suspensions from various industries and hazardous waste sites contribute to understanding effective operating parameters. Cost effectiveness is derived from the reduction or elimination of chemical aids, performance improvement of conventional mechanical separation systems, and an overall reduction in the quantity of wastes generated. Applications can be found within almost every industrial sector for a wide range of wastes and in-plant processes.

REFERENCES

1. Berry, W. F. and Justice, J. H., "Electro-Coagulation: A Process for the Future," presented at the 4th International Coal Preparation Conference and Exhibition, Lexington, KY, April 28-30, 1987.

2. Co-Ag Technology, Inc., Laboratory Report by J. H. Justice, 9/16/88, "Determine the Effect of Propagation by Mixing E/C-Treated Slurry with Various Percentages of Untreated Slurry."

3. Electro-Pure Systems, Inc., Laboratory Notebook, Data Sheets E788001 through E1288012, 1988.

4. Electro-Pure Systems, Inc., "Alternating Current Coagulation Information Summary," unpublished report, 1988.

5. Electro-Pure Systems, Inc., Laboratory Notebook Data Sheets, May '89-Confidential.

6. Goscinski, J. S., "Effects of AC Powered Electro-coagulation Related to Fine Coal Cleaning," W. F. Berry and Associates Report on Coagulator at Delta Industries Mine at Meyerdale, PA, February 8, 1982.

7. Jageline, I., Grigorovich, M. M., and Daubaras, R., "Electromechanical Treatment of Electroplating Wastes. Effect of pH of the Solution on the Elimination of Copper (+2), Zinc (+), and Chromium (+6) Ions during Electrocoagulation," *Liet. TSR Mokslu Adad. darb. Ser. B*, 1979, 65-72 (As found in *Chem. Abstrac.*, *91*, 26704m).

8. Jenke, D. R. and Diebold, F. E., "Electro-Precipitation Treatment of Acid Mine Wastewater," *Water Res.*, 1984, *18* (3), 855-859.

9. Jensen, J., Professor, Civil Engineering Department, State University of New York at Buffalo, discussion of 12/21/88.

10. Moeglich, K., "Water Purification Method," U.S. Patent 4,094,755, 1978.

11. Moeglich, K., "Water Purification Method and Apparatus," U.S. Patent 7,176,038, 1979.

12. Moeglich, K. and Hodgetts, H. L., "Water Purification Method and Apparatus," U.S. Patent 4,053,378, 1977.

13. Nikerson, F. H., "Electrical Coagulation: A New Process for Prep Plant Water Treatment," *Coal Mining and Processing*, September 1982.

14. Parekh, B. K., "The Role of Hydrolyzed Metal Ion in Charge Reversal and Flocculation Phenomena," Ph.D. Thesis, The Pennsylvania State University, 1979.

15. Parekh, B. K. and Aplan, F. F., "Flocculation of Fine Particles and Metal Ions," Annual AIME Meeting, Phoenix, AZ, 1988.

16. Plantes, W. J., "Electrocoagulator - Removal of Colloidal and Suspended Solids," Report on Mine Pond Testing, Westinghouse Electric Corporation, 1978, July/August.

17. Schwan, H. P. *et al.*, "On the Low Frequency Dielectric Dispersion of Colloidal Particles in Electrolyte Solution," *J. Phys. Chem.*, 1962, *66*, 2626.

18. Schwarz, G. A., "Theory of the Low Frequency Dielectric Dispersion of Colloidal Particles in Electrolyte Solution," *J. Phys. Chem.*, 1962, *66*, 2636.

19. Vik, E. A., Carlson, D. A., Eikum, A. S., and Gjessing, E. T., "Electrocoagulation of Potable Water," *Water Res.*, 1984, *18* (11), 1355-1360.

ELECTROCHEMICAL REACTOR SYSTEMS FOR METAL RECOVERY AND DETOXIFICATION OF INDUSTRIAL EFFLUENTS

Bernard Fleet
The Department of Chemistry
University of Toronto
Toronto, Ontario
Canada M5S 1A1

and

Toxics Recovery Systems International Inc. (TRSI)
#14-44 Fasken Drive, Rexdale
Ontario, Canada M9W 5M8

ABSTRACT

"Electrochemical reactor" is an engineering term describing an electrolytic cell that performs a useful chemical process, in this case the recovery or detoxification of metals from industrial effluent streams. Developments in this field during the last two decades have resulted in significant improvements in the performance of these systems, especially in their operating efficiency or space-time yield.

The present paper examines these advances and reviews the various configurations of commercial reactor systems ranging from low surface area, planar electrode electrowinning cells to high surface area, high-performance cathode systems designed for the treatment of very dilute waste streams.

Finally, some typical applications to pollution control in the electronics products, electroplating/surface finishing, and gold milling industries are presented.

INTRODUCTION, BACKGROUND TO POLLUTION CONTROL, ELECTROMETALLURGY, AND ELECTROCHEMICAL ENGINEERING

Pollution by toxic metals--including cadmium, copper, chromium, lead, mercury, nickel, and zinc--is generated by a wide range of manufacturing industries such as mining, metal finishing and plating, electronics, semiconductor and printed circuit board manufacturing, as well as several other industries that use metal treatment as a part of their overall manufacturing processes. Waste management practices in these industries vary widely. In some cases effluent streams are simply diluted to meet regulatory limits, but more commonly, a chemical treatment process is used where caustic soda, lime, or sulphide is added to the waste stream to convert the dissolved metals into a semisolid "sludge." After partial dewatering, this toxic metal sludge is either packed into steel drums or occasionally "fixed" with calcium sulphate or cement waste before being dumped into a land disposal site. These sites range from municipal waste dumps to "secure" chemical landfills to abandoned mineshafts or deep wells.

There is now indisputable evidence[38] that land disposal of toxic wastes is only a temporary solution to the problem. All of these storage or disposal sites leak to varying degrees, with the result that many of the toxic species find their way into natural watercourses and subsequently into the biological food chain. Increasing awareness of the risks associated with hazardous chemical waste disposal is causing governments to initiate stricter environmental legislation, which in turn is causing many industries to re-evaluate their waste management options. There are strong trends towards seeking on-site, zero discharge, resource recovery technologies[18,51].

Clearly, there are many environmental waste management problems where electrochemistry is able to offer an effective solution[5,18,37,51]. However, before these new electrochemical processes can be implemented, two important questions need to be answered. First, are the economics and environmental climate suitable to encourage industries to make the investment in new waste management strategies? Second, are alternative resource recovery technologies available, and if so, how do they compare in performance and costs with electrochemical methods?

The main aim of this review is to survey the status of the various commercial and research prototype electrochemical reactor systems for pollution control and metal recovery. This review also attempts to define the role of electrochemical reactor systems in solving some major problems of environmental waste management.

Electrometallurgy, broadly defined as the electrolytic plating or recovery of metals from solution, is among the earliest examples of applied science. Examples of crude electroplating date from early Egyptian and Assyrian times[50]; the first literary reference to electrochemistry is recorded in Pliny's Histories[2] and describes the silver plating of tin. By the 1850s, just 50 years after the introduction of Volta's electrochemical pile, electroplating was already an established practice. Electrolytic recovery of metals as an industrial process also followed closely on Faraday's and Davy's pioneering works, with the first electrowinning cells being reported in the late 1860s following the introduction of the dynamo.

The earliest recorded example of the application of electrochemical principles to the recovery of metals occurred in Czechoslovakia, in Banska Bystrica at the mine of Herrungrund, early in the 16th century[6]. This involved the recovery of copper from cupriferous mine waters by electrochemical replacement with iron. A variety of decorative copper-plated ironware was produced by this process.

ELECTROCHEMICAL ENGINEERING

Electrochemical engineering as a science dates from the mid-1960s and coincides with several advances in electrochemistry, particularly in the fields of mass transport theory, materials science, and instrumentation. The impetus given to electrochemical technology by Heyrovsky's work also should not be overlooked. The development of polarography as a technique for the study of electrode processes also provided a stimulus for many workers to expand these ideas to other fields of electrotechnology. The influence of other scientific disciplines on the development of electrochemical engineering has been summarised by Selman[46].

In conventional electrochemistry, the mechanism of the electrode process and its kinetics are often the factors of major concern. In electrochemical engineering, on the other hand, the actual mechanistic details of the process are not so important as its specificity or process efficiency. More importantly, we are concerned with the rate of the process, the current efficiency, and a measure of reactor efficiency, the space-time yield. This latter factor determines whether a process is economically or commercially viable since it can be used to compare performance of different electrode designs as well as to compare an electrochemical process with the space-time yields for alternate nonelectrochemical technologies.

The design of electrochemical reactors offers an entirely different set of challenges than those found in classical electrochemistry. In a conventional electrochemical system, one is often dealing with a microelectrode of usually uniform surface area and activity under conditions of very well defined current control--either diffusion limited or kinetically controlled or, in cases of convective control, with very well defined mass-transport conditions. In electrochemical engineering, on the other hand, one is dealing with a bulk electrode of constantly varying surface area and activity as metal is deposited. In addition, there are often difficulties in maintaining uniform potential control and current distribution over the electrode surface. It is also necessary to consider the reverse stripping process of recovering the metal after collection. Problems of scale-up, materials corrosivity, and other engineering design problems also must be addressed.

The theoretical background to electrochemical engineering is now well established[14,26,29,31,36,40-42,52]. The principles controlling the major parameters in reactor design and operation, potential/current control, potential/current distribution, mass-transport characteristics, role of electrocatalysis, and the role of electrode material, etc., are fairly well understood. The keystone undoubtedly has been the development of convective diffusion theory by Levich[31], which has led to the proper evaluation of concentration profiles in a wide range of electrolytic systems. The concept of the boundary layer defines the concentration profiles of electroactive species in the bulk solution, in the boundary diffusion layer, and at the electrode surface. The role of convective diffusion in both laminar and turbulent forced convection modes has been defined also and has led to the development of mass-transport relationships for a wide range of electrode/cell geometries.

REACTOR AND PROCESS DESIGN

The design or selection of an electrochemical reactor for a specific application or metal recovery process depends on a variety of factors. The first consideration is the process itself, the type of metal or other chemical species to be recovered, the concentration level and chemical composition of the process stream, and finally, the desired removal rate and treatment efficiency. Unless the application is well established, it is first of all necessary to investigate the electrode process. For most metal deposition processes, basic data on the electrode process may be obtained from the literature, including source references such as Pourbaix diagrams[42], which define pH-potential equilibria. In the absence of this data, laboratory bench-scale experiments may be required to define process conditions and provide data for scale-up.

In the design of electrochemical reactors, four major areas must be considered. First is the overall cell design, including electrode materials of both cathode and anode and the requirement, if any, for a diaphragm or separator. The second is the problem of potential control of the working electrode. Next is the control of current distribution over the working electrode surface area, and finally, the definition and control of the mass-transport characteristics of the system. The electrode potential of the working electrode is a complex parameter involving current distributions within the working electrode and the solution, conductivities of the bulk electrode and the process solution, and the concentrations of all reactants and products for all of the possible electrode processes. In classical electrochemical experiments, potentiostatic control is achieved by electronically comparing the voltage difference between the working and reference electrodes, and the required operating potential (set point) and feeding the difference back to a rectifier such that the rectifier provides an output through the counter electrode to maintain the desired condition. In electrochemical engineering applications, however, potentiostatic control suffers from a number of practical problems. The first is the problem of ground loops from cell currents, which can drive destructive currents through the reference electrode, effectively destroying its performance. More significantly, the reference electrode suffers from location problems since it is normally linked to the cell via a Luggin type capillary probe. It can, therefore, only monitor the potential at one small site in the working electrode, which may have very little relevance to the overall electrode potential. This approach may be useful in initial design studies for electrode potential mapping, but its real application as a control technology is minimal.

The current distribution within the electrode, which in turn controls the current efficiency of the electrode, is dependent on the mass-transport characteristics of the system and on the control of potential over the working electrode surface. Mass-transport is the parameter that controls most situations. The Nernst diffusion model defines the situation at the electrode-solution interface. Three distinct regions can be identified: the stationary diffusion layer of solution adjacent to the electrode surface; the boundary diffusion layer where, under electrolytic operating conditions, a concentration gradient and a solution flow velocity profile may be observed; and finally, the bulk layer.

REACTOR CONTROL

Since potentiostatic control of most electrochemical reactor systems is impractical, the next option is voltage control. This route, however, is also of limited value[17]. The voltage-current relationships of most cells are dominated by interelectrode conditions, electrolyte resistance, diaphragm resistance, and mode of operation, monopolar or bipolar, etc. Typically, cells are on the order of 4 to 10 volts for monopolar operation, whereas most working electrode potentials are in the range 0 to 1.0 volts. Clearly, cell voltage has some limitations for controlling working electrode potential, especially when the influence of concentration and temperature changes are considered. In most cases, especially in the case of metal recovery from dilute solutions, constant current control is the most useful operating approach from both a practical and theoretical standpoint. Since the electrode potential cannot be controlled in a "potentiostatic" sense, it can be controlled by a combination of controlled current and controlled mass-transport. For a typical current voltage curve for a mass-transport controlled process operating under constant current conditions, the electrode potential will be given by the point at which the imposed current value intersects the i-E curve. Mass-transport characteristics are, therefore, critically important in controlling electrode potential. For a given electrode-cell geometry, a well-defined range of mass-transport and imposed current conditions will achieve the desired electrode process efficiency. At the same time, there are often significant problems in controlling mass-transport conditions within a reactor. This is especially the case with three-dimensional electrodes for which mass-transport conditions in the reactor can change dramatically during the course of a process. For example, in a metal deposition process, electrode surface and effective area of electrode will change continuously and in many cases significantly influence hydrodynamic performance.

The other major critical area of advancement in electrochemical engineering has been in materials science. The materials requirements for electrochemical reactors are often particularly onerous because of the corrosivity of the process media and the extreme reactivity of some electrogenerated species. The development of novel electrocatalysts, especially the introduction of the dimensionally stable ruthenium-iridium oxide based DSA anodes, has played a vital role in the development of commercially viable reactor systems. The development of membrane/separator materials with much improved mechanical stability, ion-permeability, and conductivity also has played a key role.

ENGINEERING PARAMETERS

In electrochemical engineering, three parameters are often used to define the performance of an electrode or electrochemical reactor: current efficiency, process efficiency, and electrode/cell space-time yield. In some cases, particularly with high surface electrode cells, an additional parameter--the percentage conversion or removal per pass--may also be specified.

The process efficiency (η) is defined as moles of metal removed as a function of initial metal concentration in the process stream.

$$\eta = \frac{\text{moles of metal removed}}{\text{moles of metal in process stream}} \cdot 100$$

The current efficiency (ϕ) is defined as the ratio of cell current used to deposit metal on the working electrode as a fraction of the total current passed through the cell,

$$\phi = \frac{\text{theoretical current for metal deposited}}{\text{total charge consumed}} \cdot 100$$

The space-time yield parameter provides a measure of electrode/cell performance, allowing comparison, for a given electrolytic process, between different electrode configurations and also between electrochemical systems and competitive, nonelectrochemical technologies. The space-time yield term (Y_{ST}) is defined for an electrode as

$$Y^{E}_{ST} = \frac{A^{s}i}{czF} \text{ and for a cell,}$$

$$Y^{C}_{ST} = Y^{E}_{ST} \frac{1}{1 + V_A/V_B}$$

where A_s is the specific electrode area, ϕ the current efficiency, C the concentration change during one solution pass through the reactor, and V_A and V_B the volumes of anode and cathode compartments. The importance of the foregoing discussion is that, in the design of electrochemical reactors for metal removal, the major objective is to maximise the space-time yield and thus minimise the size and capital cost of the system for a given process or level of metal recovery.

ELECTROCHEMICAL REACTORS FOR METAL RECOVERY

APPLICATION AREAS

Three major application areas of electrochemical reactors in metal recovery can be defined:

- Electrowinning of metals from ores and primary sources
- Electrorefining of metals from aqueous solutions or molten salts
- Electrolytic recovery of metals from waste sources and industrial effluents (electrochemical detoxification).

Although the cell design requirements for electrorefining are usually quite specific, many cell designs used for both electrowinning and electrochemical effluent treatment are often very similar, and in many cases, the applications overlap. With the exception of the electrolytic processes for aluminium, magnesium, and sodium, applications of electrochemical applications of electrowinning have mainly been directed to the recovery of copper, nickel, and the precious metals. However, electrolytic processes have been developed for a wide range of metals(3), but due to the world depression in metal values, only a small fraction of these processes is being developed or operated presently on a commercial scale.

CLASSIFICATION OF REACTOR DESIGNS

The plethora of electrochemical reactor designs are, in many cases, difficult to classify. Various approaches have been attempted(23,33); one approach is based on defining the movement of process solution in relation to the direction of current flow through the working electrode (Figure 1). This definition describes three main modes of process flow: a "flow-by" mode in which solution flows past the surface of the electrode, and two flow-through modes, "flow-through parallel to current" and "flow-through perpendicular to current." A simpler classification can be based on working electrode geometry, i.e., whether the cell/reactor employs a planar (two-dimensional) or bulk electrode (three-dimensional). Two-dimensional reactor systems are typically planar electrode, low surface area designs, whereas three-dimensional designs mostly comprise high surface area systems with extended electrode surfaces typically based around porous electrode materials, packed particles, or fibres. Whereas most current research and development efforts in electrolytic metal recovery seem to be directed towards high surface area reactors, it should be noted that these two classifications of reactor system should not be viewed as competitive. Rather, they should be viewed as a spectrum of available reactor

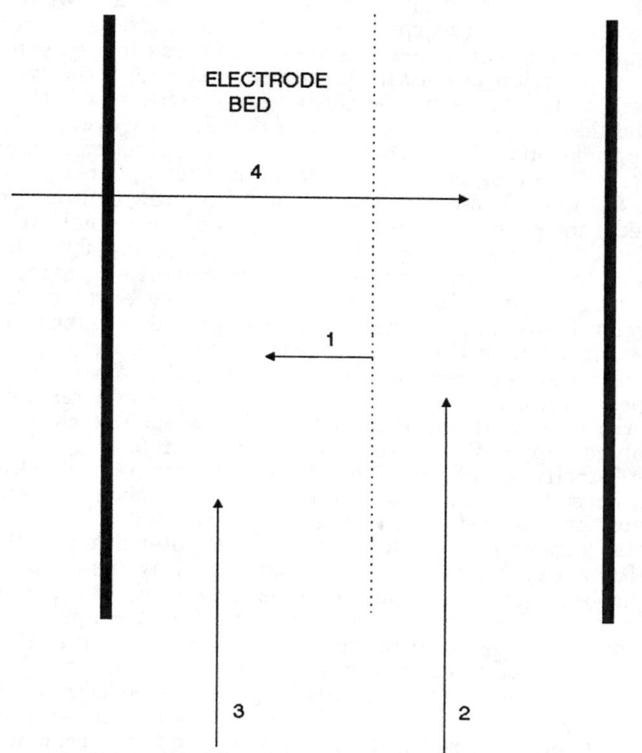

Figure 1. Mass Transport Modes in Electrochemical Reactors
1. Current Flow
2. Flow-by
3. Flow-Through Perpendicular to Current Flow
4. Flow-Through Parallel to Current Flow

and working electrode designs from which the optimum reactor and process design can be selected for a given application. A review of the major designs of reactor systems, most of which have found application in metal removal from industrial wastewaters, is presented below.

TWO-DIMENSIONAL REACTOR SYSTEMS

Classical electrochemical reactor designs invariably evolved from direct scale-up of simple laboratory electrolysis experiments. The most common example of this concept is the tank cell, in which an array of electrodes are immersed in a plastic or metal tank. More sophisticated versions involve forced convection, rotating or moving electrodes, and a wide range of plate and frame or filter press types of cells.

In the two-dimensional class of reactors, three main types will be considered: tank cells, plate and frame cells, and rotating electrode systems.

TANK CELLS

Tank cells are one of the simplest and most popular designs of cell for both electroorganic and inorganic processes. Commercial designs, available in a wide range of sizes and electrode areas, can operate both in the monopolar or bipolar mode. They usually function as an undivided cell with a single electrolyte/process stream; the incorporation of membrane/separators is inconvenient, and in cases where cathode and anode chambers must be separated, a plate and frame construction is usually preferred. Although the widest commercial use of tank cells has been in electroorganic synthesis[9,21], they also have found application in pollution control and metal recovery, mostly for electrowinning applications. The Lancy Cell[30] is a typical example; it comprises a planar electrode cell of modular construction offering a range of cathode areas from 0.5 to 5.0 m^2. A similar type of modular tank cell design with a membrane-isolated anode compartment and planar stainless steel cathodes has been developed by TRSI Systems[45]. The main application for this type of reactor system is in the electrowinning of high-concentration process streams such as spent plating baths, etchants, and ion-exchange eluates. Typically these cells can economically treat process streams to provide effluents in the 100 to 300 ppm range, after which they may require further processing by a high surface area reactor. Another example of a tank cell used as a resource recovery system is the Capenhurst Cell[20], which has been designed for regeneration of etchants in the printed circuit board industry. These spent etchants usually contain the dissolved copper(II) as well as the reduced form of the oxidant etchant, typically iron(II) from ferric chloride or Cu(I) from cupric chloride. In the Capenhurst process the cell recovers copper at the cathode, whereas in the anode compartment the active etching species is regenerated.

A bipolar version of the tank design, the Bipolar Stack Cell, consists of an assembly of parallel, planar electrodes separated by insulating spacers. Flow of electrolyte between electrodes may be either by natural or forced circulation. This design of cell is easy to construct since it simply comprises a stack of alternating electrodes and spacers with electrical connections being made to the two end electrodes. The number of electrodes in a stack typically ranges from 10 to 100. Process flow is usually by gravity feed so that the cell has none of the complex hydraulics and plumbing features of plate and frame cell designs. One example of this design, the bipolar trickle tower reactor, consists of a regular array of bipolar perforated carbon discs separated by an insulating mesh. The original work[12] on this design described the removal of metals and simultaneous cyanide destruction from cyanide containing rinse streams; the concept is now being developed by the Metelec Company (Hastings, UK).

PLATE AND FRAME CELLS

One of the most popular cell designs, primarily for large-scale electroorganic synthesis and to a lesser extent for metal recovery, is the plate and frame design, also known as the filter press. The basic design incorporates parallel plate electrodes, separated by insulators, gaskets, or diaphragms and mounted on tie rods or a filter press (Figure 2). Process flow is usually achieved by either series or parallel flow through internal gasketing, which minimises solution bypass. This arrangement clearly offers some advantages over the simpler tank cells, since the hydrodynamic characteristics of the cell are much more clearly defined. The plate and frame concept has proved very popular for electroorganic synthesis due to its flexibility of design with the wide range of working electrode materials, anodes and membrane-separators, variable hydrodynamics and operating conditions, including monopolar or bipolar and divided or undivided cell operation.

One of the most versatile plate and frame cells is the SU Cell developed by Electrocell AB in Sweden[7]. Based mainly on the work of the Lund group, Electrocell has developed a range of reactor designs from the laboratory-scale microcell up to a several square meter, pilot-scale version. The range of interchangeable electrodes, membranes, and frame materials allows for versatility in applications that range from electroorganic synthesis through metal recovery to chloralkali processes.

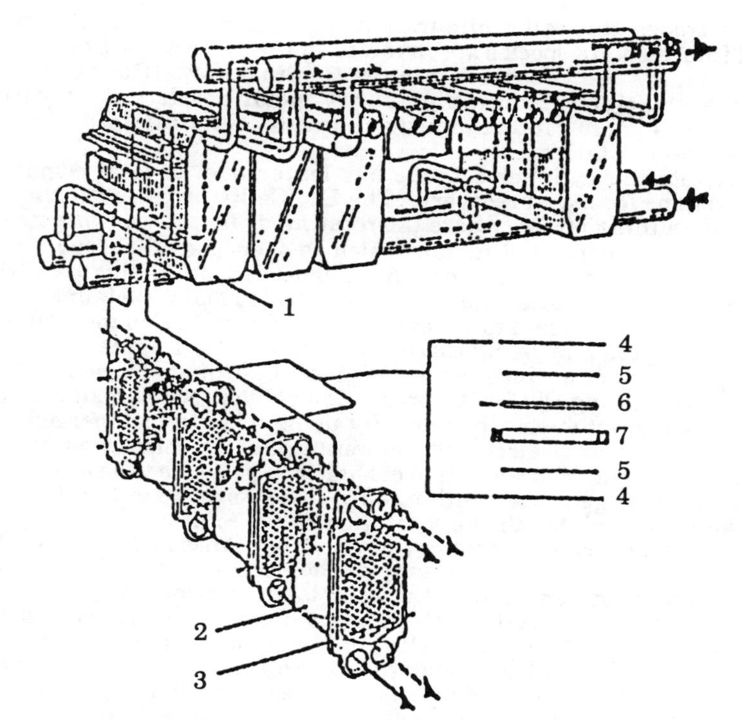

Figure 2. Plate and Frame Design (Courtesy of Electrocell AB). 1. Module; 2. Membrane; 3. Electrode Element; 4. Membrane; 5. Frame, Inner; 6. Electrode; 7. Frame, Outer

MASS TRANSPORT IN PLANAR ELECTRODE CELLS

The limited metal removal rates for planar electrodes has led to a variety of approaches for enhancing mass transport in this type of cell. These have included mechanical stirring, solution forced flow, and gas sparging. Rotating the working electrode also has proved to be one of the most practical routes for enhancing mass transfer. Other methods have tried to increase electrode surface area by roughening or even the use of wire mesh electrodes; these routes approach the domain of three-dimensional electrodes. Although not strictly a two-dimensional cell design, the Chemelec Cell(8) developed by BEWT Engineering (UK) uses an array of expanded mesh cathodes with alternate noble metal coated, DSA-type planar anodes in an undivided cell arrangement. Based on the original concept(32) developed at the Electricity Research Council (UK), the unique feature of this design is that it uses a fluidised bed of inert glass ballotini to promote mass transport. This cell has been successfully applied to a range of metal recovery and pollution control applications in the metal finishing, printed circuit board, and photographic industries.

ROTATING CELLS

Rotating the working electrode has been one of the most obvious ways of increasing mass transfer rates, particularly in metal recovery applications. One attractive feature of this approach is the ability to adjust the cathode rotation speed to the shear stress of the electrode to dislodge the deposited metal in powder form. Mechanical scraping also has been used in addition to direct shearing of dendritic deposits. By analogy to the analytical rotating disc and ring-disc electrodes, this cell configuration should also have application where short-lived intermediate species are involved in the process, for example, in the electrolytic generation of oxidants or reductants for effluent treatment. Although the ability to produce high-purity metal powders is an attractive feature of this design, in routine industrial practice rotating electrodes do have some mechanical limitations.

A rotating cylinder cathode has featured in several commercially developed systems for metal recovery. The Eco Cell(19) was one of the earliest designs based on this concept, where the rotating drum cathode was scraped by a wiper to dislodge deposited metal, which is then passed through a hydro-cyclone to produce a metal powder. A cascade version of this cell with six chambers separated by baffles was also demonstrated and was claimed to reduce a 50 ppm copper input stream to 1.6 ppm in the output(54). Commercial development of this design is now being carried out by Steetley Engineering (UK). A laboratory prototype of an ingeneous design of rotating disc electrode cell has been described by Tenygl(49) for the production of hydrogen peroxide for potable or industrial wastewater sterilization. By using a partly submerged rotating disc array, a thin film of electrolyte is maintained in contact with air, thus enabling a much higher concentration of hydrogen peroxide to be formed than in the alternative homogeneous solution route. The peroxide in the thin solution film is continuously fed into the bulk solution by the rotation process.

The pump cell concept(22), devised by Jansson and coworkers in Southampton (UK), is another variant of the rotating cell. In the simplest version the process stream enters the thin layer between the rotating disc cathode and the stationary cell body. The electrolyte is accelerated to high mass transfer rates and the cell becomes self-priming. In metal recovery applications the deposited metal film is discharged from the cell in the form of fine powder. It is also relatively straightforward to scale up this cell design to produce a multi-plate bipolar stack; a 500-amp version of this cell has been tested. Another version of the rotating electrode concept has been developed by Gotzellman KG (Stuttgart, FDR). In this design, known as the SE Reactor Geocomet Cell(24), a concentric arrangement of rod-shaped cathodes rotates inside inner and outer anodes. The deposited metal is dislodged by the rotating rods and falls to the bottom of the cell. The anodic destruction of cyanides in metal plating process rinses is also possible with this arrangement.

THREE-DIMENSIONAL REACTOR SYSTEMS

The development of high-performance, high surface area electrode reactor systems has undoubtedly been one of the most active research and development areas in electrochemical engineering. The demand for systems with high space-time yields has been driven by the need for economic metal recovery systems for both pollution control and resource recovery applications.

It should be noted at the outset that the distinction between two-dimensional and three-dimensional reactor systems is often far from rigid, with many designs, such as wire mesh and expanded metal electrodes, falling in between the two types.

The unifying feature in all of these designs is that the cell chamber is filled or partly filled with the working electrode material. The enhancement in mass-transport characteristics obtained with these designs in comparison to conventional two-dimensional planar reactors has had dramatic consequences for the commercial development of electrochemical engineering. First, the orders of magnitude increase in space-time yields obtained with many of these designs has resulted in substantial decreases in capital and operating equipment costs for a given process. This in turn has made the electrochemical processes more competitive with alternative, nonelectrochemical technical routes. Second, they opened up important new application areas, such as low-level metal recovery, which were either technically or economically impractical with the far less efficient two-dimensional cells. Some trade-offs were involved, particularly in the area of potential control and uniformity of current distribution. Despite early difficulties in maintaining the desired potential and current distribution in three-dimensional electrodes, the leverage in improved space-time yields more than compensates for this limitation. With further research in the areas of better computer modelling techniques and advanced process control systems, even this limitation may be overcome eventually.

THE FLUIDISED BED ELECTRODE

The invention of the fluidised bed cell by the University of Newcastle group(1) both marked an important landmark and created a major stimulus to the field of electrochemical engineering. Despite the elegance of this cell design, it was the inherent limitations of the fluidised bed concept that provided a focus for the development of alternative high surface area reactors.

The principle of the fluidised bed cell is shown in Figure 3. The conductive electrode particles are contacted by a porous feeder electrode, while the process stream causes fluidisation of the electrode bed. The main limitation to the design is that fluidisation of the bed causes loss of electrical contact between the particles, resulting in an extensive ohmic drop within the cell so that uniform potential and current distribution is virtually impossible to maintain. The original design by Goodridge *et al.*(1), however, illustrates why the cell was initially so attractive. The electrode had a specific area of 200 m^2/m^3 and was thus able to support large currents at an effective current density of around 0.01 A/cm^2. Despite these advantages, it was more than 15 years from the initial concept before any significant commercial demonstration of the concept was achieved.

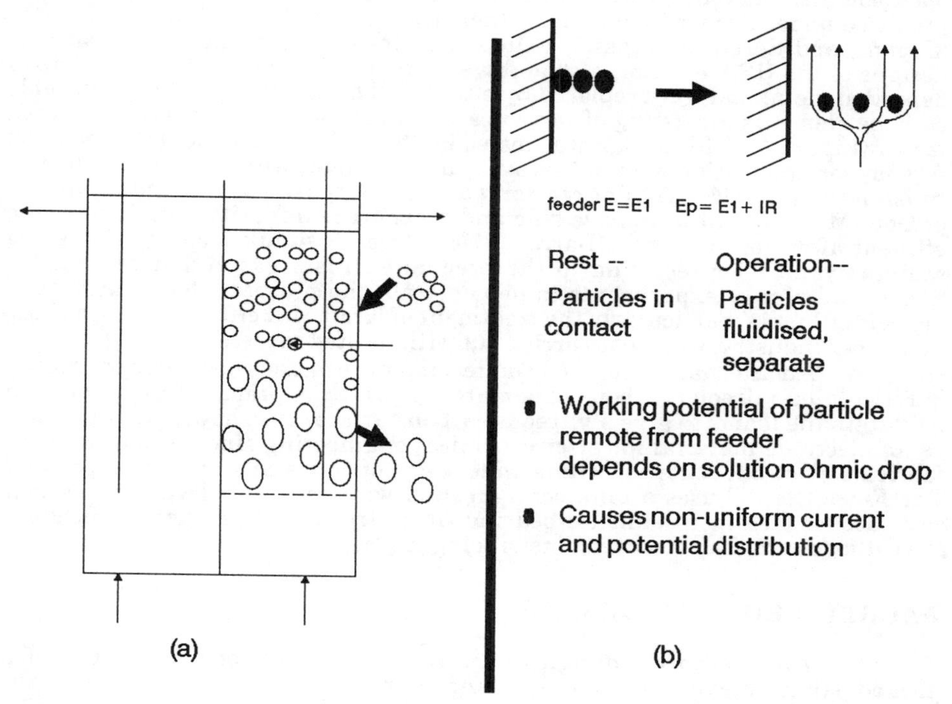

Figure 3. Schematic of Fluidised Bed Reactor
(a) With Amplified View of Electrode Particles Close to Feeder
(b) And Effect of Fluidisation on Potential Distribution

Most of the commercial development of the fluidised bed cell has been carried out by Akzo Zout Chemie in the Netherlands[53]. The Akzo cell design overcame some of the scale-up limitations of the original design by the use of a large number of rod feeders to the cylindrical, 0.35-m diameter, cathode bed. The cell also contained six symmetrical rod anodes encased in cylindrical diaphragms. Applications to copper removal from chlorinated hydrocarbon waste and mercury from brine have been demonstrated by Akzo. The technology is currently licensed and is being developed by the Billiton Group of Shell Research, also in the Netherlands[4].

The Chemelec Cell described earlier[7] is also an example of the use of the fluidised bed concept to promote mass transport. In this case, however, the working electrode is an expanded metal mesh and the inert fluidised bed simply acts to enhance mass transport.

CONTIGUOUS BED ELECTRODE DESIGNS

Many of the problems of fluidised bed cells appear to have been overcome by the development of three-dimensional contiguous bed reactors[15,48]. Evolving in many cases from porous electrode designs used in battery or fuel cell applications, these electrode designs are characterised by very high specific surface areas and space-time yields. At the same time, the ability to control potential and current distribution through the electrode bed is far better than in the earlier fluidised bed designs.

The first demonstration of the use of this electrode principle to metal recovery and pollution control applications was the use of carbon fibres as a high-performance cathode[15]. Subsequently, this electrode concept was developed commercially by HSA Reactors in Canada[25] and is currently being produced under licence by Baker Brothers (Stoughton, Massachusetts), Hitachi (Japan), and Sorensen (Kassel, FDR). The initial plate and frame cathode designs of the HSA Reactor, although very efficient, proved difficult to strip of deposited metal and were replaced by a tank cell design that used demountable cathode elements consisting of woven carbon fibre supported on a metal screen feeder (Figure 4). This design was tested by the U.S. Environmental Protection Agency for application to metal recovery and cyanide destruction in the metal-finishing industry[44]. A pilot reactor was also demonstrated for the treatment of Gold Mill effluent to remove zinc and other trace impurity metals from the effluent after gold recovery (Barren Bleed stream), simultaneously liberating sodium cyanide for recycling to the pregnant stream extraction process[34]. Some preliminary experiments to demonstrate the prospects for treatment of organic effluents, particularly the treatment of Kraft Mill effluent from the pulp and paper industry, were also carried out with limited success.

An alternative electrode material for contiguous bed electrode design is reticulated vitreous carbon. This material, which is prepared by pyrolysing polyurethane foams to give a vitreous carbon surface, has shown great promise as an electrode material for electrochemical engineering since it is possible to fabricate electrode material with a wide range of pore size and void proportion. The Retec Cell[13] uses a tank configuration with reticulated vitreous carbon cathodes for treatment of metal-bearing rinse streams. After metal collection, the cathodes are then removed for stripping in a separate cell.

PACKED BED CELL DESIGNS

An alternative design to the contiguous electrode is the use of a packed particulate bed of electrode material. Kreysa and Reynvaan[27,28] have studied the design of packed bed cells for optimal recovery of trace metals. The Enviro Cell[11], based on their work, consists of a packed bed of carbon granules and has been applied mainly to metal removal from industrial process streams. These authors also point out the desirability of altering the cathode bed geometry to compensate for depletion effects. A different design of packed bed

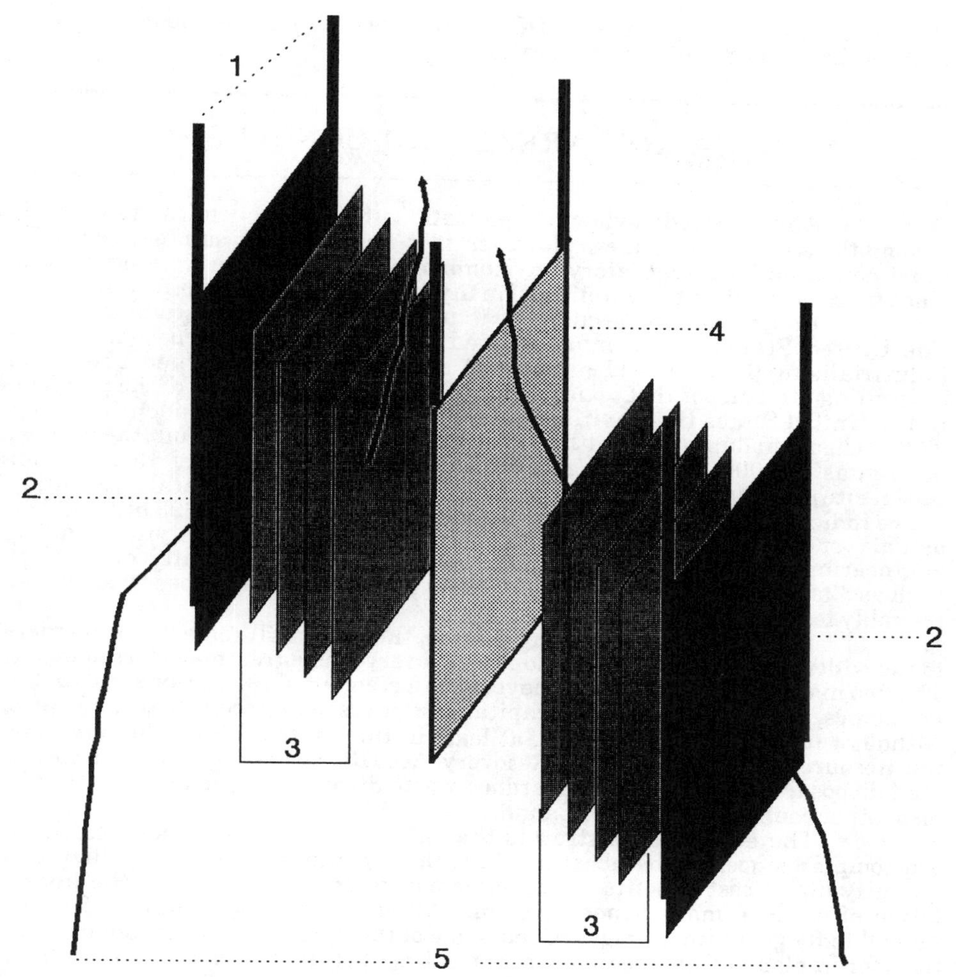

Figure 4. High Surface Area Reactor
1. **Cathode/Anode Current Feeders**
2. **Stainless Steel or Titanium Mesh Cathode Feeder/ Support**
3. **Layered Carbon Fibre Cloth Cathode (stitched to cathode feeder)**
4. **Anode (DSA type or steel)**
5. **Solution Flow Path**

reactor has been developed by the Nanao Kogyo Company[35]. The major features of their cell are the packed particulate bed cathode with central cathode feeder and symmetrical anodes separated by a diaphragm. The cell also uses the bipolar mode of operation. An interesting design of static three-dimensional electrode is shown by the Swiss-Roll Cell[43,55]. As the name implies, this design uses thin flexible sheets of cathode, separator, and anode rolled around a common axis and inserted into a cylindrical cell. Flow is axial to the bed, and the small interelectrode gap provides for a low cell voltage. A similar version of this concept was developed by Du Pont[55] and was known as the extended surface electrode or ESE Cell. This system was demonstrated at the pilot scale for metal removal from wastewaters.

A modified version of the SNDC Cell[47] has also been described. It uses a packed bed electrode for wastewater metal removal, but unlike most other packed bed designs, the SNDC reactor uses the flow-by mode as in conventional plate and frame type operation.

APPLICATION AREAS AND CONCLUSIONS

Any detailed review of applications of electrochemical reactors is beyond the scope of this paper. However, it is worthwhile to summarise briefly some points on both regulatory environmental issues and process economics, since these both affect the viability of many of these potential applications.

Environmental regulatory issues are an extremely complex topic. The United States, most industrialised nations[39], and even some newly industrialising Third World nations[16] are extremely concerned about the continuing land disposal of toxic wastes. To appreciate the scale of the problem, in the United States the Environmental Protection Agency has identified over 27,000 chemical dump sites, although some independent studies put the number as high as 400,000. The U.S. government has already allocated $9 billion for initial site cleanup, although a recent assessment from the General Accounting Office indicated that the real cost of cleanup could be as high as $23 billion. The actual cost of reclamation at each site, including site surveys, analytical studies, engineering design, legal costs, public hearings, etc., is currently running as high as $2,000 to $3,000 per ton. The initial costs for dumping these wastes was probably less than $100 per ton!

Despite these compelling figures, there are still some major barriers to the wider implementation of resource recovery waste treatment technologies. The two major barriers are (a) the level of enforcement of regulations and (b) the economics, especially the higher capital cost of resource recovery technologies. Although it would now seem that, at least in the United States, the impact of the Resource Conservation and Recovery Act (RCRA) is rapidly eliminating land disposal as an option for hazardous waste disposal, this is far from true in most other countries including Canada.

The economic constraints that affect waste management decisions are complex since, in the absence of regulatory issues, it is often difficult to quantify direct cost benefits or economic paybacks associated with the installation of a waste management system. Although the major driving force is undoubtedly government regulations, some of the economic and strategic incentives for implementing effective waste treatment strategies can be summarised briefly:

- The effect of increasingly stringent environmental regulations, especially decreasing permissible discharge limits, leading to an increased level of surveillance of industries' waste management practices
- The virtual elimination, in many parts of the United States, of land disposal as an option for hazardous waste disposal, a trend that is slowly being observed worldwide
- The increasing value of many metals and process chemicals, providing justification for implementation of resource recovery and recycling systems
- The rising costs of process water and higher sewer discharge costs, providing justification for use of wastewater treatment to recycle the bulk of process water, often with the added advantage of generating high-purity, deionised plant process water
- An increasing awareness by many industries of the direct economic benefits of waste management in terms of improved product quality and reliability, as well as improved worker safety and operating conditions.

Assuming that, for most metal pollution generating industries, some type of waste management system is mandatory, then the waste management strategy to be adopted should be based solely on an evaluation of the technical performance and economics of the various competing waste management technologies and strategies. As a concluding example, we can consider a case study describing the integrated use of electrolytic recovery in a "zero-sludge" resource recovery waste management system for the printed circuit board industry(10). Figure 5 shows a schematic of a system for treating the various metal-bearing streams from a printed circuit board plant. This shows the use of ion exchange to treat dilute metal-bearing waste streams and electrolytic recovery for treating metal concentrates such as etchants, plating baths, and the eluate from the ion exchange columns. There are two major features of this process. First, no toxic metals are discharged as the final product from the electrolytic recovery units is a metallic sheet that can be resold or recycled. Second, although the plant has a total wastewater useage of 250,000 gal/day after removal of the toxic metals followed by demineralization, over 90 percent of this process water can be recycled as high-purity, deionised water. Although the capital cost of this system at around $1.7M is between two and three times that for a conventional waste management approach involving chemical precipitation land disposal of heavy metal sludges, the total annualised operating costs of both routes are quite similar(10). This results in part from the higher operating costs of conventional treatment, including labour and process water useage. The major cost that affects conventional treatment is the rapidly escalating cost of sludge disposal, which is currently running from $200 to $400 per ton.

Although the above data were compiled at a California site for the State of California, they should translate to other geographic areas and environmental climates.

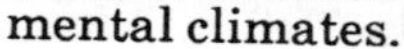

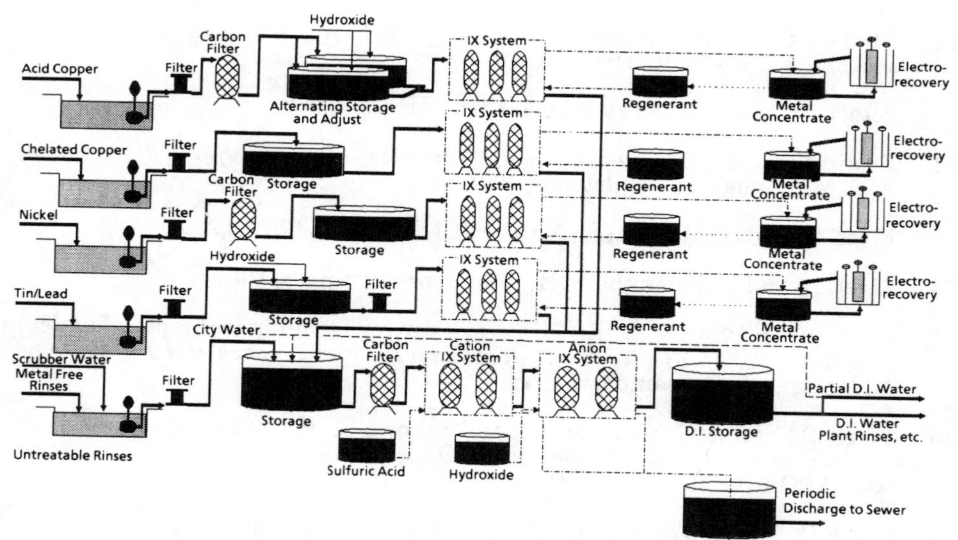

Figure 5. Integrated Ion-Exchange Electrolytic Recovery Zero-Discharge Waste Management System for the Printed Circuit Board Industry

REFERENCES

1. Backhurst, J. R., Coulson, J. M., Goodridge, F., Plimley, R. E., and Fleischmann, M., *J. Electrochem. Soc.*, 1969, *116*, 1600.

2. Bailey, K. C., *The Elder Pliny's Chapters on Chemical Subjects*, Part II, London, 1932, 60-61.

3. Barbier, M., in Ouellette, R. P., King, J. A., Cheremisinoff, P. N., eds., *Electrotechnology*, Ann Arbor Science, Ann Arbor, MI, 1978, 239-342.

4. Billiton, N. V., Arnhem, The Netherlands.

5. Bockris, J.O'M., ed., *Electrochemistry of Cleaner Environments*, Plenum, New York, 1972.

6. Brown, E., *Philosophical Transactions*, 1670, *5*, 1042-1044.

7. Carlsson, L., Sandegren, B., Simonsson, D., Rihovsky, M., *J. Electrochem. Soc.*, 1983, *130*, 342.

8. Chemelec, BEWT (Water Engineers), Alcester, UK.

9. Danly, D. E., *Emerging Opportunities for Electroorganic Process*, Marcel Dekker, New York, 1984.

10. Davis, G. A., Judd, R. L., Fleet, B., Small, C. E., Piasecki, B., and Moeller, M., "Waste Reduction Strategies for the Printed Circuit Board Industry," Report to Department of Health Services, Alternative Technology, Division, State of California, October, 1987.

11. Deutsche Carbone Akttiengesellschaft, Frankfurt, German FDR.

12. Ehdaie, S., Fleischmann, M., Jansson, R.E.W., and Alghaoui, A. E., *J. Appl. Electrochem.*, 1982, *12*, 59.

13. Eltech Systems Corporation, Chardon, OH.

14. Fahidy, T. Z., Mohanta, S., *Advances in Transport Processes*, A. S. Mujumdar, ed., Halstead, 1980, 83.

15. Fleet, B. and Das Gupta, S., *Nature 263*, 1976, No. 5573,122.

16. Fleet, B. and Gunasingham, H., Singapore Business, 1985.

17. Fleet, B. and Small, C. E., "Process Control Systems in Electrochemical Engineering," in *Electrochemical Reactors: Their Science and Technology*, M. Ismail, ed., Pergamon, in press.

18. Fleet, B., Small, C., Cardoza, B., and Schore, G., *Performance and Costs of Alternatives to Land Disposal*, E. T. Oppelt, B. L. Blaney, W. F. Kemner, eds., APCA Publications, Pittsburgh, 1987, 170-180.

19. Gabe, D. R. and Walsh, F. C., *J. Appl. Electrochem.*, 1983, *13*, 3.

20. Hillis, M. R., *Trans. Inst. Met. Finish.*, 1979, *57*, 73.

21. Jansson, R.E.W., *Chem. Eng. Sci.*, 1980, *35*, 1979.

22. Jansson, R.E.W. and Tomov, N. R., *Chem. Eng.*, 1977, *316*, 867.

23. Kalia, R. K., Weinberg, N. L., and Fleet, B., *Impact of New Electrochemical Technologies on the Demand for Electrical Energy*, Contract CEA 410-U-479, Canadian Electrical Association, Montreal, 1986.

24. Kammel, R., *Metalloberfleche*, 1982, *5*, 194.

25. Kennedy, I.F.T. and Das Gupta, S., *Proceedings*, First Annual Conf. on Advanced Pollution Control for the Metal Finishing Ind., EPA-600/8-78-010, Cincinnati, OH, 1978, 49.

26. King, C.J.H., Amer. Inst. Chem. Eng. Symposium Series, 1981, *204*, 46.

27. Kreysa, G., *Chem. Ing. Tech.*, 1970, *50*, 332.

28. Kreysa, G. and Reynvaan, C., *J. Appl. Electrochem.*, 1982, *12*, 241.

29. Kuhn, A. T., ed., *Industrial Electrochemical Processes*, Elsevier, Amsterdam, 1971.

30. Lancy International Inc., Zelienople, PA.

31. Levich, V. G., *Acta Physicochem. (URSS)*, 1942, *17*, 257; *Physicochemical Hydrodynamics*, Prentice-Hall, New York, 1962.

32. Lopez-Cacicedo, C. L., Br. Patent 1,423,369, 1973.

33. Marshall, R. J. and Walsh, F. C., *Surface Tech.*, 1985, *24*, 45.

34. Mohanta, S., Fleet, B., Das Gupta, S., and Jacobs, J., "Evaluation of a High Surface Area Electrochemical Reactor for Pollution Control in the Gold Industry," Department of Supply and Services, Contract # DS 0477K204-7-EP58, Ottawa, 1981.

35. Nanao Kogyo Co. Ltd., Yokohama, Japan.

36. Newman, J. S., *Electrochemical Systems*, Prentice-Hall, New York, 1973; *Electroanalytical Chemistry*, A. J. Bard, ed., Marcel Dekker, New York, 1973, 187-352.

37. Ouellette, R. P., King, J. A., Cheremisinoff, P. N., eds., *Electrotechnology*, Ann Arbor Science, Ann Arbor, MI, 1978.

38. Piasecki, B., ed., *Beyond Dumping: New Strategies for Controlling Toxic Contamination*, Quorum Books, Westport, 1984.

39. Piasecki, B. and Davis, G. A., *America's Future in Toxic Waste--Lessons from Europe*, Greenwood Press, 1988.

40. Pickett, D. J., *Electrochemical Reactor Design*, Elsevier, Amsterdam, 1979.

41. Pletcher, D., *Industrial Electrochemistry*, Chapman and Hall, London, 1983.

42. Pourbaix, M., *Atlas of Electrochemical Equilibria in Aqueous Solutions*, Pergamon, Oxford, 1966.

43. Robertson, P. M. and Ibl, N., *J. Appl. Electrochem.*, 1977, *7*, 323.

44. Roof, E., *Proceedings*, Third Annual Conf. on Advanced Pollution Control for the Metal Finishing Ind., EPA-600/2-81-028, Cincinnati, OH, 1982.

45. SCADELEC Technical Note, TRSI Systems Inc., Rexdale, Ontario.

46. Selman, J. R., Amer. Inst. Chem. Eng. Symposium Series, 1981, *204*, 77.

47. Simonsson, D., *J. Appl. Electrochem.*, 1984, *14*, 595.

48. Sioda, R. E. and Keating, K. B., in *Electroanalytical Chemistry*, A. J. Bard, ed., Marcel Dekker, New York, 1982, 12.

49. Tenygl, J., personal communication.

50. Tournier, M., "Galvano," 1964, *33*, 324; cited in Dubpernel, G., *Selected Topics in the History of Electrochemistry*, The Electrochemistry Society, Princeton, 1978, *78* (6), 1.

51. TRSI Systems and Cal-Tech Management Associates, *Waste Reduction Strategies for the Printed Circuit Board Industry*, California Department of Health Services, Contract 85-00173, Final Report, 1987.

52. *Tutorial Lectures in Electrochemical Engineering and Technology*, Amer. Inst. Chem. Eng. Symposium Series, 1981, *204*, 77.

53. Van der Heiden, G., Raats, C.M.S., and Boon, H. F., *Chem. and Ind. (London)*, 1978, *13*, 465.

54. Walsh, F. C., Gardner, N. A., and Gabe, D. R., *J. Appl. Electrochem.*, 1982, *12*, 229.

55. Williams, J. M., U.S. Patent 3,859,195, 1975.

ULTRASONIC INTERACTIONS DURING SLUDGE DEWATERING

Dr. Nagabhusan Senapati
Battelle
505 King Avenue
Columbus, Ohio 43201-2693
(614) 424-4506

INTRODUCTION

Ultrasonic energy has been demonstrated to be effective in enhancing the mechanical dewatering of a variety of sludges by the electroacoustic dewatering (EAD) process. EAD is a patented process[1] developed at Battelle. This process is based on the synergistic effects of dc electric field, ultrasonic field, and mechanical pressure (or vacuum) on a suspension to increase the rate and the degree of solid/liquid separation.

Ultrasonic energy interacts with liquids and suspended solids to change their surface and bulk properties based on a wide range of mechanisms[4,6]. This paper highlights our present understanding of the mechanisms of ultrasonic interaction during the EAD process for sludges.

ELECTROACOUSTIC DEWATERING OF SLUDGE

Electroacoustic dewatering (EAD) is a process of enhancing conventional pressure or vacuum dewatering or filtration process by the simultaneous application of a dc electric field and an ultrasonic field. As discussed in previous publications[5,6], the water in solid/liquid suspension is present as free or bulk water and in bound form in capillaries, open pores, and on the surfaces of the suspended particles. Conventional filtration processes are very effective in removing the free water, but not the bound water. The EAD process has been demonstrated to help remove bound water by several mechanisms of synergistic interaction of the electric and the ultrasonic fields with the solid/liquid suspension. The effect of the ultrasonic field on the rate and degree of sludge dewatering during the EAD process is very intriguing and is discussed here.

Figure 1 shows the results of batch EAD tests on Hertogenbosch sludge. The initial solids were about 16.4 weight percent. This sludge was subjected to an electric voltage of 35 V, and the ultrasonic power (U) was varied from 0 to 20 W. Total dewatering time was 2 minutes. The test equipment and the test procedures are described in detail in another publication[2]. The electric field alone (i.e., U = 0) was effective in increasing the solids from 16.4 to 24.2 percent. The final solids percent appears to increase monotomically from 24.2 to 30.2 percent with the increase in ultrasonic power from 0 to 20 W. The rate of increase appears to be almost constant with increase in ultrasonic power.

Figure 2 shows the results of a similar batch test on Ridderkerk sludge. The initial solids weight percent was 16, and the EAD residence time was 3 minutes. When a 60-V field was applied, the solids percent increased from 16 to 24 with no ultrasonic power. Similar to the previous example, the final solids percent appears to increase from 24 to 26 with the increase in ultrasonic power from 0 to 20 W. However, not only the change in percent solids is significantly lower than the previous example, but also the rate of increase appears to be decreasing with ultrasonic power.

Figure 3 shows the results of yet another batch test on Columbus Southerly waste-activated sludge. For this sludge the final solids percent is not affected by ultrasonic power up to 10 W. Between 10 and 20 W, the final solids percent appears to assume a bell-shaped curve with the highest value for an ultrasonic power of 15 W.

These three examples illustrate that the effect of ultrasonic power on the rate and the degree of dewatering is a strong function of the type of sludge. We have developed a hypothesis to explain these phenomena. The ultimate goal of this exercise is to develop, if possible, a predictive model to estimate the EAD performance of sludges. This hypothesis is based on our understanding of the mechanisms of interaction of ultrasound with solid/liquid suspension.

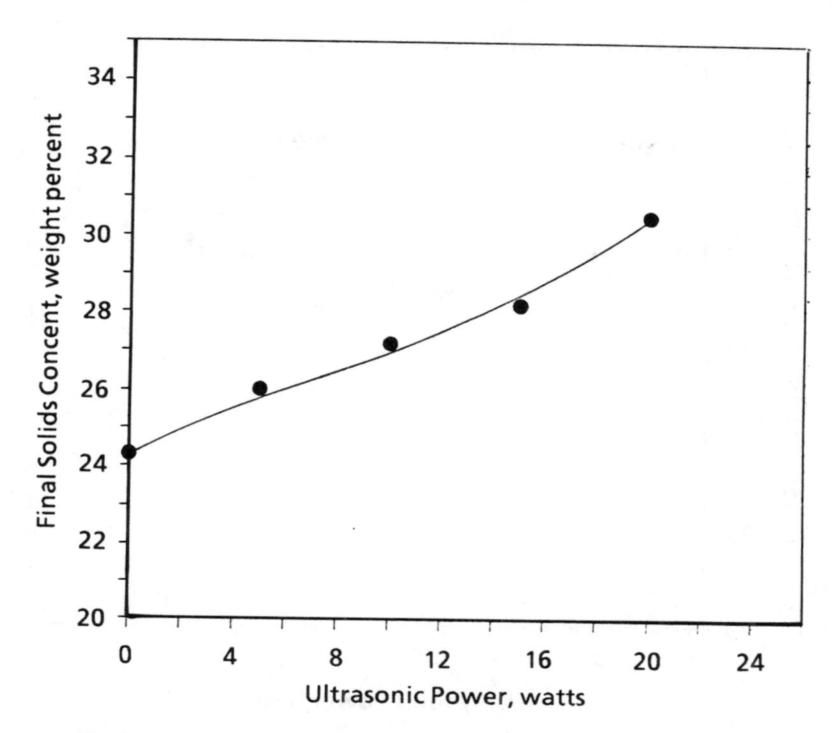

Figure 1. Effect of Ultrasonic Power on the Final Solids Content of Hertogenbosch Sludge at 35 V, 2 Minutes Dewatering Time, Initial Solids Weight Percent = 16.4

MECHANISMS

The degree of effectiveness of the ultrasonic field on sludges during EAD is product dependent. There are many mechanisms of interaction of an ultrasonic field with a solid/liquid suspension[2]. On one hand, ultrasonic energy is known to agglomerate particles suspended in a liquid. On the other hand, if the ultrasonic intensity (i.e., power per unit area) is increased above the cavitation threshold in that medium, the ultrasonic energy is known to homogenize or mix the suspension. For cake-like structures, removal of liquid trapped or bound in open pores, capillaries, and between the interstitial spaces depends on many variables including

- Percent of bound water
- Degree of nucleation to promote rectified diffusion
- Degree of shear thinning of the suspension
- The absolute value of the dynamic viscosity at zero frequency and at the operating ultrasonic frequency.

Most practical sludges are highly complex materials with complex bulk and surface properties. Although these physical, chemical, and electric properties are complex and difficult to measure, an analysis of the possible mechanisms suggests a few of these properties of cake-like materials that are most likely to correlate with performance. Two of these properties related to the acoustic behavior of the material are the dynamic complex shear modulus (G) and the dynamic viscosity (η).

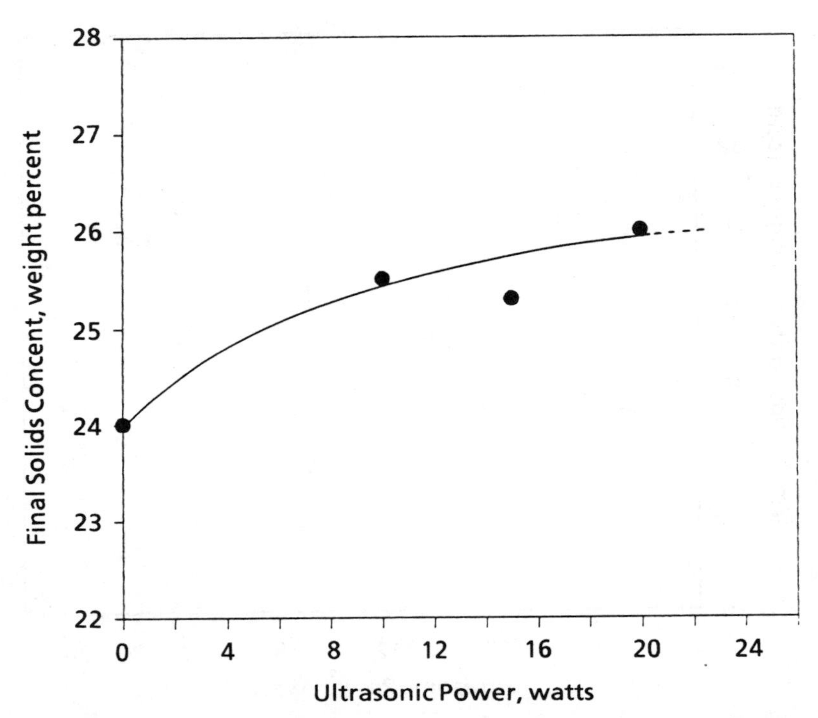

Figure 2. Effect of Ultrasonic Power on the Final Solids Content of Ridderkerk Sludge at 60 V, 3 Minutes Dewatering Time, Initial Solids Weight Percent = 16.0

The acoustic impedance of the medium is strongly related to the dynamic complex modulus. The complex shear modulus can be written as

$$G = G' + j\,G'' \tag{1}$$

The propagation of elastic waves is related to G', and the absorption is related to G". High value of G" indicates high absorption of ultrasonic waves in the medium and therefore a low value of penetration. In other words, a relatively high value of G" indicates that the penetration of ultrasonic energy into the cake will be limited, and therefore, only a part of the cake will be subjected to the optimum level of ultrasonic intensity for effective enhancement of dewatering.

The dynamic viscosity of most sludges is a strong function of the frequency and can be represented as

$$\eta \omega^{a} = c \tag{2}$$

where
- η = dynamic viscosity
- ω = angular frequency = 2πf
- f = frequency in cycles/second
- a = exponent
- c = constant.

From Darcy's relation[3] the rate of flow of a liquid (Q) under a potential gradient is inversely proportional to the viscosity:

$$Q \propto \frac{1}{\eta} \tag{3}$$

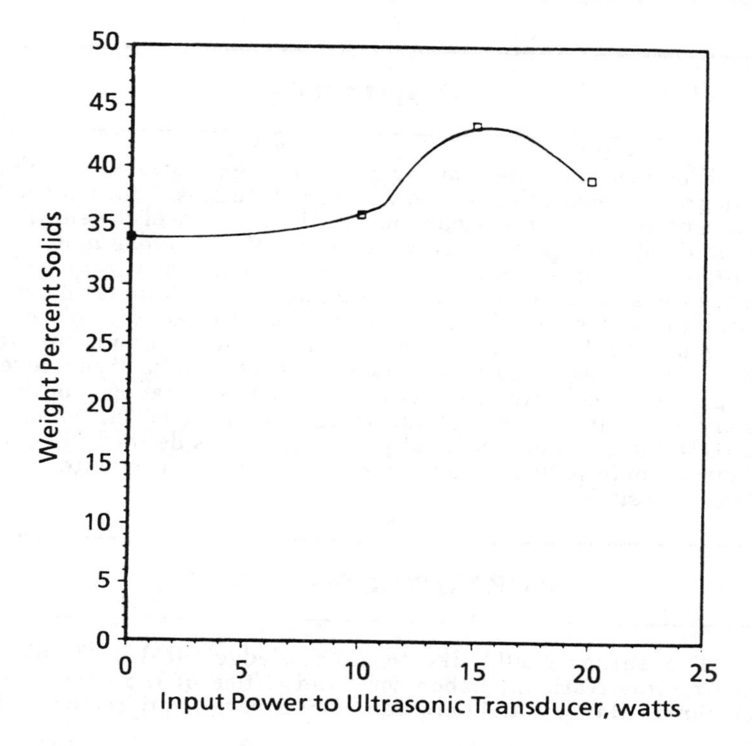

Figure 3. Effect of Input Power to the Ultrasonic Transducer on Weight Percent Solids for Southerly Waste-Activated Sludge at 50 V, pH 7, 15 in. Vacuum

However, η is not a constant. Most sludges appear to be thixotropic, i.e., η decreases with the rate of shear. For most sludges, the exponent (a) (equation (2)) is positive and is often >1. Therefore, application of ultrasonic energy is most likely to decrease the effective viscosity and therefore increase the rate and the degree of dewatering. From these considerations, it appears that the effectiveness of ultrasonics during EAD should be related strongly to the exponent (a), especially if shear thinning is a major mechanism of EAD effectiveness.

Logarithmic differentiation of equation (2) gives

$$\frac{d\eta}{\eta} + a\,\frac{d\omega}{\omega} = 0 \qquad (4)$$

or

$$a = -\left\{\left(\frac{d\eta}{d\omega}\right) / \left(\frac{\eta}{\omega}\right)\right\} \qquad (5)$$

Note that (a) is a dimensionless parameter that embodies both shear thinning property (i.e., dη/dω), the viscosity η and the frequency ω. From the data on the dynamic viscosity (η(ω)) measured by a mechanical spectrometer, the value of (a) can be estimated at a frequency of 100 radians/sec.

For most sludges (a) is positive since dη/dω is negative. A higher value of (a) is most likely to indicate higher ultrasonic effectiveness during EAD. Preliminary observations strongly indicate that G" and (a) are likely to be two parameters of a sludge that are measurable and can indicate the degree of effectiveness of ultrasonics to enhance dewatering during EAD. We are in the process of further verifying this hypothesis and developing a theory of ultrasonic effectiveness, which will be reported in future publications.

SUMMARY

Electroacoustic dewatering has been demonstrated to enhance the rate and degree of dewatering of a wide range of sludges. The additional solids weight percent due to the ultrasonic field in the presence of the electric and the pressure fields appears to be a strong function of the sludge material. Since most sludges are very complex material, it has been very difficult to characterize the sludge to identify a cause and effect relationship between the sludge and the potential effectiveness of the ultrasonic field to enhance dewatering during EAD. From the potential mechanisms of ultrasonic interaction with solid/ liquid suspensions, a hypothesis has been developed to correlate performance with the physical properties that can be measured. Preliminary study indicates that the imaginary part of the complex shear modulus (G") and a nondimensional parameter (a) as defined in equation (5), are two measurable physical parameters that can be correlated with EAD performance of a sludge.

ACKNOWLEDGMENTS

The author would like to acknowledge the U.S. Department of Energy's Argonne National Laboratory and Office of Industrial Programs, Ashbrook-Simon-Hartley, and Bergmann for supporting projects on EAD.

REFERENCES

1. Muralidhara, H. S., Parekh, B. K., and Senapati, N., "Solid-Liquid Separation Process for Fine Particle Suspension by an Electric and Ultrasonic Field," U.S. Patent No. 4561953, December 31, 1985.

2. Muralidhara, H. S., Senapati, N., and Beard, R. B., "A Novel Electro Acoustic Separation Process for Fine Particle Suspensions," *Advances in Solid-Liquid Separation*, H. S. Muralidhara, ed., Battelle Press, 1986.

3. Purchas, D. B., "Solid/Liquid Separation Technology," Uplands Press Ltd., 1981.

4. Senapati, N., "Mechanisms of Ultrasonic Interaction During Electro-Acoustic Dewatering," Fifth International Drying Technical Symposium, Cambridge, MA, 1986.

5. Senapati, N. and Muralidhara, H. S., "Dewatering by Electro-Acoustic Techniques," Proceedings of the Engineering Foundation Conference, Palm Coast, FL, January 10-15, 1988.

6. Senapati, N., Muralidhara, H. S., and Beard, R., "Ultrasonic Interaction in Electro Acoustic Dewatering," British Sugar plc, Technical Conference, Eastbourne, England, 1988.

SCALE-UP OF ELECTROACOUSTIC DEWATERING OF SEWAGE SLUDGES

S. P. Chauhan
Battelle
505 King Avenue
Columbus, Ohio 43201-2693
(614) 424-4812

and

H. W. Johnson
Ashbrook-Simon-Hartley
P.O. Box 16327
Houston, Texas 77222
(713) 449-0322

ABSTRACT

The electroacoustic dewatering (EAD) process is a patented process being commercialized by Battelle. The process utilizes electric and ultrasonic fields to obtain improved solid/liquid separation. The scale-up of the EAD process for belt pressing of sewage sludge is discussed in this paper. A number of sewage sludge samples were first tested in a 7.6-cm-I.D. batch EAD system to determine the relevant parameters and the extent of additional dewatering possible with EAD. Then extensive parametric studies were carried out on two sludges to design a 1.2-m belt width EAD postdewatering prototype unit. The prototype was tested with four mixed undigested sludges. The results confirmed the commercial viability of EAD for belt presses. The results from the prototype also showed general agreement with batch dewatering data.

INTRODUCTION

Most of the municipal wastewater treatment facilities today are concerned about excessive cost of sludge disposal. One of the key contributors to sludge disposal costs is the excessive moisture (low solids levels) in the sludges after conventional mechanical dewatering. For example, the typical solids content of a mixed undigested sludge after centrifugation or belt pressing is 15 to 20 percent. The excess water adds to cost of landfill both because of its added volume and because of the need to blend lime. For the incineration alternative, more than half of the water in the filter cake must be evaporated using supplemental fuel. And for the composting alternative, the wet feed must be blended with dried and partially composted sludge and sawdust to achieve the desired solids level in the composting reactor, which adds to the size of composting equipment. It is therefore desirable to further dewater the filter cake without resorting to evaporation.

To address the need for enhanced mechanical dewatering of sewage and other wastewater treatment sludges as well as various products, Battelle is developing a process called electroacoustic dewatering (EAD), for which three patents[1] have been issued and others are pending. The process received an IR-100 award in 1986. This paper will focus on the scale-up of EAD from small batch to commercial-sized belt pressing for sewage sludge dewatering application.

EAD PROCESS BACKGROUND

The EAD process utilizes conventional mechanical processes, such as belt presses, screw presses, plate and frame filters, and vacuum disc or drum filters, in combination with electric and acoustic (e.g., ultrasonic) fields. The dominant mechanism of dewatering enhancement is electroosmosis through the application of DC electric field. Electroosmosis is caused by an electrical double layer of negative and positive ions formed at the solid/liquid interface and is characterized by a zeta potential[2,5]. In the case of sewage sludge, the zeta potential at natural pH of sludge is negative, resulting in movement (pumping) of water to a cathode.

The liquid flow rate (Q) and energy consumption (E_{sp}) in idealized electroosmosis, at a given solids content of filter cake, can be expressed as follows[9]:

$$Q \propto \frac{DZV}{\mu L} \qquad \text{Eq. (1)}$$

$$Q \propto \frac{DZI}{\mu\lambda} \qquad \text{Eq. (1A)}$$

$$E_{sp} \propto \frac{\mu\lambda V}{DZ} \qquad \text{Eq. (2)}$$

where

Q = flow rate of water per unit area
D = dielectric constant of water
Z = zeta potential
V = voltage across filter cake
μ = viscosity of water
L = cake thickness
I = current
E_{sp} = energy use per unit of filtrate removed
λ = electrical conductivity of cake.

The flow rate is thus proportional to zeta potential and voltage gradient (V/L). The energy use is directly proportional to voltage and conductivity and inversely proportional to zeta potential. High electrical conductivity results in high energy use and waste of energy by resistive heating. The excessive heating can, for some materials, limit the allowable voltage gradient and therefore the dewatering rate. The practical range in our work with sludges has been below 10 milliohm/cm. These equations assume open-ended capillaries with regular surfaces and uniform flow of current across filter cake.

The electroosmosis in a filter cake quickly declines with loss of the liquid continuum from anode to cathode as dewatering proceeds. However, in the EAD process, the use of an ultrasonic field helps electroosmosis by consolidating the filter cake and releasing inaccessible liquid, which help maintain a liquid continuum(5). In addition, ultrasonics can aid electroosmotic and mechanical dewatering through other mechanisms(5).

The EAD process has been tested on over 75 suspensions in a variety of equipment designed to simulate a variety of commercial filters. The applications include dewatering of municipal and wastewater treatment sludges, process effluent sludges, food products, wood pulp, coal and minerals slurries, and clay suspensions(1-3,5-7).

The initial efforts on EAD of sewage sludges were concentrated on vacuum-assisted filtration since it was envisioned that EAD could be easily retrofitted into a horizontal vacuum belt filter. The results of batch vacuum EAD tests and continuous EAD tests in a 10-cm belt width, vacuum belt unit are reported elsewhere(1). However, a prototype design effort indicated some difficulties in coupling ultrasonics in such a system. It was considered easier to implement EAD in a belt filter press mode. The results of pressurized EAD systems are given in this paper.

BATCH PARAMETRIC STUDIES

A 7.6-cm-diameter pressurized batch EAD unit, shown in Figure 1, was used to obtain design data for a continuous EAD prototype. The sample is placed between a polymeric filter, which is supported by a perforated plate cathode, and a floating anode. The ultrasonics can be applied in several ways; one method is shown in Figure 1. A downward force was applied on the floating anode to press the sludge at 0.14 to 0.70 bar (2 to 10 psi).

The tests were conducted on a waste-activated sludge (WAS) produced at the Ridderkerk plant in the Netherlands, which had been flocculated and predewatered by a belt press to a nominal 16 percent dry solids (DS) level. The effect of the key EAD parameters on final solids content (i.e., average rate of dewatering) and specific energy use (i.e., kWh/kg filtrate removed) are discussed below.

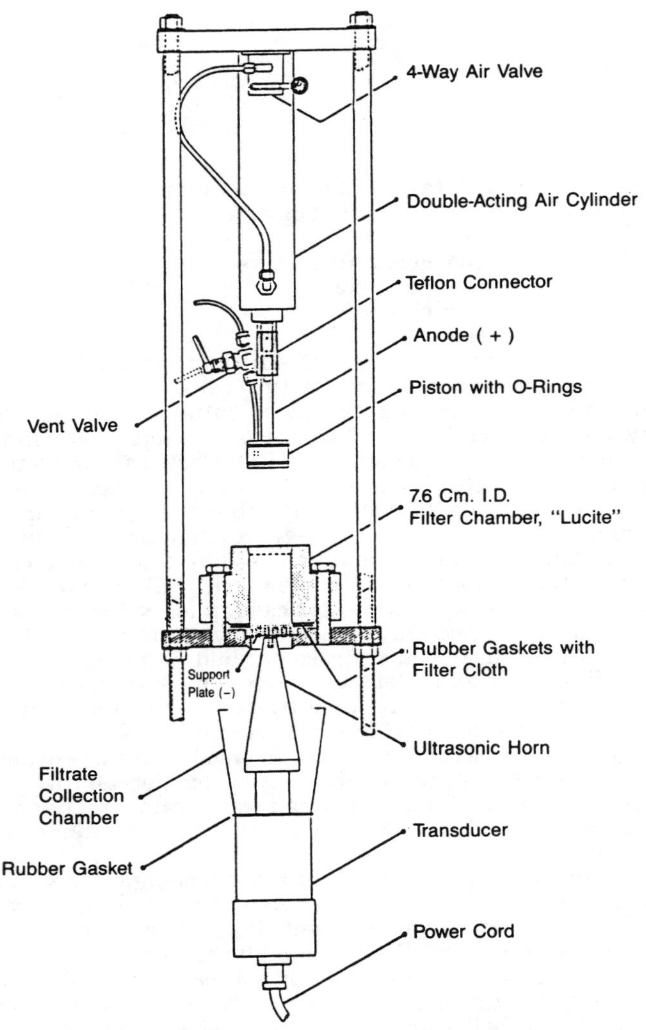

Figure 1. Schematic of Pressurized Batch EAD Unit

VOLTAGE AND TIME

The effect of applied DC voltage and residence time (dewatering time) on final cake solids is shown in Figures 2 and 3. As expected, the cake solids content increases with increasing voltage and residence time. A large number of voltage and residence time combinations can be used to achieve the desired cake solids. Data on specific energy use plotted in Figure 4 show that lower voltage and longer time give lower energy use than higher voltage and shorter time to obtain a desired cake solids content. For example, the specific energy use values to achieve 28 percent cake solids content read from the three curves in Figure 4 at a constant solids loading of 2.1 kg DS/m^2 are

Residence Time, min	Voltage, volt	Specific Energy Use, kWh/kg
3	70	0.25
4	60	0.19
5	50	0.16

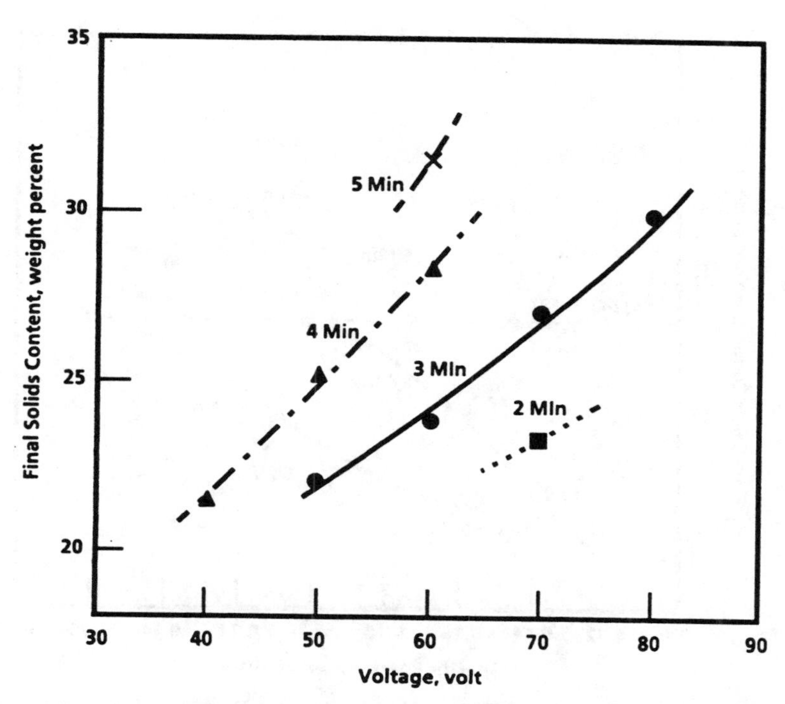

Figure 2. Effect of Voltage on Final Solids at Several Residence Times (WAS; Batch Tests; Voltage Only)

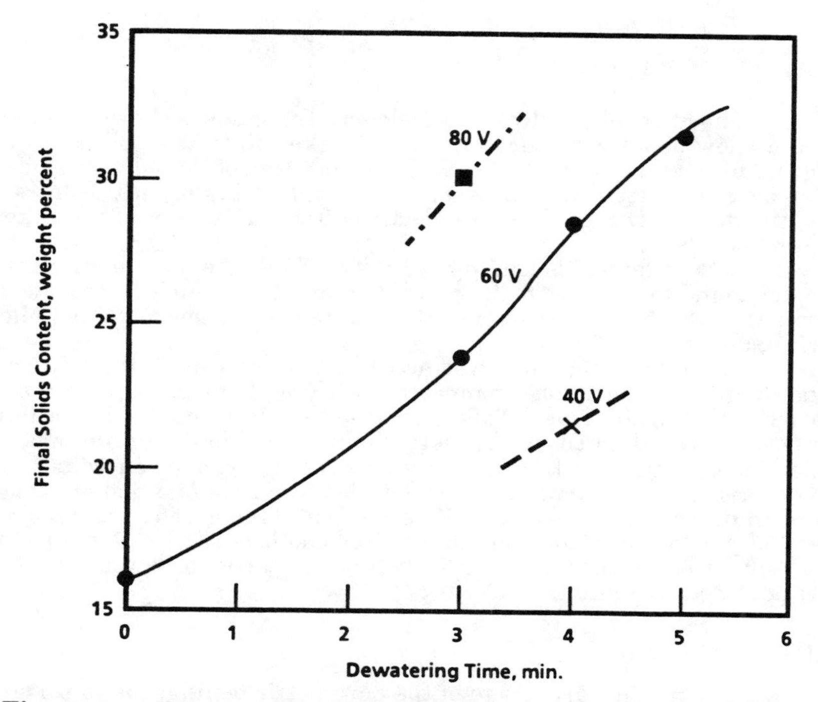

Figure 3. Effect of Dewatering Time on Final Solids at Several Voltages (WAS; Batch Tests; Voltage Only)

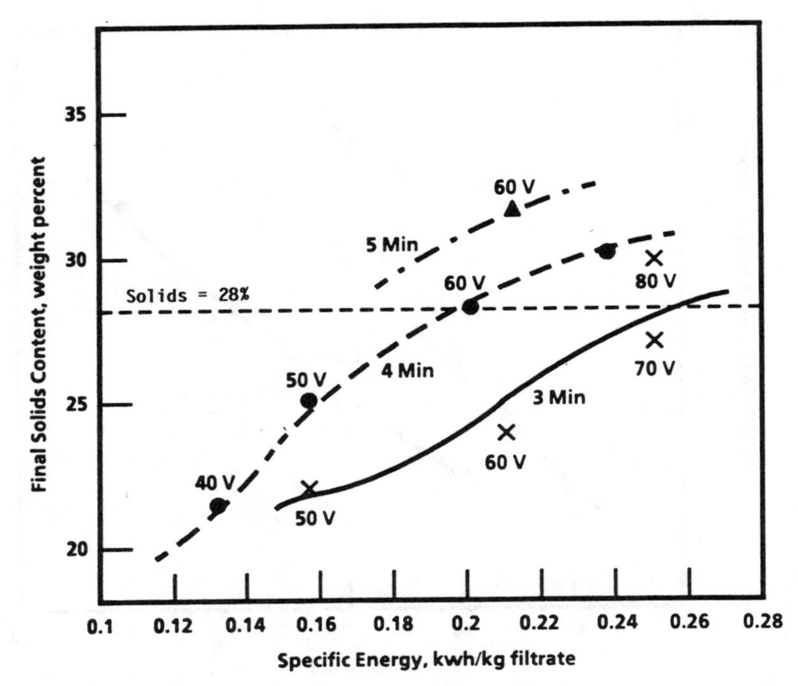

Figure 4. Effect of Time, Voltage, and Final Cake Solids on Specific Energy Use (WAS; Batch Tests; Voltage Only; Solids Loading at 2.1 kg DS/m^2)

ULTRASONIC FIELD

The effect of an ultrasonic field on cake solids is shown in Figure 5. The data show about 2 percent increase in cake solids at a solids loading of 2.1 kg DS/m^2 with a power input of 20 W. It has been observed with sludges as well as several other colloidal suspensions that at higher ultrasonics field levels, the dewatering performance declines from disruption of the cake consolidation process.

At a lower solids loading of 1.0 kg DS/m^2, the maximum ultrasonic effect was found to increase to 3.2 percent cake solids, which represents about 30 percent of the total increase in solids content from the combined effect of electric and ultrasonic fields.

The ultrasonic effect was accompanied by a more than 10 percent decline in specific energy use compared with use of the electric field only to achieve the same solids level. This is because the ultrasonic field, at relatively lower power levels than DC power, helps maintain a liquid continuum between the electrodes, thus resulting in a higher current for a given voltage. The higher current, according to equation (1A), leads in turn to a higher rate of filtration. In this manner the rate of filtration can be increased by increasing the current only. On the other hand, for electroosmotic-only filtration, increasing the current will require increasing the voltage proportionally, thus requiring a higher power consumption.

SOLIDS LOADING

During the early stage of the parametric testing, the importance of solids loading (or cake thickness) became obvious; this is shown in Figures 6 and 7. In Figure 6, for example, the final cake solids at 60 volts and 3 minutes

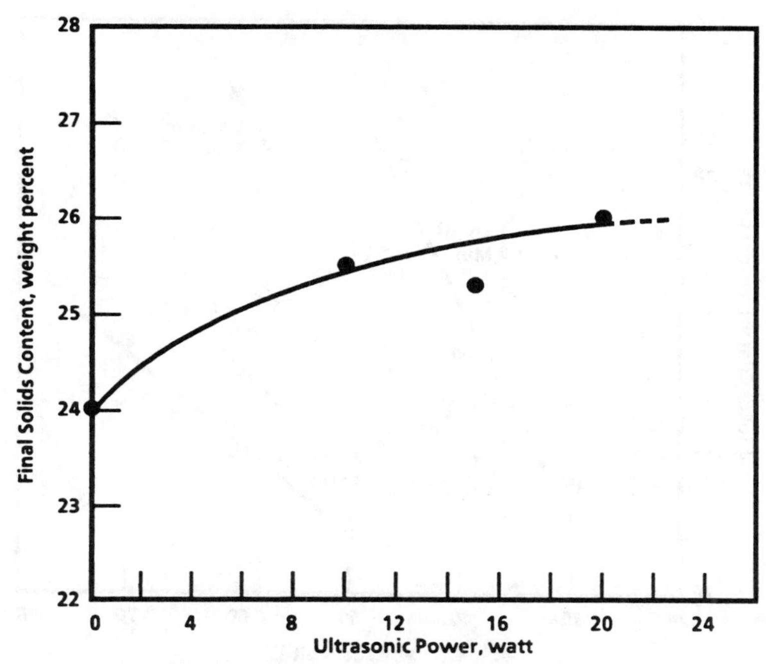

Figure 5. Effect of Ultrasonic Power, at Generator Outlet, on Final Cake Solids (WAS; Batch Tests; Solids Loading at 2.1 kg DS/m^2)

at solids loading of 2.1 and 1.6 kg DS/m^2 are 25.2 and 37.4 percent, respectively. The specific energy use for a given final cake can be reduced by as much as half by lowering the solids loading, as shown in Figure 7. It can be shown from the dewatering rate energy use equations for electroosmosis given earlier that the specific energy use is proportional to the second power of solids loading at a constant final cake solids and a constant residence time. However, this relationship may change on superimposition of ultrasonics. The data in Figure 7 at a residence time of 3 minutes and solids loading values of 1.6 and 2.1 gave the following relationship:

$$E_{sp} \propto (\text{solids loading})^{1.84} \quad . \qquad \text{Eq. (3)}$$

For a moving belt system with constant throughput, residence time must be adjusted in proportion to solids loading to maintain the machine throughput and therefore the overall dewatering rate. Under these conditions, it can be shown that the specific energy is proportional to solids loading, i.e., proportional to initial cake thickness at a given initial solids level.

A few tests were performed to determine the effect of initial cake solids. The results from two such tests given in Table 1 show that the specific energy use is lower at the lower initial solids, but the total energy use per ton of dry solids is much higher. This result, which agrees well with EAD in other test programs, indicates that EAD should be utilized after as much mechanical predewatering as possible.

Limited tests were performed to determine the effect of pressure. The results shown in Table 2 indicate very little effect of pressure in the range of 0.17 to 0.34 bar.

The results of the parametric tests were used to design an EAD belt filter prototype described in the next section.

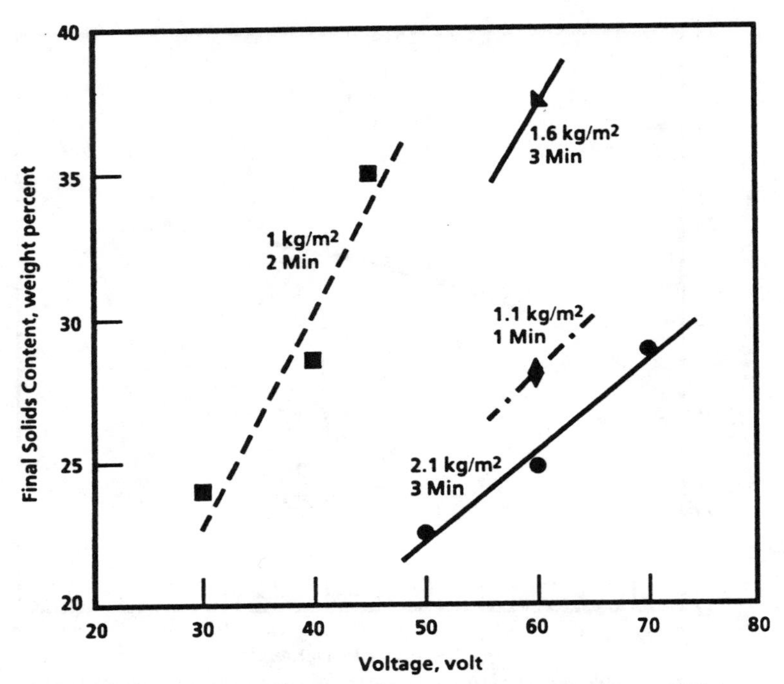

Figure 6. Effect of Voltage, Solids Loading, and Time on Final Cake Solids (WAS; Batch Tests; Optimum Ultrasonic Level)

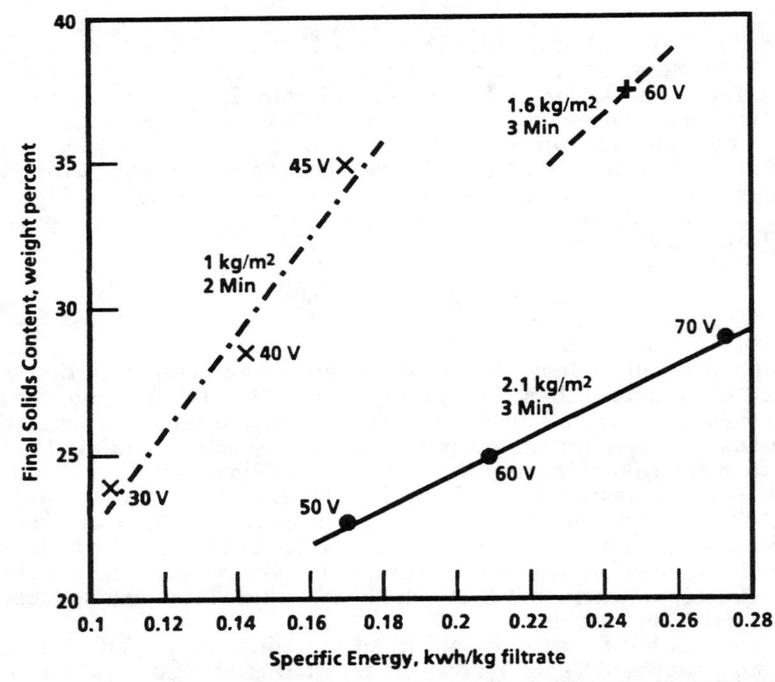

Figure 7. Effect of Solids Loading, Time, Voltage, and Final Cake Solids on Specific Energy Use (WAS; Batch Tests; Optimum Ultrasonic Level)

Table 1. Effect of Initial Solids on Energy Use for WAS (Batch Tests)

	Initial Solids, %	
	16.0	18.7
D.C. Volt	60	60
Residence Time, min	3	3
Solids Loading, kg DS/m^2	2.1	2.5
Final Solids, %	23.8	26.5
Increase in Cake Solids, %	7.8	7.8
Specific Energy Use, kwh/kg filtrate	0.111	0.119
Predicted Total Energy Use for 28% Final Solids, kwh/ton DS	297	211

Table 2. Effect of Pressure on Dewatering of WAS (2 kg DS/m^2; 60 Volts; 15 W Ultrasonics; Batch Tests)

Initial Solids, %	Pressure, bar	Final Solids, %	Energy Use, kwh/kg filtrate
15.5	0.17	25.2	0.208
15.5	0.34	24.8	0.208

COMMERCIAL PROTOTYPE DESIGN

To exploit the EAD process commercially for sludge dewatering, it is possible to adapt EAD to a variety of mechanical dewatering concepts, such as belt filter presses, screw presses, and plate and frame presses. Among these, belt filter presses were considered the most applicable. Therefore, a belt press EAD design effort was initiated several years ago. First, a 0.5-m belt width press was retrofitted with EAD in several locations along the dewatering path. It was concluded, as was also supported by batch tests, that the most efficient use of EAD was at the end of mechanical dewatering(3). The 0.5-m unit was also used to identify a very low electrolytic corrosion rate anode as well as concepts for efficient coupling of electric and ultrasonic fields into the filter cake.

The results of batch parametric tests and the 0.5-m EAD retrofitted system were used to design a 1.2-m belt width EAD postdewatering unit. The principle is to feed partially dewatered (normally 15 to 20 percent solids) sludge into the belt press EAD unit and then remove approximately 50 percent of the remaining water. Such a postdewatering unit could be an integral part of a belt press fed with 1 to 3 percent solids sludge or it could be a stand-alone unit as shown schematically in Figure 8. In the latter case, the predewatering method is not limited to belt pressing.

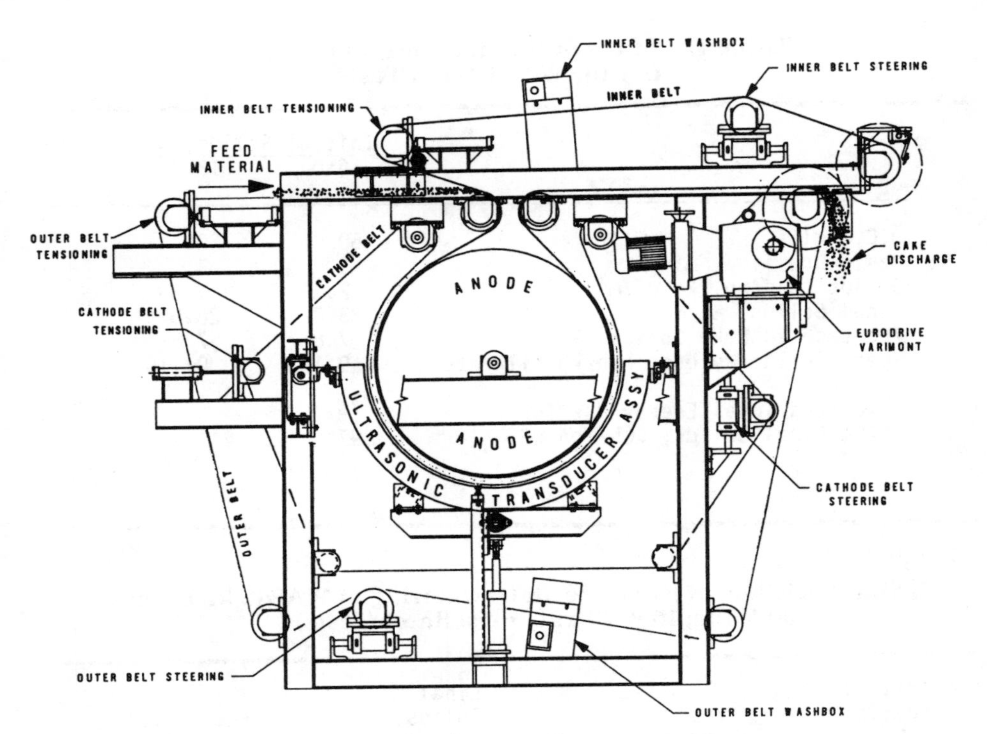

Figure 8. EAD General Arrangement

The partially dewatered sludge is pumped or screw-conveyed into a specially designed feed chute that provides a uniform feed cake thickness. The sludge is then sandwiched tightly between an outer polymer belt under medium tension and an inner polymer belt under low tension and travels around a large roll of 1.3 m diameter. In this manner, the sludge experiences a pressure in the range of 0.14 to 0.28 bar (2 to 4 psi). The large roll is used as the anode, and a moving wire-mesh belt on the outside of the outer polymer belt serves as the cathode, as shown in Figure 9. An ultrasonic transducer plate system couples ultrasonic energy with the sludge through the cathode belt. The electrical current passing through the belts, combined with the acoustic field and mechanical pressure, removes water from the sludge. The dewatered cake is then removed by scraper blades.

The anode and the ultrasonic transducer plate are the result of 5 years of design effort. The anode is of a special construction and is coated to minimize electrochemical corrosion and dissolution. A simple design also isolates the anode from the frame of the EAD machine, which is grounded.

A unique ultrasonic transducer design with a novel coupling system makes it possible to couple ultrasonic energy into the filter cake through belts moving in a circular path. To date, a curved ultrasonic transducer plate has never been designed and manufactured.

The inner, outer, and cathode belts are steered and tensioned independently. The hydraulic system allows for the proper adjustment of tensioning for each belt independently. Provisions are made for backwashing all these belts. The overall system for prototype testing is mounted on three separate skids, as shown in Figure 10.

The cathode belt has a wrap angle of about 240 degrees, whereas the ultrasonic system covers 180 degrees approximately towards the end of the EAD application zone. The reason for such a design is that ultrasonics is more

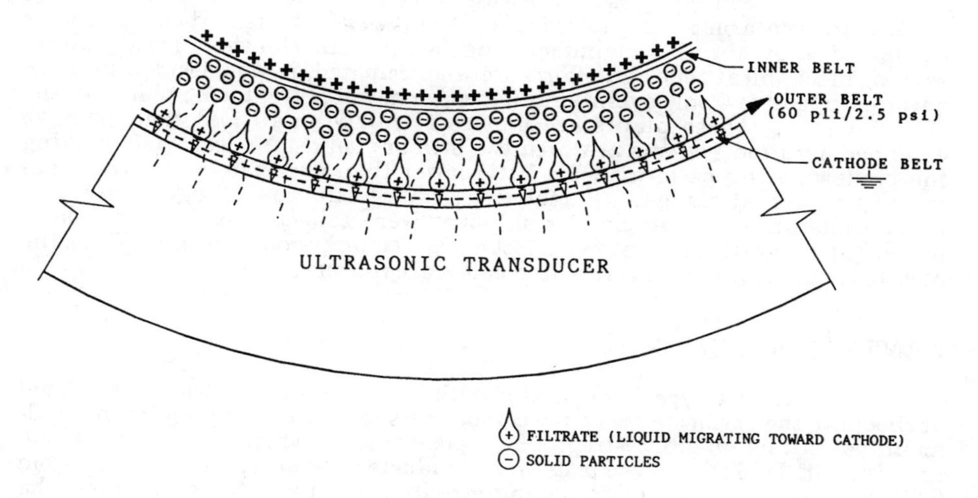

Figure 9. Belt Press EAD Process Concept

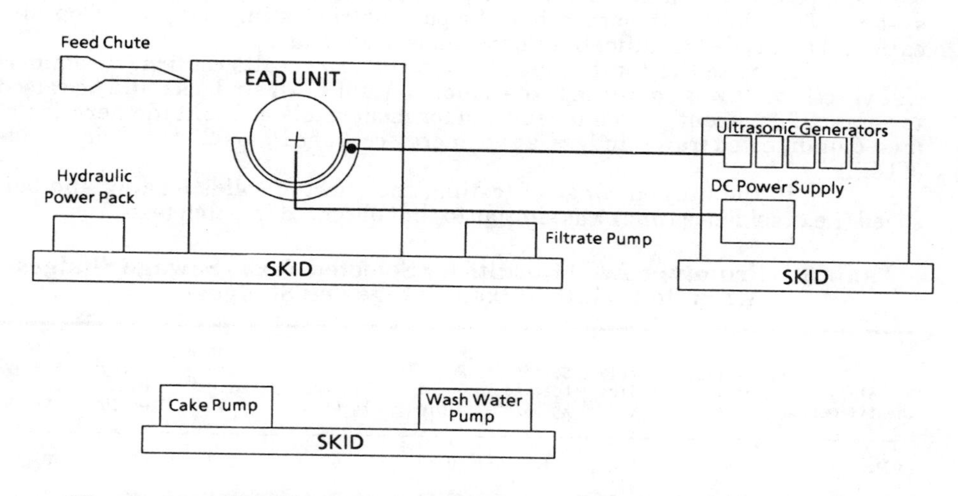

Figure 10. Skid-Mounted Belt Press EAD Prototype System

effective in dewatering when the cake is more consolidated. In fact, applying it when the cake is very thin may disrupt electroosmotic dewatering. With a 1.3-m-diameter roll and a 240-degree wrap angle, the total EAD contact length along the belt is 2.7 m. Thus, EAD prototype residence times of 1.7 to 8.3 minutes can be achieved for sludge dewatering by varying the belt speed from 1.6 to 0.33 m/min. For a particular application, the belt speed can be determined by the prototype and batch tests and thus adjusted.

COMMERCIAL PROTOTYPE RESULTS

The prototype has been used to obtain mechanical and EAD process performance data on predewatered mixed undigested sludges. These tests were conducted at Ashbrook's manufacturing facilities in Houston, Texas. Since several mechanical redesign efforts were anticipated, it was decided to truck dewatered sludge from several municipal wastewater treatment plants in the Houston area. These plants are generally fairly small and tend to have extended aeration systems in which the primary solids are destroyed, making the predewatering performance a little lower than expected for conventional mixed undigested sludges. Furthermore, due to the low sludge production rates, sometimes the samples for shipping were collected over several days, which has a significant negative effect on EAD energy consumption. Thus, the results for the four sludges presented below are conservative.

GENERAL RESULTS

The prototype was operated with four sludges. The tests were aimed at checking the performance of key components to modify and prepare the skid-mounted unit for on-site testing. The typical results, which are not optimized, are given in Table 3. These tests were conducted at belt speeds ranging from 0.58 to 1.20 m/min and solids loading ranging from 1.6 to 3.0 kg DS/m^2. The voltage in the tests was selected to achieve a current near the maximum possible value of 750 amps since the DC power supply was somewhat undersized. And, although the batch data showed energy efficient operation at solids loadings of around 1.0 kg DS/m^2 and belt speeds above about 1.5 m/min, such tests were not possible with the current design of the sludge feeding and distribution system. This limited the breadth of the parametric testing and process optimization. The needed modifications have since been made.

Despite the limitations of the tests, the results confirmed commercial viability. It was shown that the solids content of all sludges could be raised to at least 26 percent, which is required for moderately efficient incineration of mixed undigested (raw) sludges, at an energy cost of $19 to $27/ton of dry solids (TDS).

In the limited range of testing, the effect of solids loading and belt speed (i.e., residence time) was similar to that observed in batch tests.

Table 3. Prototype EAD Results for Selected Texas Sewage Sludges (1.2-m Belt Width; Mixed Undigested Sludges)

Sludge Identification	Initial Solids %	Final Solids %	Solids Loading kg DS/m^2	Belt Speed m/min	Throughput kg DS/hr	Increase in Solids[a] %	EAD Energy Use kwh/TDS	EAD Energy Cost[b] $/TDS
La Marque	16.8	26.6	1.5	0.68	63	9.8	433	21.7
Atoscacita	18.3	26.4	2.4	0.58	88	8.1	540	27.0
Kingwood	19.8	23.4	1.9	1.20	142	3.6	322	16.1
	20.6	25.9	2.5	0.80	124	5.3	380	19.0
	19.5	28.8	1.9	0.80	94	9.3	445	22.2
Memorial Municipal Utility District	17.2	27.6	2.9	0.58	105	10.4	523	26.1

(a) Percent final solids less percent initial solids.
(b) Electricity cost: 0.05 dollars/kwh.

The filter cake was fairly uniform across the belt width and discharged easily. The rise in cake temperature at the point of discharge was limited to about 55 C, which was made possible by a special design feature. This will help minimize any odor problems from heating the sludge.

The anode surface remained free of scratches after over 100 hours of operation and did not develop any deposits during the tests, which were typically 5 hours long.

EFFECT OF ULTRASONICS

As in the batch tests, the rate of dewatering as well as the current, at a fixed voltage gradient, increased on application of ultrasonics at an optimum level. The optimum ultrasonic level varied from 1.7 to 6.0 kW, depending on the sludge and the operating conditions. At higher ultrasonic energy levels, the performance declined to as low as or lower than performance with electroosmosis only. The results for two sludges are shown in Table 4.

In the case of Memorial sludge, the additional energy consumption by ultrasonics was only 3.5 percent, but it increased the filtration rate by 17 to 34 percent (normalized for voltage gradient). This resulted in an 8 to 20 percent reduction in specific energy use. Similarly, for Kingwood sludge, the filtration rate per unit voltage gradient increased 68 percent and the specific energy declined 18 percent, even though the solids loading for ultrasonics plus electric field test was higher. In the Kingwood case, the current per unit voltage gradient, which is an indication of degree of liquid continuum between the electrodes, was 23 percent higher when 6 kW ultrasonics was superimposed on 46 volts of electric field.

Additional work is continuing to determine the sludge properties that control the magnitude of ultrasonic effect in terms of enhanced dewatering. However, there is an additional key benefit of ultrasonics for all sludges. Apparently, during electroosmosis, some metal hydroxide (e.g., calcium hydroxide) deposits form as a result of the electrolytic reactions accompanying electroosmosis. These deposits can blind the cathode belt and retard current conduction. The coupling of ultrasonic energy through the cathode belt apparently keeps the belt clean. Or, if necessary, the belt can be periodically cleaned at higher ultrasonic energy levels using the same ultrasonic system.

Table 4. Effect of Ultrasonics for EAD Prototype

Sludge	Cake Thickness cm	Voltage Gradient volts/cm	Ultrasonic Power kw	Current amps	Filtration Rate kg/hr	Increase in Solids[a] %	Specific Energy kwh/kg Filtrate
Memorial Municipal Utility District	1.68	40.0	0	780	185	7.8	0.331
	1.73	37.0	1.7	780	229	10.4	0.264
	Repeat on a Second Sample						
	1.75	36.5	0	780	190	7.3	0.311
	1.83	35.0	1.7	790	213	8.5	0.287
Kingwood	0.74	62.4	0	765	117	8.9	0.353
	0.94	48.9	6.0	740	154	9.2	0.298

(a) Percent final solids less percent initial solids.

SCALE-UP RATIOS

As mentioned above, the results from the 1.2-m belt width prototype EAD unit were comparable to those from the 7.6-cm-diameter batch EAD unit, except on occasions when cake feeding or distribution was made difficult by very low solids loading or a high belt speed. Most of the tests were carried out at a belt speed of 0.80 m/min, which corresponded to a residence time of 3.4 minutes. The results of a typical prototype test and a batch simulation are given in Table 5. As shown, the batch unit closely simulates the prototype performance in terms of increase in solids content and DC electrical energy use.

The main difference in the batch unit is with respect to the efficiency of ultrasonics. In the batch, the optimum ultrasonic level for most sludges is around 20 W (Figure 5) output of the generator. This corresponds to 12 kW output for the prototype, for which the optimum was in the range of 1.7 to 6 kW input to the generator. Thus, a scale-down factor from generator output in batch to generator input in prototype is 7 to 2 for an average of about 5. For this reason, the batch data in this paper use one-fifth of the ultrasonic energy at the outlet of the generator in calculating specific and total energy consumption.

The voltage requirement in the prototype unit does appear to be somewhat higher in the presence of an additional (inner) polymer belt. Attempts are now under way to reduce the resistance of all belts.

Table 5. Comparison of 1.2-m Prototype and Batch Tests for Kingwood, Texas, Sludge

	Prototype	Batch
Recorded Data		
EAD Residence Time, min	3.4	3.5
Cake Pressure, kg/cm^2	0.18	0.27
Initial Cake Weight, g	NA	45.4
Feed Rate, kg DS/hr	481	NA
DC Voltage, volts	46	40
DC Current, amps	740	1.19
Ultrasonic Power, watts	6000[a]	20[b]
Initial Solids, %	19.5	20.1
Final Solids, %	28.8	29.0
Calculated Data		
Solids Loading, kg DS/m^2	1.9	2.0
Increase in Solids, %	9.3	8.9
Specific Energy, kwh/kg Filtrate	0.300	0.287[c]
Total Energy, kwh/TDS	445	396
Current Density, amps/volts m^2	5.7	6.6

NA Not Applicable.
(a) Input to generator.
(b) Output of generator.
(c) Using one fifth of output of ultrasonic generator.

EAD ECONOMICS

Using the unoptimized prototype data, we looked at the potential economic impact of using EAD for incineration and composting. The analysis revealed commercial viability of EAD for both applications as discussed below.

INCINERATION

The material balances and relative energy costs for conventional and EAD processes applied to sludge incineration are shown in Figure 11. For EAD, an energy consumption figure of 0.22 kWh/kg filtrate is used, which is the same as for unoptimized tests with La Marque sludge, 15 percent less than for unoptimized tests with Memorial, and 25 percent less than for tests with Kingwood sludge. This is a reasonable upper estimate based on the batch data for solids loading of 1.0 kg/m^2, which resulted in an energy consumption of about 0.14 kWh/kg (Figure 7). These lower solids loading will be tried after modification of the sludge feeding and distribution.

The excess fuel cost for incineration is based on a heating value of sludge of about 18.6 MJ/kg (8,000 Btu/lb, dry basis) and incineration in a multiple hearth furnace[8]. For such a sludge, autogenous combustion can be achieved at about 28 percent solids in sludge. A comparison of the two cases shows a savings of $1,717/day for 50 TDS/day plant. The anticipated cost for EAD units using unoptimized data for throughput calculations is estimated to be $3.0 to $3.5 million. It is also anticipated that the capital and nonenergy operating cost of an incinerator will decline substantially when a 28 percent sludge rather than an 18 percent sludge, which represents 35 percent less feed volume and 44 percent less water, is fed to the incinerator. Thus, EAD is attractive for incineration of sludge.

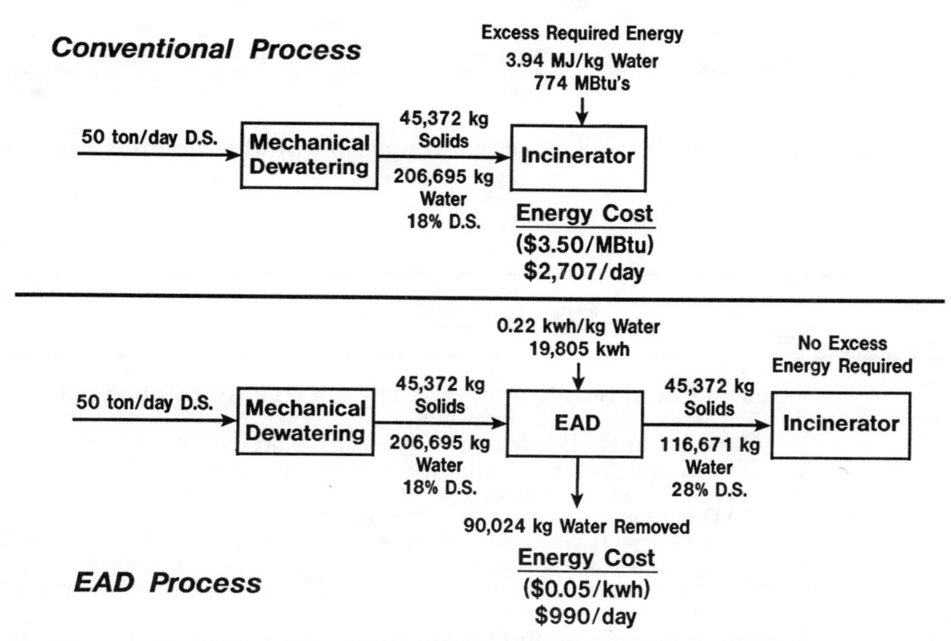

Figure 11. Material and Energy Balances for Sludge Incineration

COMPOSTING

The material and energy balances for composting are shown in Figure 12. In composting, part of the sludge is recycled and mixed with wet sludge and sawdust to prepare a 40 percent infeed. By using EAD, the total volume requirement for composting is reduced. This reduces the number of composting reactors required and therefore the capital cost by 53 percent. This corresponds to a capital cost savings of $27 million for a 50 TDS/day plant. The capital cost for EAD to achieve additional dewatering, on the other hand, is in the range of $3.0 to $3.5 million only. The EAD energy cost is also about 15 percent lower than the composting energy cost. Thus, EAD is quite attractive for composting.

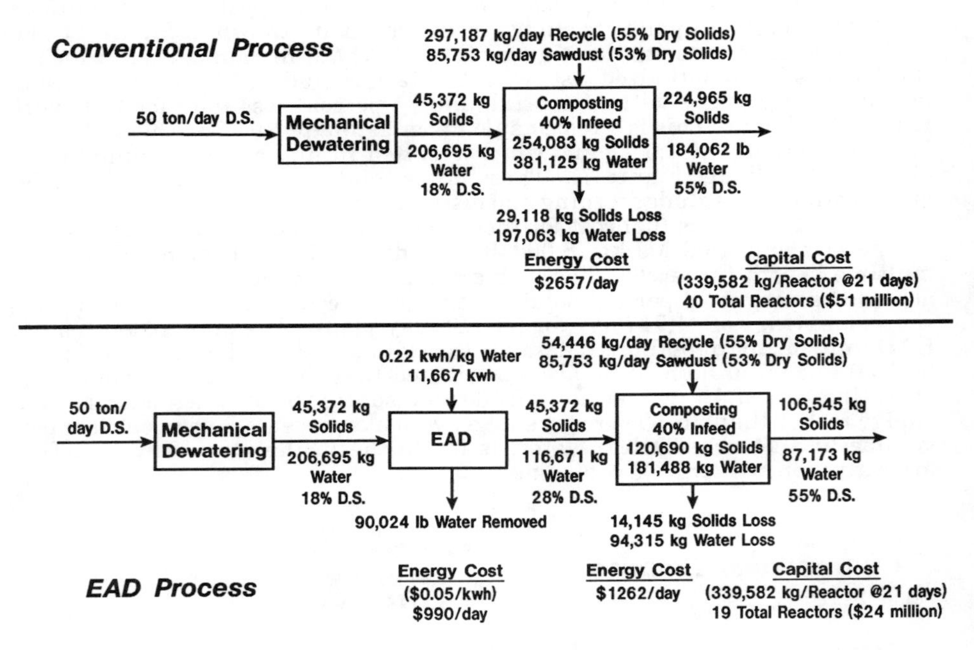

Figure 12. Material and Energy Balances for Sludge Composting

CONCLUSIONS

Based on initial results from the testing of an EAD belt press prototype, the following conclusions can be drawn:

1. Continuous EAD is feasible for a variety of sewage sludges.
2. The solids content of sludges can be raised economically to a level to allow self-sustained combustion in an incinerator.
3. EAD is economically attractive for composting.
4. For energy-efficient operation, a maximum amount of mechanical predewatering and a low solids loading should be used.
5. The ultrasonics help improve the efficiency of electroosmotic dewatering.
6. The prototype performance can be easily simulated in batch tests.

ACKNOWLEDGMENTS

The authors wish to thank B. C. Kim, Dr. N. Senapati, C. Criner, Dr. H. S. Muralidhara, and B. Jirjis of Battelle and D. Gamlen, T. Eason, and T. Bartoni of Ashbrook-Simon-Hartley, who helped obtain the data for this paper. Thanks are also due to M. Corrigan of DOE/OIP, C. V. Pearson of DOE/ANL, and A. Rozendaal of Bergmann b.V. of the Netherlands for funding part of the developmental effort.

REFERENCES

1. Chauhan, S. P., Muralidhara, H. S., and Kim, B. C., "Electroacoustic Dewatering Process for POTWs (Sewage Sludges)," *Proceedings of the National Conference on Municipal Treatment Plant Sludge Management*, held in Orlando, FL, May 1986.

2. Chauhan, S. P., Muralidhara, H. S., Kim, B. C., Senapati, N., Beard, R. E., and Jirjis, B. F., "Electroacoustic Dewatering (EAD)--A Novel Process," paper presented at 1987 Summer AIChE Meeting, Minneapolis, MN, August 1987.

3. Chauhan, S. P., Kim, B. C., Muralidhara, H. S., Senapati, N., and Criner, C. L., "Scale-Up of Electroacoustic Dewatering (EAD) Process for Food Products," Paper No. 48a, presented at 1989 Summer AIChE Meeting, Philadelphia, PA, August 1989.

4. Muralidhara, H. S., Parekh, B. K., and Senapati, N., "Solid-Liquid Separation Process for Fine Particle Suspensions by an Electric and Ultrasonic Field," U.S. Patent Nos. 4,561,953 (1985), 4,747,920 (1988), and Serial No. 400,296 (1989).

5. Muralidhara, H. S., Senapati, N., and Beard, R. E., "A Novel Electroacoustic Separation Process for Fine Particle Suspensions," *Advances in Solid-Liquid Separation*, H. S. Muralidhara, ed., Battelle Press/Royal Society of Chemistry, Chapter 14, 1986, 335-374.

6. Muralidhara, H. S., Senapati, N., Ensminger, D., and Chauhan, S. P., "A Novel Electroacoustic Separation Process for Fine Particle Suspensions," *Proceedings of World Filtration Congress IV*, held in Ostende, Belgium, April 1986.

7. Muralidhara, H. S., Chauhan, S. P., Senapati, N., Beard, R. E., Jirjis, B. F., and Kim, B. C., "Electroacoustic Dewatering (EAD)--A Novel Approach for Food Processing and Recovery," *Separation Science and Technology*, 1988, *23* (12 and 13), 2143-2158.

8. U.S. EPA, "Sludge Treatment and Disposal," U.S. EPA Report No. EPA 625/1-79-011, September 1979.

9. Yukawa, H., Kobayashi, K., and Hakoda, H., "Study of the Performance of Electrokinetic Filtration Using Rotary Drum Vacuum Filter," *J. Chem. Engr. Japan*, 1980, *13* (5), 390-396.

PARTICLE SEPARATIONS BY THE ELECTROCOAL PROCESS

Luis Borzone, Robin R. Oder,
J. T. Mang, and S.-M.B. Chi
EXPORTech Company, Inc.
P.O. Box 588
New Kensington, Pennsylvania 15068-0588
(412) 337-4415

INTRODUCTION

EXPORTech Company, Inc. (ETCi) has developed a novel electrostatic method to separate particulates using emulsions of water in organic liquid. The separation is based on differences in surface characteristics that allow particles to associate with the appropriate liquid component of the emulsion. Since emulsified water droplets are micron sized, the emulsion provides a large interfacial area for particle contact. Line frequency electric fields on the order of 1 to 2 kV/cm rms are used to speed coalescence of the water phase. The water phase and the hydrophilic particles are separated by sedimentation. The organic phase and the hydrophobic particles are separated after the breaking of the emulsion. Low HLB surfactants are employed to promote selective particle wetting and to enhance particle separations in the coalescer.

EXPERIMENTAL

The coalescence phenomenon has been observed visually using a microscope and in experiments carried out in a batch cell. In the microscopic tests, a few drops of emulsion were placed between two wire electrodes. Different electric fields were applied and the coalescence modes photographed. Batch cell experiments were carried out using blends of kaolin clay (hydrophilic component) and low ash bituminous coal (hydrophobic component) to demonstrate particle separation.

A schematic diagram of the batch cell is shown in Figure 1. The cell consists of a 2-1/2-in.-diameter Plexiglas® cylinder with a working volume of 100 cc. Inside the cell, two circular brass screen electrodes are placed horizontally, separated 5 mm apart. The emulsion, prepared using a Tekmar SDT 182En Tissumizer at 20,000 rpm, was poured slowly into the cell, over deionized water, whose level was kept constant at about 5 mm below the electrodes.

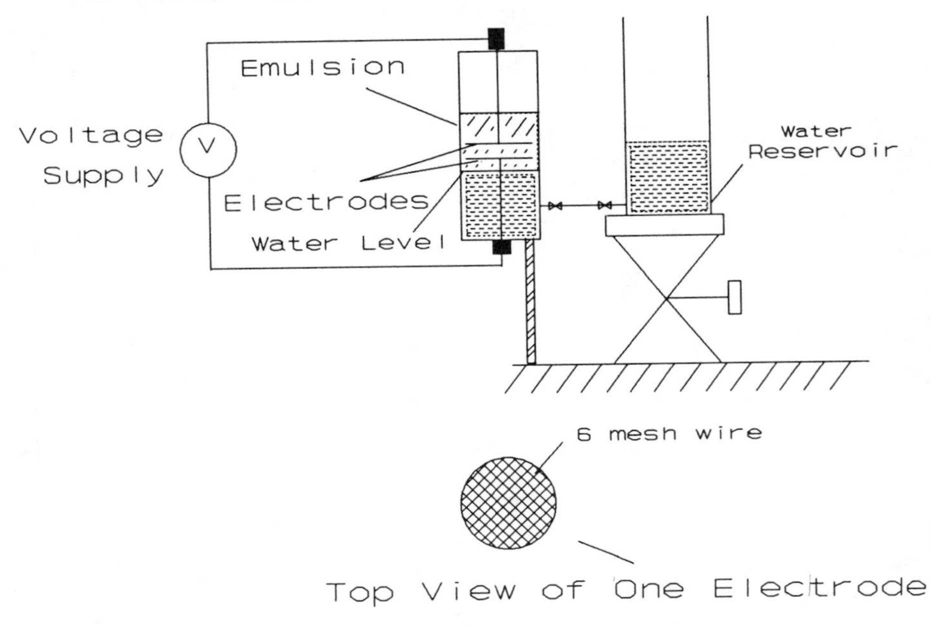

Figure 1. Batch Coalescer Cell

A 60 Hz ac signal, generating electric fields up to 2 kV/cm rms, was applied between the electrodes to initiate coalescence. Slow mixing was provided to circulate the emulsion. Voltages and currents to the cell were measured. After coalescence, the upper organic phase was recovered and analyzed for water, solids, and ash content. The same analysis was carried out for the aqueous phase, including sediments. The variables measured were emulsion formulation, solids, and water content of the emulsion, as well as coalescence time, voltage, and particle size.

Two proprietary emulsifiers were chosen on the basis of the ability to prepare stable emulsions containing high levels of solids and water. The first one (emulsifier C) is a mixture of sorbitan monooleate, nonylphenol, and ethylene oxide. The other (emulsifier T) is an ABA-type block copolymer of castor oil and nonylphenol plus some bonding ethylene oxide molecules. Both promote flocculation of the hydrophilic mineral matter with the water in the preparation of the feed to the electrocell.

RESULTS

Response to the electric field was found to be dependent upon the type of emulsifier used. Emulsions prepared with emulsifier C chained in the presence of the electric field, whereas emulsions prepared with use of emulsifier T coalesced into filamentary or tubular structures. These two modes of coalescence are shown in the photographs of Figure 2. The same modes appear in the presence of solids. The particular coalescence mode has an effect on the hydrophobic component distribution: chaining generates intense electric field gradients that attract the hydrophobic polarizable component. In the case of tubular coalescence, no gradients are formed and the hydrophobic particles remain dispersed in the organic phase. In the batch tests, this second mode of coalescence translated into better separations.

Coalescence times ranging from 5 to 15 minutes were tested in an emulsion containing 60 g of oil, 16.7 g of deionized water, 2 g of coal, 2 g kaolin, and 1 g of emulsifier. For an electric field of 1.2 kV/cm, kaolin rejection was 88 percent in the case of emulsifier T and 75 percent for emulsifier C. The ash rejection increased to 95 percent and 87 percent, respectively, after 15 minutes of coalescence time. Settling of the coalesced water droplets is included in that time. The present effort is dedicated to the testing of shorter coalescence times followed by settling in the absence of electric fields; this will reduce power consumption significantly.

The water content of the emulsion is important to the separation efficiency (Figure 3). Besides improving kaolin rejection, an increase in water content is followed by an increase in temperature, an effect that becomes important for water contents in excess of 15 percent. This phenomenon may be associated to an increase of chaining or tubing extending from one electrode to another, short circuiting the cell. This phenomenon is being investigated by varying electrode spacing and mixing inside the cell, two operations that are commonly used in commercial electrostatic coalescers to reduce chaining effects.

An increase in field strength also results in augmented separation efficiency, as shown in Figure 4. There is, however, a practical limit at fields on the order of 1.2 kV/cm, above which no further separation is obtained. For the emulsifier T, the maximum kaolin rejection was of the order of 96 percent.

A practical application of the coalescer separation was accomplished using Upper Freeport coal. This coal has an ash content of 20 percent and sulfur content of 1.7 percent. The emulsion was formulated as 60/x/4/1 (grams) = oil/water/coal/emulsifier. The field was applied for 20 minutes. Results of the effect of water content are presented in Figure 5 and show a clear increase in separation performance with an increase in water content. Other variables relevant to coal separation using emulsions are the effect of feed ash, coal prep-

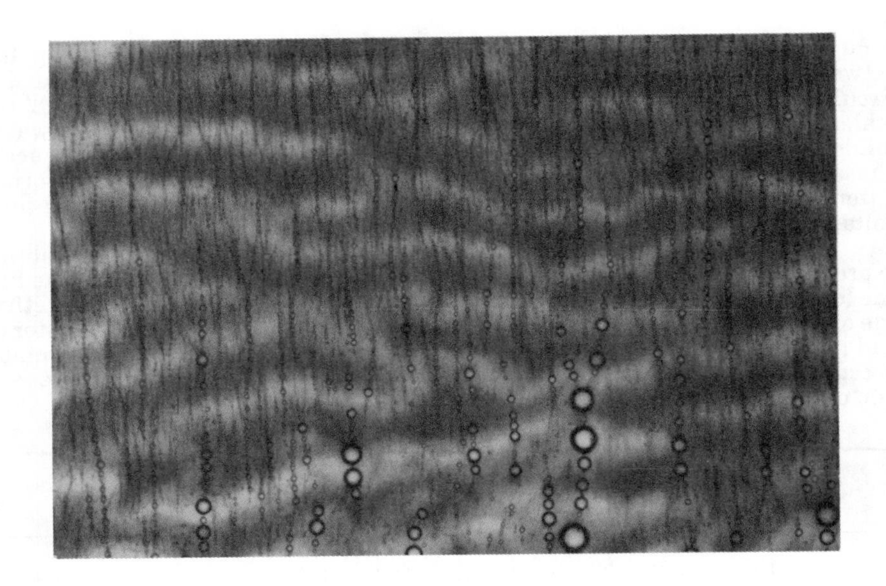

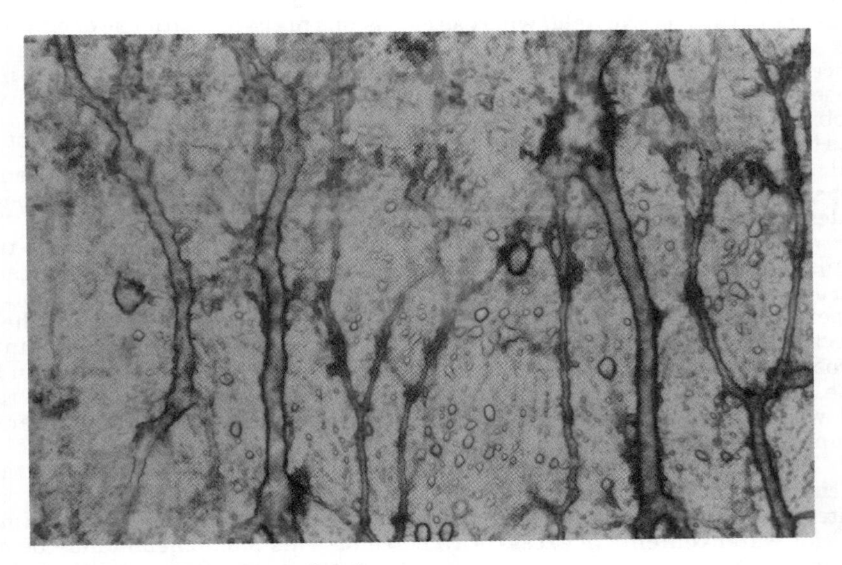

Figure 2. Emulsion Coalescence

(a) Chaining Mode (Emulsifier C, 8.84 μ/cm)

(b) Filamentary Mode (Emulsifier T, 8.84 μ/cm)

aration, voltage frequency, particle size, and field strength.

Significant differences in separation efficiency were found when the coal was ground in water rather than oil. Water grinding resulted in better separation. The effects of field frequency and strength showed the presence of an optimum occurring at 200 Hz and 1.8 kV/cm, respectively. At this field strength, an ash rejection of 82 percent with a combustible matter recovery of 82 percent was achieved. The best results were achieved with particle sizes of the order of 6 μ.

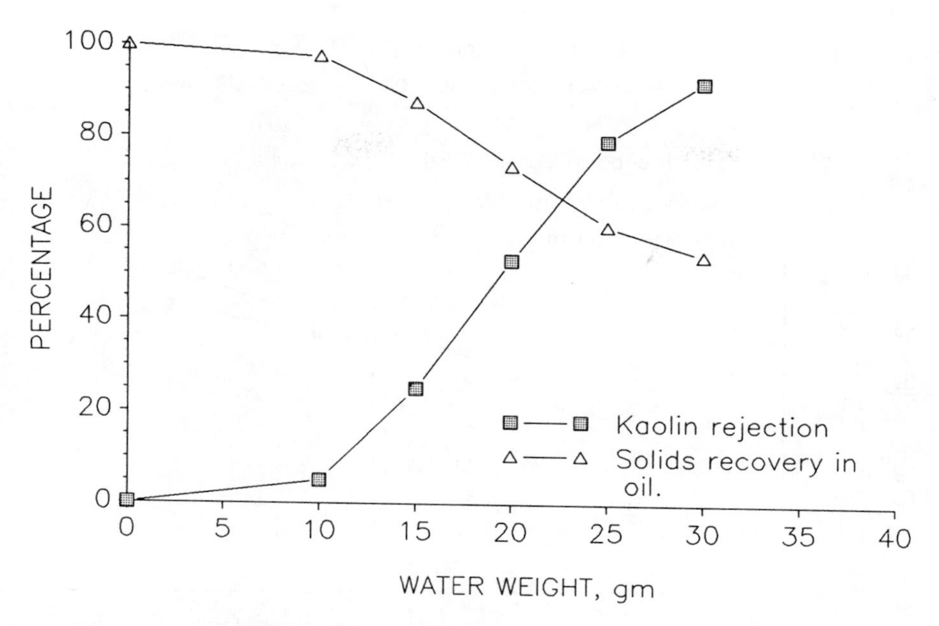

Figure 3. Effect of Water Content on the Separation of a Blend of Coal and Kaolin Clay Using Emulsifier T

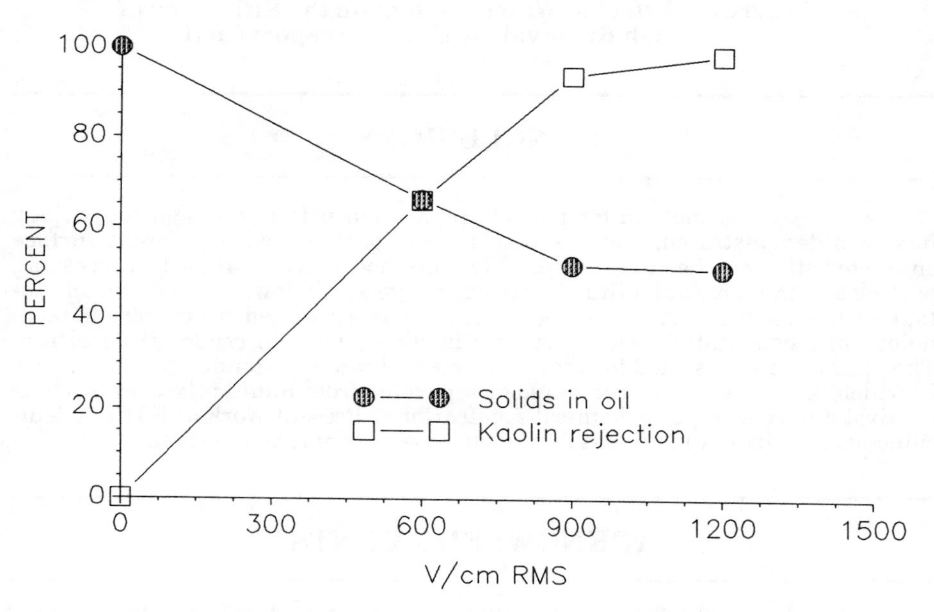

Figure 4. Effect of Electric Field Strength on the Separation of Blend of Coal and Kaolin Clay Using Emulsifier T

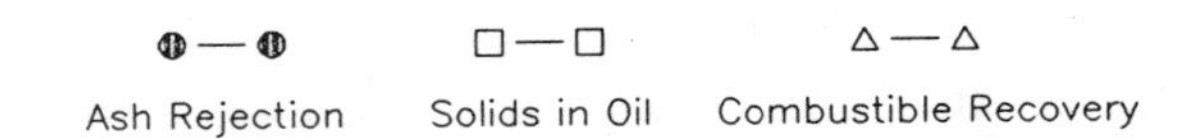

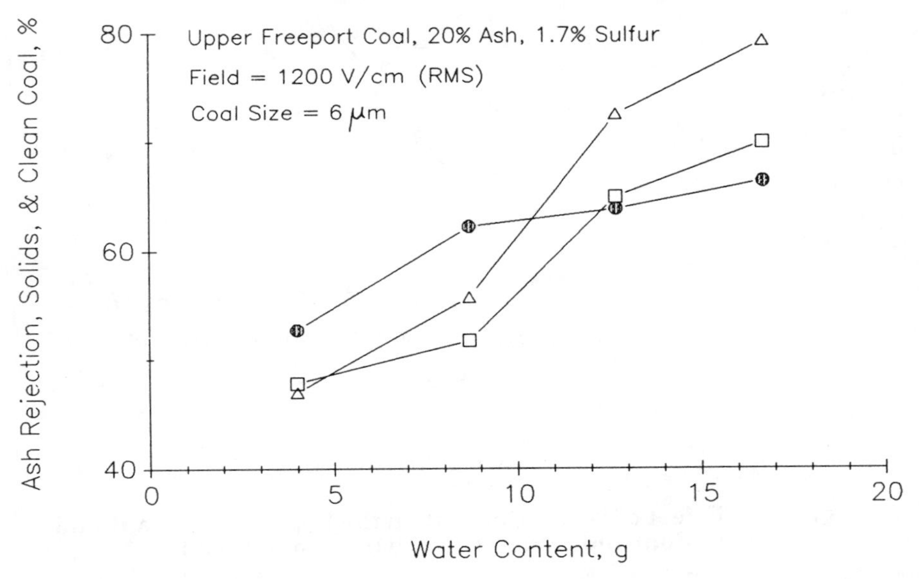

Figure 5. Effect of Water Content on the Efficiency of Ash Removal for Upper Freeport Coal

CONCLUSIONS

A novel method for particle separation using emulsion technology has been demonstrated, and the separation of particles with different surface characteristics has been successful. The method is well suited to processing particles in the nominal minus 100 μ size range and below, and has the advantage of low capital and operating costs, as it is estimated from present technology of electrostatic coalescence that is widely used in crude oil desalting. The method is well suited for coal cleaning and can be extended to other areas in which solids have to be recovered or separated from liquid mixtures, such as catalyst recovery in petrochemical applications. Present work at ETCi is dedicated to optimizing the process for the purpose of commercialization.

ACKNOWLEDGMENTS

The results presented in this paper are based upon work supported by the National Science Foundation under Grant ISI-87-01187.

MEMBRANES AND MEMBRANE PRESSES

In the United States alone, the current level of spending on pollution control is estimated at $70 billion per year, almost $50 billion of which is spent by industry. At the same time, because of the increasing demand on the extraction and processing of energy and material sources, a great deal of pollutants are generated. Hence, there is a dire need to develop more efficient pollution control methods.

Advanced membrane filtration processing is playing a vital role in the treatment of municipal and industrial wastewater. Membrane processes have proven very effective in decreasing levels of such toxic materials as arsenic to less than 0.05 mg/L and heavy metals to less than 0.1 mg/L.

Membrane technology will be seriously considered by a number of industries such as metal finishing, electronics, and food processing. Over the past 30 years, economic and environmental factors have accelerated the application of membrane processes as an alternative to evaporation or distillation. The evaporation processes are energy intensive and generate considerable thermal pollution, whereas membrane separations, which do not require a change in state of the solution, lower energy costs and release no emissions. An increase in the world market for membranes from $10 million/year before 1960 to over $1 billion currently is a strong indication of the increased use of membranes in the process industry.

Membrane fouling is a common phenomenon observed during the operation of any membrane system. Membrane fouling leads to reduction in flux. Dr. Muralidhara's paper describes an electromembrane technology that utilizes electrokinetic and flow properties to enhance the efficiency of commercial membranes. Mr. Cartwright of C3 International provides a number of case studies in which membrane systems have been successfully employed.

Plate and frame filter presses are experiencing a comeback brought about by recent developments in automatic cake discharge and fabric washing. Membrane plates are also being developed to further enhance productivity. Dr. Mayer of Du Pont has written an interesting case study describing the use of such an approach. Dr. James will overview the separation technologies demonstrated under the U.S. Environmental Protection Agency's SITE program.

ELECTROMEMBRANE TECHNOLOGY: A NOVEL APPROACH FOR ANTIFOULING

H. S. Muralidhara and S. V. Jagannadh
Battelle
505 King Avenue
Columbus, Ohio 43201-2693
(614)424-5018

ABSTRACT

Membrane performance is affected mainly by two factors--fouling and concentration polarization. One approach to limiting membrane fouling is external field application. Battelle has been investigating an external field approach that utilizes electrokinetic and flow properties to enhance the efficiency of commercial membranes. An important focus of the work has been to minimize the power consumption and electrode costs for adapting the technology to existing membrane materials and basic module designs. This approach has the potential to minimize the reduction in membrane flux that is commonly observed after a period of operation. Potential application of this technology in environmental issues such as wastewater treatment and waste sludge treatment is presented and compared with controls, using a conventional flat plate module. Preliminary results showed that electromembranes (EMBs) performed 15 to 50 percent better than non-EMBs.

INTRODUCTION

During the past few years, membrane separations have become more widely used, replacing the conventional concentration techniques. They are becoming an increasingly important tool for separating and concentrating a variety of feedstocks ranging from oil/water emulsions to sewage sludges. The efficient use of membrane technology is, however, hindered by two factors--namely, fouling and concentration polarization.

The negative impact of membrane fouling is roughly estimated at $500 million on a yearly basis[22]. Fouling involves the adsorption or trapping of material being transported (foulant) across the membrane. The foulant may adsorb very strongly to the membrane surface and, in some cases, chemically react with the polymeric membrane. Thus, in some cases, fouling can be a chemical reaction phenomenon.

The second factor, concentration polarization, is the accumulation of the solute species at the upstream surface of the membrane. This hydrodynamic/diffusion phenomenon can be overcome by high velocity, if such an operation can be tolerated. The effect always will be involved during membrane processing because of the fundamental limitations of mass transfer and the existence of a boundary layer.

Both fouling and concentration polarization reduce membrane flux (permeate flow rate per unit area) and thus decrease the useful life of membranes. This results in frequent cleaning or replacement of the membranes. Thus, the two factors clearly can limit the application of membranes in industrial sectors.

The various approaches to limit membrane fouling can be grouped into four categories: boundary layer (or velocity) control[5,9,14]; turbulence inducers/generators[2]; membrane modification and materials[10,13]; and external fields. These approaches were discussed in detail in a recent article by Muralidhara and Huffman[17].

The general approach of using external fields (e.g., electrical, gravitational, ultrasonic, magnetic) address most of the limitations of the above-mentioned other approaches. The external fields can be developed and implemented independent of the velocity field on the membrane and the type of the membrane material. This approach should be independent of the feed stream, as long as certain electrokinetic properties such as electrical conductivity and zeta potential are present in the feed stream. Of course, such properties can be introduced into the feed if warranted and feasible from a product end-use and/or an economic standpoint. Battelle's electromembrane (EMB) technology is based on the external field approach. Typical developments and investigations of this approach by others are given in Table 1. The references and areas mentioned in the table suggest that the external field approach applies to a wide variety of materials including colloids, proteins, and oils.

Battelle's EMB technology employs an external electric field to minimize the membrane fouling. This technology differs from the other electric field approaches in the placement of the electrodes. The novel placement of electrodes is expected to maintain higher fluxes through the membrane for a lower energy input. We tested the performance of electromembranes for a different variety of feed materials. The application of EMB technology to acid whey, sewage sludge, and oil/water emulsion and the resulting performances are discussed in the present paper.

Table 1. Examples of External Field Membrane Approach

Topic or Area	Reference
Development of an electrofilter	Beechold, 1926; Manegold, 1937
Combination of membrane and an electrical field for colloids	Bier, 1959
Electrical field with hollow fiber module for chromatography	Reis and Lightfoot, 1976
Cross-flow electrofilter for clay and oil suspensions	Henry *et al.*, 1977
High frequency vibration and excitation during hyperfiltration	Herrmann, 1982
Cross-flow electroultrafiltration	Yukawa *et al.*, 1983
Prevention of protein and paint fouling using an electric field	Mullon *et al.*, 1985
Modeling of electroultrafiltration	Radovich and Bennam, 1985
Optimization of an electrical field for cross-flow microfiltration	Wakeman and Tarleton, 1986
Turbulence promoting agents to minimize polarization prevention of ultrafiltration	Rios *et al.*, 1987
Prevention of protein clogging using electrophoresis	Hong, 1988
Flux decline in the presence of electric field	Rios *et al.*, 1988
Prevention of microfiltration flux decline with electricity	Wakeman, 1988
In situ electrolytic membrane cleaning and restoration	Bowen *et al.*, 1989

EXPERIMENTAL METHOD

For the results reported herein, a Millipore Minitan plate and frame unit was used during all the experimental trials. Both micro- and ultrafiltration membranes of 60 cm^2 membrane area were tested. The membrane materials used were polyvinylidene difluoride (PVDF) for microfiltration (MF) and polysulfones for ultrafiltration (UF). Platinum electrodes were connected to a dc power supply. The feed concentration was increased during the experiments by recycling the retentate. A control experiment (without EMB) was run parallel to all the EMB experiments, and the performance of membranes with and without electromembranes was compared.

PRELIMINARY EXPERIMENTAL RESULTS

A summary of experimental results for the three feed materials is given in Table 2. The preliminary data show that the permeate flux from EMBs is about 15 to 50 percent higher than the flux from control membranes or non-EMBs. Further, it was noticed at the end of runs that the EMBs have less solid deposits on the membrane surface compared to the control membranes.

Table 2. Performance of EMB Technique Versus Conventional Technique for Various Suspensions

		lb/hr-ft^2		
		Initial Flux	Final Flux	Duration
Acid Whey Polysulfone 10K UF				
	Control	0.0017	0.00068	25 hr
	EMB	0.0043	0.0021	
Sewage Sludge PVDF 0.22μ				
	Control	0.02	0.0	8 hr
	EMB	0.02	0.01	
Vegetable Oil/Water PVDF 0.22μ				
	Control	5.3	1.66	6 hr
	EMB	6.14	2.6	

ACID WHEY RESULTS

Experiments using acid whey were conducted over a period of 25 hours. Polysulfone membranes with a molecular weight cut-off (MWCO) of 10,000 were used. Figure 1 shows the normalized permeation rate as a function of filtration time. The control membrane started with an initial permeate flux of 0.0017 lb/hr/ft^2; at the end of 26 hours, the flux was 0.00068 lb/hr/ft^2--a decrease in flux of 72 percent. On the other hand, during the same period, the EMB permeate flux decreased from 0.0043 to 0.0021 lb/hr/ft^2--an improvement of 15 percent over control. Both control and EMB experiments were conducted under an identical set of conditions using the same feed.

SEWAGE SLUDGE RESULTS

Figure 2 shows the variation of normalized permeation rate as a function of filtration time for domestic sewage sludge. The sludge contains fine particles as well as proteins and lipid matter. The typical solids concentration

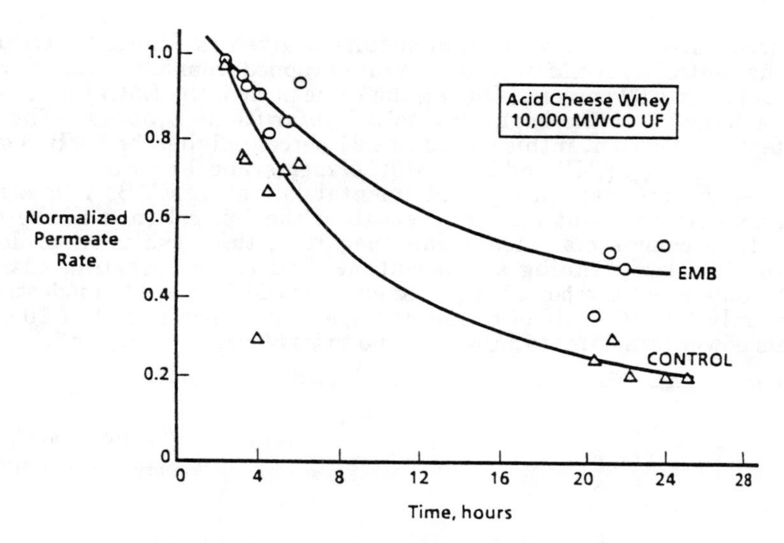

Figure 1. Results--Acid Cheese Whey

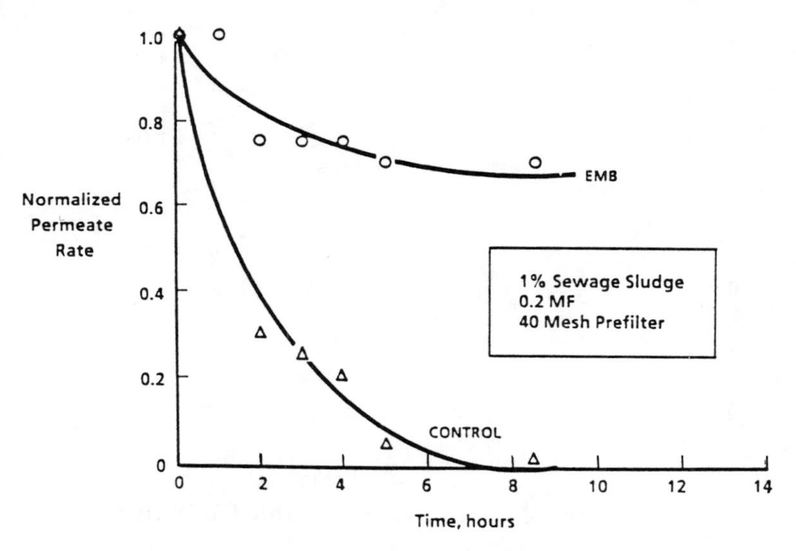

Figure 2. Results--Sewage Sludge

during experimentation was about 1 percent. PVDF membrane of 0.22 μ pore size was used. The figure shows that the permeate flux for the control (non-EMB) system declined dramatically by 70 percent in the first 2 hours, and the membrane was virtually blinded at the end of 8 hours. On the other hand, for the same period of time, the permeate flux for the EMB dropped only by 30 percent. This demonstrated the advantages of EMB technology in controlling fouling with a complex mixture such as sewage sludge. This result was also validated with a sludge feed containing 2 to 3 percent solids.

VEGETABLE OIL/WATER EMULSION RESULTS

The experiments using vegetable oil/water emulsions were conducted using a 0.22-μ pore size PVDF membrane for a period of 6 hours. Variation of normalized flux as a function of time is given in Figure 3. Over this period, the control (non-EMB) permeate rate dropped from 5.3 to 1.66 lb/hr/ft^2, a decline of nearly 69 percent. During the same period, the EMB flux changed from 6.14 lb/hr/ft^2 at the beginning to 2.6 lb/hr/ft^2 at 6 hours. The total permeate vol collected over this period was 21 percent higher for EMB (3 gal/ft^2 of membrane area for EMB and 2.37 gal/ft^2 of membrane for control).

Thus, the preliminary experimental data using EMBs always maintained permeate fluxes at higher levels than the control (non-EMB) membranes. In a commercial membrane operation, this results in prolonged membrane life, less cleaning and downtime, and lower operating costs. It should be noted that the power requirements for EMB are quite modest, ranging from only 1 to 10 W/ft^2 of membrane area. At a power cost of \$0.05 per kWh, this power translates to a cost of \$0.05 to \$0.50 per 1,000 ft^2/hr.

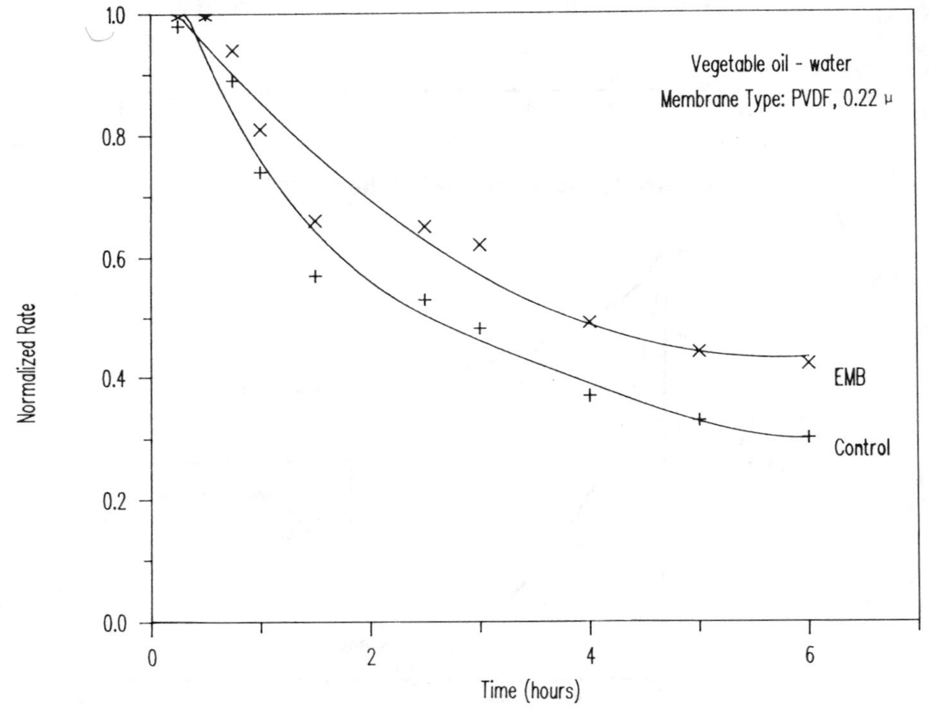

Figure 3. Results--Vegetable Oil/Water

FEATURES AND BENEFITS

Some of the features of electromembranes can be explained, to some extent, using a model developed by Maulik[12]. Maulik's model is based on modified Darcy's equation that contains an additional term to account for electrokinetic effects. This model specifically includes an electric term. Whenever a dc electric field is imposed, the following phenomena occur--electrophoresis, electroosmosis, electrolysis, and Joule's heating. The major contributing phenomena to EMB are not known completely at this point. However, several hypotheses are available.

Maulik's model takes into account both electroosmosis and electrophoresis. The model equation was developed by adding these electric terms to Darcy's equation. The model equation can be obtained from Maulik (1971).

According to the model, a plot of permeate vol (v) against the reciprocal of filtration rate yields a straight line. The decrease in the slope of the resulting straight line is indicative of decreasing filter cake buildup. The decrease in the intercept values corresponds to the decrease in filter medium resistance.

A fit of Maulik's model to vegetable oil/water emulsion data is given in Figure 4. The EMB data followed a single straight line, whereas two straight line fits were necessary for control data. During the first hour of operation, the straight line for control data was parallel to the line for EMB data, and after 1 hour, the slope for control straight line increased significantly compared with the slope for the EMB data. Comparison of slope suggests that the control (non-EMB) membrane plugged faster than the EMB; however, both membranes have about the same rate of plugging up to 1 hour. Thus, Maulik's model shows that the EMBs have a lower rate of plugging and, thus, maintain fluxes at higher levels for longer periods of time. The expected benefits of EMB technology are given in Table 3.

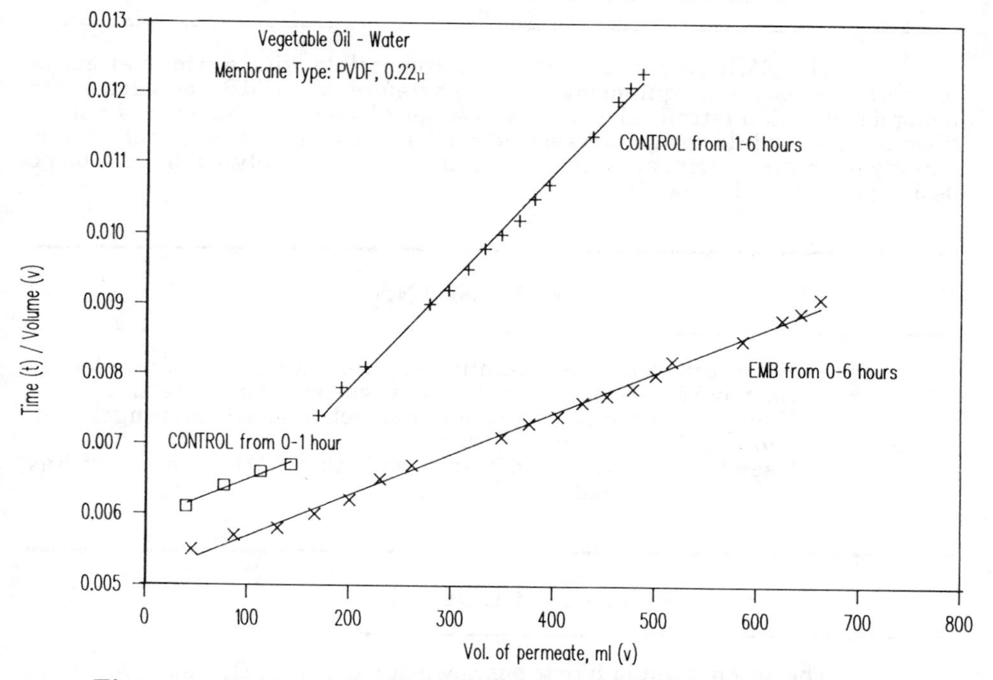

Figure 4. A Fit of Maulik's Model to Vegetable Oil/Water Data

Table 3. Features and Benefits of EMB

Feature	Benefit
Higher rates	Less membrane area or cost Less backflushing or cleaning Longer membrane life
Commercial membranes	Limited regulatory approval Existing application materials
Retrofit design	Existing hardware and designs Rapid commercialization

PROCESS SCALE-UP

On the basis of promising bench-scale results such as those given previously, the EMB process is being evaluated now on a larger scale. At present, the technology is being adapted to a 0.7-m^2 plate and frame design to handle 30 to 120 gal/hr of feed; spiral wound development is also planned for the near future. More detailed application testing is expected to be initiated for the food and beverage, biotechnology, and electronic industries in the near future.

COMMERCIALIZATION POTENTIAL

The EMB process appears to be compatible with a wide variety of membrane separation equipment types. Therefore, the EMB approaches are being developed to retrofit plate and frame, spiral wound, tubular, and hollow fiber types of module designs. As stated earlier, plate and frame retrofitting is already in progress. Other design specifications are still evolving, but the basic design principles are now clear.

CONCLUSIONS

- Flux decline has been minimized through the use of an electric field by 15 to 50 percent with a wide variety of materials.
- Energy requirement using our approach is small and ranges from 1 to 10 W/ft^2 of membrane surface.
- Use of commercial membranes with this EMB technology has been demonstrated.

ACKNOWLEDGMENTS

The authors would like to acknowledge Dr. W. J. Huffman, Mr. B. F. Jirjis, and Battelle Columbus Division management for supporting this work, and Laurie Peters for typing the manuscript.

REFERENCES

1. Beechold, H., "Ultrafiltration and Electro-Ultrafiltration," in *Colloid Chemistry*, J. Alexander, ed., The Chemical Catalog Company, 1926, Vol. I.

2. Belfort, G., "Membrane Modules: Comparison of Different Configurations Using Fluid Mechanics," *J. Membr. Sci.*, 1988, *35*, 245-270.

3. Bier, M., ed., *Electrophoresis*, Academic Press, New York, NY, 1959, Vol. I, 263.

4. Bowen, R. W., Kingdom, R. S., and Sabuni, H.A.M., "Electrically Enhanced Separation Processes: The Basis of In Situ Intermittent Electrolytic Membrane Cleaning (IIEMC) and In Situ Electrolytic Membrane Restoration (IEMR)," *J. Membr. Sci.*, 1989, *40*, 219-229.

5. Brian, P.L.T., "Concentration Polarization in Reverse Osmosis Desalination with Variable Flux and Incomplete Salt Rejection," *Ind. Chem. Fund.*, 1965, *4*, 438-445.

6. Henry, J. D., Lawler, L. F., and Kuo, C.H.A., "A Solid/Liquid Separation Process Based on Cross Flow and Electrofiltration," *AIChE Journal*, 1977, *23* (6), 851-859.

7. Hermann, C. C., "High Frequency Excitation and Vibration Studies on Hyperfiltration Membranes," *Desalination*, 1982, *42*, 329-338.

8. Hong, J., "Electrophoresis Stops Ultrafiltration Membrane Clogging," Dept. of Chemical Engineering, Illinois Institute of Technology, Chicago, IL, personal communication, March, 1988.

9. Huffman, W. J., "The Effect of Forced and Natural Convection During Ultrafiltration of Protein-Saline Solutions in Thin, Horizontal Channels," Ph.D. Dissertation, Clemson University, Clemson, SC, 1970.

10. Light, W. G., "Contending With Chlorine Attack of RO Membranes," Fifth Annual Membrane Technology/Planning Conference, 1987, 233-239.

11. Manegold, E., "The Effectiveness of Filtration, Dialysis, Electrolysis, and Their Intercombinations as Purification Processes," *Trans. Faraday Soc.*, 1937, *33*.

12. Maulik, S. P., "Physical Aspects of Electro-Filtration," *Env. Sci. and Tech.*, 1971, *5* (9), 771-776.

13. Michaels, A. S., personal communication to W. J. Huffman, 1987.

14. Michaels, A. S., "Progress in Separation and Purification," *Ultrafiltration*, E. S. Perry, ed., Interscience Publishers, New York, NY, 1968, Vol. I, 297-298.

15. Mullon, C., Radovich, J. M., and Behnam, B., "A Semiempirical Model for Electroultrafiltration-Diafiltration," *Separation Science and Technology*, 1985, *20* (1), 63-72.

16. Muralidhara, H. S., "The Combined Field Approach to Separations," *Chem. Tech.*, 1988, 224-235.

17. Muralidhara, H. S. and Huffman, W. J., "Electromembrane Technology--A Novel Approach for Antifouling," Sixth Annual Membrane Technology/Planning Committee, 1988.

18. Radovich, J. M. and Behnam, B., "Steady State Modelling of Electroultrafiltration at Constant Concentration," *Sep. Sci. Technology*, 1985, *2* (4), 315.

19. Reis, J.F.G. and Lightfoot, E. N., "Electropolarization Chromatography," *AIChE Journal*, 1976, *22* (4), 770-785.

20. Rios, G. M., Rakotoarisoa, H., and Tarado de la Fuente, B., "Basic Transport Mechanisms of Ultrafiltration in the Presence of Fluidized Particles," *J. Membrane Sci.*, 1987, *34*, 331-343.

21. Rios, G. M., Rakotoarisoa, H., and Tarado de la Fuente, B., "Basic Transport Mechanisms of Ultrafiltration in the Presence of an Electric Field," *J. Membrane Sci.*, 1988, *38*, 147-159.

22. Smolders, C. A. and Boomgard, T. H., Guest Editorial in *J. Membr. Sci.*, 1989, *40*, 121.

23. Wakeman, R. J., "Prevention of Flux Decline in Electrical Microfiltration," Special Issue on Combined Field Techniques for Dewatering, H. S. Muralidhara and N. C. Lockhart, eds., in *Drying Technology Journal*, 1988, *6*, 547-573.

24. Wakeman, R. J. and Tarleton, E. S., "Modelling Cross Flow Electro- and Micro-Filtration," Proc. 4th World Filtration Congress, Ostende, Belgium, April, 1986.

25. Yukawa, H., Shimora, K., Suda, A., and Maniwa, A., "Cross Flow Electroultrafiltration for Colloidal Solution of Protein," *J. Chem. Eng. Jpn.*, 1983, *16* (4), 305.

WASTE MINIMIZATION AND RECYCLING WITH MEMBRANE FILTER PRESSES

E. Mayer
E. I. du Pont de Nemours & Co., Inc.
Engineering Department Louviers 1359
P.O. Box 6090
Newark, Delaware 19714-6090

ABSTRACT

Resource Conservation and Recovery Act (RCRA) regulations have imposed severe restrictions on land disposal of sludges, whether hazardous or not. In fact, recent land-ban restrictions have further exacerbated this situation to the extent that best demonstrated available technology should be practiced to meet the Toxic Characteristic Leaching Procedure regulatory levels for priority pollutants[4]. As a consequence, solids for disposal must be drier, and more widespread waste minimization and recycle techniques are being practiced. This paper examines a relatively new technique, membrane filter press dewatering, in light of these RCRA regulations and highlights specific examples. Comparisons to conventional filter presses, belt presses, and centrifuges are made where appropriate. Examples of waste recycling using this new technology are also discussed.

INTRODUCTION

As a consequence of recent Resource Conservation and Recovery Act (RCRA) regulations, severe restrictions have been applied to land disposal of wastes, whether hazardous or not. In fact, recent land-ban restrictions have further exacerbated this situation to the extent that best demonstrated available technology (BDAT) should be practiced to meet the Toxic Characteristic Leaching Procedure (TCLP) regulatory levels for priority pollutants[4]. As a result, solids for disposal must be drier, and more widespread waste minimization and recycle techniques are being practiced[11]. These drier solids must now pass the modified 50-psig "Paint Filter Test" for land disposal[11], which generally precludes sludges and cakes produced by centrifuges, rotary drum vacuum filters (RVFs), and belt filter presses--the "big three" dewatering techniques most widely practiced throughout all waste disposal industries[8,9]. This paper examines a relatively new technique, membrane filter press dewatering, in light of these RCRA regulations and compares it with conventional techniques. Specific case histories are cited in which improved waste has been formed that either is less hazardous or can be recycled for some intrinsic value.

HISTORY

Filter presses at up to 225 psig dewatering pressure have had a tremendous resurgence in the past 10 years with the introduction of polypropylene (PP) recessed chamber and membrane plates, primarily because of drier cakes and lower disposal costs compared with the "big three"[9,12]. The old manual plate-and-frame design is now outmoded, especially since most plants are becoming more automated. In addition, filter presses generally are simpler mechanically and do not have many moving parts (and those that do are low speed). Steward's[12] discussion revealed that membrane filter press dewatering can significantly reduce cake moisture, and thus disposal costs, particularly when compared with belt presses and vacuum filters. However, the filter's main component, the membrane plate, was not discussed.

The membrane (or diaphragm) plate was invented by Weber of Ciba-Geigy[16] in 1964 to improve fine-pigment cake washing, not for waste dewatering[2]. Flexible membranes (or diaphragms), which are an integral part of the plates, are inflated with either air or water up to 225 psig pressure to squeeze and dry the cakes. These membrane plates assure dry, firm cakes regardless of fill volume (provided at least one-third of recessed chamber is filled) and regardless of pump pressure or pumping time (i.e., standard recessed chamber plates require a 225-psig pump and long pumping times to develop dry cakes between their fixed volume recesses)[9].

The early membrane plates in the late 1970s usually were made of rubber over a steel plate (web) for support (i.e., Fletcher & Moseley in England; Hoesch, Hansen, and Rittershaus and Blecher [R&B] in Germany; and Ishigaki and Kurita in Japan). These were generally cumbersome and tended to crack, extrude, or have the rubber membrane separated from the steel web. In some cases they were limited to only 125 psig. As a result, the all-PP "one-piece" molded membrane plate was developed by Lenser in Germany in the middle 1970s[7] and I was introduced to it in 1976. It had a center-feed eye, a number of stay bosses for support between adjacent plates (which are still used today), and vertical filtrate drainage grooves. Extensive pilot testing in the next 2 years demonstrated the clear superiority of these membrane plates over the old rubber-style plates, as well as significantly better dewatering when compared with centrifuges, nutsches, plate-and-frame presses, and vacuum filters (both RVF and horizontal belt type). The only negatives found were wash-water channeling and membrane plate failures at the stay boss and center-feed eye areas (as high as 50 percent plate failures). Since the entire membrane plate had to be discarded once membrane rupture occurred, the final full-scale presses used half Lenser PP and half Hoesch rubber/PP membrane plates. The subsequent track record clearly showed the Lenser PP membrane plates to be more reliable, especially since they could operate up to 225 psig, whereas the Hoesch plates were limited to 125 psig. In addition, Lenser made some improvements that drastically reduced their failure rate, particularly better quality control and a "pipped" filtrate drainage surface (i.e., round nubs supported the filter cloth)[2]. Furthermore, rubber cannot be used in food, pharmaceutical, and vegetable oil applications, so the Lenser PP membrane plates were preferred. Thus, I was firmly convinced then that PP membrane plates were the new emerging BDAT of high-pressure dewatering, especially for applications involving products.

In 1980, I had an extensive literature search conducted on membrane press dewatering; only four references[3,5,13,14] plus the EPA report[15] referred to rubber membrane plates, and none referred to PP membrane plates. Even so, Heaton[5] pointed out that the value of membrane plate filtration is that 75 percent of the cake, formed in about half the time, can then be compressed reasonably quickly with the membranes (diaphragms) to form drier cakes than from conventional presses; hence, membrane plates increase capacity and produce drier cakes. In addition, some controversy existed concerning the usefulness of membrane plates (i.e., compare Heaton and Cherry[3,5]). However, subsequent publications[2,9,13,15,17] unequivocally supported the value and utility of membrane plates. In fact, the exhaustive study with Blue Plains sewage sludge[15] showed the superiority of membrane presses (with Japanese membrane plates marketed by Envirex, USA) over other conventional dewatering techniques. This study never generated much interest in the waste industry, and few membrane presses were installed. The first installation of significant size was at Daytona Beach, Florida, where Lenser PP center-feed membrane plates were used. These had many failures resulting from maloperation, poor sludge conditioning, and of course, the poor plate design using a center feed with stay bosses[2,9]. As a consequence, conversion to Hansen rubber/PP membrane plates has been made. These have no stay bosses. Earlier minor installations with only a few chambers also failed, and most are shut down now or converted to standard recessed presses.

As a result of these many center-feed Lenser PP membrane plate failures in the waste industry, most companies have decided to remain with standard recessed plates and high-pressure pumping. However, in 1982 Klinkau introduced a corner-feed, all PP membrane plate without stay bosses[6], which eliminated the previously mentioned wash-water channeling, as well as the membrane failures at the stay bosses and the center-feed eye. The corner-feed design also resulted in better cake washing, slightly drier cakes, and better cake release[9]. In addition, Klinkau also developed a "mixed-pack" arrangement in which membrane plates are alternated with standard recessed plates to reduce costs and halve the number of membrane plates that can fail (e.g., previous designs used all membrane plates). This corner-feed design was introduced into the U.S. market in 1983 at the Williamsburg Filtration Conference[1]. Acceptance was slow, since Lenser had a strong foothold in the U.S. market. The discussion[8] suggests when center-feed versus corner-feed plates should be used, but I have found that corner-feed membrane plates can be used in all cases with superior results, and the case histories cited herein will support this assertion.

CASE HISTORIES

Steward's examples[12] clearly show the value of 225-psig membrane plate presses compared with standard 100-psig recessed chamber plate presses, belt presses, and in one case, a conventional vacuum filter, particularly with respect to cake volume reduction and disposal costs. Pressure driving forces are also tabulated for the various devices, and membrane filter presses are clearly much higher, which accounts for their drier, denser cakes. However, I have found routinely that a 225-psig membrane squeeze pressure results in drier cakes than a 225-psig standard recessed press, even if the 225-psig pumping is continued beyond the normal cutoff point and the entire press plus feed eyes are filled with cake. Typically, the membrane press produces 2 to 8 percent higher solids cakes (depending on sludge characteristics), which I believe is due to more uniform cake compression that eliminates the soft center portion of standard recessed cakes where the two formed cakes meet. (This also can be shown by laboratory piston press testing[17] or by a modified Passavant R-Meter test.) Nevertheless, it is not clear whether these examples[12] are from tests or from actual installations, or whether the cakes were washed to remove hazardous materials prior to discharge. This cake washing ability of membrane filter presses is precisely why they are better suited for hazardous wastes and why Du Pont has installed many state-of-the-art units to produce nonhazardous, or at least less toxic, cakes. (See Netzsch, Inc.[10] for description of this state-of-the-art press.)

PLANT CLARIFIER UNDERFLOW SLUDGE

Many tests on an 8 percent solids sludge from the main plant waste treatment clarifier clearly showed the superiority of membrane filter presses; e.g., the driest cakes were 46 percent without polymer and 51 percent with polymer. In addition, the membrane press was the only unit that could produce dry, firm, landfill-ready cakes without the need for filter aid or polymer, which reduces operating costs significantly (see Table 1). The membrane press option was also the least costly alternative. Subsequent to these tests, the plant could no longer dispose of this sludge in an off-site landfill, so in the interim before the membrane presses are installed, they contracted with a waste handler. Their standard recessed press with a 225-psig Abel pump could produce only about 40 percent solids cakes that were soft, spongy, and difficult to handle. These results prompted the plant to expedite the membrane press project, and three systems have been ordered.

ACIDIC INORGANIC WASTE

A low-pH sludge basin had to be closed per RCRA, so an extensive program was undertaken to determine the best dewatering method, particularly since the plant desired to wash and neutralize the cakes for recycle. The results in Table 2 show that the membrane press produces much clearer

Table 1. Clarifier Underflow Sludge Dewatering

Device	Polymer Dose, lb/ton	Filter Aid Dose, g/L	Annual Costs, $M	Cake, % solids	Rate, gpm/ft2	Capital Cost, $M	Pass Mod. Paint Test?
Centrifuge	23	0	630	<30	Impractical		No
Oberlin APF	3	0	80	20	Impractical		No
Schneider APF	46	0	1,260	20	0.17	500	No
	46	5	1,620	24	0.11		
Verti-Press	23	0	630	35	0.12	800	Yes
Filter Press							
• Std. Recessed	0	0	0	37	0.02	400*	?
	20	0	540	44	0.02		
• Membrane	0	0	0	46	0.03	450*	Yes
	20	0	540	51	0.04		

* With feed pumps.

Table 2. Acidic Inorganic Waste Dewatering

Device	Polymer Dose, lb/ton	Solids Capture, %	Filtr. TSS, ppm	Cake, % solids	Washing	Capital Cost, $MM	Pass EP Toxicity?	Pass Paint Test?
Vacuum Filter	6-8	95	9,000	45	Fair	10.8	?	No
Centrifuge	6	95	2,000	60	None	1.2	No	?
Membrane Filter Press	0	100	<100	65	Exc.	1.0	Yes	Yes

effluent; achieves better capture; produces much firmer, drier cakes; and most importantly, results in excellent washed cakes suitable for landfilling or recycle that pass the standard EP toxicity leaching tests. As a consequence, a complete membrane press system has been installed and will begin operation soon. Note that the membrane press alternative was also least costly.

EP toxicity tests found that washed cakes above pH 3 are acceptable for landfilling. The membrane press was the only device that could wash the cakes sufficiently to greater than pH 3 (typically cakes were in the 5 to 7 pH range).

POND DREDGINGS ACID WASTE

A membrane filter press did not require polymer flocculant (a significant cost savings); produced a filtrate clean enough to be recycled without further processing; achieved better capture; produced much firmer, drier cakes; and again, produced excellent washed cakes suitable for landfilling at the lowest capital cost. Table 3 presents some typical results.

HEAVY METAL LAGOON SLUDGE

Du Pont was faced with a very high cost to close a waste sludge lagoon per RCRA regulations. In this case, only a membrane plate press was investigated because operating centrifuges at the plant produce sloppy cakes and extensive treatment was necessary to recover the solids for sale to a metal reclaimer. Of 16 schemes studied, the membrane plate press was the only one that both decommissioned the lagoon and recovered the solids for sale. This

Table 3. Pond Dredgings Acid Waste Dewatering

Device	Polymer Dose, lb/ton	Solids Capture, %	Filtr. TSS, ppm	Cake, % solids	Washing	Capital Cost, $MM	Pass Paint Test?
Vacuum Filter	3	>99	500	54	Fair	4	No
Centrifuge	30	96	700	40-50	None	1.2	No
Membrane Filter Press	0	100	<100	65	Exc.	1.0	Yes

membrane press has been operating for over a year, 24 hr/day, with absolutely no instances of sloppy cake production. In fact, the cakes have been quite acceptable to the reclaimer. The only operating problems have been some membrane plate failures (bulges only) due to overheating, lack of sufficient cake (one instance only), repeated membrane squeeze water pump failures (replaced pump with another type), and cloth life shortened by chemical attack (replaced with a similar cloth of different fiber type). Cake release has been outstanding throughout the year with only occasional cloth washing required when a slimy feed slurry batch was inadvertently made. Filter aid and polymer addition are not needed. In short, the membrane press has met expectations and is believed to be the only dewatering technique that could have accomplished this difficult task.

Table 4 presents some typical results.

Table 4. Heavy Metal Lagoon Sludge Waste Dewatering

Condition	% Solids	Volume, ft^3/day	% Volume Reduction	Saleable?
Membrane Filter	0.5	85,000	--	No
Membrane Filter Press Cake (at only 140 psig)	65	340	99.6	Yes

CERAMICS PLANT CLARIFIER UNDERFLOW SLUDGE

This existing plant uses two rotary vacuum precoat filters (RVPFs) to dewater the plant's total solids waste from a clarifier, which discharges to a nearby river. Although this waste is nonhazardous, the RVPFs produce sloppy cakes that do not pass the EPA Paint Filter Test, and they require nearly $250M annual filter aid cost. Both a centrifuge and a screw press were tried on this waste without much success. As a consequence, I evaluated a membrane filter press, which demonstrated better filtrate quality (without filter aid addition), acceptable dewatering so two small presses would be required, and much drier cakes without filter aid (Table 5). As a consequence, a project has been initiated to install two small membrane presses (for greater flexibility to handle variable waste flows), and as an added bonus, the membrane press cake (without filter aid) can be sold as a raw material to an offsite reclaimer.

ORGANIC WASTE DEWATERING

Per RCRA regulations, a large organic waste sludge basin required closure by November 1988. As a consequence, the entire plant's wastes had to

Table 5. Clarifier Underflow Sludge Waste Dewatering

Device	Filter Aid Cost, $M/yr	Cake, % solids	Filtrate TSS, ppm	Pass Paint Filter Test?
Existing RVPFs	250	15-20	200	No
Membrane Filter Press	0	30	<50	Yes

be collected and dewatered to produce dry cakes suitable for secure landfilling. Bench-scale testing showed that a vacuum filter, centrifuge, and belt press were all unsuitable. As a consequence, only an automatic pressure filter (APF), a standard recessed filter press, and a membrane filter press were piloted (Table 6). The results clearly demonstrated the superiority of the membrane filter press because of drier cakes, cleaner filtrate that could be disposed of without further treatment, lower capital cost, and the capability to wash the cakes thoroughly. This press has been operating very successfully for 6 months and has never produced a sloppy cake. This completely automatic, state-of-the-art membrane press has generated much interest, both within Du Pont and in the larger waste-dewatering field (see Table 5)(10). It is completely DCS controlled from a remote control room (about a mile away) and has a small local manual control panel that can override the DCS for maintenance or process optimization purposes. It has an automatic cloth washer system because some wastes are oily and sticky, but it has been used only twice in the past year. Cake release from the state-of-the-art cloths has been outstanding regardless of waste characteristics, and the same set of cloths has been in use for a year. These cloths also contain laminated rubber necks (by Micronics, Inc., of Dover, New Hampshire) that have eliminated feed eye pluggage and improved cake release from the corner-feed channel. The only problems encountered with this installation were about 20 Klinkau plate failures that resulted from weld failure in the corner-feed eyes at 225 psig, which was caused by incorrect feed eye location, and higher than threshold limit value (TLV) limits of certain vapors. These problems were corrected by complete plate set replacement by Klinkau (old plates, however, were used in another installation that operated at lower squeeze pressure) and by lengthening the wash cycle. This decreased the toxic vapor emissions to well below the TLVs. In short, this installation has met or exceeded all our expectations and is a state of the art that the waste disposal industry can duplicate for years to come.

Table 6. Organic Waste Sludge Dewatering

Device	Polymer Dose, mg/L	Annual Cost, $M	Cake, % solids	Cake Density, lb/ft³	Filtr. TSS, ppm	Capital Cost, $M	Cake Washing?	Pass Paint Filt. Test?	Cake Oil Extr.?
Sludge Basin	0	0	0.3	--	--	--	--	No	--
APFs	100	270*	52	85	450	600	No	Yes	Yes
Standard Recessed Filter Press	0	10**	51	85	250	320	No	Yes	Yes
Membrane Filter Press	0	10**	56	95	<100	300	Yes	Yes	No

* Includes disposable media cost required with these APFs.
** Assumes two sets of cloths per year.

CONCLUSIONS

The brief membrane plate history, as well as the case histories presented here, demonstrates the clear superiority of membrane filter press dewatering over the "big three" devices (vacuum filters, belt presses, and centrifuges). A state-of-the-art membrane press is introduced; its wide applicability to waste dewatering, as well as waste minimization and recycle, has been shown.

REFERENCES

1. Augerot, B. J., paper presented at Filtration Society Conference, Williamsburg, VA, November, 1983.

2. Avery, L., "Latest Development of Diaphragm Plate Filter Presses and Other Variable Volume Filters," paper presented at International Technical Conference on Filtration and Separation, Ocean City, MD, March 21-24, 1988.

3. Cherry, G. B., *Filtration and Separation*, 1978, July/August, 313-316.

4. *Federal Register*, June 13, 1986, Vol. 51, No. 114.

5. Heaton, H. M., *Filtration and Separation*, 1978, May/June, 232-238.

6. Klinkau & Co., GmbH, Marktoberdorf, West Germany, Bulletin, "Klico Membrane Plate," August, 1982, plus subsequent bulletins.

7. *Lenser Bulletin LM-O1*, "Lenser PP-Membraneblock," and earlier bulletins, August, 1987.

8. Mayer, E., paper presented at International Technical Conference on Filtration and Separation, Ocean City, MD, March 21-24, 1988.

9. Mayer, E., "New Trends in SLS Dewatering Equipment," *Filtration News*, 1988, May/June, 24-27.

10. Netzsch, Inc., "High Performance Membrane Filter Press Systems," Bulletin #400/11-87, November, 1987.

11. Sell, N. J., *Pollution Eng.*, August, 1988, 44-49.

12. Steward, C. R., "Minimizing Waste, Utilizing - High Pressure Membrane Filter Presses," paper presented at ASME Northwest Hazardous Waste Conference and Exhibition '88, Tacoma, WA, October 24-27, 1988.

13. Svarovsky, L., *Chemical Engineering*, July 2, 1979, 62-76; July 16, 1979, 93-105; July 30, 1979, 69-78.

14. Tiller, F. M. and Crump, J. R., *CEP*, October, 1977, 65-75.

15. U.S. Department of Commerce, "Evaluation of Dewatering Devices for Producing High-Solids Sludge Cake," NTIS Report PB80-11153, August, 1979.

16. Weber, O., Ciba-Geigy, U.S. Patent 3,289,845.

17. Young, I. M., Purdey, J. A., and Bradbury, N., in *Solid/Liquid Separation Equipment Scaleup*, 2nd Edition, D. B. Purchas and R. J. Wakeman, eds., Uplands Press, Ltd., 1986, 446-506.

MEMBRANE SEPARATION TECHNOLOGIES IN POLLUTION CONTROL: CASE HISTORIES

P. S. Cartwright, P.E.
Cartwright Consulting Company
8324 16th Avenue South
Bloomington, Minnesota 55425
(612) 854-4911

INTRODUCTION

The membrane separation technologies of microfiltration, ultrafiltration, reverse osmosis, and electrodialysis all possess characteristics that make them attractive for industrial pollution control applications. These include

- Continuous process, resulting in automatic and uninterrupted operation
- Low energy utilization involving neither phase nor temperature changes
- Modular design--no significant size limitations
- Minimum number of moving parts with low maintenance requirements
- No effect on the form or chemistry of the contaminant
- Discrete membrane barrier to ensure physical separation of contaminants
- No chemical addition requirements.

Cartwright Consulting Co. is engaged in the investigation of membrane separation technologies with primary emphasis on waste treatment and food and chemical processing applications. This investigation usually takes the form of an analysis of the stream to be treated in the context of disposal restrictions and reuse or recovery options. After selecting the appropriate membrane polymer and device configuration, an applications test characterizes the stream and obtains design data. If necessary, an on-site pilot test is run to gather long-term operating data. From these tests, it is possible to design and provide all of the necessary engineering for construction or purchase of a total treatment system.

This paper summarizes the membrane technologies commercially available, outlines test procedures, and cites case histories illustrating the use of membrane technologies in effluent treatment applications.

REVIEW OF TECHNOLOGIES

Figure 1 depicts the mechanism of microfiltration. Generally, microfiltration involves the removal of particulate or suspended materials ranging in size from 0.1 to 10.0 μ (1,000 to 100,000 angstroms).

Figure 2 depicts ultrafiltration, which is used to separate materials in the 0.001- to 0.1-μ range (10 to 1,000 angstroms). Basically, ultrafiltration is used to remove dissolved materials, whereas the suspended solids are removed by microfiltration.

Figure 3 illustrates reverse osmosis, which typically separates materials less than 0.001 μ (10 angstroms in size). Reverse osmosis offers the advantage of rejecting ionic materials that are normally small enough to pass through the pores of the membrane. As with ultrafiltration, reverse osmosis is used to remove dissolved materials.

Figure 4 depicts electrodialysis, which combines permeable membranes with an anode and cathode to effect separation. Instead of driving pure water through the membrane and leaving contaminants behind, as in the case of the other membrane processes, electrodialysis utilizes membranes that selectively allow either the cationic or anionic solute to pass through in response to the electrical charges imposed by the anode and cathode.

Figure 5 depicts a general schematic for the membrane processes of microfiltration, ultrafiltration, and reverse osmosis. In these technologies, an important engineering design consideration is recovery, which is defined as the permeate flow divided by the feed flow; in other words, the percentage of the feed flow pumped through the membrane. Typically, for effluent treatment

Microfiltration

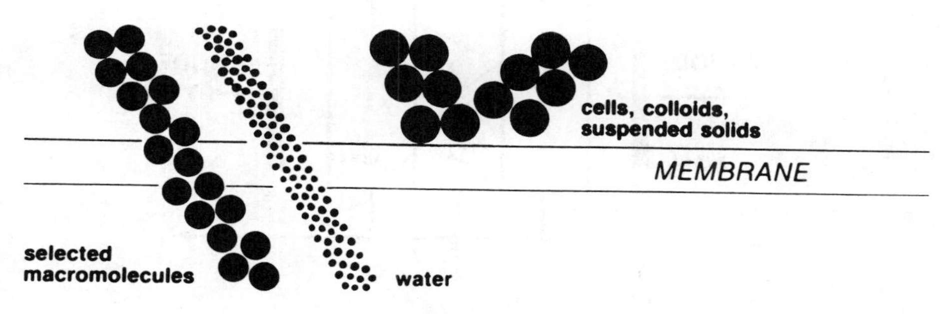

Figure 1. Microfiltration

Ultrafiltration

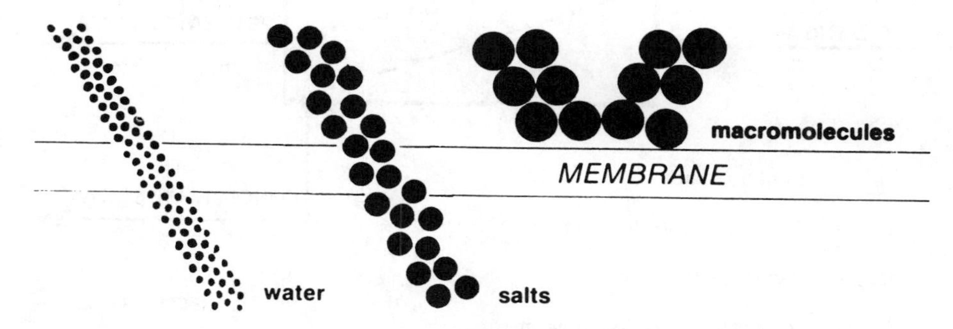

Figure 2. Ultrafiltration

Reverse Osmosis

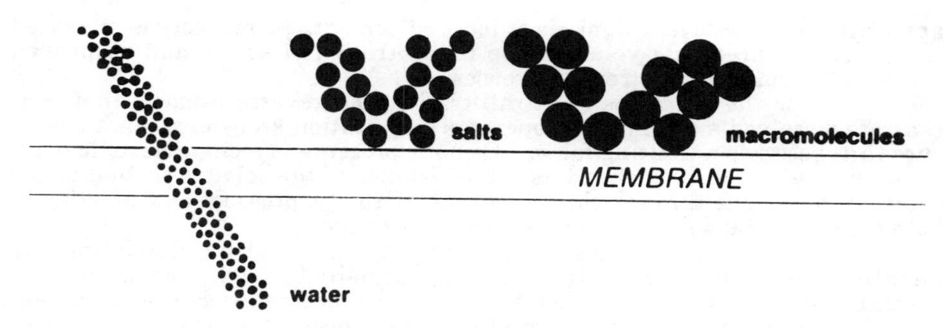

Figure 3. Reverse Osmosis

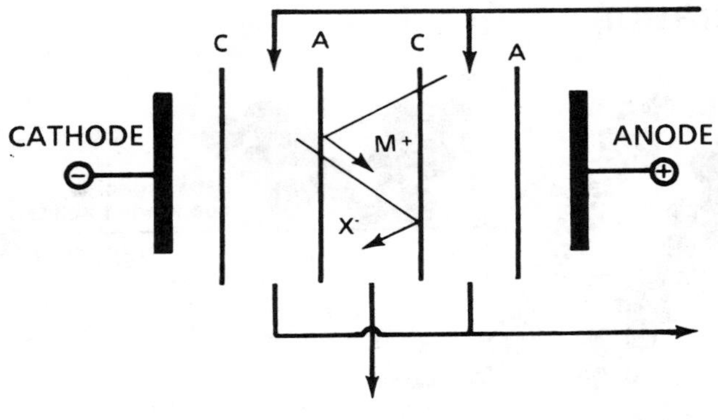

Figure 4. Electrodialysis

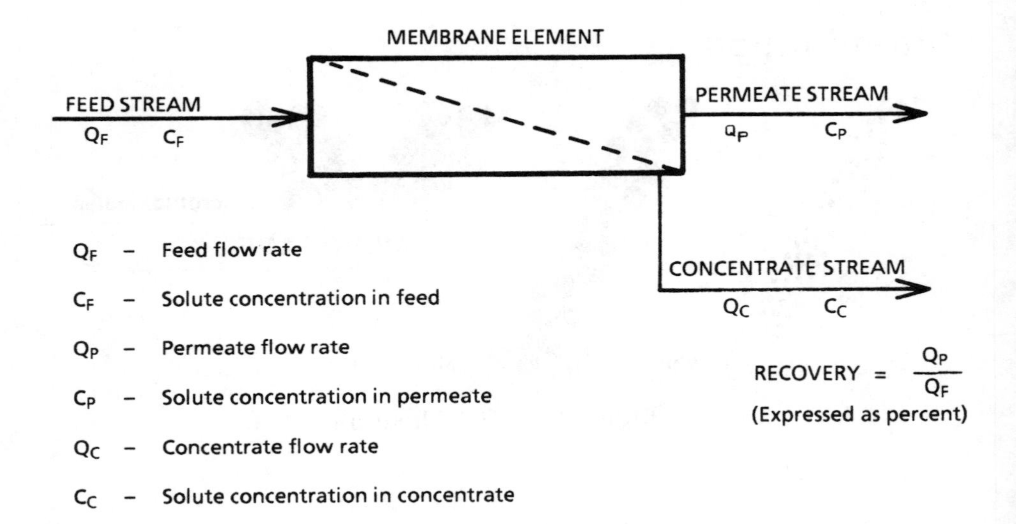

Figure 5. Microfiltration, Ultrafiltration, and Reverse Osmosis Schematic

applications, the recovery figure is at least 90 percent. As recovery is increased (to decrease concentrate volume), the concentration of solute and suspended solids in the concentrate stream increases.

For the processes of ultrafiltration and reverse osmosis that concentrate dissolved materials, a property of the solution known as osmotic pressure ($\Delta\pi$) becomes a limiting factor. Osmotic pressure is a characteristic of all solutions and is loosely defined as the resistance of the solvent portion of the solution to passage through the membrane. Osmotic pressure is a function of both the particular solute as well as its concentration.

The recovery of the system can be controlled by restricting the quantity of flow in the concentrate stream, normally through the use of a concentrate valve. As recovery is increased, with the resulting decrease in concentrate flow, the concentration of solute in the concentrate stream increases, resulting in an increased osmotic pressure.

No membrane is perfect in that it rejects 100 percent of the solute on the feed side; this solute leakage is known as passage. Expressed as percent passage, the actual quantity of solute that passes through the membrane is a function of the concentration of solute on the feed side. Under high recovery conditions, the concentration of solute on the feed side is markedly increased, therefore, the actual quantity of solute passing through the membrane also increases. Because most effluent applications demand that, in addition to a minimum concentrate volume, the permeate quality be high enough to allow reuse or to meet discharge regulations, the catch-22 predicament of permeate quality decreasing as recovery is increased can impose design limitations. Additionally, the increased osmotic pressure resulting as recovery is increased also imposes a design limit. Generally, pumping pressures in excess of 1,000 psi are impractical for most applications.

Because of the propensity of suspended or precipitated materials to settle out on the membrane surface and plug the membrane pores, turbulent flow conditions must be maintained (Reynolds numbers in excess of 2,000). For high recovery systems, this usually requires recycling a significant percentage of the concentrate back to the feed side of the pump. The addition of this portion of the concentrate stream into the feed solution obviously increases the dissolved solids concentration, further increasing osmotic pressure.

In an ideal system, all of the contaminants to be removed will be separated by the membrane and exit in the concentrate stream. As the recovery is increased, the concentration of contaminants in the concentrate stream will increase dramatically. Table 1 summarizes this increasing concentration factor as a function of system recovery, and Figure 6 illustrates this graphically.

All of these factors--recovery, osmotic pressure, permeate quality, recycling, etc.--serve to underscore the value of testing the specific waste stream as thoroughly as possible. Because effluent streams often vary in analysis as a function of time, it is important that either a composite or a worst case sample be obtained for test purposes.

Table 2 compares membrane technologies in several important areas.

Table 1. Concentration Factor Versus Recovery

$$C_C \approx \frac{C_F}{1\text{-Recovery}} = X\ C_F$$

$X \equiv$ Concentration Factor

Recovery	X
33%	1.5
50%	2
67%	3
75%	4
80%	5
90%	10
95%	20
97 1/2%	40
98%	50
99%	100

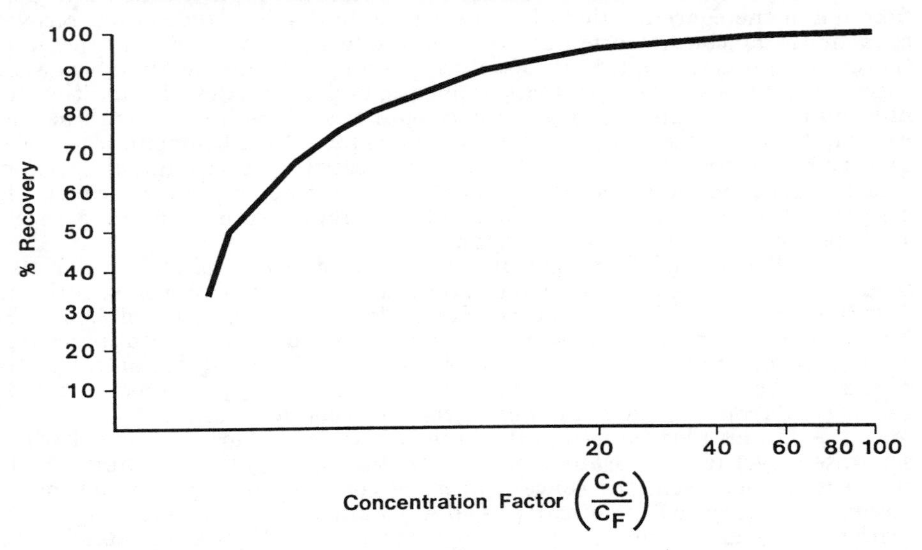

Figure 6. Effect of Recovery on Increase in Solute Concentration in Concentrate Stream

Table 2. Membrane Separation Technologies Compared

Feature	Micro-filtration	Ultra-filtration	Reverse Osmosis	Electro-dialysis
Suspended Solids Removal	yes	yes	yes	no
Dissolved Organic Removal	no	yes	yes	no
Dissolved Inorganic Removal	no	no	yes	yes
Concentration Capabilities	high	high	moderate	high
Permeate Purity	high	high	high	moderate
Energy Usage	low	low	moderate	moderate
Membrane Stability	high	high	moderate	high
Operating Cost ($/1000 gal.)*	.50-1.00	.50-1.00	1.00	1.00

*feed rate

One of the major advantages of these technologies over other pollution control technologies is the potential for direct on-site recycling. The contaminant can be dewatered or concentrated without chemical or physical change, enabling it to be returned directly to the process. When a pollutant cannot be directly recycled, membrane processes can be utilized to concentrate it to make off-site recycling more attractive. In either case, the resulting purified water may be reused.

Each membrane separation technology has strengths and weaknesses when compared with the others; this creates opportunities to combine them and take advantage of a particular strength in a specific application.

The following case histories detail applications of membrane separation technologies in pollution control applications.

EXAMPLE I

This example utilizes electrodialysis reversal (EDR) to concentrate the stream from a reverse osmosis system used to dewater an agricultural wastewater effluent. A solar pond ultimately receives the effluent and, without the dewatering features of reverse osmosis and electrodialysis, would overflow. Osmotic pressure effects limit the maximum concentration of the reverse osmosis concentrate stream to 31,000 mg/L. The flow rate of the stream at this concentration is still higher than the evaporation rate of the solar pond, so the electrodialysis unit, which is unaffected by osmotic pressure, is used to increase the concentrate (to as much as 150,000 mg/L) and further dewater the stream. Some of the EDR brine stream is also used to regenerate the softener ,which is pretreatment for the reverse osmosis system. Feed flow is up to 33,000 gal/day, with overall product water recovery as high as 90 percent. Figure 7 illustrates this installation.

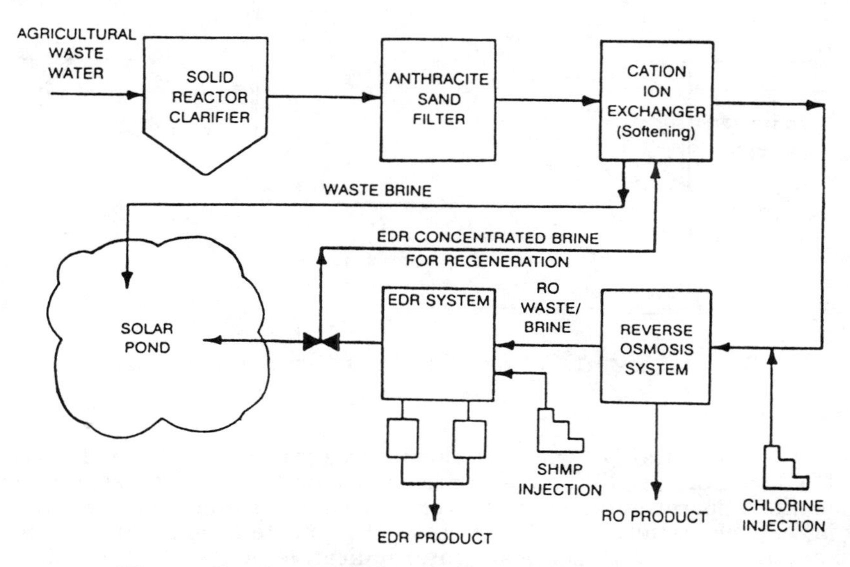

Figure 7. Reverse Osmosis and Electrodialysis

EXAMPLE II

This company manufactures ester-based synthetic hydraulic fluids. Wastewater produced from the condensation reactions contains a potpourri of organic chemicals, including glycols, amines, polyols, fatty acids, esters, and anhydrides.

This waste stream, at a volume of approximately 200 gal/day, is now sent to a hazardous waste treatment facility at a current cost of $1/gallon. In addition to the economic considerations, the company is concerned about the perpetual responsibility associated with off-site removal of hazardous wastes.

If the contaminants can be concentrated into a sufficiently small volume, they can be more easily treated or removed. Tight ultrafiltration was considered to accomplish this concentration step with the following goals in mind:

- The permeate must meet local discharge regulations.
- The membrane materials must be resistant to the concentrated chemicals produced during the dewatering operation.

Application tests were run on a worst case sample using the equipment illustrated in Figure 8. Because the waste stream contains many low molecular weight organic materials, it was decided to use reverse osmosis membranes, which have molecular weight cut-off (MWCO) properties of 100 to 200 daltons.

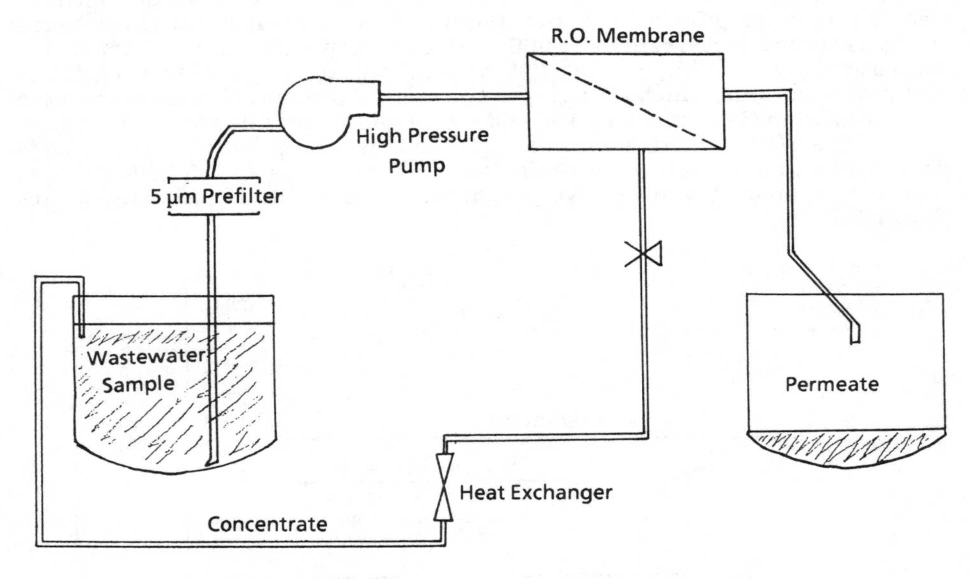

Figure 8. Applications Test System

The initial test was performed with a spiral-wound thin film composite (TFC) membrane. TFC membranes are the first choice for waste treatment applications operating at high recoveries because of their high rejection and low MWCO characteristics. In this case, the TFC test membrane exhibited a permanent loss of flux (permeate rate) indicative of chemical attack. As a result of discussions with the membrane manufacturer, it was concluded that the alkylated diphenylamine compounds in the waste stream reacted with the membrane polymer to produce this result. Because of the similar polymeric structure of the other available TFC membranes, it was decided not to test any other manufacturer's TFC product.

The second test involved a high-rejection cellulose acetate (CA) membrane. The results of this testing are summarized in Table 3. The low rejection of COD can probably be attributed to the fact that most of these COD chemicals have molecular weights below 200 daltons.

Testing is currently under way to determine how best to handle the COD contaminants. Initial studies are directed at using activated carbon in either powdered or granular form to adsorb the COD.

Table 3. High Rejection CA Membrane Test Results

	Feed	PERMEATE 90%	95%
TDS (ppm)	226	17	27
% rejection		92.5%	88.1%
COD (ppm)	17,000	13,000	14,000
% rejection		23.5%	17.7%
Kjeldahl Nitrogen (ppm)	49	4	5
% rejection		91.8%	89.8%
Oil/Grease (ppm)	193	27	36
% rejection		86.0%	81.3%
Osmotic Pressure ( )	---	253 psig	253 psig
Permeate Rate (gph)	---	6.6	6.6

EXAMPLE III

This company produces automotive catalytic converters from a slurry of aluminum, cerium, and nickel. The slurry is dewatered in a filter press and the filtrate discharged to the local Publicly Owned Treatment Works (POTW).

The discharge stream has been consistently out of compliance in the following parameters: total dissolved solids, BOD, COD, acetate, ammonia, and aluminum. The company also desires to recover as much of the wastewater as possible. The total quantity of this effluent stream is 25,000 gal/day, continuous.

Reverse osmosis was selected as the technology to evaluate in this application. The company believed that, if the contaminants could be concentrated enough, they could be recycled to the process, and with the water reuse, this would result in a true zero discharge application.

A worst case sample was submitted for testing in the apparatus illustrated in Figure 8. A high-rejection TFC spiral-wound membrane was used and samples obtained at three recoveries--85, 90, and 95 percent. Table 4 summarizes the data from this test and compares the permeate quality with the acceptable discharge limits. From these data, it is apparent that, even at 95 percent recovery, only the acetate ion exceeds this discharge limit; however,

Table 4. TFC Membrane Test Results

Parameter	MCL*	Feed	Permeate Concentration (mg/l) 85% Recovery	90% Recovery	95% Recovery
TDS	1000	1330	232	393	576
BOD	204	514	49	70	98
COD	500	631	61	108	161
Acetate	100	539	41	81	120
Ammonia (as N)	30	25	25	21	21
Aluminum	2	2	<.01	<.01	<.01

*Maximum Contaminant Level acceptable for discharge to POTW.

it was believed that, by keeping the pH of the stream elevated (above 9), the acetate will exist primarily in ionic form and thus be more highly rejected.

This application test was followed by a 3-month pilot test using a full-sized membrane element of the same type used in the application test.

During the pilot studies, a change in the production process resulted in higher suspended solids loadings in the effluent stream, which caused fouling problems in the reverse osmosis membrane. Cross-flow microfiltration is currently being evaluated as a pretreatment process, and initial results appear very encouraging.

EXAMPLE IV

This aerospace manufacturer uses deionized water to wash out tanks that had contained nitrogen tetroxide (N_2O_4). The only contaminants in this waste stream are nitrite (NO_2^-) and nitrate (NO_3^-) ions. They wish to concentrate these contaminants into the smallest volume possible for off-site disposal and reuse the recovered treated water.

The total daily volume to be treated is 5,000 gal/day. The maximum concentration of nitrite and nitrate expressed as nitrogen does not exceed 1,000 ppm, and the waste stream has a pH of approximately 2.

Cartwright Consulting Co. received a contract to evaluate reverse osmosis in this application. It was determined that, if the permeate quality would meet the proposed drinking water standards of 1 ppm nitrite and 100 ppm nitrate, this stream could either be reused or discharged to the local POTW.

Because both nitrite and nitrate are weakly ionized salts, it was decided to neutralize this stream with sodium hydroxide to a minimum pH of 6.0. A literature search was made to investigate the osmotic pressure of sodium nitrate, and Table 5 illustrates these data.

Table 5. Data for the System ($NaNO_3 \cdot H_2O$) at 25 C

Molality	Mole fraction × 10^3	Weight % solute	Osmotic pressure (lb/in^2)	Density of solution (g/cm^3)	Molar density (mole/cm^3 × 10^2)	Kinematic viscosity (cm^2/sec × 10^2)	Solute diffusivity (cm^2/sec × 10^5)
0	0	0	0	0.9971	5.535	0.8963	1.568
0.1	1.798	0.8429	66	1.0027	5.529	0.8958	1.443
0.2	3.590	1.6718	130	1.0082	5.523	0.8950	1.427
0.3	5.375	2.4869	192	1.0137	5.517	0.8943	1.414
0.4	7.154	3.2886	253	1.0191	5.510	0.8937	1.407
0.5	8.927	4.0772	314	1.0245	5.504	0.8941	1.403
0.6	10.693	4.8531	374	1.0297	5.497	0.8960	1.399
0.7	12.453	5.6165	434	1.0351	5.492	0.8977	1.394
0.8	14.207	6.3677	494	1.0401	5.484	0.8997	1.389
0.9	15.955	7.1071	553	1.0453	5.477	0.9016	1.384
1.0	17.696	7.8350	613	1.0503	5.470	0.9036	1.379
1.2	21.160	9.2569	730	1.0603	5.456	0.9138	1.371
1.4	24.600	10.6356	846	1.0700	5.442	0.9241	1.362
1.6	28.016	11.9731	963	1.0796	5.427	0.9342	1.353
1.8	31.408	13.2711	1078	1.0892	5.414	0.9443	1.345
2.0	34.777	14.5314	1193	1.0984	5.399	0.9544	1.336
2.5	43.096	17.5275	1476	1.1210	5.363	0.9835	1.325
3.0	51.273	20.3206	1759	1.1405	5.317	1.0141	1.318
3.5	59.312	22.9308	2038	1.1635	5.291	1.0576	1.310
4.0	67.216	25.3754	2311	1.1836	5.256	1.1020	1.303
4.5	74.987	27.6696	2586	1.2027	5.220	1.1459	1.296
5.0	82.631	29.8270	2861	1.2210	5.185	1.1892	—
5.5	90.149	31.8594	3146	1.2390	5.151	—	—
6.0	97.545	33.7775	3439	1.2560	5.116	—	—

An applications test was run using the apparatus illustrated in Figure 8; the data from this test are summarized in Table 6. TFC membranes from two different manufacturers were evaluated in this test, and from the data, it is obvious that membrane A is superior to membrane B. Considering the concentration effects of the high recoveries, the actual rejection of sodium nitrate was in the range of 96 to 97 percent.

Table 6. Applications Test Data

		Permeate Quality							
		90% Recovery				95% Recovery			
Parameter	Feed	A	% rej	B	% rej	A	% rej	B	% rej
pH	6.0	5.8	---	5.7	---	6.4	---	6.3	---
Sodium	1000 ppm	750 ppm	25%	830 ppm	17%	380 ppm	62%	840 ppm	16%
Nitrate (as N)	603 ppm	350 ppm	42%	422 ppm	30%	184 ppm	70%	444 ppm	26%

Because the lowest nitrate concentration in the permeate stream was still well above the proposed maximum contaminant level, it was determined that a two-pass reverse osmosis system would be required. Figure 9 illustrates the system design for the two-pass reverse osmosis system. Note that 95 percent (approximately 4,700 gal/day) of the waste stream can be reused with this design. The estimated cost for the system is $70,000 to $90,000.

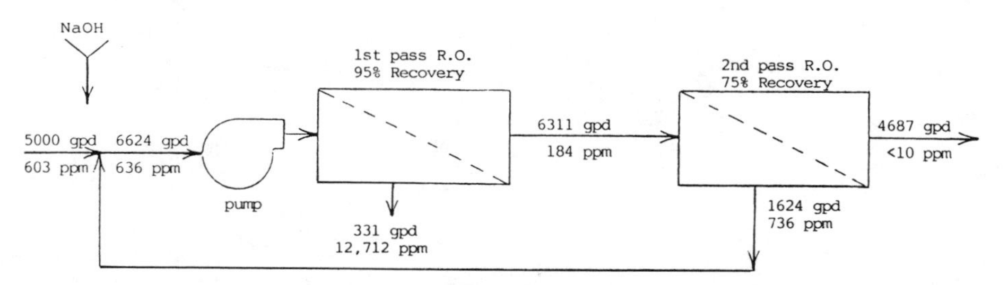

Figure 9. Nitrite, Nitrate Treatment System

CONCLUSIONS

This paper has illustrated some of the practical applications of membrane separation technologies to pollution control. Although they may not be optimum in all applications, and testing is required in virtually every application, these technologies have much to offer and are beginning to be recognized for the advantages they can provide.

NOMENCLATURE

AAS	atomic absorption spectrometry
AC/EC	alternating current electrocoagulation
ACS	American Chemical Society
AFS	atomic fluorescence spectrometry
APF	automatic pressure filter
API	American Petroleum Institute
BAT	best available technology
BCF	bioconcentration factor
BDAT	best demonstrated available technology
BOD	biochemical oxygen demand
BOM	biodegradable organic matter
c.c.c.	critical coagulation concentration
c.s.c.	critical stabilization concentration
CA	cellulose acetate
CB	cacodylate buffer
CDHS	California Department of Health Services
CEC	cation exchange capacity
CEGB	Central Electricity Generating Board [UK]
CERCLA	Comprehensive Environmental Response, Compensation and Liability Act
CF	cationized ferritin
COD	chemical oxygen demand
CSIRO	Commonwealth Scientific and Industrial Research Organization [Australia]
CST	capillary suction time
CVAAS	cold vapor atomic absorption spectrometry
DAF	dissolved air flotation
DDC	diethyl dithio carbamate
DEPH	diethylhexylphthalate
Dfl	Dutch guilder
DNP	2,4-dinitrophenol
DPASV	differential pulse anodic stripping voltammetry
DS	dry solids
DW	dry weight
EAD	electroacoustic dewatering
EC	European Community
EDR	electrodialysis reversal
EMB	electromembrane
EPA	U.S. Environmental Protection Agency
ES	emission spectrometry
ESD	electroacoustic soil decontamination
ESR	electron spin response
ET-AAS	electrothermal atomic absorption spectrometry
ETCi	EXPORTech Company, Inc.
FID	flame ionization detector
GC	gas chromatography
gfd	gallons per square foot per day
HGMS	high gradient magnetic separation
HMDE	hanging mercury drop electrode
HPOAS	high-purity oxygen-activated sludge
IAF	induced air flotation
ICP	inductively coupled plasma emission spectrometry
IDMS	isotope dilution mass spectrometry
IEMR	in-situ electrolytic membrane restoration
IFT	interfacial tension
IIEMC	in-situ intermittent electrolytic membrane cleaning

INAA	instrumental neutron activation analysis
ISV	in-situ vitrification
K	potassium
KOH	potassium hydroxide
LCP	local control panel
MeCl	methylene chloride
METHA	mechanical treatment of dredged material
MF	microfiltration
MGM	magnetic gravimetric separation
MHS	magnetohydrostatic separation
MIBK	methyl isobutyl ketone
MS	mass spectrometry
MTARRI	MTA Research Resources, Inc.
MWCO	molecular weight cut-off
N	nitrogen
NAF	nozzle air flotation
NAPL	nonaqueous phase liquid
NPDES	National Pollution Discharge Elimination System
OM	organic matter
P	phosphorus
PBB	polybrominated biphenyl
PCB	polychlorinated biphenyl; polychlorobiphenyl
PCP	pentachlorophenol
POTW	publicly owned treatment works
PP	polypropylene
PTFE	polytetrafluoroethane
PVC	polyvinyl chloride
PVDF	polyvinylidene difluoride
pzc	point of zero charge
RCRA	Resource Conservation and Recovery Act
RF	radio frequency
RNAA	neutron activation analysis with radiochemical separation
RO	reverse osmosis
RR	ruthenium red
RVF	rotary drum vacuum filter
RVPF	rotary vacuum precoat filter
SAIC	Science Applications International Corporation
SCAQMD	South Coast Air Quality Management District
SCC	Sludge Conditioning Controller
SCR	silicon-controlled rectifier
SITE	Superfund Innovative Technology Evaluation
SSMS	spark source mass spectrometry
SURTEK	(a company)
SVC	semivolatile component
SVH	semivolatile hydrocarbon
SVI	sludge volume index
TCE	trichloroethylene
TCF	thin cake filtration
TCLP	Toxic Characteristic Leaching Procedure [of EPA]
TDS	total dissolved solids
TEM	transmission electron microscopy
TFC	thin film component
TKN	total Kjeldahl nitrogen
TLV	threshold limit value
TOC	total organic carbon
TRI	Texas Research Institute
TSD	treatment, storage, and disposal
TSS	total suspended solids
TTUSA	Toxic Treatments (USA) Inc.
TWAS	thickened waste-activated sludge
UF	ultrafilter; ultrafiltration

USLE	Universal Soil Loss Equation
VOC	volatile organic compound
WAS	waste-activated sludge
WTC	Wastewater Treatment Centre (Environment Canada)
XRF	X-ray fluorescence spectrometry

INDEX